Lightweight Polymer Composite Structures

Lightweight Polymer Composite Structures

Design and Manufacturing Techniques

Sanjay Mavinkere Rangappa,
Jyotishkumar Parameswaranpillai,
Suchart Siengchin, and Lothar Kroll

CRC Press
Taylor & Francis Group
Boca Raton London New York

CRC Press is an imprint of the
Taylor & Francis Group, an **informa** business

First edition published 2020
by CRC Press
6000 Broken Sound Parkway NW, Suite 300, Boca Raton, FL 33487-2742

and by CRC Press
2 Park Square, Milton Park, Abingdon, Oxon, OX14 4RN

CRC Press is an imprint of Taylor & Francis Group, LLC

ISBN: 978-0-367-19920-3 (hbk)
ISBN: 978-0-367-54162-0 (pbk)
ISBN: 978-0-429-24408-7 (ebk)

Typeset in Times
by Deanta Global Publishing Services, Chennai, India

Contents

Preface

Polymers are lightweight materials that are used for many advanced applications such as automobile, construction and building, aerospace, toys, household appliances, etc. However, during its operation in many applications, polymer materials are exposed to harsh environments, high load, friction, wear and tear, etc. Scientists have established that the incorporation of fillers has significantly improved the thermomechanical, electrical, and other physical properties of polymer composites. Thus, to withstand the external forces during use, micro and nanofillers including both natural and synthetic fibers such as carbon nanotubes, carbon black, graphene, nanoclay, nanosilica, plant fibers, glass fibers, carbon fibers, carbon mats, plant fiber mats, etc., were added to the thermoplastics or thermosetting plastics. Moreover, many fillers aid additional functionality to the polymers. For example, when carbon-based fillers such as carbon nanotubes, carbon nanofibers, graphene, etc., are used with polymers, it imparts conductivity to the polymers. Similarly, magnetic particles are incorporated in polymers to aid shape memory properties. Recently, polymer composites have been the preferred option for making components for automobile, aerospace, and marine applications due to their lightweightedness, that enhances the energy and fuel efficiency, thus making the entire industry more environmentally friendly.

Significant progress has been made in the area of lightweight polymer composite structures in the last decade. This has led to a surge in the number of papers, patents, and reviews on the subject. Therefore, we believe it is important to edit a book on *Lightweight Polymer Composite Structures: Design and Manufacturing Techniques*. Eminent scholars have contributed the chapters for the book. We hope that this book will be very informative for the scientist, academic staff, faculty members, researchers, and students working in the area of lightweight polymer composite structures.

The book consists of 13 chapters that shed light on lightweight polymer composite structures. Chapter 1 focuses on the synthesis of graphene, its properties, and highlights the properties of graphene-based thermoplastic and thermosetting composites. The authors emphasized the importance of dispersion of graphene platelets in the polymer matrix for the enhancement of the properties. Chapter 2 discusses various processing techniques that are commonly used for the manufacturing of polymer composites. Chapter 3 describes the necessity of biodegradable and biocompatible polymer composites for biomedical and packaging applications. Chapter 4 reviews the recent developments in polymer composites including ceramics, woody materials, and metal particles. Chapter 5 focuses on lightweight composites for aerospace, marine, and automobile applications. Different processing methods such as hand lay-up, spray-up, vacuum bagging, vacuum infusion, filament winding, resin transfer molding, prepreg, etc., used for the manufacturing of polymer components are highlighted. Chapter 6 gives an overview of using more than one reinforcing filler in the polymer matrix to introduce new properties. Chapter 7 highlights the importance of designing and modeling of polymers for its effective utilization in real-time applications. Chapter 8 gives an overview of the recent developments in smart composites.

The efficiency and usability of traditional polymer composites can be improved by introducing additional functionalities. Chapters 9 and 10 discuss the use of carbon and glass fiber in thermoset and thermoplastic composites, respectively. Chapter 11 reviews the influence of nanofillers on the thermomechanical, electrical, and physical properties of thermoset and thermoplastic composites. Chapter 12 gives an overview of the broad spectrum of applications of thermoplastic and thermosetting polymer composites. The last chapter, Chapter 13, highlights the importance of life cycle analysis on the assessment of the environmental impact of polymer composites.

The editors are thankful to the authors for their contributions and the Taylor & Francis editorial team for their support and guidance.

Editors

Dr. Sanjay Mavinkere Rangappa (Thailand)
Dr. Jyotishkumar Parameswaranpillai (Thailand)
Prof. Dr.-Ing. habil. Suchart Siengchin (Thailand)
Univ.-Prof. Dr.-Ing. habil. Prof. h. c. Dr. h. c. Prof. Lothar Kroll (Germany)

Editor Biographies

Dr. Sanjay Mavinkere Rangappa, received a B.E. (Mechanical Engineering) from Visvesvaraya Technological University, Belagavi, India in 2010, M. Tech (Computational Analysis in Mechanical Sciences) from VTU Extension Center, GEC, Hassan, in 2013, Ph.D. (Faculty of Mechanical Engineering Science) from Visvesvaraya Technological University, Belagavi, India in 2017, and Post Doctorate from King Mongkut's University of Technology North Bangkok, Thailand, in 2019. He is a Life Member of Indian Society for Technical Education (ISTE) and an Associate Member of Institute of Engineers (India). He is a reviewer for more than 50 international journals and international conferences (for Elsevier, Springer, Sage, Taylor & Francis, Wiley). In addition, he has published more than 85 articles in high-quality international peer-reviewed journals, 20+ book chapters, one book, 15 books as an editor, and also presented research papers at national/international conferences. His current research areas include natural fiber composites, polymer composites, and advanced material technology. He is a recipient of DAAD Academic exchange-Project-related Personnel Exchange Program (PPP) between Thailand and Germany to Institute of Composite Materials, University of Kaiserslautern, Germany. He has received a Top Peer Reviewer 2019 award, Global Peer Review Awards, from Web of Science Group powered by Publons.

Dr. Jyotishkumar Parameswaranpillai is currently working as a research professor at Center of Innovation in Design and Engineering for Manufacturing, King Mongkut's University of Technology North Bangkok. He received his Ph.D. in Polymer Science and Technology (Chemistry) from Mahatma Gandhi University. He has research experience in various international laboratories such as Leibniz Institute of Polymer Research Dresden (IPF) Germany, Catholic University of Leuven, Belgium, and University of Potsdam, Germany. He has published more than 100 papers in high-quality international peer-reviewed journals on polymer nanocomposites, polymer blends and alloys, and biopolymer, and has edited five books. He received numerous awards and recognitions including the prestigious Kerala State Award for the Best Young Scientist 2016, INSPIRE Faculty Award 2011, and the Best Researcher Award 2019 from King Mongkut's University of Technology North Bangkok.

Prof. Dr.-Ing. habil. Suchart Siengchin is President of King Mongkut's University of Technology North Bangkok (KMUTNB), Thailand. He received his Dipl.-Ing. in Mechanical Engineering from University of Applied Sciences Giessen/Friedberg, Hessen, Germany in 1999, M.Sc. in Polymer Technology from University of Applied Sciences Aalen, Baden-Wuerttemberg, Germany in 2002, M.Sc. in Material Science at the Erlangen-Nürnberg University, Bayern, Germany in 2004, Doctor of Philosophy in Engineering (Dr.-Ing.) from Institute for Composite Materials, University of Kaiserslautern, Rheinland-Pfalz, Germany in 2008, and Postdoctoral Research from Kaiserslautern University and School of Materials Engineering, Purdue University, USA. In 2016, he received the habilitation at the Chemnitz University in Sachen, Germany. He worked as a lecturer for the Production and Material Engineering Department at The Sirindhorn International Thai–German Graduate School of Engineering (TGGS), KMUTNB. He has been full Professor at KMUTNB and became the President of KMUTNB. He won the Outstanding Researcher Award in 2010, 2012, and 2013 at KMUTNB. His research interests are in polymer processing and composite material. He is Editor-in-Chief of *KMUTNB International Journal of Applied Science and Technology* and the author of more than 150 peer-reviewed journal articles. He has participated with presentations in more than 39 international and national conferences on Materials Science and Engineering topics.

Prof. Lothar Kroll received a Diploma in Mechanical Engineering in 1987 and a Doctorate in Engineering (Dr.-Ing.) in 1992 from Institute for Technical Mechanics, Clausthal University of Technology, Germany as well as the a habilitation in the field of Lightweight Construction and Material Mechanics in 2005 from Dresden University of Technology, Germany. Since 2006, Prof. Kroll is Full Professor of Lightweight Structures and Polymer Technology at the Institute of Lightweight Structures at Chemnitz University of Technology, Germany. Starting with ten academic staff, he has actively established one of the largest professorships in Germany with currently three endowed professorships, nine research areas, and more than 200 employees. Prof. Kroll won the Medal of Honor of Chemnitz University of Technology in 2011, the Medal of Honor of Opole University of Technology, Poland, in 2014, and the Outstanding International Academic Alliance Award at KMUTNB in 2015. In 2017, he was awarded with the Honorary Doctorate, Doctor of Philosophy (Ph.D.) in Production Engineering at KMUTNB. In 2018, he was made Professor of Technical Science by the President of the Republic of Poland. Prof. Kroll is Editor-in-Chief of *Technologies for Lightweight Structures* (open access journal), Deputy Editor-in-Chief of *Journal of Achievements in Materials and Manufacturing Engineering* (JAMME), Poland, on the Editorial Board of *Composites Theory and Practice* (research journal of the Polish Society of Composite Materials) as well as of *Applied Science and Technology* (international journal of KMUTNB). Furthermore, he is on the Editorial Advisory Board of MECHANIK (Miesiecznik Naukowo-Techniczny, Polska) and on the Editorial Key Reviewers Committee of *Archives of Materials Science and Engineering* (AMSE), Poland. He is author of more than 150 peer-reviewed journal articles, in particular on the topics of manufacturing technologies for fiber- and fabric-reinforced lightweight structures, calculation of lightweight structures under mechanical, thermal, and medial load, construction and manufacturing of active composites with integrated electronics, as well as merging of material-specific manufacturing processes to technologies suitable for mass-production of lightweight structures.

Contributors

V. Acar
Atatürk University
Erzurum, Turkey

Özay Aksoy
İzmir Katip Çelebi University
İzmir, Turkey

Felipe Cerdas
Chair of Sustainable Manufacturing & Life Cycle Engineering
Institute of Machine Tools & Production Technology (IWF)
Technische Universität Braunschweig
Braunschweig, Germany
and
Open Hybrid Lab Factory e.V.
Wolfsburg, Germany

M. Chandrasekar
School of Aeronautical Sciences
Hindustan Institute of Technology and Science
Tamil Nadu, India

P. K. Chattopadhyay
Department of Leather Technology
Government College of Engineering and Leather Technology (Post-Graduate)
Maulana Abul Kalam Azad University of Technology
West Bengal, India

Naga Srilatha Cheekuramelli
Polymer Science and Engineering Division
CSIR-National Chemical Laboratory
Pune, India
and
Academy of Scientific and Innovative Research (AcSIR)
Ghaziabad, India

Narayan Chandra Das
Rubber Technology Center
Indian Institute of Technology
Kharagpur, India

Tushar Kanti Das
Rubber Technology Center
Indian Institute of Technology
Kharagpur, India

Mousumi Deb
Advanced Polymer Laboratory
Department of Polymer Science and Technology
Government College of Engineering and Leather Technology (Post-Graduate)
Maulana Abul Kalam Azad University of Technology
West Bengal, India

Antal Dér
Chair of Sustainable Manufacturing & Life Cycle Engineering
Institute of Machine Tools & Production Technology (IWF)
Technische Universität Braunschweig
Braunschweig, Germany
and
Open Hybrid LabFactory e.V.
Wolfsburg, Germany

Alperen Doğru
Ege University, İzmir Katip Çelebi University
İzmir, Turkey

Baijayantimala Garnaik
Polymer Science and Engineering Division
CSIR-National Chemical Laboratory
Pune, India
and
Academy of Scientific and Innovative Research (AcSIR)
Ghaziabad, India

Sabyasachi Ghosh
Rubber Technology Center, Indian Institute of Technology
Kharagpur, India

Johanna Sophie Hagen
Chair of Sustainable Manufacturing & Life Cycle Engineering
Institute of Machine Tools & Production Technology (IWF)
Technische Universität Braunschweig
Braunschweig, Germany
and
Open Hybrid LabFactory e.V.
Wolfsburg, Germany

Christoph Herrmann
Chair of Sustainable Manufacturing & Life Cycle Engineering
Institute of Machine Tools & Production Technology (IWF)
Technische Universität Braunschweig
Braunschweig, Germany
and
Open Hybrid LabFactory e.V.
Wolfsburg, Germany

Naman Jain
Meerut Institute of Engineering & Technology
Meerut, India

Gobinath Velu Kaliyannan
Department of Mechanical Engineering
Kongu Engineering College
Tamil Nadu, India

Alexander Kaluza
Chair of Sustainable Manufacturing & Life Cycle Engineering
Institute of Machine Tools & Production Technology (IWF)
Technische Universität Braunschweig
Braunschweig, Germany
and
Open Hybrid LabFactory e.V.
Wolfsburg, Germany

M. Batıkan Kandemir
İzmir Katip Çelebi University
İzmir, Turkey

M. Karmakar
Advanced Polymer Laboratory
Department of Polymer Science and Technology
Government College of Engineering and Leather Technology (Post-Graduate)
Maulana Abul Kalam Azad University of Technology
West Bengal, India

S. Kiran
Polymer Science and Engineering Division
CSIR-National Chemical Laboratory
Pune, India
and
Academy of Scientific and Innovative Research (AcSIR)
Ghaziabad, India

T. Senthil Muthu Kumar
Center for Composite Materials
Department of Mechanical Engineering
Kalasalingam Academy of Research and Education
Tamil Nadu, India

Dattatraya Late
Physical and Materials Chemistry Division
CSIR-National Chemical Laboratory
Maharashtra, India
and
Academy of Scientific and Innovative Research (AcSIR)
Ghaziabad, India

Manas Mahapatra
Advanced Polymer Laboratory
Department of Polymer Science and Technology
Government College of Engineering and Leather Technology (Post-Graduate)
Maulana Abul Kalam Azad University of Technology
West Bengal, India

Madhushree Mitra
Department of Leather Technology, Government College of Engineering and Leather Technology (Post-Graduate)
Maulana Abul Kalam Azad University of Technology
West Bengal, India

H. Mondal
Advanced Polymer Laboratory
Department of Polymer Science and Technology
Government College of Engineering and Leather Technology (Post-Graduate)
Maulana Abul Kalam Azad University of Technology
West Bengal, India

N. Mohd Nurazzi
Department of Aerospace Engineering
Faculty of Engineering
Universiti Putra Malaysia
Selangor, Malaysia

Avinash Parashar
Indian Institute of Technology
Roorkee, India

N. Rajini
Center for Composite Materials
Department of Mechanical Engineering
Kalasalingam Academy of Research and Education
Tamil Nadu, India

Sanjay Remanan
Rubber Technology Center
Indian Institute of Technology
Kharagpur, India

M. R. Sanjay
Natural Composites Research Group Lab
Academic Enhancement Department
King Mongkut's University of Technology North Bangkok
Bangkok, Thailand

T. P. Sathishkumar
Department of Mechanical Engineering
Kongu Engineering College
Tamil Nadu, India

K. Senthilkumar
Center for Composite Materials
Department of Mechanical Engineering
Kalasalingam Academy of Research and Education
Tamil Nadu, India

M. Ö. Seydibeyoğlu
İzmir Katip Çelebi University
İzmir, Turkey

Suchart Siengchin
Department of Mechanical and Process Engineering
The Sirindhorn International Thai–German Graduate School of Engineering
King Mongkut's University of Technology North Bangkok
Bangkok, Thailand

M. K. Singh
Indian Institute of Technology Mandi
Mandi, India

Vinay K. Singh
G. B. Pant University of Agriculture and Technology
Pantnagar, India

N. R. Singha
Advanced Polymer Laboratory
Department of Polymer Science and Technology
Government College of Engineering and Leather Technology (Post-Graduate)
Maulana Abul Kalam Azad University of Technology
West Bengal, India

M. Sütçü
İzmir Katip Çelebi University
İzmir, Turkey

E. Teke
Izmir Katip Çelebi University
İzmir, Turkey

Akarsh Verma
Indian Institute of Technology
Roorkee, India

Deepak Verma
Department of Mechanical Engineering
Graphic Era Hill University
Dehradun, India

N. Verma
Indian Institute of Technology Mandi
Mandi, India

S. Zafar
Indian Institute of Technology Mandi
Mandi, India

1 Lightweight Graphene Composite Materials

Akarsh Verma, Naman Jain, Avinash Parashar, Vinay K. Singh, M. R. Sanjay, and Suchart Siengchin

CONTENTS

1.1 INTRODUCTION

1.1.1 Synthesis of Graphene

Micromechanical exploitation of graphene from the graphite is one of the simplest and oldest methods to extract graphene [1–4], as shown in Figure 1.1. In 2005, Novoselov et al. [5] extracted the two-dimensional (2D) crystallites by rubbing graphite crystal against another surface like drawing chalk on the blackboard; but, only some of the amount of flakes were present. In micromechanical exploitation, graphene 2D layers are peeled off by scotch tape, and after that deposited on SiO_2. But major disadvantages of this method are its non-scalability and uneven production in a small area.

Epitaxial growth is another technique used for the synthesis of graphene [6, 7], as shown in Figure 1.2. In this technique, SiC crystals are annealed at a high temperature of about 2000 K in ultra-high vacuum. At this elevated temperature, silicon desorption occurs from the uppermost surface of SiC, resulting in yielding of multilayer graphene. In 2004, Berger et al. [6] fabricated the graphene ultrathin films by thermal desorption of Si from 6H-SiC single crystal face. After, the hydrogen etching of the surface was done, and then samples were heated to 1000°C to remove the oxides. Graphene layers were controlled by limiting heat treatment either by

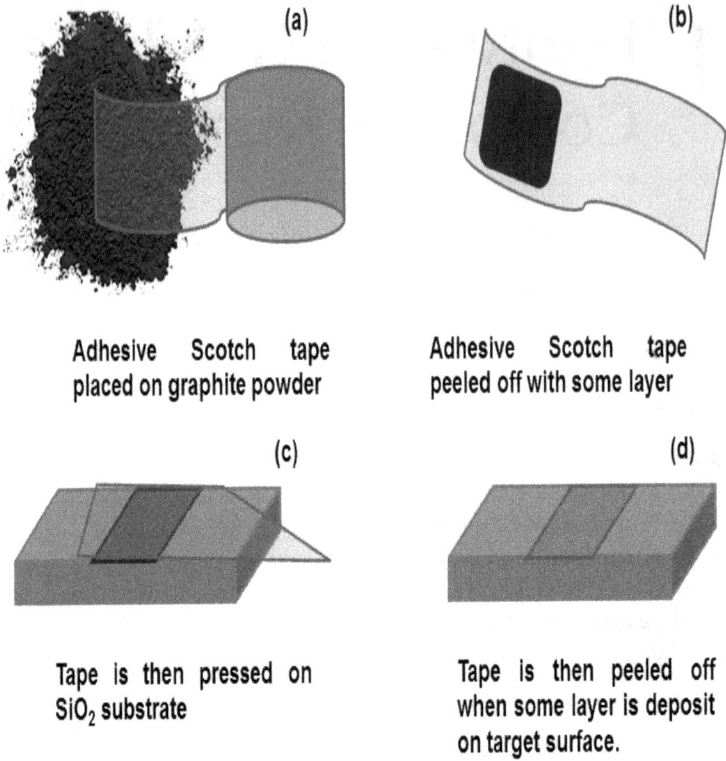

(a)
(b)
Adhesive Scotch tape placed on graphite powder
Adhesive Scotch tape peeled off with some layer
(c)
(d)
Tape is then pressed on SiO2 substrate
Tape is then peeled off when some layer is deposit on target surface.

FIGURE 1.1 Micromechanical exploitation of the graphene.

temperature or time. The epitaxial growth technique yielded larger area compared to the exfoliation method, but still the size is not large enough for those required in electronic applications.

Researchers also employed the wet chemical approach to synthesize graphene from graphene oxide [8–10], as shown in Figure 1.3. In this process, graphite is first converted into graphite oxide by nitric and sulphuric acid treatment. Then, rapid evaporation of intercalant is done at high temperatures, followed by a ball milling process. Stankovich et al. [8] in 2007 prepared graphene by the wet chemical approach. Firstly, graphene oxide (GO) was prepared by the hummers method from pure nature graphite. Then 10 mg of GO was added into 100 ml water, which yielded in yellow-brown inhomogeneous dispersion. Ultra-sonication of dispersion was done to clear the particulate matter. After that, hydrazine hydrate was added and the resulting solution was heated to 100°C for 24 hours under condenser, which reduced GO to a black solid. This method is more versatile than the above two methods, but it also has certain limitations such as poor control of number of layers and partially oxidized graphene is obtained that may alter the electronic and mechanical properties.

The most effective technique to synthesize the graphene sheet is the chemical vapor deposition (CVD) approach [11–15], as shown in Figure 1.4. In this technique, hydrocarbon gases such as methane and hydrogen are heated to a 1000°C temperature into a furnace. A thin layer of transition metallic which is deposit to substrate is

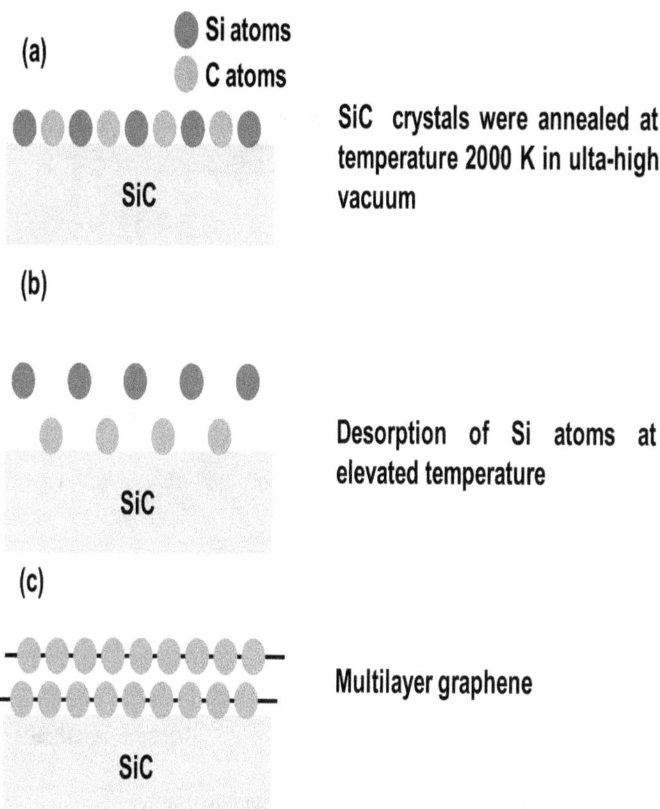

FIGURE 1.2 Graphene synthesis by the epitaxial growth.

used to catalyze the decomposition gas. The metallic layer gradually gets absorbed or diffused, dissociating carbon atoms depending upon the metal. Many researchers have used different metallic catalysts such as Ru, Ir, Pd, Ni, Cu, etc. Mainly, two types of mechanisms occur in the CVD process:

(1) The decomposed carbon atoms are absorbed by the metallic catalyzer and then upon subsequent cooling (with decrease in temperature solubility of carbon decreases) carbon atoms precipitate on the surface of metal to form graphene. This mechanism occurs in those metals which have strong interaction with carbon atoms e.g., Ni.
(2) The decomposed carbon atoms remain or diffuse into metallic catalyzer surface and are then incorporated into graphene. This mechanism occurs in those metals which do not form carbide such as Cu.

1.1.2 Properties of Graphene

Graphene is made up of two-dimensional hexagonal rings like honeycomb lattice with carbon atoms. In graphene, carbon atoms are covalently bonded with

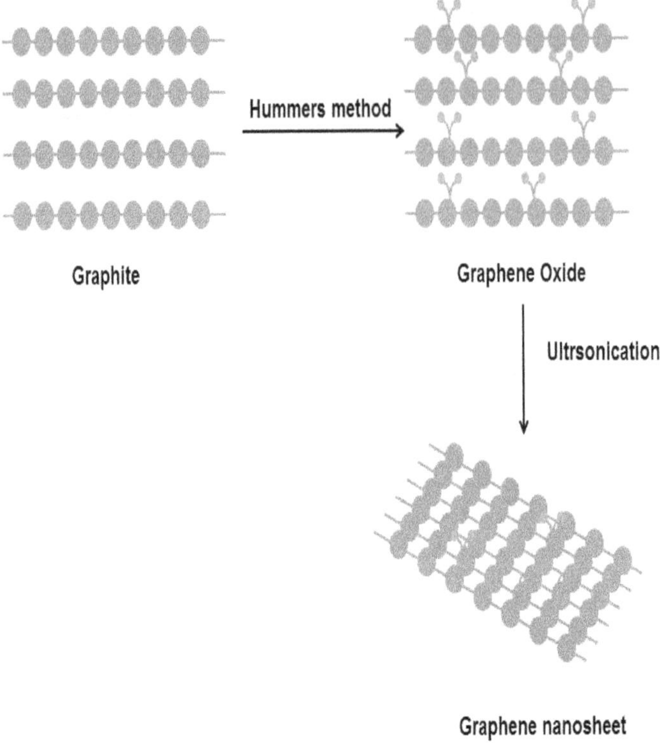

FIGURE 1.3 Wet chemical approach for synthesizing the graphene.

each other with sp^2 hybridization state. Each carbon atom in graphene is bonded with three neighboring carbon atoms, although it has the tendency to make a bond with the fourth atom also. A molecular model of graphene can be visualized as the planar benzenoid hexagonal rings such that peripheral hydrogen atoms are saturated. When these graphene layers are stacked up to form a three-dimensional structure, it is known as graphite (which is commonly used in day-to-day life such as pencils, batteries, and more). Each carbon atom of graphene lies on the surface, due to which there is more interaction with the surrounding atoms. High surface area to volume ratio, elevated stiffness, and distinct electrical and thermal properties of graphene have gained researcher interest as well as their potential application to the area of nanotechnology. At room temperature, the mobility of electrons in graphene is very high due to its two-dimensional structure, that further results in high electrical and thermal conductivity; but on the other hand, there is no band gap in graphene, which restricts its application in the electronics sector. Nowadays, researchers are working to overcome the above problem in many ways such as applying the electrical field or by doping. As per the study of Feng et al., [16] energy gap and binding energy of graphene are strongly affected by the size, compared with the stacking sequence and number of layers. Broad energy gaps in the range 1–2.5 eV have been obtained by π-π interaction between graphene finite size sheets. Another way to open the energy gap is to pattern the graphene

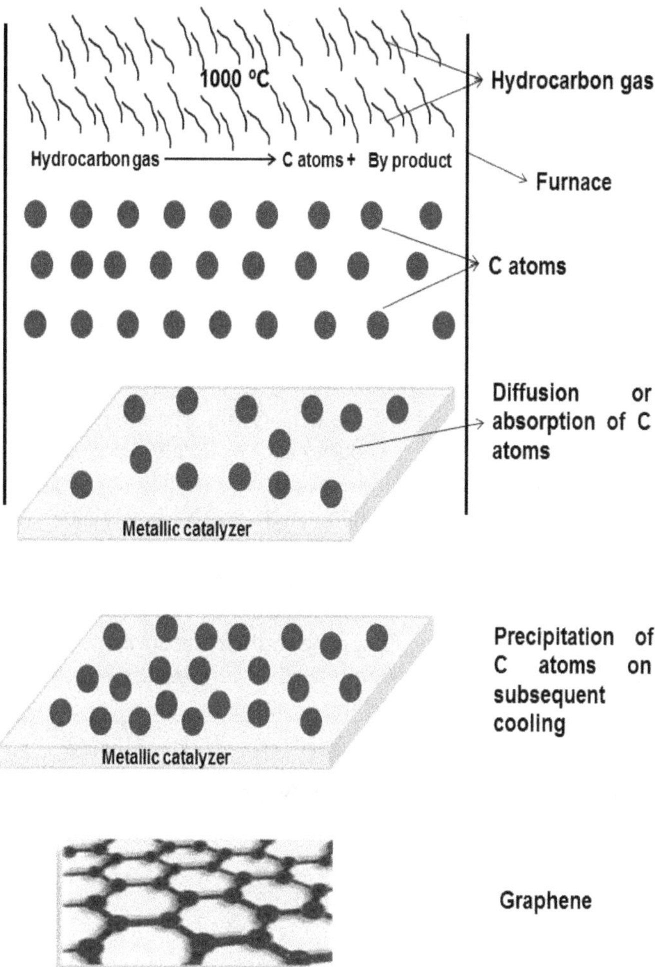

FIGURE 1.4 Chemical vapor deposition approach to synthesize the graphene.

into a narrow ribbon in which carriers are confined into a quasi-one-dimensional system and energy gap varies inversely with the ribbon width [17]. Another alternative may be hydrogenation of the graphene. In 2010, Samarakoon and Wang [18] applied bias voltage across the hydrogenated graphene sheet, which resulted in continuous tuning of energy gap. In 2009, Pereira et al. [19] studied the effect of mechanical strain on the electronic structure of graphene. They observed that to generate a gap, minimum threshold must be applied along the specific lattice direction. Gui et al. [20] also investigated the graphene structure under planer strain. Their results found that zero band gaps were obtained with applied symmetrical strain, while asymmetrical strain results in opening of Fermi level energy gap. When strain was applied along the C-C bonds, maximum energy gap of about 0.486 eV was obtained, whereas the maximum gap of about 0.170 eV was obtained when the strain was applied perpendicular to the C-C bonds.

In parallel to the electrical properties of graphene, graphene-based composites also have fascinating mechanical properties. The Young's modulus E^{3D} on the material can be calculated in terms of shear energy γ_s by the simple equation $E^{3D} = \frac{\sigma_x}{\epsilon_x} = \frac{1}{V_0}\left(\frac{\partial^2 \gamma_s}{\partial \epsilon_x^2}\right)_{E_0}$ where V_0 is the volume of material. Whereas in graphene, in-plane stiffness is defined as compared to E^{3D}, which is calculated by the equation $E^{3D} = \frac{\sigma_x}{\epsilon_x} = \frac{1}{A_0}\left(\frac{\partial^2 \gamma_s}{\partial \epsilon_x^2}\right)_{E_0} = c_0 E^{3D}$. Lee et al. [21] in 2008 evaluated the elastic properties of graphene membranes through atomic force microscopy by nano-indentation. Young's modulus of graphene was found to be 1 Terapascal (TPa) with fracture strength of 42 Nm^{-1}. In 2009, Faccio et al. [22] determined the mechanical properties of graphene nanoribbons by applying the density functional theory. Young's modulus in the present investigation comes out to be 0.96 TPa for graphene and 1.01 TPa for carbon nanoribbons. Other researchers have also studied the mechanical properties of graphene and single walled carbon nanotubes (SWCNT)/nanoribbons, which are presented in Table 1.1.

Due to these extraordinary properties, many researchers have employed graphene as reinforcement in the metal-based matrix to enhance the mechanical and electrical properties of materials. Graphene in different forms such as nanosheets, nanoplatelets, nanoflakes, etc. are being used as reinforcement materials in metal

TABLE 1.1
Mechanical Properties of Graphene and Nanoribbons

Graphene (potential used)	E^{3D}	M	References
Graphene (ab initio)	0.96	0.17	[22]
Graphene	0.799	—	[23]
Graphene (ab initio)	1.02	0.149	[24]
Graphene (ab initio)	1.11	—	[25]
Graphene (Brenner-non-minimized potential)	1.012	0.245	[26]
Graphene (Brenner-minimized potential)	0.669	0.416	[26]
Graphene (Brenner)	0.694	0.412	[27]
Graphene (true model)	1.11	0.45	[28]
Carbon nanotubes/nanoribbons (technique used)			
SWCNT (Brenner)	0.694	—	[29]
SWCNT (empirical model)	0.97	0.28	[30]
SWCNT (MM)	0.213–2.08	0.16	[31]
SWCNT (experimental)	0.32–1.47	—	[32]
SWCNT (MD)	0.7	—	[33]
SWCNT (vibrations)	1.0	0.25	[34]
SWCNT (experimental)	0.81–1.13	—	[35]
SWCNT (ab initio)	0.8–1.05	—	[36]
SWCNT (5,5)–(ab initio)	1.05	—	[36]
SWCNT (5,5)–(ab initio)	1.01	—	[22]

matrix composites by many researchers [37]. Yang et al. [38] fabricated the Ag/graphene composites. To fabricate the composite, graphene oxide is prepared by the hummer's method from graphite and then spark plasma sintering of GO is done to obtain the high-quality graphene (HQG). In FTIR spectroscopy, peaks for bond C-O and C=O were removed after spark plasma treatment. Ball milling of Ag with HQG at different content is done at 400 rpm for 8 hours with 1% stearic acid as the controlling agent. The Vickers hardness is maximum at 1% content of HQG and 42% higher with value of 72 HV, compared with other composites. Wang et al. [39] fabricated the aluminum-based composites reinforced with nano-graphene sheet. The wet aqueous solution approach has been employed to fabricate the graphene nanosheets (GNs). Tensile strength of Al/GNs composite is 249 MPa which shows 62% improvement compared to Al. Kim et al. [40] strengthened the Ni and Cu metals by incorporating single-layer graphene. Graphene here was synthesized using the CVD approach, which is to transfer metal by poly(methyl methacrylate) (PMMA) support to fabricate Ni-Graphene (with repeating layer spacing of 70, 125, and 200 nm) and Cu-graphene (with repeating layer spacing of 100, 150, and 300 nm). Ultra-high compressive strength of 1.5 GPa and 4 GPa were obtained during compressive testing. Rashad et al. [41] fabricated the magnesium composite reinforced with graphene to enhance the mechanical properties by powder metallurgy. With respect to sole magnesium, Young's modulus, fracture strain, and yield strength improved by 131%, 74%, and 49.5%, respectively. Bustamante et al. [42] fabricated the aluminum and graphene nanoplatelets (GNPs) composites through mechanical milling (for different amounts of time) process in which graphene is obtained through a graphite exfoliation process. Results show that 2 hours of sintering composites as compared to 5 hours results in a higher value of hardness. With increase in milling time hardness of Al/GNPs composites increases, due to synthesis of finer grain structure. Bastwros et al. [43] studied the effect of ball milling time on mechanical properties Al6061/graphene composites through solid state sintering. Graphene is synthesized through graphite exfoliate by modified Brodie's process. Flexural strength of the composites increases with increase in bill milling time and is found to be twice that of Al6061 or 90 minutes ball milling, compared with Al6061. Jeon et al. [44] employed a friction stir process to manufacture Al/graphene metal matrix composite in which graphene is synthesized by the wet chemical approach. Graphene and friction stir process slightly improve the ductility of Al, while reducing thc tensile strength.

1.1.3 Graphene-Based Polymer Composites

Graphene-based polymer composites have widely been recognized as promising materials with a large number of applications in diverse fields such as automobiles, electronic industries, defense, aerospace, etc. [45–50]. In order to get the maximum benefit out of these composites, fabricating individual graphene sheets inside the polymer matrices is a very crucial factor; since the graphene sheets tend to agglomerate inside the polymer matrices and form stress concentration sites that lead to early failure of the composite material [51, 52]. Moreover, other factors that may affect the applications, performance, and properties of these mentioned composites are the types of graphene used with their intrinsic properties, quantity of wrinkling inside its domain, its network structure inside the polymer matrix, and their interfacial interaction and dispersion state within the matrix.

Moving on to the polymer resins section, thermosetting polymer resins have also been applied in several diverse applications such as coatings, adhesives, automotive components, and aerospace structures; but numerous applications have been limited because of low stiffness, strength, toughness, thermal performance, and electrical conductivity of these polymer resins [53–60]. For example, basic composite aircraft structures could be much lighter if we could somehow increase the polymer resin toughness and conductivity or use some light fibers [61–64]. For this purpose, graphene has been shown to have a positive impact on enhancing significantly the thermosetting polymer matrices toughness, stiffness, and electrical and thermal performances [65–67]. For instance, Galpaya et al. [68] conducted a comparative study to compare the mechanical properties improvement generated by graphene (two-dimensional structure in the form of nanosheets) and functionalized graphene (in the form of graphene oxide) with those of other different nanofillers such as one-dimensional carbon nanotubes, titanium dioxide, aluminum oxide, and silicon dioxide. They stated that an enhancement of 65% in the fracture toughness values was observed when graphene of 0.125% was added to the mixture. For the same quantitative enhancement in the properties, a 30-fold higher weight fraction of aluminum oxide, a 120-fold higher weight fraction of silicon dioxide, and a 60-fold higher weight fraction of titanium oxide was needed. As for the carbon nanotubes, fracture toughness value increased by 48% and this was also achieved when a four-fold higher weight fraction was mixed with the polymers. On analyzing, despite these impressive properties, still the graphene reinforced thermosetting polymers are yet to find a suitable place in the market for large-scale applications; this is mainly due to graphene's high cost, no standardization and reproducibility of graphene composite structure and the lack of dispersion techniques in matrix domain experimentally. Several challenges exist for developing lightweight, cost-effective, high performance, and multi-material composite structures. Firstly, fabricating a pristine large graphene sheet experimentally is impossible to date, as topological defects such as grain boundaries, dislocations, and Stone–Thrower–Wales defects get induced in the domain during large-scale production [69–74]. Their functionalization is also difficult, since we can tailor the properties by placing various chemical functional groups on the two-dimensional graphene sheet [75–79]. The most common graphene production technique is the mechanical exfoliation technique but to date it is not scalable to industrial application. Next, the highly brittle (low ductility) nature of graphene composites can lead to catastrophic failure in certain cases. The literature has reported a wide range of properties for these composites; hence, standardization is required [80]. Lastly, no reliable commercial finite element software (unable to capture the joining and bonding phenomenon) exists to model the graphene-based composites that can be further utilized for the structural applications.

1.2 LARGE-SCALE PRODUCTION OF GRAPHENE-BASED COMPOSITE MATERIALS

Graphene is an allotrope of carbon, with a hexagonal structure and sp^2 hybridized carbon atoms. It has exceptional mechanical and thermal properties; thus, it is a promising candidate to be the ultimate nanofiller [81]. But, its real time commercial applications are still to be implemented as the scientific community lacks the knowledge of physical chemistry of material and its dispersion characteristics under hot melt conditions during fabrication. Several parameters such as the stacking sequence, orientation, and ordering of the nanofiller play an important role in determination of the mechanical and thermal properties of composite formed.

Graphene-based composites are manufactured using the following techniques [82]:

1. Chemical vapor deposition technique that is extremely costly, requires high process temperature, and has moderate scalability.
2. Chemical exfoliation (it has the potential for large-scale production of low-cost graphene).
3. Micromechanical exfoliation, which is also exceedingly expensive and is to date suitable for only small-scale production.
4. Chemical reduction of graphene oxide (in raw form) leads to formation of highly defective, low-in-purity graphene and also large amounts of wastes are produced.
5. Epitaxial growth on silicon carbide that requires very costly substrates and high processing temperature.

To date, graphene produced by mechanical exfoliation (with deposition on particular substrates) has fascinating properties; however, when we want to apply graphene to the field of lightweight polymer composites, the only concern is the cost and scalability (regardless of defect location and purity in the domain). Graphene nanoflakes that are produced using this technique lead to the formation of agglomeration, whereas the 2D graphene nanosheets have comparatively low chances of agglomerating in the polymer matrices. Challenges associated with cheap graphene production are selecting an appropriate material and applying it for the right application. Tailoring and optimizing the graphene-based composites production technique (selection of raw materials, process parameters, and the particles dispersion) depends on which application you intend the composites should perform. A major thought that must be taken into consideration for successfully completing the fabrication of graphene lightweight composites on a large-scale is the extent to which we can extrapolate the laboratory-based results to predict the properties on a massive scale. In laboratories, we are confined to performing experiments on pristine graphene that has only two to four layers but in the actual case graphene (or any other 2D nanomaterial) sheets are likely to be distributed randomly and with a large number of layers [83]. The number of layers (of graphene) and the inter-layer spacing between them is also an important factor in determination of the mechanical properties, as the monolayer and bilayer of graphene showcased the same mechanical strength but from the inclusion of third layer on, this property starts deteriorating [84]. Interestingly, when the number of graphene layers exceeds seven, the load-carrying efficiency reduces to more than half of that of when we used monolayer graphene. The more convenient and probably the best method to produce graphene nanocomposites up to now is through the chemical reduction of graphite oxide in defined conditions. The disadvantage associated with this method is that the reduction gives a highly defected graphene that probably would worsen the mechanical and thermal aspects.

In addition, some major questions need to be answered about graphene's surface modification with polymers, such as whether a covalent or a non-covalent bond between their interfaces should be present and what is its dispersion behavior. Recently, many manufacturing techniques have been developed to produce very high-quality graphene-based nanocomposites, namely solution blending, electro-polymerization, melt compounding, latex mixing, and in situ polymerization [85, 86]. For the stronger covalent bonding, there are "grafting to" methods such as nucleophilic ring-open and substitution reactions, esterification, and amidation reactions, whereas "grafting from" methods include ring-opening polymerization, atom transfer radical polymerization, in situ condensation polymerization, and reversible addition fragmentation chain transfer polymerization. From the non-covalent modification point of view, weak interactions dominate between graphene and polymer interface that are hydrogen bonding, electrostatic interactions, and π-π interactions [87]. Switching to the polymers section, thermoplastics (easier to recycle, lighter, and impact resistant) have an upper hand over the thermoset polymers. They tend to get hard when cooled but retain their plasticity; they re-melt and can be reshaped by heating above their melting temperature; moreover, they harden in a short time and at low temperatures, and thus can be rapidly produced in short cycle times. Since, graphene produced by recent techniques contains surface defects, so one can use these defects as high-energy adsorption sites for polymer chains. Lastly, to avoid the agglomeration condition and assure a uniform dispersion of graphene within the polymer matrix, solution mixing via ultrasonic process is mostly utilized. After manufacturing step comes the joining part, which is expensive; also, the performance is not up to the mark for these types of composites. Moreover, for the lightweight composites, lower temperature and hybrid joining techniques are required that itself is a topic for research. The latest joining techniques that are being used for joining composites are mechanical fastening (riveting and folding), adhesive bonding, and ultrasonic welding. Especially for the structural applications, adhesive bonding has a certain advantage in that it successfully connects two dissimilar materials without re-melting them or altering their mechanical properties individually.

1.3 MODELING AND SIMULATION OF GRAPHENE-BASED LIGHTWEIGHT COMPOSITE MATERIALS

Until now, there have been lots of theoretical studies based on the graphene-based nanocomposites. Mainly, the research scientists have focused on the fracture and deformation pattern of the present nanocomposites [88]; but, modeling (algorithm and computational tools) for high-performance structural applications and for severe loading conditions (crash and dynamic analysis) is still a fantasy. There have been mainly two characterization techniques that may be used to model the graphene-based nanocomposites and these are the finite element method (continuum model) and molecular dynamics (focus on the molecular level interactions among atoms/chemical mechanisms between the nanofiller and matrix; they are limited to a very small simulation box due to the constraint of computational cost) [89]. Still, these techniques lack integration between them and we cannot use them for extrapolating the real structural applications. Also, the mechanical and thermal studies are performed at a fixed temperature; in contrast, the temperature value varies and therefore, a temperature-dependent model needs to be investigated. Focusing on the continuum modeling, they are further subdivided into the Mori–Tanaka model, the Halpin–Tsai model, and the rule of mixtures [90]. For example, Cho et al. [91] used the Mori–Tanaka approach to reveal the mechanical properties of nanocomposites with randomly distributed graphene sheets. Recently, there have also been some multi-scale models suggested (combining both the molecular and continuum models). For instance, Montazeri and Rafii-Tabar [92] created a multi-scale finite element model to evaluate the mechanical properties of graphene-polymer nanocomposites, whereas Parashar and Mertiny [93] proposed a multi-scale model for buckling phenomenon in graphene-polymer composites. Generally, the point to initiate in a modeling is to select a material and their constitutive law.

Some of the most highlighted literature articles in the field of molecular dynamics are as follows: Lu et al. [94] exemplified graphene with high-density polyethylene nanocomposites through molecular dynamics (MD) simulations and a Drieding potential force field. In these simulations, the hydrocarbon chains were modeled with the aid of a united atom model that treats individual hydrocarbon sub-units as a solitary entity. In the meantime, Rahman and Foster [95] calculated the mechanical properties of polyethylene (PE)-graphene nanosheet composite materials and revealed that the randomly-oriented graphene nanosheet possessed the best properties. It was elucidated that a common neighborhood region in the PE matrix helps in increasing a tough non-local interaction, whereas the scattered allocation of non-local region in the neighborhood atoms leads to relatively weaker interactions. They also concluded that the crazing and presence of voids were among the chief reasons for failure of material to occur. In molecular dynamics simulations work conducted by Nikkhah et al. [96], polyethylene was functionalized with various sets of chemical functional groups which include the cyano, carboxylic, isocyanato, hydroxy, NH_2, and ethylamino groups. It was observed that the atoms within and greater than 3Å distance from the graphene sheet tend to both repulse and attract, respectively. The presence of chemical functional groups with heavy atoms on the polymer allow a large number of atoms to be there in the attractive region (>3Å) and the remaining few in the repulsive region (<3Å). This modification in surface helps in maintaining the work of adhesion to be higher than the separation work. Among all these functional groups, the presence of cyano group with the polyethylene led to in the highest increase in interfacial adhesion. Moreover, the work of adhesion increased with the increase in surface group density because of the reduction in surface atomic roughness at the interface, whereas the separation work decreased except for the amino, oxo, and carboxy functional groups. In 2015, Bačová et al. [97] focused attention on the edge-functionalization (with carboxylic and hydrogen functional groups) of graphene, reinforced in polyethylene matrix (nonpolar matrix) and polyethylene oxide (polar matrix). They examined the static and dynamic properties around graphene and found that the induction of dynamic heterogeneities in nanocomposite leads to the retardation in chain dynamics. Dewapriya et al. [98] conducted the atomistic and continuum-based analysis to investigate the interfacial fracture between hydrogen functionalized graphene-polyethylene nanocomposites. They stated that the interfacial properties depend on equilibrium spacing between graphene and PE matrix and the cohesive energy of the system. They also predicted that the critical spacing between graphene and polyethylene increases with the percentage of hydrogen atoms. This increase in spacing with hydrogen atoms was attributed to the repulsion acting between polymer and hydrogen as the adatoms. On the other hand, they have also predicted a decrease in cohesive stresses with the increase in percentage of adatoms. Their work concluded an overall degradation in the interfacial strength occurs with the increase in hydrogen adatoms.

1.4 ADVANCED GRAPHENE-BASED LIGHTWEIGHT COMPOSITE MATERIALS

Recently, the conventional composites used for the structural applications have begun to be replaced by hierarchical advanced graphene-polymer composites. Particularly, glass fiber—graphene nanoplatelets reinforced polypropylene composites have showcased that the combined effect of two different fillers of various size scales (nano and microscale) has the capacity to significantly improve the mechanical properties, whereas the nanofiller dispersion throughout the matrix prompted a stronger interface formation between nanoparticles and the matrix [99]. In addition, graphene nanoplatelets—carbon fiber reinforced poly-arylene ether nitrile composites have shown more enhanced mechanical properties than the host polymer matrix and combined it with individual particles [100]. For efficient use, these innovative graphene-based hybrid composites must be economical and should be prepared by environmentally simple viable techniques (such as melt mixing). To manufacture these two-phase composites, firstly, prepare the graphene nanoplatelets and polymer resin mixture. Then paint it layer-by-layer on the concerned fiber plies. Finally, place a vacuum bag over the composite preparation bench and apply curing for one day at room temperature followed by a high temperature curing (at approximately 100°C) for 4–5 hours. These materials are in high demand in the automobile sector (for interior as well as exterior parts) as they are lightweight (low density) and have higher stiffness values. Thus, one can expect them to withstand high structural loads at a relatively low temperature. Lightweight graphene composites are in high demand due to their weight reduction property (increases fuel efficiency in automobiles, reduces carbon emissions, and may be used for electromagnetic interference shielding [101–114]), and their capacity to be recycled, as well as renewable nature. These graphene-based lightweight structures have also found a place in the electronics (especially sensors) sector [115–123] and microwave absorbing applications, etc. [124–130].

ACKNOWLEDGMENT

Monetary support from the All India Council for Technical Education (AICTE) and the Department of Science and Technology (DST), India is gratefully acknowledged. The corresponding author would also like to thank the United States-India Educational Foundation (USIEF) for giving him the Fulbright-Nehru Doctoral Research (FNDR) Fellowship."

CONFLICTS OF INTEREST

There are no conflicts of interest to declare.

REFERENCES

1. Novoselov, K.S., Geim, A.K., Morozov, S., Jiang, D., Katsnelson, M.I., Grigorieva, I., Dubonos, S. and Firsov, A.A., 2005. Two-dimensional gas of massless Dirac fermions in graphene. *Nature*, *438*(7065), p.197.
2. Novoselov, K.S., Geim, A.K., Morozov, S.V., Jiang, D., Zhang, Y., Dubonos, S.V., Grigorieva, I.V. and Firsov, A.A., 2004. Electric field effect in atomically thin carbon films. *Science*, *306*(5696), pp.666–669.
3. Novoselov, K.S., Fal, V.I., Colombo, L., Gellert, P.R., Schwab, M.G. and Kim, K., 2012. A roadmap for graphene. *Nature*, *490*(7419), p.192.
4. Geim, A.K., 2009. Graphene: Status and prospects. *Science*, *324*(5934), pp.1530–1534.
5. Novoselov, K.S., Jiang, D., Schedin, F., Booth, T.J., Khotkevich, V.V., Morozov, S.V. and Geim, A.K., 2005. Two-dimensional atomic crystals. *Proceedings of the National Academy of Sciences of the United States of America*, *102*(30), pp.10451–10453.
6. Berger, C., Song, Z., Li, T., Li, X., Ogbazghi, A.Y., Feng, R., Dai, Z., Marchenkov, A.N., Conrad, E.H., First, P.N. and De Heer, W.A., 2004. Ultrathin epitaxial graphite: 2D electron gas properties and a route toward graphene-based nanoelectronics. *The Journal of Physical Chemistry. Part B*, *108*(52), pp.19912–19916.
7. Berger, C., Song, Z., Li, X., Wu, X., Brown, N., Naud, C., Mayou, D., Li, T., Hass, J., Marchenkov, A.N. and, Conrad, E.H., 2006. Electronic confinement and coherence in patterned epitaxial graphene. *Science*, *312*(5777), pp.1191–1196.
8. Stankovich, S., Dikin, D.A., Piner, R.D., Kohlhaas, K.A., Kleinhammes, A., Jia, Y., Wu, Y., Nguyen, S.T. and Ruoff, R.S., 2007. Synthesis of graphene-based nanosheets via chemical reduction of exfoliated graphite oxide. *Carbon*, *45*(7), pp.1558–1565.
9. Pichon, A., 2008. Graphene synthesis: Chemical peel. *Nature Chemistry*, pp.1755–4330.
10. Eda, G., Fanchini, G. and Chhowalla, M., 2008. Large-area ultrathin films of reduced graphene oxide as a transparent and flexible electronic material. *Nature Nanotechnology*, *3*(5), p.270.
11. Li, X., Cai, W., Colombo, L. and Ruoff, R.S., 2009. Evolution of graphene growth on Ni and Cu by carbon isotope labeling. *Nano Letters*, *9*(12), pp.4268–4272.
12. Li, X., Zhu, Y., Cai, W., Borysiak, M., Han, B., Chen, D., Piner, R.D., Colombo, L. and Ruoff, R.S., 2009. Transfer of large-area graphene films for high-performance transparent conductive electrodes. *Nano Letters*, *9*(12), pp.4359–4363.
13. Reina, A., Jia, X., Ho, J., Nezich, D., Son, H., Bulovic, V., Dresselhaus, M.S. and Kong, J., 2008. Large area, few-layer graphene films on arbitrary substrates by chemical vapor deposition. *Nano Letters*, *9*(1), pp.30–35.
14. Reina, A., Thiele, S., Jia, X., Bhaviripudi, S., Dresselhaus, M.S., Schaefer, J.A. and Kong, J., 2009. Growth of large-area single-and bi-layer graphene by controlled carbon precipitation on polycrystalline Ni surfaces. *Nano Research*, *2*(6), pp.509–516.
15. Sutter, P.W., Flege, J.I. and Sutter, E.A., 2008. Epitaxial graphene on ruthenium. *Nature Materials*, *7*(5), p.406.
16. Feng, C., Lin, C.S., Fan, W., Zhang, R.Q. and Van Hove, M.A., 2009. Stacking of polycyclic aromatic hydrocarbons as prototype for graphene multilayers, studied using density functional theory augmented with a dispersion term. *The Journal of Chemical Physics*, *131*(19), p.194702.
17. Han, M.Y., Özyilmaz, B., Zhang, Y. and Kim, P., 2007. Energy band-gap engineering of graphene nanoribbons. *Physical Review Letters*, *98*(20), p.206805.

18. Samarakoon, D.K. and Wang, X.Q., 2010. Tunable band gap in hydrogenated bilayer graphene. *ACS Nano*, *4*(7), pp.4126–4130.
19. Pereira, V.M., Neto, A.C. and Peres, N.M.R., 2009. Tight-binding approach to uniaxial strain in graphene. *Physical Review. Part B*, *80*(4), p.045401.
20. Gui, G., Li, J. and Zhong, J., 2008. Band structure engineering of graphene by strain: First-principles calculations. *Physical Review. Part B*, *78*(7), p.075435.
21. Lee, C., Wei, X., Kysar, J.W. and Hone, J., 2008. Measurement of the elastic properties and intrinsic strength of monolayer graphene. *Science*, *321*(5887), pp.385–388.
22. Faccio, R., Denis, P.A., Pardo, H., Goyenola, C. and Mombrú, A.W., 2009. Mechanical properties of graphene nanoribbons. *Journal of Physics: Condensed Matter*, *21*(28), p.285304.
23. Coluci, V.R., Dantas, S.O., Jorio, A. and Galvao, D.S., 2007. Mechanical properties of carbon nanotube networks by molecular mechanics and impact molecular dynamics calculations. *Physical Review. Part B*, *75*(7), p.075417.
24. Kudin, K.N., Scuseria, G.E. and Yakobson, B.I., 2001. C 2 F, BN, and C nanoshell elasticity from ab initio computations. *Physical Review. Part B*, *64*(23), p.235406.
25. Van Lier, G., Van Alsenoy, C., Van Doren, V. and Geerlings, P., 2000. Ab initio study of the elastic properties of single-walled carbon nanotubes and graphene. *Chemical Physics Letters*, *326*(1–2), pp.181–185.
26. Reddy, C.D., Rajendran, S. and Liew, K.M., 2006. Equilibrium configuration and continuum elastic properties of finite sized graphene. *Nanotechnology*, *17*(3), p.864.
27. Arroyo, M. and Belytschko, T., 2004. Finite crystal elasticity of carbon nanotubes based on the exponential Cauchy-Born rule. *Physical Review. Part B*, *69*(11), p.115415.
28. Reddy, C.D., Rajendran, S. and Liew, K.M., 2005. Equivalent continuum modeling of graphene sheets. *International Journal of Nanoscience*, *4*(04), pp.631–636.
29. Zhang, P., Huang, Y., Geubelle, P.H., Klein, P.A. and Hwang, K.C., 2002. The elastic modulus of single-wall carbon nanotubes: A continuum analysis incorporating interatomic potentials. *International Journal of Solids and Structures*, *39*(13–14), pp.3893–3906.
30. Lu, J.P., 1997. Elastic properties of carbon nanotubes and nanoropes. *Physical Review Letters*, *79*(7), p.1297.
31. Shen, L. and Li, J., 2004. Transversely isotropic elastic properties of single-walled carbon nanotubes. *Physical Review. Part B*, *69*(4), p.045414.
32. Yu, M.F., Files, B.S., Arepalli, S. and Ruoff, R.S., 2000. Tensile loading of ropes of single wall carbon nanotubes and their mechanical properties. *Physical Review Letters*, *84*(24), p.5552.
33. Sammalkorpi, M., Krasheninnikov, A., Kuronen, A., Nordlund, K. and Kaski, K., 2004. Mechanical properties of carbon nanotubes with vacancies and related defects. *Physical Review. Part B*, *70*(24), p.245416.
34. Yoon, J., Ru, C.Q. and Mioduchowski, A., 2005. Terahertz vibration of short carbon nanotubes modeled as Timoshenko beams. *Journal of Applied Mechanics*, *72*(1), pp. 10–17.
35. Wu, Y., Huang, M., Wang, F., Huang, X.H., Rosenblatt, S., Huang, L., Yan, H., O'Brien, S.P., Hone, J. and Heinz, T.F., 2008. Determination of the Young's modulus of structurally defined carbon nanotubes. *Nano Letters*, *8*(12), pp.4158–4161.
36. Bogár, F., Mintmire, J.W., Bartha, F., Mező, T. and Van Alsenoy, C., 2005. Density-functional study of the mechanical and electronic properties of narrow carbon nanotubes under axial stress. *Physical Review. Part B*, *72*(8), p.085452.
37. Kim, Y., Lee, J., Yeom, M.S., Shin, J.W., Kim, H., Cui, Y., Kysar, J.W., Hone, J., Jung, Y., Jeon, S. and Han, S.M., 2013. Strengthening effect of single-atomic-layer graphene in metal–graphene nanolayered composites. *Nature Communications*, *4*, p.2114.

38. Yang, Y., Ping, Y., Gong, Y., Wang, Z., Fu, Q. and Pan, C., 2019. Ag/graphene composite based on high-quality graphene with high electrical and mechanical properties. *Progress in Natural Science: Materials International* 2(4), p. 384–389.
39. Wang, J., Li, Z., Fan, G., Pan, H., Chen, Z. and Zhang, D., 2012. Reinforcement with graphene nanosheets in aluminum matrix composites. *Scripta Materialia, 66*(8), pp.594–597.
40. Kim, Y., Lee, J., Yeom, M.S., Shin, J.W., Kim, H., Cui, Y., Kysar, J.W., Hone, J., Jung, Y., Jeon, S. and Han, S.M., 2013. Strengthening effect of single-atomic-layer graphene in metal–graphene nanolayered composites. *Nature Communications, 4*, p.2114.
41. Rashad, M., Pan, F., Hu, H., Asif, M., Hussain, S. and She, J., 2015. Enhanced tensile properties of magnesium composites reinforced with graphene nanoplatelets. *Materials Science and Engineering: Part A, 630*, pp.36–44.
42. Pérez-Bustamante, R., Bolaños-Morales, D., Bonilla-Martínez, J., Estrada-Guel, I. and Martínez-Sánchez, R., 2014. Microstructural and hardness behavior of graphene-nanoplatelets/aluminum composites synthesized by mechanical alloying. *Journal of Alloys and Compounds, 615*, pp.S578–S582.
43. Bastwros, M., Kim, G.Y., Zhu, C., Zhang, K., Wang, S., Tang, X. and Wang, X., 2014. Effect of ball milling on graphene reinforced Al6061 composite fabricated by semi-solid sintering. *Composites Part B: Engineering, 60*, pp.111–118.
44. Jeon, C.H., Jeong, Y.H., Seo, J.J., Tien, H.N., Hong, S.T., Yum, Y.J., Hur, S.H. and Lee, K.J., 2014. Material properties of graphene/aluminum metal matrix composites fabricated by friction stir processing. *International Journal of Precision Engineering and Manufacturing, 15*(6), pp.1235–1239.
45. Das, T.K. and Prusty, S., 2013. Graphene-based polymer composites and their applications. *Polymer-Plastics Technology and Engineering, 52*(4), pp.319–331.
46. Verdejo, R., Bernal, M.M., Romasanta, L.J. and Lopez-Manchado, M.A., 2011. Graphene filled polymer nanocomposites. *Journal of Materials Chemistry, 21*(10), pp.3301–3310.
47. Tang, L.C., Zhao, L. and Guan, L.Z., 2017. 7 Graphene/polymer composite materials: Processing, properties and applications In *Advanced Composite Materials: Properties and Applications* (Bafekrpour, E., ed.) (pp. 349–419). Sciendo Migration.
48. Dhand, V., Rhee, K.Y., Kim, H.J. and Jung, D.H., 2013. A comprehensive review of graphene nanocomposites: Research status and trends. *Journal of Nanomaterials, 2013*, p.158.
49. Mukhopadhyay, P. and Gupta, R.K. eds., 2012. *Graphite, Graphene, and Their Polymer Nanocomposites*. CRC Press.
50. Huang, X., Qi, X., Boey, F. and Zhang, H., 2012. Graphene-based composites. *Chemical Society Reviews, 41*(2), pp.666–686.
51. Stankovich, S., Dikin, D.A., Dommett, G.H., Kohlhaas, K.M., Zimney, E.J., Stach, E.A., Piner, R.D., Nguyen, S.T. and Ruoff, R.S., 2006. Graphene-based composite materials. *Nature, 442*(7100), p.282.
52. Kuilla, T., Bhadra, S., Yao, D., Kim, N.H., Bose, S. and Lee, J.H., 2010. Recent advances in graphene based polymer composites. *Progress in Polymer Science, 35*(11), pp.1350–1375.
53. Verma, A. and Singh, V.K., 2018. Mechanical, microstructural and thermal characterization of epoxy-based human hair–reinforced composites. *Journal of Testing and Evaluation, 47*(2), pp.1193–1215.
54. Verma, A., Gaur, A. and Singh, V.K., 2017. Mechanical properties and microstructure of starch and sisal fiber biocomposite modified with epoxy resin. *Materials Performance and Characterization, 6*(1), pp.500–520.

55. Verma, A., Negi, P. and Singh, V.K., 2018. Physical and thermal characterization of chicken feather fiber and crumb rubber reformed epoxy resin hybrid composite. *Advances in Civil Engineering Materials*, *7*(1), pp.538–557.
56. Verma, A., Negi, P. and Singh, V.K., 2018. Experimental investigation of chicken feather fiber and crumb rubber reformed epoxy resin hybrid composite: Mechanical and microstructural characterization. *Journal of the Mechanical Behavior of Materials*, *27*(3–4), pp.3–4.
57. Verma, A., Joshi, K., Gaur, A. and Singh, V.K., 2018. Starch-jute fiber hybrid biocomposite modified with an epoxy resin coating: Fabrication and experimental characterization. *Journal of the Mechanical Behavior of Materials*, *27*(5–6), pp.5–6.
58. Verma, A., Negi, P. and Singh, V.K., 2019. Experimental analysis on carbon residuum transformed epoxy resin: Chicken feather fiber hybrid composite. *Polymer Composites*, *40*(7), pp.2690–2699.
59. Verma, A., Budiyal, L., Sanjay, M.R. and Siengchin, S., 2019. Processing and characterization analysis of pyrolyzed oil rubber (from waste tires)-epoxy polymer blend composite for lightweight structures and coatings applications. *Polymer Engineering and Science*, *59*(10), pp.2041–2051.
60. Verma, A., Baurai, K., Sanjay, M.R. and Siengchin, S., 2020. Mechanical, microstructural, and thermal characterization insights of pyrolyzed carbon black from waste tires reinforced epoxy nanocomposites for coating application. *Polymer Composites*, *41*(1), pp.338–349.
61. Verma, A., Singh, V.K., Verma, S.K. and Sharma, A., 2016. Human hair: A biodegradable composite fiber–a review. *International Journal of Waste Resources*, *6*(206), p.2.
62. Verma, A., Singh, C., Singh, V.K. and Jain, N., 2019. Fabrication and characterization of chitosan-coated sisal fiber–Phytagel modified soy protein-based green composite. *Journal of Composite Materials*, DOI: 0021998319831748.
63. Verma, A. and Singh, V.K., 2016. Experimental investigations on thermal properties of coconut shell particles in DAP solution for use in green composite applications. *Journal of Materials Science and Engineering*, *5*(3), p.1000242.
64. Jain, N., Verma, A. and Singh, V.K., 2019. Dynamic mechanical analysis and creep-recovery behaviour of polyvinyl alcohol based cross-linked biocomposite reinforced with basalt fiber. *Materials Research Express*. DOI: 10.1088/2053-1591/ab4332.
65. Loh, K.P., Bao, Q., Ang, P.K. and Yang, J., 2010. The chemistry of graphene. *Journal of Materials Chemistry*, *20*(12), pp.2277–2289.
66. Qin, W., Vautard, F., Drzal, L.T. and Yu, J., 2016. Modifying the carbon fiber–epoxy matrix interphase with graphite nanoplatelets. *Polymer Composites*, *37*(5), pp.1549–1556.
67. Ramanathan, T., Abdala, A.A., Stankovich, S., Dikin, D.A., Herrera-Alonso, M., Piner, R.D., Adamson, D.H., Schniepp, H.C., Chen, X.R.R.S., Ruoff, R.S. and, Nguyen, S.T., 2008. Functionalized graphene sheets for polymer nanocomposites. *Nature Nanotechnology*, *3*(6), p.327.
68. Galpaya, D., Wang, M., Liu, M., Motta, N., Waclawik, E.R. and Yan, C., 2012. Recent advances in fabrication and characterization of graphene-polymer nanocomposites. *Graphene*, *1*(2), pp.30–49.
69. Verma, A. and Parashar, A., 2017. The effect of STW defects on the mechanical properties and fracture toughness of pristine and hydrogenated graphene. *Physical Chemistry Chemical Physics*, *19*(24), pp.16023–16037.
70. Verma, A., Parashar, A. and Packirisamy, M., 2018. Tailoring the failure morphology of 2D bicrystalline graphene oxide. *Journal of Applied Physics*, *124*(1), p.015102.
71. Verma, A. and Parashar, A., 2018. Reactive force field based atomistic simulations to study fracture toughness of bicrystalline graphene functionalized with oxide groups. *Diamond and Related Materials*, *88*, pp.193–203.

72. Verma, A., Kumar, R. and Parashar, A., 2019. Enhanced thermal transport across a bi-crystalline graphene–polymer interface: An atomistic approach. *Physical Chemistry Chemical Physics*, *21*(11), pp.6229–6237.
73. Singla, V., Verma, A. and Parashar, A., 2018. A molecular dynamics based study to estimate the point defects formation energies in graphene containing STW defects. *Materials Research Express*, *6*(1), p.015606.
74. Verma, A., Parashar, A. and Packirisamy, M., 2019. Effect of grain boundaries on the interfacial behavior of graphene-polyethylene nanocomposite. *Applied Surface Science*, *470*, pp.1085–1092.
75. Verma, A. and Parashar, A., 2018. Molecular Dynamics based simulations to study failure morphology of hydroxyl and epoxide functionalized graphene. *Computational Materials Science*, *143*, pp.15–26.
76. Verma, A. and Parashar, A., 2018. Molecular Dynamics based simulations to study the fracture strength of monolayer graphene oxide. *Nanotechnology*, *29*(11), p.115706.
77. Verma, A. and Parashar, A., 2018. Structural and chemical insights into thermal transport for strained functionalized graphene: A molecular dynamics study. *Materials Research Express*, *5*(11), p.115605.
78. Verma, A., Parashar, A. and Packirisamy, M., 2019. Role of chemical adatoms in fracture mechanics of graphene nanolayer. *Materials Today: Proceedings*, *11*, pp.920–924.
79. Eswaraiah, V., Sankaranarayanan, V. and Ramaprabhu, S., 2011. Functionalized graphene–PVDF foam composites for EMI shielding. *Macromolecular Materials and Engineering*, *296*(10), pp.894–898.
80. Verma, A., Parashar, A. and Packirisamy, M., 2018. Atomistic modeling of graphene/hexagonal boron nitride polymer nanocomposites: A review. *Wiley Interdisciplinary Reviews: Computational Molecular Science*, *8*(3), p.e1346.
81. Xu, Z., Zhang, Y., Li, P. and Gao, C., 2012. Strong, conductive, lightweight, neat graphene aerogel fibers with aligned pores. *ACS Nano*, *6*(8), pp.7103–7113.
82. Elmarakbi, A. and Azoti, W., 2018. State of the art on graphene Lightweighting nanocomposites for automotive applications In *Experimental Characterization, Predictive Mechanical and Thermal Modeling of Nanostructures and Their Polymer Composites* (pp. 1–23). Elsevier.
83. Chaurasia, A., Verma, A., Parashar, A. and Mulik, R.S., 2019. An Experimental and Computational Study to Analyse the Effect of h-BN nanosheets on Mechanical Behaviour of h-BN/Polyethylene nanocomposite. *The Journal of Physical Chemistry C*, *123*(32), pp.20059–20070.
84. Gong, L., Young, R.J., Kinloch, I.A., Riaz, I., Jalil, R. and Novoselov, K.S., 2012. Optimizing the reinforcement of polymer-based nanocomposites by graphene. *ACS Nano*, *6*(3), pp.2086–2095.
85. Istrate, O.M., Paton, K.R., Khan, U., O'Neill, A., Bell, A.P. and Coleman, J.N., 2014. Reinforcement in melt-processed polymer–graphene composites at extremely low graphene loading level. *Carbon*, *78*, pp.243–249.
86. Zhang, M., Li, Y., Su, Z. and Wei, G., 2015. Recent advances in the synthesis and applications of graphene–polymer nanocomposites. *Polymer Chemistry*, *6*(34), pp.6107–6124.
87. Eswaraiah, V., Sankaranarayanan, V. and Ramaprabhu, S., 2011. Functionalized graphene–PVDF foam composites for EMI shielding. *Macromolecular Materials and Engineering*, *296*(10), pp.894–898.
88. Rafiee, M.A., Rafiee, J., Wang, Z., Song, H., Yu, Z.Z. and Koratkar, N., 2009. Enhanced mechanical properties of nanocomposites at low graphene content. *ACS Nano*, *3*(12), pp.3884–3890.
89. Awasthi, A.P., Lagoudas, D.C. and Hammerand, D.C., 2008. Modeling of graphene–polymer interfacial mechanical behavior using molecular dynamics. *Modelling and Simulation in Materials Science and Engineering*, *17*(1), p.015002.

90. Hashin, Z. and Rosen, B.W., 1965. Erratum: "The Elastic Moduli of Fiber-Reinforced Materials" (Journal of Applied Mechanics, 1964, 31, pp. 223–232). *Journal of Applied Mechanics*, *32*(1), pp.219–219.
91. Cho, J., Luo, J.J. and Daniel, I.M., 2007. Mechanical characterization of graphite/epoxy nanocomposites by multi-scale analysis. *Composites Science and Technology*, *67*(11–12), pp.2399–2407.
92. Montazeri, A. and Rafii-Tabar, H., 2011. Multiscale modeling of graphene-and nanotube-based reinforced polymer nanocomposites. *Physics Letters. A*, *375*(45), pp.4034–4040.
93. Parashar, A. and Mertiny, P., 2012. Representative volume element to estimate buckling behavior of graphene/polymer nanocomposite. *Nanoscale Research Letters*, *7*(1), p.515.
94. Lu, C.T., Weerasinghe, A., Maroudas, D. and Ramasubramaniam, A., 2016. A comparison of the elastic properties of graphene-and fullerene-reinforced polymer composites: The role of filler morphology and size. *Scientific Reports*, *6*(1).
95. Rahman, R. and Foster, J.T., 2014. Deformation mechanism of graphene in amorphous polyethylene: A molecular dynamics based study. *Computational Materials Science*, *87*, pp.232–240.
96. Nikkhah, S.J., Moghbeli, M.R. and Hashemianzadeh, S.M., 2015. Interfacial adhesion between functionalized polyethylene surface and graphene via molecular dynamic simulation. *Journal of Molecular Modeling*, *21*(5), pp.1–12.
97. Bačová, P., Rissanou, A.N. and Harmandaris, V., 2015. Edge-functionalized graphene as a nanofiller: Molecular Dynamics simulation study. *Macromolecules*, *48*(24), pp.9024–9038.
98. Dewapriya, M.A.N., Rajapakse, R.K.N.D. and Nigam, N., 2015. Influence of hydrogen functionalization on the fracture strength of graphene and the interfacial properties of graphene–polymer nanocomposite. *Carbon*, *93*, pp.830–842.
99. Pedrazzoli, D., Pegoretti, A. and Kalaitzidou, K., 2015. Synergistic effect of graphite nanoplatelets and glass fibers in polypropylene composites. *Journal of Applied Polymer Science*, *132*(12).
100. Yang, X., Wang, Z., Xu, M., Zhao, R. and Liu, X., 2013. Dramatic mechanical and thermal increments of thermoplastic composites by multi-scale synergetic reinforcement: Carbon fiber and graphene nanoplatelet. *Materials and Design*, *44*, pp.74–80.
101. Zhang, H.B., Yan, Q., Zheng, W.G., He, Z. and Yu, Z.Z., 2011. Tough graphene–polymer microcellular foams for electromagnetic interference shielding. *ACS Applied Materials and Interfaces*, *3*(3), pp.918–924.
102. Yan, D.X., Ren, P.G., Pang, H., Fu, Q., Yang, M.B. and Li, Z.M., 2012. Efficient electromagnetic interference shielding of lightweight graphene/polystyrene composite. *Journal of Materials Chemistry*, *22*(36), pp.18772–18774.
103. Xu, L., Jia, L.C., Yan, D.X., Ren, P.G., Xu, J.Z. and Li, Z.M., 2018. Efficient electromagnetic interference shielding of lightweight carbon nanotube/polyethylene composites via compression molding plus salt-leaching. *RSC Advances*, *8*(16), pp.8849–8855.
104. Shen, B., Zhai, W., Tao, M., Ling, J. and Zheng, W., 2013. Lightweight, multifunctional polyetherimide/graphene@ Fe3O4 composite foams for shielding of electromagnetic pollution. *ACS Applied Materials and Interfaces*, *5*(21), pp.11383–11391.
105. Ling, J., Zhai, W., Feng, W., Shen, B., Zhang, J. and Zheng, W.G., 2013. Facile preparation of lightweight microcellular polyetherimide/graphene composite foams for electromagnetic interference shielding. *ACS Applied Materials and Interfaces*, *5*(7), pp.2677–2684.
106. Gavgani, J.N., Adelnia, H., Zaarei, D. and Gudarzi, M.M., 2016. Lightweight flexible polyurethane/reduced ultralarge graphene oxide composite foams for electromagnetic interference shielding. *RSC Advances*, *6*(33), pp.27517–27527.

107. Yuan, Y., Liu, L., Yang, M., Zhang, T., Xu, F., Lin, Z., Ding, Y., Wang, C., Li, J., Yin, W. and Peng, Q., 2017. Lightweight, thermally insulating and stiff carbon honeycomb-induced graphene composite foams with a horizontal laminated structure for electromagnetic interference shielding. *Carbon*, *123*, pp.223–232.
108. Zeng, Z., Jin, H., Chen, M., Li, W., Zhou, L., Xue, X. and Zhang, Z., 2017. Microstructure design of lightweight, flexible, and high electromagnetic shielding porous multiwalled carbon nanotube/polymer composites. *Small*, *13*(34), p.1701388.
109. Tan, Y., Luo, H., Zhang, H., Zhou, X. and Peng, S., 2016. Lightweight graphene nanoplatelet/boron carbide composite with high EMI shielding effectiveness. *AIP Advances*, *6*(3), p.035208.
110. Chaudhary, A., Kumari, S., Kumar, R., Teotia, S., Singh, B.P., Singh, A.P., Dhawan, S.K. and Dhakate, S.R., 2016. Lightweight and easily foldable MCMB-MWCNTs composite paper with exceptional electromagnetic interference shielding. *ACS Applied Materials and Interfaces*, *8*(16), pp.10600–10608.
111. Raagulan, K., Braveenth, R., Jang, H., Seon Lee, Y., Yang, C.M., Mi Kim, B., Moon, J. and Chai, K., 2018. Electromagnetic shielding by MXene-graphene-PVDF composite with hydrophobic, lightweight and flexible graphene coated fabric. *Materials*, *11*(10), p.1803.
112. Zhang, H.B., Yan, Q., Zheng, W.G., He, Z. and Yu, Z.Z., 2011. Tough graphene–polymer microcellular foams for electromagnetic interference shielding. *ACS Applied Materials and Interfaces*, *3*(3), pp.918–924.
113. Chen, Z., Xu, C., Ma, C., Ren, W. and Cheng, H.M., 2013. Lightweight and flexible graphene foam composites for high-performance electromagnetic interference shielding. *Advanced Materials*, *25*(9), pp.1296–1300.
114. Singh, A.K., Kumar, A., Haldar, K.K., Gupta, V. and Singh, K., 2018. Lightweight reduced graphene oxide-Fe3O4 nanoparticle composite in the quest for an excellent electromagnetic interference shielding material. *Nanotechnology*, *29*(24), p.245203.
115. Eda, G. and Chhowalla, M., 2009. Graphene-based composite thin films for electronics. *Nano Letters*, *9*(2), pp.814–818.
116. Konwer, S., Guha, A.K. and Dolui, S.K., 2013. Graphene oxide-filled conducting polyaniline composites as methanol-sensing materials. *Journal of Materials Science*, *48*(4), pp.1729–1739.
117. Aphirakaramwong, C., Phattharasupakun, N., Suktha, P., Krittayavathananon, A. and Sawangphruk, M., 2019. Lightweight multi-walled carbon nanotube/N-doped graphene aerogel composite for high-performance lithium-ion capacitors. *Journal of the Electrochemical Society*, *166*(4), pp.A532–A538.
118. Qin, Y., Peng, Q., Ding, Y., Lin, Z., Wang, C., Li, Y., Xu, F., Li, J., Yuan, Y., He, X. and Li, Y., 2015. Lightweight, superelastic, and mechanically flexible graphene/polyimide nanocomposite foam for strain sensor application. *ACS Nano*, *9*(9), pp.8933–8941.
119. Huang, W., Dai, K., Zhai, Y., Liu, H., Zhan, P., Gao, J., Zheng, G., Liu, C. and Shen, C., 2017. Flexible and lightweight pressure sensor based on carbon nanotube/thermoplastic polyurethane-aligned conductive foam with superior compressibility and stability. *ACS Applied Materials and Interfaces*, *9*(48), pp.42266–42277.
120. Ma, Z., Wei, A., Ma, J., Shao, L., Jiang, H., Dong, D., Ji, Z., Wang, Q. and Kang, S., 2018. Lightweight, compressible and electrically conductive polyurethane sponges coated with synergistic multiwalled carbon nanotubes and graphene for piezoresistive sensors. *Nanoscale*, *10*(15), pp.7116–7126.
121. Zeng, Z., Liu, M., Xu, H., Liao, Y., Duan, F., Zhou, L.M., Jin, H., Zhang, Z. and Su, Z., 2017. Ultra-broadband frequency responsive sensor based on lightweight and flexible carbon nanostructured polymeric nanocomposites. *Carbon*, *121*, pp.490–501.
122. Stoller, M.D., Park, S., Zhu, Y., An, J. and Ruoff, R.S., 2008. Graphene-based ultracapacitors. *Nano Letters*, *8*(10), pp.3498–3502.

123. Liu, C., Yu, Z., Neff, D., Zhamu, A. and Jang, B.Z., 2010. Graphene-based supercapacitor with an ultrahigh energy density. *Nano Letters*, *10*(12), pp.4863–4868.
124. Singh, V.K., Shukla, A., Patra, M.K., Saini, L., Jani, R.K., Vadera, S.R. and Kumar, N., 2012. Microwave absorbing properties of a thermally reduced graphene oxide/nitrile butadiene rubber composite. *Carbon*, *50*(6), pp.2202–2208.
125. Wen, B., Wang, X.X., Cao, W.Q., Shi, H.L., Lu, M.M., Wang, G., Jin, H.B., Wang, W.Z., Yuan, J. and Cao, M.S., 2014. Reduced graphene oxides: The thinnest and most lightweight materials with highly efficient microwave attenuation performances of the carbon world. *Nanoscale*, *6*(11), pp.5754–5761.
126. Li, J.S., Huang, H., Zhou, Y.J., Zhang, C.Y. and Li, Z.T., 2017. Research progress of graphene-based microwave absorbing materials in the last decade. *Journal of Materials Research*, *32*(7), pp.1213–1230.
127. Jiang, Y., Chen, Y., Liu, Y.J. and Sui, G.X., 2018. Lightweight spongy bone-like graphene@ SiC aerogel composites for high-performance microwave absorption. *Chemical Engineering Journal*, *337*, pp.522–531.
128. Song, W.L., Guan, X.T., Fan, L.Z., Zhao, Y.B., Cao, W.Q., Wang, C.Y. and Cao, M.S., 2016. Strong and thermostable polymeric graphene/silica textile for lightweight practical microwave absorption composites. *Carbon*, *100*, pp.109–117.
129. You, X., Yang, J., Feng, Q., Huang, K., Zhou, H., Hu, J. and Dong, S., 2018. Three-dimensional graphene-based materials by direct ink writing method for lightweight application. *International Journal of Lightweight Materials and Manufacture*, *1*(2), pp.96–101.
130. Wu, H. and Drzal, L.T., 2012. Graphene nanoplatelet paper as a light-weight composite with excellent electrical and thermal conductivity and good gas barrier properties. *Carbon*, *50*(3), pp.1135–1145.

2 Conventional Processing of Polymer Matrix Composites

M.K. Singh, N. Verma, and S. Zafar

CONTENTS

2.1 INTRODUCTION

Huge growth over the last two decades can be observed in the automobile and aerospace industries. These industries require specific material properties to manufacture their products. A single material can't fulfill the requirements of all properties and it is not possible to use two materials for the same component. So, composites were created, which may be altered according to the required property of the product. Composites are created by of physically mixing of two constituents. These two constituents should be chemically insoluble in each other. One constituent is known as matrix phase and another constituent is known as reinforced phase. Composites are generally used for applications which require lightweight and high-strength material. Nowadays, composites are replacing heavier metals. For example, the traditional automobile engine block is now replaced by aluminum matrix composite (Singh et al., 2015; Verma and Vettivel, 2018). The automobile and aerospace industries are always looking for material that is lightweight and strong in order to reduce the overall weight of the vehicle. The reduction of weight is important for higher fuel economy and efficiency. Polymers matrix composites are lighter than metal matrix composite (MMC). The strength of some polymers matrix composite (carbon fiber reinforced composite) is higher than MMC. The construction of Boeing 787 was accomplished using carbon fiber reinforced polymer (CFRP). The decrease in weight by 20% was observed by using CFRP instead of aluminum alloy. The applications of polymers are not restricted only to automobile and aerospace uses; polymer composite (Kevlar fiber-reinforced composite) has excellent applications in body armor. The applications of polymers can be also seen in biomedical implants. For example, hydroxyapatite-reinforced polymer composite has the ability to replace heavier biomedical implants of metals and ceramics (Liu et al., 2008). The composite should fulfill the following conditions to be called composite:

- The wt. % of constituents must be greater than 10%.
- The property of reinforcement should be five times greater than the property of another constituent.
- The addition of constituents should contribute to a significant change in the property.

The composite materials may be naturally founded or man-made. Various naturally occurring composites are wood (lignin acts as matrix material and cellulose acts as reinforced material) and bone (composite of calcium phosphate and soft collagen). Due to the limited properties of the naturally occurring composite, the man-made composites were created. The man-made composites were based on metal, ceramics, and polymers. The various traditionally used man-made polymer composites in the aircraft industry are CFRP composite, glass fiber reinforced polymer (GFRP) composite, and Kevlar fiber reinforced polymer composite. The composite materials have the following unique properties which separate them from conventional materials.

2.1.1 Advantages of Composite Materials

- High strength.
- High stiffness.
- High fatigue life.
- Resistance to wear is high.
- Degradation of composite in the environment is of higher rate.
- It can be used where thermal and electrical insulation is required.
- Energy dissipation.
- Reduction in cost, as well as weight, can be achieved.
- Composites are of higher endurance strength.
- They provide long-term resistance to severe chemical and temperature environments.

2.1.2 Major Applications of Composite Materials

- In 1969, epoxy/boron carbide composite was used to make horizontal stabilizers of F-14 aircraft and epoxy/carbon fiber composite was used to make wings and fuselage.
- The airframe of F-22 and AV-8B contains 25 wt.% of carbon fiber.
- Lear fan 2100 (business aircraft) was made by Kevlar fiber composite (Noyes, 1983).
- The roof panel of the BMW M6 is made of carbon fiber composite.
- The chassis interior and suspension of Formula 1 racing car is made of carbon fiber-reinforced composite.

The first use of composite may be observed in 1500 BC when the settlers of Mesopotamians and Egyptians were made of straw (reinforcement) and mud (matrix). Later in AD 1200, the Mongols invented the first composite bow by using animal glue, wood, and bone. The birth of the plastic era was observed in 1900. The various plastics developed in the 1900s were phenolic, vinyl, polystyrene, and polyester. However, plastic alone was not efficient for attaining better mechanical properties. The invention of fiberglass-reinforced plastic was observed during the Second World War. In the 1970s, the industry of composites began to mature. Various synthetic fibers such as Kevlar and carbon fibers were developed at that time. The composite materials are still in demand and effectively used in energy sectors such as manufacturing of wind turbine blades.

2.1.3 Classifications of Composites

Composites are mainly classified into two categories. Classifications may be done on the basis of their constituents (either matrix or reinforcement) as is shown in Figure 2.1.

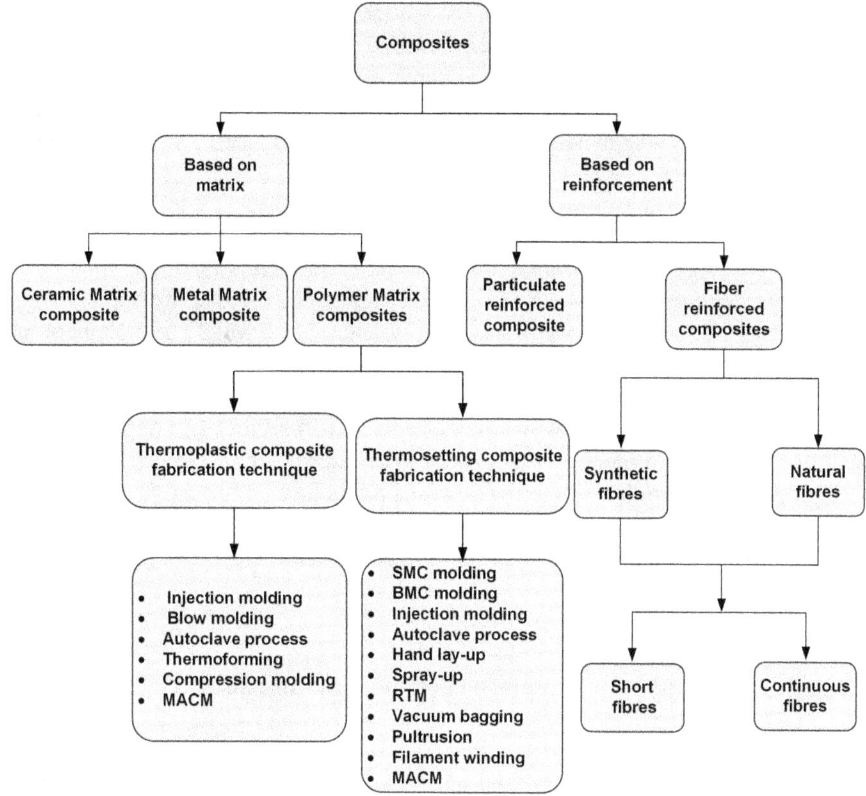

FIGURE 2.1 Classification of various composite materials.

2.1.3.1 Based on Matrix

The matrix acts as a continuous phase of the composite. The role of the matrix is to give shape to the structure. As the matrix is present in the continuous phase in the structure, it provides protection to reinforcement from the external environment. The matrix acts as a constituent of composite that first encounters the imposed force on composite. Generally, the matrix does not contribute to withstanding the imposed forces acting on it. The role of the matrix is to transfer the load to fiber. The matrix classification is discussed in the section below.

2.1.3.1.1 Ceramic Matrix Composite

The use of ceramics in the fabrication of composite is more popular due to their high resistance to temperature and corrosive environments. Ceramics are very hard and stiff, but the fracture toughness of ceramics is less than metals. The main applications of ceramics are in the manufacturing of hard machining tools, automotive engines, mining equipment, and pressure vessels.

2.1.3.1.2 Metal Matrix Composite (MMC)

Metal matrix composites are the composites in which the available matrix phase is in the form of metals. The most popular metals used in MMC are aluminum alloys, magnesium alloy, and titanium alloys. The advantage to using metals as matrix material is its higher value of elastic modulus and its resistance to elevated temperature. The various metallic materials used in composites are aluminum, aluminum zinc alloy, aluminum-lithium alloy, aluminum-magnesium alloy, copper, and titanium. The limitations of using metals in the composite are its higher density and difficulty in fabrication processes due to their high melting points.

2.1.3.1.3 Polymer Matrix Composite (PMC)

Polymers are receiving more attention over the last few decades, due to their being lightweight, featuring high extension under tension, and corrosion resistance. There are varieties of polymers matrix available to fabricate PMC. The polymers are mainly classified into categories such as thermosetting polymers and thermoplastic polymers. The thermoplastic polymers melt on heating. These types of polymers can be softened again and again by heating. The various thermoplastic polymers along with their mechanical properties and applications are shown in Table 2.1.

The use of thermoplastics matrix in the fabrication of composite is not so popular due to limited mechanical properties. The thermosetting polymers have excellent mechanical properties as compared to thermoplastic polymers. Thermosetting

TABLE 2.1
Properties of Popular Thermoplastic Polymers (Azom, 2019; Matweb, 2019)

S.No	Material	Density (g/cc)	Modulus (GPa)	Flexural yield strength (MPa)	Ultimate tensile strength (MPa)	Application
1	ABS	1.01–1.2	1.0–2.6	49.6–113.0	22.1–49.0	Instrument board, outer panel, steering wheel, and dashboard
2	Polyamides	1.11–1.17	1.3–4.2	20–150	50–90	Safety airbag in cars
3	Polyethylene	0.93–1.27	0.62–1.45	28–91	11–25	Packaging, pipes and fitting
4	Polypropylenes	0.88–2.40	0.008–8.25	20–180	9–80	Packaging and automotive Industry
5	Polystyrenes	0.013–1.18	0.303–3.35	28–106	17.9–60.7	Smoke Detectors, plastic cutlery
6	Polyether ether ketone	1.26–1.50	2.20–6.48	24.1–380	54.5–265	Bearing, piston parts and pumps
7	Polyvinyl chloride	1.13–1.85	1.12–4.83	28.0–99.9	3.74–55.9	Wiring and cables
8	Poly-lactic acid	1.00–2.47	0.08–13.8	6.0–145	14–114	Disposable goods, medical bags

TABLE 2.2
Properties of Popular Thermosetting Polymers (Matweb, 2019)

S.No.	Material	Density (g/cc)	Modulus (GPa)	Flexural yield strength (MPa)	Ultimate tensile strength (MPa)	Applications
1	Polyester	0.60–2.20	1.0–10	53.8–265	10–123	Car tires reinforcements, conveyor belts, and safety belts
2	Vinyl ester	1.03–1.95	3.72–94.5	20–150	30.3–993	Tanks and vessels
3	Polyimides	1.60	27.6	400	241	Firefighters' cloths, acoustic and thermal insulation
4	Polyurethanes	1.21	2.62	68.5–92	36.5–55.2	Building insulation, footwear
6	Melamine	1.50 –1.81	5.0–13.3	70–140	53–76	Wiring and cables
7	Urea formaldehyde	1.5	9	100	55	Disposable goods, medical bags

polymers cannot reform or remelt after curing. These polymers form three-dimensional (3D) cross-linked chains on curing. Due to this cross-linking effect, it is difficult to break the molecules of polymers. The rigidity and thermal stability of polymers increase with an increase in the cross-linking effect. The processing of thermosetting materials is easy because it is available in the form of liquid resin at room temperature. The resin provides better wettability to reinforced fibers. The popular thermosetting polymers along with their mechanical properties are presented in Table 2.2.

2.1.4 Comparison of Thermoplastic and Thermosetting Polymers

- The thermoplastic polymers are more ductile than thermosetting polymers.
- The applications of thermoplastics polymers are limited due to their lesser strength while the strength of thermosetting polymers is higher than that of thermoplastics.
- The thermoplastic polymers can be recycled again while thermosetting polymers cannot be recycled. Therefore, thermosetting polymers are hazardous to the environment.
- The impact resistance of thermoplastic polymers is higher than that of thermosetting polymers due to their high extension and energy-absorbing properties under the action of load.
- Thermosetting polymers are highly stable in a thermal environment as compared to thermoplastic polymers.
- Thermoplastic polymers can be remolded and reshaped by a cycle of heating while thermosetting polymers cannot be reshaped after curing.

2.1.5 Based on Reinforcement

The role of reinforcement is to provide stiffness, strength, and other physical properties to composite. It is well known that the mechanical properties of the composite are highest in the direction of the reinforced fiber. For example, if all fibers are reinforced in the length direction, then the composite is strongest against the tensile load in the length direction. This property of composite allows the designer to decide the percentage of fiber. If a force acts on composite materials from all directions, then the designer would prefer random fibers. The resultant properties of the composite are not the only function of properties of matrix and reinforcement, but it is also dependent on bonding and interaction between matrix and reinforcement. When the load acts on the composite, the following two phenomena may be observed i.e., delamination of fiber and fracture on matrix. Delamination of fiber may occur due to poor bonding between the matrix and the reinforcement. Fracture on matrix under the action of load may be observed due to load transfer from the matrix to the reinforcement failing. The reinforcement may be present in two forms

(i) Particulate reinforcement
(ii) Fiber reinforcement

Particulate reinforcements are present in the matrix in the form of particles. The size of the particles must be less than 0.25 μm. The strengthening of composite depends upon the wettability between matrix and reinforced particles. Particulate reinforced composites are more popular in the case of MMC. For example, graphite reinforced aluminum alloy 6061 composite may be used for wear applications due to the better lubrication properties of graphite. However, particulate reinforcement may also be used in the case of polymers. For example, hydroxyapatite reinforced polylactic acid (PLA) composite for bone scaffolds applications, carbon nanotube reinforced polyethylene composite for structural applications, and composite of ultra-high molecular weight polyethylene and hydroxyapatite may be used as an alternative material for biomedical implants (Liu et al., 2008; Arora and Pathak, 2019a, b). The various types of reinforcements used in the case of particulate reinforcements are carbides (SiC, WC, TiC, and B_4C), nitrides (AlN and Si_3N_4), oxides (SiO_2 and Al_2O_3), and. hydroxyapatite particle reinforcement may be used for biomedical applications.

Fibers are the systemic arrangement of strands or threads in one or two directions. In the case of one direction, it is known as unidirectional fiber while an arrangement in two directions is known as bidirectional fiber. The fiber may be present in the composite in various forms i.e., continuous fibers, discontinuous fibers, long fibers, and short fibers (Arora and Pathak, 2019c). The diameter of fiber varies from 5 to 20 μm. The fibers are classified as natural fibers and synthetic fibers.

2.1.5.1 Natural Fibers

These are types of fibers which are obtained from nature. The properties of the natural fiber depend on structural features and the chemical composition of the fiber. The natural fibers consist of various chemicals such as pectin, cellulose, hemicellulose, and lignin. Cellulose is responsible for the strength of natural fiber, while lignin acts

TABLE 2.3
Properties of Natural Fibers (Mohanty et al., 2005)

Fiber	Density (g/cm³)	Diameter (μm)	Tensile strength (MPa)	Young's modulus (GPa)	Elongation at break %
Flex	1.5	40–600	345–1500	27.6	2.7–3.2
Ramie	1.55		400–938	53	1.2–3.8
Cotton	1.5–1.6	12–38	287–800	5.5–12.6	7–8
Jute	1.3–1.49	25–200	393–800	13–26.5	1.16–1.5
Sisal	1.45	50–200	468–700	38	3–7
Hemp	1.47	25.500	690	70	1.6
Coir	1.15–1.46	100–460	131–220	4–6	15–40

as a binder to cellulose. The physical properties of natural fibers are presented in Table 2.3.

2.1.5.2 Synthetic Fibers

Synthetic fibers are man-made fibers used for applications requiring strong material. The mechanical properties of synthetic fibers are better than natural fibers. The properties of various synthetic fibers along with their properties are presented in Table 2.4.

2.1.5.3 Comparison of Synthetic Fibers and Natural Fibers

- The strength of synthetic fibers is higher than natural fibers.
- The cost of synthetic fibers is higher than natural fibers.
- Most strong materials are based on synthetic fiber-reinforced thermosetting polymers. The applications of natural fibers are limited to low load only. For example, carbon fiber-reinforced composite may be used in aerospace applications, where high fatigue resistance is required and Kevlar fiber-reinforced composite may be used in body armor applications. Whereas natural fiber composites are used for the fabrication of tables, chairs, roofing, photo frames, door panels, etc.

TABLE 2.4
Properties of Synthetic Fibers (Mohanty et al., 2005)

Fiber	Density (g/cm³)	Diameter (μm)	Tensile strength (MPa)	Young's modulus (GPa)	Elongation at break %
Carbon fiber	1.78	5.7	3400–4800	240–425	1.4–1.8
E glass	2.55	17	3400	73	2.5
Aramid	1.44		3000	60	2.5–3.7

2.2 PROCESSING OF POLYMER MATRIX COMPOSITES

Processing of polymer matrix composites can be discussed according to the type of polymer used in the composites. Polymers may be thermoplastic or thermosetting. The properties of both types of polymers are different; therefore, the choice of fabrication/processing method can also vary. In some of the cases, the choice of the fabrication/processing method may be the same. In the case of common composites, the fabrication method for both polymers is discussed in Section 2.3 and 2.4. Before discussion of the fabrication process, it is necessary to know the factors which affect the fabrication of polymer composites. Below, some important factors are discussed in detail.

2.2.1 Degree of Cure

For more efficient use of the mold, post-curing is an important process. The post-curing is done by removal of the cured specimen from the mold after dimensional stability of the specimen in the mold is attained and the specimen has been placed in an oven at elevated temperature. This is done to increase the cross-linking of the specimen. Cross-linking leads to an increase in desired physical and mechanical properties. The pot curing is mainly done on polyester and other thermoset polymers. Post-curing is important for those polymers which do not cure completely at room temperature.

2.2.2 Viscosity

The viscosity of resin may be defined as the resistance to flow under shear stress. Fluids with low molecular weight i.e., water, oil, etc. have low viscosities and they are readily able to flow. Whereas, molten polymers have high molecular weight, and as a result, the viscosities of molten polymers are high and they flow under the higher value of shear stress. Temperature and shear rate are the two important factors on which viscosity depends. For all fluids, the viscosity decreases with increasing temperature. The viscosity of the lower molecular weight fluids is not affected by the shear rate, whereas in the case of higher molecular weight fluid, it either increases (shear-thickening) or decreases (shear thinning). Molten polymers are generally shear-thinning fluid; therefore, their viscosity decreases with increasing intensity of shearing. The starting material for a thermoset resin is a low-viscosity fluid. However, the viscosity of thermosets increases with proper curing; as a result, it transforms into a solid mass. By varying the curing properties, different levels of viscosity of the resin can be achieved. It is observed that as the time and temperature of curing increases, the viscosity of the thermoset increases. It is interesting to note that at an earlier stage of curing, viscosity increases at a low rate. Whereas, after attaining a particular threshold value, the viscosity increases rapidly (Mallick, 2007). The time at which the viscosity of the thermoset increases and after this point, thermosets are no longer workable and this is called gel time. Therefore, the gel time is an important parameter during the fabrication of thermoset composites.

2.2.3 Resin Flow

For the proper wetting of the fiber reinforcement, it is necessary that the flow of resin should be easy inside the mold. Therefore, the selection of resin should be in such a way that its viscosity should not increase too rapidly during the curing. In the case of too rapid an increase in viscosity, the wettability of the fiber decreases, causing voids and poor interfacial bonding or adhesion. The flow of resin through fiber reinforcement has been modeled using Darcy's equation (see Equation 2.1) for porous media (Zeng and Grigg, 2006). This equation helps in finding out the flow rate of resin per unit area through fiber using the pressure gradient, which causes the resin flow to occur. Darcy's equation (see Equation 2.1) for one-dimensional flow in x-direction can be written as:

$$q = -\frac{k}{\mu}\left(\frac{dP}{dx}\right) \tag{2.1}$$

where q = volumetric flow rate of resin per unit area (m/s) in x-direction
k = permeability
μ = viscosity (Ns/m^2)
$\frac{dP}{dx}$ = negative pressure gradient (N/m^3) in direction of flow

Equation 2.1 has been used by many researchers for the modeling of resin flow from prepregs in bag-molding process and mold filling in resin transfer molding (RTM). During modeling, it is assumed that the size of pores that have to be filled is uniform and the porous medium is isotropic. Equation 2.1 is used very well to predict the resin flow rate in the fiber direction. Whereas in the case of transverse direction flow, Equation 2.1 is not valid.

2.2.4 Consolidation

Good resin flow and compaction are the two necessary parameters for the consolidation of the composite. During composite fabrication, consolidation of matrix and reinforcement is necessary, otherwise, a variety of defects will occur in the composites which include voids, resin-rich areas, and interply cracks. Good resin flow is not the only condition necessary for the consolidation; proper compaction pressure is also required. The compaction pressure helps in squeezing out the trapped air or volatiles, as liquid resin flows through the fiber network, suppresses voids, and attains a uniform fiber volume fraction.

2.2.5 Gel-Time Test

The gel-time test is the main method to determine the curing of the thermoset composites. In the gel-time test, a measured amount (10 g) of a thoroughly mixed resin–catalyst combination is poured into a standard test tube. The rise in temperature of the mixture is monitored as a function of time with the help of thermocouple. During this procedure, the test tube is suspended into the water bath with a constant temperature of 82°C. As the curing begins, the liquid resin mixes with the catalyst and after some time it transforms into a gel-like mass. This gel-time of the particular resin is noted and used for further fabrication processes.

2.2.6 Shrinkage

During curing, shrinkage of the resin takes place because of the solidification as well as thermal contraction. The shrinkage may be linear or volumetric. Curing shrinkage mainly takes place because of the rearrangement of polymer molecules into a more compact structure during curing reaction. The thermal shrinkage occurs during the cooling period that follows the curing reaction and may take place both inside and outside of the mold. The volumetric shrinkage for vinyl ester resins are in the range of 5–12% and for cast-epoxy resin, it is in the range of 1–5%.

With the addition of fillers or fibers, the volumetric shrinkage decreases. Considering unidirectional fibers, the reduction in the longitudinal direction is more than when compared to that of the transverse direction. The shrinkage rate of polyester or vinyl ester is high; therefore, to reduce the shrinkage, addition of some additives is required. Some of these additives are thermoplastic polymers i.e., polymethyl acrylate, polycaprolactone, polyethylene, and polyvinyl acetate. These additives are also called low-profile agents. The reason behind this may be that the thermoplastics have a neutral effect on cross-linking and forms in the second phase in cured resin.

2.2.7 Voids

Voids are the most influential defect among various defects present in the composite laminates. The creation of voids is mainly due to the presence of air in fiber, which cannot be removed during wetting. Another reason for void formation may be entrapped air bubbles or volatiles entrapped in the liquid resin. Sometimes, the voids are formed due to the solvents used for viscosity control, moisture, or chemical contamination, which may remain dissolved in the resin during elevated curing temperature. There are various techniques used to remove the void such as creating a vacuum or applying compaction pressure during curing. The presence of large-volume fractions of voids in the composite laminate reduces the level of desired mechanical properties. Also, these voids may lead to absorption of moisture in humid environments. Thus, the properties of the composites may be distorted.

2.3 MANUFACTURING TECHNIQUES OF THERMOPLASTIC COMPOSITES

2.3.1 Injection Molding

Injection molding is a widely used technique to manufacture different composites. It is also successfully used to manufacture the thermoset composites. The term injection is used because we are injecting the molten plastic through the nozzle into the die. It is used to manufacture an incredible variety of parts, ranging between 5 g and 85 kg (Mazumdar, 2002). Typically, it is used for large-volume production.

The injection molding technique is used with both types of polymer i.e., thermoplastics and thermosets. The technique of injection molding of thermoplastic/thermoset composite is the same as the injection molding of neat thermoplastic/thermoset. For both manufacturing processes i.e., that of neat polymer and polymer composites, the same tool is required.

In the injection molding process, a fixed amount of raw material is poured into the mold cavity through a heated screw barrel as shown in Figure 2.2. The raw material used for thermoplastic/thermoset composite materials is in the form of pellets or granules. These pellets or granules are obtained by extruding the different mixture of thermoplastic and reinforcement with different weight percentages. After extruding the thermoplastics with different reinforcement, the pelletizer is used to make the pellets of different sizes as per the requirement. In most of the requirements, the size of pellets is 10 mm. The reinforcements within the thermoplastic or thermosets are natural fiber, synthetic fiber, ceramics, dust, etc. The size of the chopped fiber is in the range of 0.2 to 6 mm long. During the extruding process, the size of the fiber reduces to 0.1 to 0.5 mm because of fiber crack during screw crushing load (Ville et al., 2013). The thermoplastics which are used to make pellets are polyethylene (PE), polypropylene (PP), high-density polyethylene (HDPE), polyphenylene sulfide, polyethylene terephthalate (PET), nylon, etc. In the case of the thermosets, the materials used to make pellets are polyester, bakelite, polyurethane, resin, phenolic, vinyl ester, etc. (Yao et al., 2018).

The arrangement of piston in the form of screw and cylinder is used to feed the mold cavity from hopper to mold cavity (see Figure 2.2). Heating coils are used to heat the feeding pellets or powder according to their melting point or glass transition temperature. For the sufficient feeding pressure, heated pellets or powder were passed through the nozzle, attached at the end of the barrel. The selection mold cavity was based on the type of material that was going to be molded as well as the material's thermal properties. For the easy removal of parts, split molds are used. In some cases, the cooling system is used for quick removal of the molded parts. The cycle time of the process is between the range of 20 to 60 s. In most of the cases, injection molding is of an automated nature. The major segment of the part

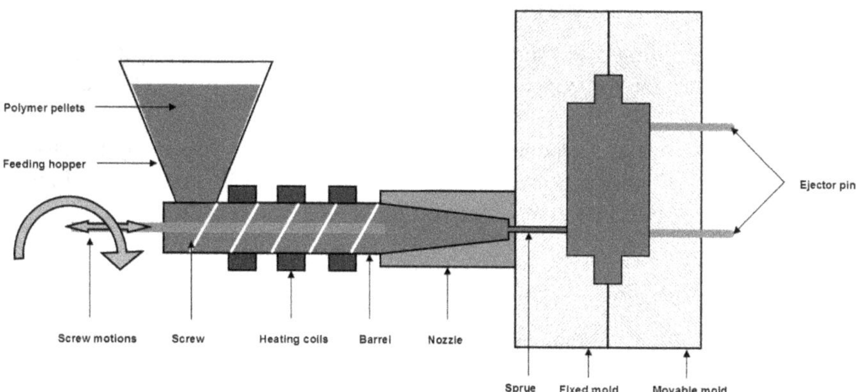

FIGURE 2.2 Injection molding setup.

fabrication cycle time is occupied by the cooling of the parts. Parts that are used to fabricate may be injected either into the single-cavity mold or into the multi-cavity mold. It can be concluded that the injection molding process can be achieved in four stages: clamping, injection, cooling, and ejection.

2.3.1.1 Application

Injection molding is generally used to produce high-volume products. Some of the products manufactured using neat polymers are power tools of small size, tubes of sink, parts of sewing machines, pocket combs, wire spools, bottle caps, packaging, mugs, soap casings, toys, toothbrushes, electrical plug fuses, medical devices, including valves and syringes, etc. Other applications related to reinforced composites are computer parts, automotive dashboards, sprockets, equipment housings, and more.

2.3.1.2 Benefits of Injection Molding

- The process cycle time of injection molding is in the range of 20 to 60 s, which leads to a higher production rate than that of other methods.
- For small complex parts, close tolerance is achieved.
- Neat and near-neat shape products are obtained since the in-process waste is negligible.
- There is no need for finishing and trimming because the product obtained from this process has a good surface finish.
- The geometry of complex shapes can be achieved in one shot.
- Repeatability of the parts is much better as compared to the other manufacturing technique.
- Injection molding leads to low production cost because of low labor cost and in-process waste is negligible.
- The range of the production volume is very high from 5 g to 85 kg.

2.3.1.3 Limitations of Injection Molding

- Equipment required during this technique is of higher cost.
- The cost of installing the injection molding setup is very high.
- The frequency of die changing is limited because of the higher cost of tool change.
- Injection molding is not suitable for the prototype parts. Rapid prototyping is preferable for the conception of a new product and to decide the final product.
- The quality of parts produced cannot be determined immediately because of so many process variables i.e., injection pressure, back pressure, melting point of feed polymer, temperature of mold, and size of pellets, etc.

2.3.2 Blow Molding

The concept of blow molding derived from the glassblowing process. During the 1880s there was a patent published which explains the method of extruding a celluloid polymer into a parison and blow molding a part. Earlier methods and techniques to deform plastics into various shapes were very crude and not suitable for mass

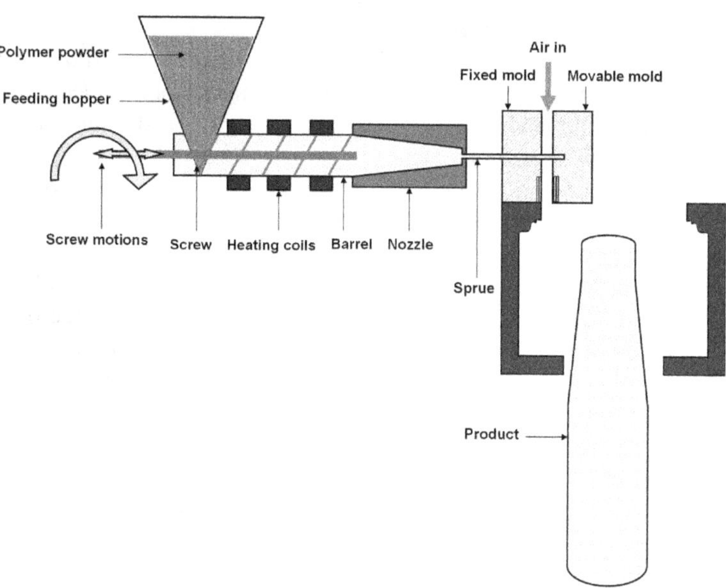

FIGURE 2.3 Schematic of blow molding.

production. It wasn't until the late 1930s that the first commercial machines were used to develop blow-molded bottles. The major element that was explored in the early years was the development of high-density polyethylene and low-density polyethylene for blow molding and consumer packages. After this, there was an explosion of blow-molded products and equipment in both Europe and North America.

Blow molding is used to mold hollow parts of a symmetric shape. The size of the product produced through the blow mold is small as compared to the product produced through rotational molding. Blow molding is basically used to mold plastic materials. The principal behind blow molding is the pressurized air is blown through the gap between the fixed mold and movable mold. Blow molding is a two-step process; in the first step the desired shape of plastic is molded and in the second step pressurized air is blown into the molded plastic and the final product is obtained as shown in Figure 2.3. Sometimes, there is the problem of sticking of the component into the mold. For that purpose, ejector is used for proper removal of the component. The process parameters required for blow molding are type of parison to be made, melting temperature of the plastic materials, blow ratio, air pressure, and cooling time. The processing cycle time of blow molding is 60–140 s. The typical materials used in blow molding are low-density polyethylene (LDPE), HDPE, PP, PET, polyvinyl chloride (PVC), co-polyester, nylon, polycarbonate, polystyrene, ethylene-vinyl alcohol (EVOH), ethylene-vinyl acetate (EVA), thermoplastic elastomers (TPE), etc.

Blow molding can be classified into three types.

All main differences are based on the method of forming the parison; either by injection molding or extrusion, the method of movement between the parison and blow molds, which may be stationary, shuttling, linear, or rotary, and the size of the parison.

(i) Extrusion blow molding (EBM)

In EBM, the polymer is melted and solid extrude specimen of that melt is extruded through a die to form a tube. Two halves of a cooled mold are then closed and pressurized air is injected through a needle, and entered in the shape of the mold for producing the hollow part. After the hot plastic has cooled sufficiently, the mold is opened and the part is removed.

EBM consists of two methods of extrusion, continuous and intermittent. In the continuous process, the parison is continuously extruded and the mold moves to and away from the parison. The plastic is accumulated by extruder in a chamber, then is forced through the die to form the parison. The molds are typically stationary under or around the extruder. Examples of the continuous process are with rotary wheel machines and continuous extrusion shuttle machines. Intermittent extrusion machines can be reciprocating screw or accumulator head. Various factors are considered when selecting between the processes and the size or models available. Examples of parts made using the EBM process include many hollow products, such as bottles, industrial parts, toys, automotives, appliance components, and industrial packaging.

(ii) Injection blow molding (IBM)

In IBM, the polymer is injected into a core within a cavity to form a hollow tube called a preform. The preforms rotate on the core rod to mold. This process is a feasible technique for making bottles. The process is divided into three steps: injection, blowing, and ejection, all done in an integrated machine. Parts come out with accurate finished dimensions and with being capable of holding tight tolerances with no extra material in the formation, it is highly efficient. Examples of parts made using IBM are pharmaceutical bottles, medical parts, and cosmetic and other consumer product packages.

(iii) Injection stretch blow molding (ISBM)

The ISBM process is similar to the IBM process described above, in that the preform is injection molded. The molded preform is then presented to the blow mold in a conditioned state, but before the final blowing of the shape, the preform is stretched in length as well as radially. The typical polymers used are PET and PP, that have physical characteristics that are enhanced by the stretching part of the process. This stretching gives the final part improved strength and barrier properties at much lighter weights and better wall thicknesses than with IBM or EBM but not without some limits such as handled containers, etc. ISBM can be divided into a one-step and two-step process.

In the case of a one-step process, the same machine is used to make preform and bottle blowing simultaneously. This can be done in three or four station machines (injection, conditioning, blowing, and ejection). This process and related equipment can handle small to high volumes of various shapes and sizes of bottles.

In the case of a two-step process, the plastic is first molded into the preform using an injection molding machine separate from the blow molder. These preforms are produced with the necks of the bottles, including threads on the open end of the closed-end hollow preform. These preforms are cooled, stored, and fed later into a reheat stretch blow-molding machine. Before obtaining the final product, the preforms are preheated using typical infrared heaters and blown using high-pressure air in the blow mold.

2.3.2.1 Application

- Garden, lawn, and household items
- Construction industry products
- Components of automobile which are under the hood
- Jars, bottles, toys, containers, etc.
- Ducting
- Bulk containers for industries
- Parts required in the medical field, i.e., bottles, tubes, etc.
- Tankers to store fluid
- Components of various appliances

2.3.2.2 Advantages of Blow Molding

- The production rate is very fast in addition to the tooling cost being low.
- Complex parts can be molded easily.
- The generation of wastes is very low.
- Hollow parts of large size can be produced but not as large as in the case of rotational molding.
- Produced parts can be recycled.
- The production level is high with this molding process.
- It offers the benefits of automation.

2.3.2.3 Limitations of Blow Molding

- It is highly dependent on petroleum.
- Blow molding has a severe impact on the environment.
- It is limited to the hollow form of product.
- Thick parts cannot be manufactured.
- Low strength of the fabricated products.

2.3.3 Autoclave Process

Autoclave molding is an alteration of pressure-bag and vacuum-bag molding. This technique is basically used to apply heat and pressure during curing. With the help of higher heat and pressure during autoclave curing, a void-free and denser product is obtained. Before autoclaving, laminates are stacked layer-wise in a mold and stitched at some places to avoid any type of relative movement. After stacking, the whole assembly is placed in a vacuum bag to remove entrapped air between the laminates. The detailed schematic of the autoclave molding process can be seen in Figure 2.4. After removal of the entrapped air, the whole assembly is put into an autoclave. To achieve consolidation, a proper amount of heat and pressure can be applied as per requirement. After completion of the consolidation, composite parts are taken

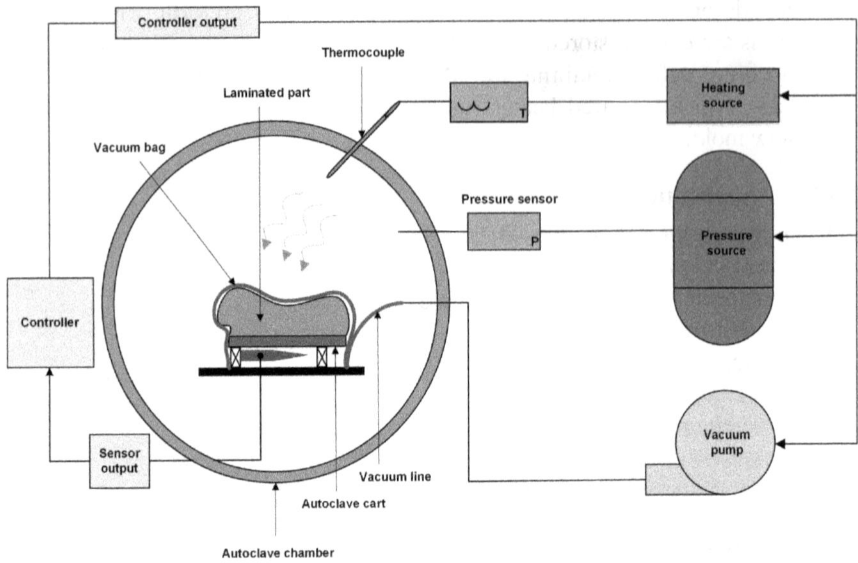

FIGURE 2.4 Illustration of autoclave processing.

out from the mold. In the autoclave molding process, good polymer composites are obtained with enhanced interfacial bonding. To avoid the sticking of the composites to the mold, release gel is applied. The raw materials used in these techniques are matrix i.e., epoxy, unsaturated polyesters, polyvinyl esters, polyester, phenolic resin, polyurethane resin, and thermoplastic resins and reinforcement i.e., glass fiber, carbon fiber, aramid fiber, natural fiber, etc. All of these fibers may be unidirectionally or bidirectionally oriented and these fibers may be continuously or randomly oriented.

The processing cost of autoclave heating is higher than oven heating, therefore, it is used when there is the necessity of isostatic pressure on the complex shape. In the case of smaller parts, cycle time effectively reduces when the parts are pressed after heating. For other applications, processing pressure is not required because of the use of steam, which acts as a heating as well as a pressing medium. The temperature of steam depends on the amount of pressure required for curing. Vulcanization of rubber is one of the examples of this technique (Ang et al., 2015). Similar to autoclave processes, hydroclave processes are used for exceptional requirements. In hydroclave processes, pressurized water at higher temperature is used for curing. The phase of water is liquid despite higher temperature because of high pressure. Examples of components cured using hydroclave processes are the nozzles of rocket engines and the nosecones of missiles. Hydroclave processes involves high equipment cost and higher risk during operation.

2.3.3.1 Application

Autoclave curing is mostly used to fabricate components with higher specific strength. These components may be spacecraft, marine ship, missiles, aircraft parts, military weapons, etc.

2.3.3.2 Advantages of Autoclave Process

- The autoclave process has the flexibility of curing composites with a higher-volume fraction of reinforcement.
- Thermoplastic and thermosets can both be processed using an autoclave process.
- Composites obtained in this process have higher uniformity and better interfacial adhesion between matrix and reinforcement.
- During the autoclave process entrapped air is removed because of vacuum, which leads to the fabrication of void-free products.
- Fibers are wetted completely because of vacuum during the processing of composites.
- Better bonding during the use of cores and inserts.

2.3.3.3 Limitations of Autoclave Process

- The size of part produced depends on the size of autoclave.
- The cost of autoclave processing is high.
- The production rate is low.
- It requires skilled labor.
- In autoclave molding various defects are observed i.e., delamination, debonding, voids, resin that is too rich, pores, and lack of resin (Ang et al., 2015).

2.3.4 Thermoforming

Thermoforming is a process in which plastic sheets are formed with the application of heat and pressure. The thermoplastic sheet is placed horizontally over the surface of the mold and clamped using a holding device as shown in Figure 2.5. The sheet

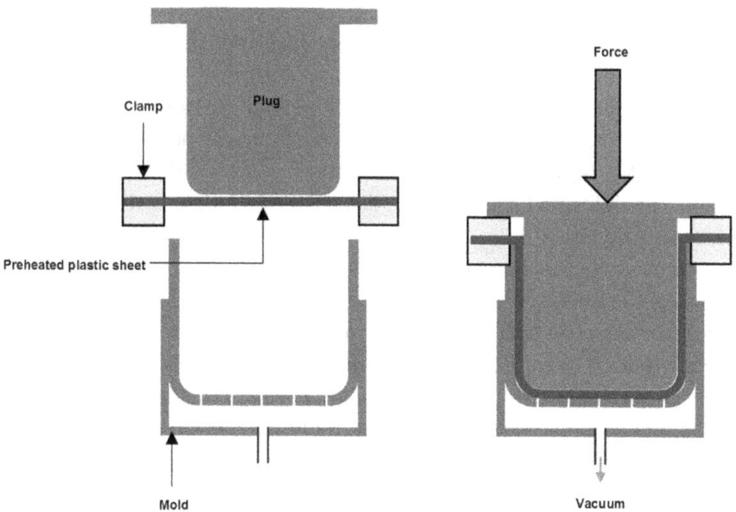

FIGURE 2.5 Schematic of thermoforming.

is heated up to glass transition temperature using a heater. The thermostat sets the heater temperature. The cooled air is provided when the temperature becomes high in the mold. The thermoplastic sheet softens with the effect of heat and is pressed into the mold surface by air pressure or by any other suitable means. The softened sheet deforms into the mold shape and removed after gradual cooling of mold. The mold cavity is opened and the thermoformed part is released. Some plastics require air cooling to become rigid quickly because the thermal conductivity of some plastics is low. The excess material is removed by secondary operations. Excess material can mix again with unused plastic, and can be again deformed into plastic sheets. Thin and thick sheets can be deformed easily. The different types of thermoplastic materials which can be processed using thermoforming process are polypropylene, acrylonitrile butadiene styrene (ABS), acrylic (PMMA), HDPE, polystyrene (PS), LDPE, cellulose acetate, and PVC.

Thermoforming setup usually consists of the heaters, clamping unit, mold, and air cooling system. The molds should be cleaned after each and every cycle, as materials in the mold can cause a change in the shape of the finished goods. There are mainly three different types of thermoforming processes depending upon the pressure required i.e., pressure forming, vacuum forming, and matched die forming.

(i) Vacuum Forming

Vacuum forming is the technique in which vacuum pressure is used to form the thermoplastic sheet into a certain shape. The placement of thermoplastic sheet is done on the mold surface and is fixed by clamping unit. The sheet is heated up to its softening temperature. The sheet is heated until it is softened and after that, vacuum is applied quickly. A surge tank is provided for the quick pull out of the air. When the vacuum is created, the sheet conforms to the shape of the mold cavity. The formed part is cooled and then ejected from the mold cavity.

(ii) Pressure Forming

This process is the same as vacuum forming. In this process, the air pressure required is higher than vacuum forming. The preheated plastic sheet is placed on the mold surface, and then air pressure is quickly applied over the sheet. The pressure developed between the pressure box and softened sheet is high. Due to high pressure, the sheet deforms quickly in the mold. The deformed sheet is placed in the mold for some time for cooling.

(iii) Matched Die Forming

Matched die forming is also known as mechanical forming. In match die forming, the mold consists of two parts i.e., the punch and the die. The thermoplastic sheet is heated up to the softening temperature. The preheated sheet is placed onto the surface of the mold and punch pressure is applied on the hot sheet. The presented air in between the die and sheet is removed by application of vacuum pump, and therefore the sheet deforms into the mold shape. The formed part is cooled and ejected from the mold cavity. The important process parameters of the thermoforming process are heating time, heating temperature, vacuum pressure, mechanical pressure, air pressure, cooling time, and ejection mechanism.

2.3.4.1 Application

This process is used for manufacturing of components for various applications, such as automotive parts, food packaging, building products, trays, and aircraft windscreens. Thick parts of gauge may be used as cosmetics on permanent structures like medical equipment, trucks, electrical and electronic equipment, material handling equipment, spas and shower enclosures, air ducts, vehicle door and dash panels, refrigerator liners, printer enclosures, utility vehicle beds, office furniture, and plastic pallets. Thin gauge parts are primarily used to package or contain a food item, disposable cups, containers, lids, blisters, and clamshells.

2.3.4.2 Advantages of Thermoforming

- Thermoforming can be adopted as per the requirements of design of product.
- It is useful in developing prototypes.
- Good dimensional stability.
- Low production cost.
- The cost of installation of this technique is very little.
- Thermal stress generated during this process is less than that of injection and compression molding on comparison.

2.3.4.3 Limitations of Thermoforming

- Poor surface finish.
- It is difficult to mold ribs and bosses.
- The thickness of part may be non-uniform.
- Materials used during this technique are limited.
- All parts need to be trimmed.
- Plastic sheets of higher thickness cannot be achieved.

2.3.5 Compression Molding

Compression molding is a popular technique used to manufacture various composite products. In compression molding, closed molds with high pressure is used to fabricate composites. The closed molds which have to be matched are generally made of metal. There are two plates in compression molding i.e., base plate and upper plate as shown in Figure 2.6, where the upper plate is movable and the base plate is stationary. The charge consisting of reinforcement and matrix is placed in the metallic mold and the whole assembly is kept in between the compression mold. For adequate bonding between matrix and reinforcement, proper application of heat and pressure takes place for a fixed period of time. Due to the proper application of heat and pressure on the mold, charge flows into the cavity and takes the cavity shape with high dimensional accuracy of the specimen. The temperature required to cure the composite may vary according to the type of polymer and it may happen either at room temperature or at a higher temperature. When the curing process is completed, composites are removed from the mold and processed further. If concentrating on the principle of working, it can be said that it is a type of press (oriented vertically) consisting of two halves i.e., top and bottom.

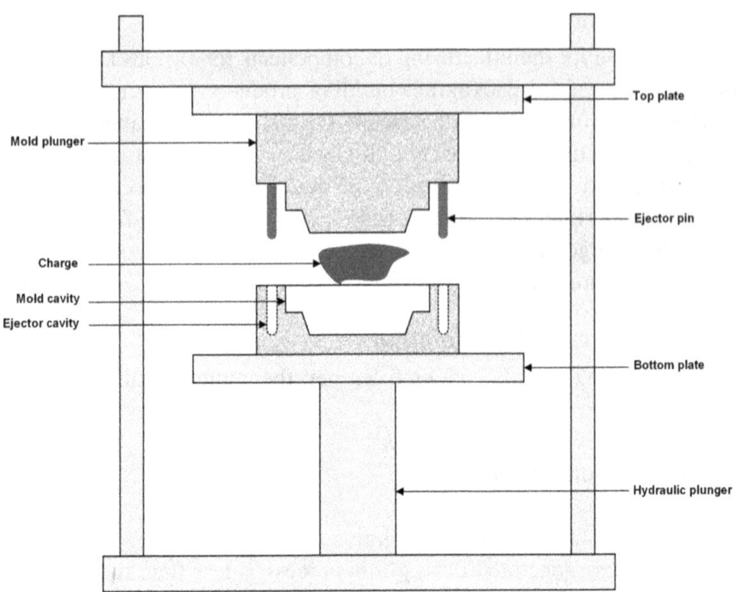

FIGURE 2.6 Setup of compression molding.

For the most part, in compression molding pressure is applied to the mold using a hydraulic system. During compression molding, the main aim of the controlling parameter is to develop a higher-quality product with desired properties. The controlling parameters are mainly compression pressure, temperature of mold, and time of application of pressure. All these three parameters are critical to optimization for the achievement of a better-quality product. If the applied compression pressure is not sufficient, then the interfacial bonding of the composite will decrease, whereas in the case of pressure application that is too high, fiber breakage and resin leakage takes place. Similarly, in the case of temperature that is too low, thermoplastics will not flow properly and the problem of wetting takes place and in the case of temperature that is too high, degradation of matrix or reinforcement takes place. Now, if considering the factor of time, then when given insufficient time, various manufacturing defects will occur. Other process parameters of compression molding i.e., closing rate of two halve plates, mold wall heating, and de-molding time also affect the properties of products.

Rubber molding is one of the widely used production methods in compression molding. It can be used to produce products from low to medium volumes. Additionally, variety of part size and materials can be produced. It is very efficient in producing bulky parts i.e., seals, O-rings, gaskets, etc. There is no excess consumption of rubber in this technique, like in the cases of injection and transfer molding techniques. If considering the principle of all molding processes, it can be observed that compression molding is used everywhere. Materials that can be used in compression molding are polyurethane, polyethylene, neoprene, nylons, polycarbonates, acrylics, polyesters, fluoroelastomers, fluorosilicone, etc.

2.3.5.1 Application

- This method can be used to process both the polymer types i.e., thermosets and thermoplastics.
- This method is better for high and medium hardness products. This technique has the capability to process even harder materials also.
- The largest range of typical products can be manufactured using this method, i.e., gaskets, diaphragms, seals, bearings, golf balls, bushings, shoes soles, roof, life gates, battery trays, fenders, automobile panels, bumpers, hoods, spoilers, air deflectors, furniture, kitchen bowls and trays, buttons, large containers, dinnerware, recreational vehicle body panels, medical equipments (ultrasound equipments), printed circuit board and fuel cells.
- It is able to process all material from bulky compounds to very hard material.
- Almost every industry uses this technique, from industry to aerospace.

2.3.5.2 Advantages of Compression Molding

- Lowest installation cost.
- Shortest tool setting time.
- Low internal stresses.
- Capability to process very stiff materials.
- Less wastage of raw materials.
- Ability to process large and thin parts.
- Provision of flexibility of molding of various part sizes.
- Low shrinkage.
- More cavities per mold are possible as lower molding pressure is required.
- Suitable for low- to medium-scale production.
- Suitable for large- and medium-sized parts.
- Parts of thick cross-section can be obtained.

2.3.5.3 Limitations of Compression Molding

- Exact weighed material is required for processing.
- Products obtained can show more dimensional inconsistency.
- It produces large parting lines.
- Requires secondary operation to remove the flash.
- In this technique, labor requirements are higher but the process can be automated to some extent.

2.4 MANUFACTURING TECHNIQUES OF THERMOSET COMPOSITES

2.4.1 Sheet Molding Compound (SMC) Molding

SMC is used to fabricate larger parts. Instead of the same composition of SMC and BMC, the mixing process is a little different in the case of SMC. A batch-blending process is used to mix the constituents in the case of BMC. In the case of SMC,

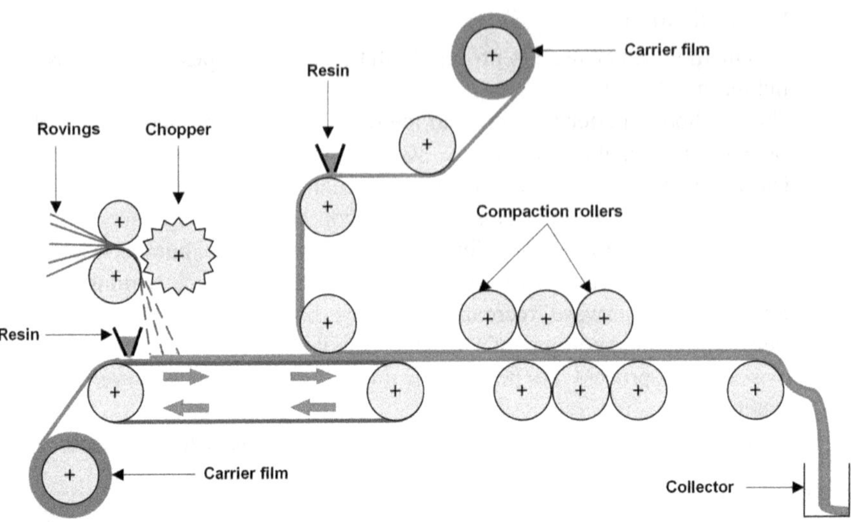

FIGURE 2.7 Illustration of SMC molding.

a pre-initiated resin/filler mixture is poured on the moving sheet of polyethylene film as shown in Figure 2.7. This pre-initiated resin/filler mixture is layered up to 0.6 cm thick. Reinforcements in the form of chopped fibers with length in the range of 3–7 cm are sprinkled onto the resin/filler layer. These chopped fibers are then covered with another polyethylene sheet. As a result a sandwich is formed, consisting of reinforcement in the middle and resin/filler mixture at the top, and enclosed by the polyethylene film. For the wetting of fibers, the above-formed sandwich is passed between the rollers. Finally, the entire mixture is rolled to an appropriate size for easy handling and storage.

The length of fiber can be longer in SMC because the mixing system involves less movement of fiber in the mold in SMC. After the SMC process, manufactured rolls are taken to a molding station and unrolled for cutting purpose as per desired lengths. The length and area of cut are the same as the size of part to be molded. Before placing the SMC in mold, polyethylene sheets are removed from the SMC sandwich. As per the thickness of the desired part, SMC sheets are placed in the mold layer by layer. For variable thickness parts, strips of SMC can be used. After placing the SMC sandwich into the mold, it is closed properly and the curing process is started. During the curing process, the movement of SMC materials within the mold is started and it covers the entire mold surface. Hence, there must be enough resin movement such that proper binding takes place. This technique is mostly used to manufacture large parts e.g., the body of a corvette (General Motors) and airplane wings.

Fillers are added during SMC molding to increase the dimensional constancy of the composite, reduce the volumetric and linear shrinkage, and finally to reduce overall cost. For the purpose of viscosity increment, thickener is used. The role of added styrene during curing is to increase the cross-linking. To avoid the premature curing of the composite, inhibitor is used. Mold release gel is used to remove the cured composites easily from the mold. Currently, SMC is in great demand in automotive industries because of its low-cost, high-volume production capabilities. The range of thickness of SMC is very vast and the maximum thickness is up to 6 mm. For the feed of compression molding, SMCs are cut into the rectangular piece and using this piece different shapes are obtained. For the proper curing of the SMC during compression molding, proper pressure and temperature should be applied such that the charge/ SMC spread into the mold cavity. Compression molding during the SMC process takes cycle time in the range of 1–4 mins.

SMC is commonly available in three types according to the form of fiber used:

(a) **SMC-R:** R denotes SMC is fabricated using short fibers in random orientation. To reveal the amount of short fiber used in SMC, the percentage of short fiber is mentioned after the R i.e., SMC-R30 has 30% short fiber.
(b) **SMC-CR:** C denotes continuous unidirectional fiber and R is the random short fiber. The method of representation of fiber percentage is the same as in the case of SMC-R. For example, SMC-C25R30 has 25% continuous unidirectional fiber and 30% randomly oriented short fiber.
(c) **XMC:** X represents X pattern of continuous fiber mixed with randomly oriented fibers. The angle between the crossed fiber is in the range of 5 to 7°.

2.4.1.1 Application

The SMC technique is used to manufacture various products from different industries e.g., bonnet of automobile, electrical and electronics components, structural components, furniture, sanitary ware, etc. In addition to the above applications, it is also used to manufacture baths, arenas, spas, cinemas, and stadium seating.

2.4.1.2 Advantages of SMC

- Near neat shape can easily be produced.
- High rate of production achieved.
- Owing to its flexibility in processing, many parts can be processed simultaneously.
- It is a high-volume and low-cost production technique.
- Excellent reproduction of parts can be achieved.
- Owing to its lesser dimensional requirement and ability to consolidate many parts into one, it helps in weight reduction.

2.4.1.3 Limitations of SMC

- The higher weight fraction of fiber in composites cannot be achieved using the SMC technique.
- SMC parts have lower stiffness and strength because of lesser fiber concentration and short length fibers and their isotropic distribution (Wulfsberg et al., 2014).

2.4.2 BMC Molding

BMC is sometimes also called dough molding compound (DMC). It is obtained by mixing the unsaturated polyester resin with a filler, initiator, and reinforcement

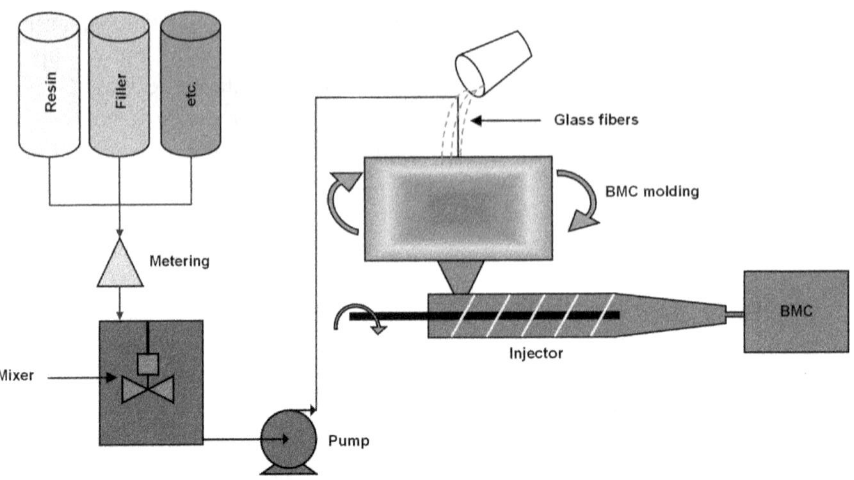

FIGURE 2.8 Schematic of BMC molding.

fibers and extruded in the form of log or rope. The length of the extruded part is decided as per the requirements. Generally, the mixture used for BMC is also called premix. The temperature during the mixing process is maintained at room temperature to avoid any curing. For proper mixing of the fibers with the fillers and resin, some agitation was provided as shown in Figure 2.8. The obtained product after mixing is stored at low temperature to extend its shelf life. The typical composition of BMC components is 25% resin and styrene mixture, 45% filler, 30% reinforcement. This composition can vary as per the requirements for different applications. The length of the fiber reinforcement varies between 6–12 mm. The strength of the BMC composite is lower than the SMC composites because of the lesser volume fraction of the fiber reinforcement and lesser length of the fibers.

BMC materials have a dough-like structure, which is helpful to varying the chopping length and the materials getting the shape during compression. Therefore, compression molding is the primary molding technique for BMC. Thus, BMC is helpful in molding large components through automation and reduces the cycle time. It can be used to mold highly complex parts easily. In compression molding, SMCs and BMCs are the most common raw materials. This method is mostly used in automobile industries for the production of various products e.g., body panels, electrical components, grills, trim, air-conditioning ducts, and various semi-structural members. One of the inherent limitations of BMC is that the movement of materials is limited within the mold. Generally, during compression molding, the charge of BMC is placed in the center of the mold and as the closing mold presses the charge, it flows toward the edge of mold. Thus, there should be a proper selection of resin, filler, and reinforcement. As the viscosity of resin increases, the flow of material will be hindered and this could lead to separation of the components. Therefore, there is a limitation of size with BMC molding; in most cases, the maximum size of part can be 16 in. (40 cm) (Mallick, 2007).

2.4.2.1 Application

- SMC is compression molded to obtain one- and two-piece panels of automotives.
- It can be used in the manufacturing of fenders, panels for roof, etc.
- Two-piece panels are manufactured using BMC. These panels are used as closer panel in hoods, doors, and deck lids.
- Other example of compression-molded SMC products are radiator supports, skid plates, pickup box components, and military drop-boxes.
- SMC can be used to produce enclosure of some electrical components, e.g., outdoor lamps, fuses, streetlights, switches, etc. (Mazumdar, 2002).

2.4.2.2 Advantages of BMC

- Dimensional control is better.
- BMC charges are resistant to flame and are non-melting in nature.
- Products obtained from BMC molding is resistant to chemical stains, environment, and corrosion.
- Products obtained from BMC molding is of higher mechanical and physical properties.
- Specific strength of obtained product is excellent.
- Products manufactured through BMC molding reveals higher dielectric strength and are of good electrical insulation.
- There is an increase in the acoustic properties of the manufactured component.
- Final product obtained is of near net shape.
- In BMC molding, variable designs of product can be manufactured and the color of product is also stable.

2.4.2.3 Limitations of BMC

- The movement of charge material within the mold is limited.

2.4.3 Hand Lay-Up

The simplest curing method among the methods is hand lay-up. The capital investment in the process is minimum compared to other methods. The processing steps in hand lay-up are very simple. Before starting the fabrication of composites, the application of release gel takes place on the mold for proper removal of fabricated composites. For enhancing the surface finish of the final product, plastic sheets are applied on the top and bottom surface of the mold plate. Stacking of alternate layers of prepreg matrix and fiber reinforcement (chopped or mat form) takes place within the mold as per requirements. After that, the mixture of resin and hardener is poured on an already stacked surface. For proper distribution of the mixture on the stacked surface brush is used. After that, a second layer of the mat is placed and with the help of a roller, slight pressure is applied on the mat-polymer layer to remove the trapped air. This process is repeated for each layer of polymer and mat until the required layers are stacked. At last, thin plastic sheet is placed on the top surface and release gel

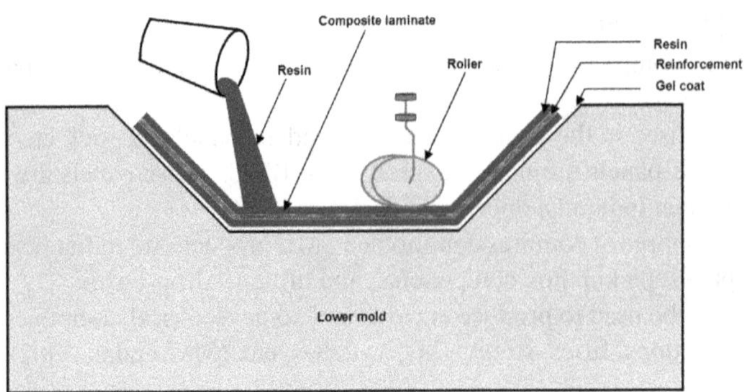

FIGURE 2.9 Illustration of hand lay-up technique.

is applied on the inner surface of the top mold plate. For proper interfacial bonding, a suitable amount of pressure is applied. For curing at room temperature or at higher temperatures, the whole setup is placed in a predefined environment. After the completion of curing, the final product is removed from the mold and is further processed as per requirements. The detailed setup of hand lay-up can be seen in Figure 2.9. The time of curing for each polymer varies as per their gel time period. Like in the case of epoxy-based composite, it will take 24–48 hours for curing at room temperature. The main limitation of this technique is that it can be used to manufacture thermoset-based composites. Matrix material used for composite fabrication using hand lay-up technique can be polyvinyl ester, unsaturated polyester, epoxy, polyester, phenolic resin, polyurethane resin, etc. The reinforcements are glass fiber, aramid fiber, carbon fiber, natural plant fibers (sisal, coir, nettle, flax, hemp, banana, etc.) in the form of mat of chopped.

2.4.3.1 Application

Hand lay-up is widely used to manufacture bulkier products e.g., boat hulls, wind turbine blades, architectural molding, automotive parts, deck, etc.

2.4.3.2 Advantages of Hand Lay-Up

- The cost of establishment is very little when compared to another manufacturing process.
- Tooling cost is very low because the working temperature is room temperature.
- There is no complexity in this process.
- Flexibility in variation of reinforcement and matrix materials.
- Higher-volume fraction of fibers can be used and longer length of fiber can be used in this process.

2.4.3.3 Limitations of Hand Lay-Up

- When comparing this process with the fiber-reinforced plastic pultrusion process, filament winding, and pulwinding process, it is found that the efficiency of production, cycle time, and speed are lesser.

- During the processing of composite material, hazardous factors and safety should be considered.
- The hand lay-up technique is good for resin of lower viscosity.
- Quality of the final product mainly depends on the skills of laborers.
- Generation of voids takes place in the product because of the non-uniform distribution of resin.
- There may be inclusion of foreign particles in the composite material.

2.4.4 Spray-Up

Spray-up technique is an extension of hand lay-up technique. Mainly a spray gun is used in this technique to spray the pressurized mixture of resin and chopped fibers. Generally, glass roving is used as fiber reinforcement which passes through a spray gun where it is chopped with a chopper gun. In this technique, matrix material is either mixed with reinforcement or simultaneously sprayed one after one. For the proper removal of the product, releasing gel is sprayed on the surface of the mold. To remove entrapped air from the sprayed lay-ups, a roller is rolled over the sprayed material. This spray-up material along with the mold is placed for curing either at room temperature or at a higher temperature. The curing time of composites may vary, which depends on the type of polymer. A detailed schematic of the spray lay-up process can be seen in Figure 2.10. Spray lay-up technique is used to fabricate the low load-bearing composites e.g., fairing of trucks, small boats, bathtubs, etc. There is no limitation of fiber volume fraction and also no limitation on part size. Commonly used matrix materials in this technique are epoxy, polyvinyl esters, polyurethane resin, unsaturated polyester, phenolic resin, and polyester and

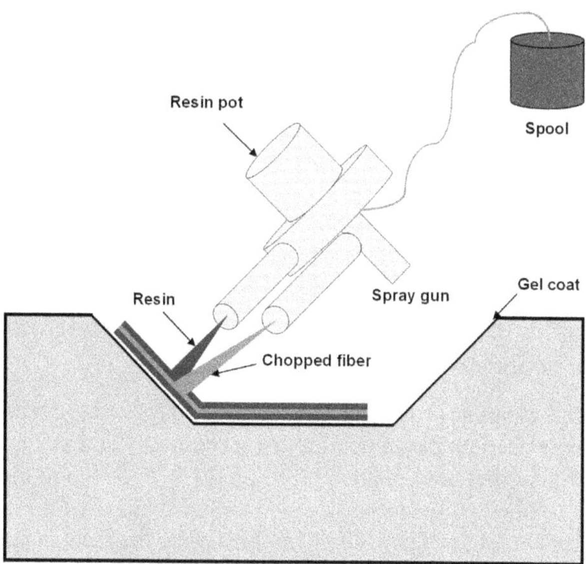

FIGURE 2.10 Schematic of a spray-up composite fabrication process.

reinforcements commonly used are glass fiber, aramid fiber, carbon fiber, natural plant fibers (sisal, hemp, banana, flax, coir, cotton, nettle, jute, etc.). These fibers can be used in the form of flakes, particle fillers, chopped fibers, etc.

2.4.4.1 Application

There are various applications of the spray-up composite manufacturing process varying from custom parts to medium-volume quantities. Some of these applications are simple enclosures, structural panels of lightweight, e.g., caravan bodies, shower trays, bathtubs, truck fairings, and some small dinghies. Other applications are commercial products e.g., swimming pools, duct and air handling equipment, storage tanks, boat hulls, and furniture components.

2.4.4.2 Advantages of Spray-Up

- Widely used for many years.
- Small to medium-volume products are easily fabricated.
- The manufacturing cost for small products is much less.
- Cost and cycle time of composite manufacturing is less.
- Tooling cost and material cost are less tha that of other techniques.

2.4.4.3 Limitations of Spray-Up

- The composites obtained from this technique are very heavy because of the high amount of resin present in the composite, which is heavier in weight.
- Only short fibers are used in this technique, which limits the mechanical properties attainable of the fabricated composite materials.
- Viscosity of resin should be low such that it can be sprayed easily. Due to this, the thermal and mechanical properties of the product are compromised.
- The styrene used to avoid sudden curing is of harmful nature.
- High structural requirement parts cannot be fabricated using this technique.
- The control of fiber volume fraction and thickness of composite material is difficult in this technique.
- The whole process and the accuracy of the obtained product depend on the operator's skill.
- Surface finish of the obtained product can be varied on both sides.
- The lack of dimensional accuracy and repeatability are the major limitations of this technique.

2.4.5 Resin Transfer Molding

RTM is a closed molding process. As the name indicates, the resin is transferred over the already placed reinforcement in this technique. Reinforcement in the form of either woven mat or strand mat is placed on the surface of the lower half mold. A release gel is applied on the mold surface for easy removal of the composite. The mold is properly closed and clamped. After clamping, the resin is pumped into the mold through a vent and air is displaced through another vent. Uniform flow of resin can be attained by using a catalyst as an accelerator and vacuum. After that, the whole setup is placed for curing and the cured product is removed. Figure 2.11 shows a detailed schematic of RTM.

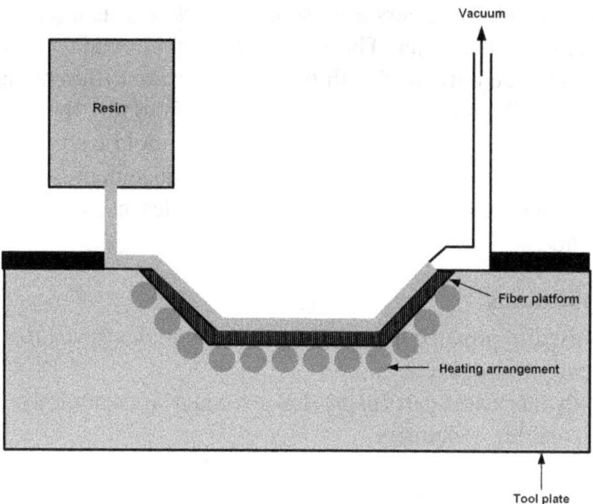

FIGURE 2.11 Illustration of RTM technique.

In RTM, hard and soft mold can be used as per the requirement of run duration. Molding materials for soft molding may be polymers, whereas for hard mold, steel and aluminum can be used. The cost of mold varies as per the mold life requirement of mold. The cycle time of this technique can be reduced by automation. Generally, simple parts are easily fabricated but for production of complex shape, preformed fiber reinforcements are used. Resin transfer molding mainly depends on the viscosity of the resin. The time of injection of resin mainly depends on its viscosity. For the proper flow of resin, high pressure is required. Due to this high pressure, displacement of fiber takes place, which is called fiber wash.

RTM is used to fabricate the composites which are of high-end use. The RTM technique produces high-strength composites that are used for aerospace applications. With the advancement in technology of RTM, it is used to produce lightweight products with excellent mechanical properties.

The RTM technique mainly uses the below materials:

Matrix: methyl methacrylate, polyester, phenolic resin, epoxy, polyvinyl ester.
Reinforcement: carbon fiber, aramid fiber, glass fiber, natural fibers (sisal, coir, banana, flax, nettle, hemp, etc.). Reinforcements can be used in woven mat form, chopped form, and unidirectional mat form. Fillers can also be added to the resin for the enhancement of surface finish and fire retardancy.

RTM mainly consists of five components that govern the processing of composites. These components are

(i) Container for resin and catalyst
(ii) Pump for pumping of resin
(iii) Chamber to mix resin and catalyst
(iv) Injector to inject the resin
(v) Mold at which curing takes place

In RTM, two separate containers are used. The first one is a resin container and the other is a catalyst container. The size of the resin container is larger than the catalyst container. The outlets of both the container are different and converge in a mixing chamber. The role of the mixing chamber is to mix the resin and catalyst properly. After that, this mixer is injected into the mold cavity with the help of the injector. For the purpose of curing, heating arrangement is provided into the lower half mold. During curing some gases will be released, which can be removed through the vent.

2.4.5.1 Applications

- RTM is used to produce hollow shapes and complex structures like motor casing, engine covers, etc.
- Commonly fabricated parts using this technique are automotive body parts, bathtubs, and big containers.

2.4.5.2 Advantages of RTM

- The process is very efficient.
- Suitable for complex shapes.
- Both surfaces of parts produced have good surface finish.
- RTM allows the flexibility to change the combination of reinforcement materials and their orientations.
- Cycle time in the case of RTM can be reduced by using a heating device at the lower half of the mold.
- RTM can be manually controlled, semi-automated, or fully automated as per the requirements.
- Products obtained from this technique have uniform thickness as a function of mold cavity.
- Dimensional tolerance is very good in this process.
- Other attachments can easily be fitted in RTM.
- In RTM low injection pressure is sufficient for resin injection.
- Near neat shapes are obtained and due to this material wastage is reduced.
- The production rate can be increased by automating the process.
- Composite with higher fraction of fiber volume and low void numbers is obtained.
- Since RTM resin is placed in an enclosure, health and environmental safety are good.
- Because of automation requirement of RTM, the need for labor is reduced.
- Better reproducibility.
- Relatively low clamping pressure and ability to induce inserts.

2.4.5.3 Limitations of RTM

- The size of the composites is limited because of the mold cavity.
- The tooling cost is high with this technique.
- Limited reinforcing materials can be used because of resin flow and saturation limitation.

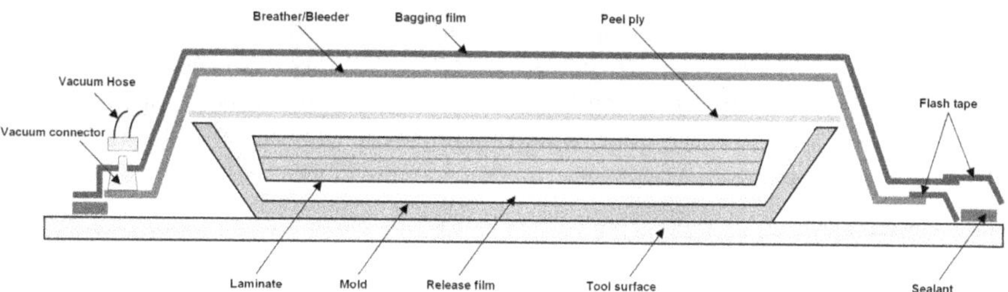

FIGURE 2.12 Schematic of vacuum bagging technique.

- The matched tool used in this technique is expensive and the capability to withstand higher pressure is not good.
- RTM is mostly limited to fabricating smaller products.
- There may be the possibility of improper resin flow which creates un impregnated area and results in scrap of higher cost.

2.4.6 Vacuum Bagging

This process is nearly the same as the wet lay-up technique; the main difference is the presence of the vacuum. The vacuum is created in a vacuum bag, in which composite is fabricated. With the help of vacuum in the vacuum bag, entrapped air, excess resin, and gases are removed. The vacuum bag may be of nylon or polyvinyl alcohol (PVA) as shown in Figure 2.12. Generally, vacuum is created on the mold with the help of a vacuum bag. Curing is performed either at room temperature or at a higher temperature. The suction of air from the vacuum bag at atmospheric pressure helps in the compaction of composite layers; thus higher quality of composite is obtained. The vacuum bagging technique is basically the extension of the wet lay-up technique where pressure is applied to the stacked laminates in order to achieve consolidation. Vacuum pressure can be achieved by sealing the mold with a plastic film and vacuum pump. The vacuum pump is used to extract the air from the vacuum bag. Polymers used in vacuum bagging are generally phenolic and epoxy. Other polymers e.g., vinyl ester and polyester create the problem of unnecessary extraction of styrene during vacuum pumping. Reinforcements used may be synthetic and natural fibers, ceramic particles, etc. For proper wet out of the fiber reinforcement, the consolidation pressure should be high.

Steps involved in vacuum bagging are

(i) Release film is applied on the top surface of prepreg. For the extraction purpose of excess resin, entrapped air, and volatiles, the release film is made perforated.
(ii) Breather is placed at the top surface of the release film to absorb moisture or excess resin. Breather is like a porous bleeder.
(iii) Barrier film is applied on the top surface of the bleeder. It is similar to the release film; the only difference is that it is not perforated.

(iv) At last, the vacuum bag is applied. The material of the vacuum bag is elastomer or polyamide. After this, seal the vacuum bag with the help of seal tape. The whole mold can be sealed when it is porous. Mostly, the width of seal tape is 1.27 to 2.54 cm, which is a rubbery material that sticks to the mold and vacuum bag. A vacuum hose is connected to the vacuum bag for the purpose of vacuum creation.

2.4.6.1 Applications

The major applications of vacuum bagging bag molding are to fabricate parts of cruising boats, core-binding in production boats, race car components, furniture, musical instruments, wind turbine blades, and model boats.

2.4.6.2 Benefits of Vacuum Bagging

- Compared to wet lay-up technique, it can mold composites of higher reinforcement percentage.
- Void content in the products is less when using this technique, compared to that found when using the wet lay-up technique.
- The wetting of fibers is better due to the pressure.
- Because of the enclosed atmosphere, health and safety concern is maintained well.

2.4.6.3 Limitations of Vacuum Bagging

- Fabrication cost is more because of extra adds-in and bagging materials.
- Skilled labor is required.
- Proper mixing and control of the resin require skilled labor.

2.4.7 Pultrusion

Pultrusion is a continuous automated closed molding process. Its basic mechanism is the same as that of the metal extrusion process. The only difference is that in the case of pultrusion, material is pulled through the die, whereas in the case of extrusion, material is pushed through a die. The reinforced material, in this case, is continuous fiber or mat, which passes through the resin tank. The reason behind the passing of fiber into the resin tank is to attain proper wetting of fiber into the resin. After the wetting of fiber into the resin, it is passed through the hot die to attain the desired shape. Heated die also helps in the curing of composite. After the curing, composite is pulled out with the help of a gripper. Lastly, pultruded profile is cut using the inbuilt cutter at the end of the pulling mechanism (see Figure 2.13). For the variation of composite, some fillers are added into the resin. To remove the excess resin, a hot die is sufficient but sometime pre-former is also used. Pre-former acts as a squeezer, which squeezes extra resin from the fiber. The pultrusion process is mostly used for thermoset polymers and the product obtained is of uniform cross-section. Due to the uniformity of product, distribution and alignment of fiber are good and the impregnation of resin is also good. The production rate in the pultrusion process is high but the variation of the cross-section is hard to achieve. The setup cost of the pultrusion

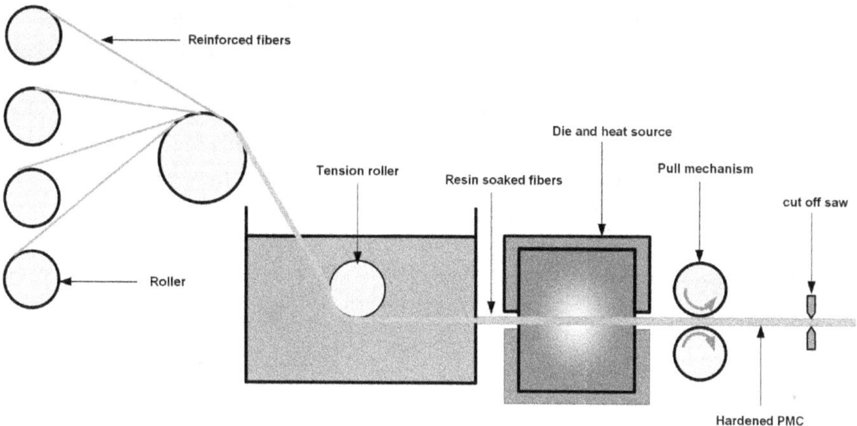

FIGURE 2.13 Setup of the pultrusion process.

process is less than other complex molding processes. In the pultrusion process, there are mainly six components. These components are

(i) Fiber creels
(ii) Pre-former
(iii) Impregnation system for resin
(iv) Dies for shaping
(v) Mechanism for pulling
(vi) Cutter to cut the composite

For uniform and controlled tension, creel should be properly located during the pultrusion process. To maintain the continuity of the roving of fibers a backup creel is provided behind the running creel. Shape and size of creel are decided on the basis of the number of roving packages and the distance between strands to be maintained. The role of the preform plate is crucial in the pultrusion system because it aligns the reinforcement before feeding to the hot die. The quality of the pultrusion system depends mainly on the preform plates. For the wetting of fiber resin, a bath tank is used and its capacity depends on the amount of resin to be handled. The resin tank may be incorporated with the heating system to enhance the wetting of fibers, but this may decrease the life of the tank. In most cases, a dip bath system is used to impregnate the fiber, which is also called an open bath system. In a dip bath system, fibers come into contact with resin with the help of creel and this wetted fiber comes out with the help of guided bars. Heating dies are the key components of the pultrusion system, which help give the shape and curing of composites. After the curing of composites, the pulling system comes into action in a continuous manner. A cut off saw is the last unit of the pultrusion system, which gives the desired size to the composites. Sometimes, there may be a requirement of some coolant or lubricant; for this purpose water is used. This type of saw is called a wet saw. It helps in flushing dust and debris during the cutting of the composites. Another type of cutter is a dry cutter, which uses a diamond blade and with which there is no requirement of lubricant and coolant (see Figure 2.13). The reinforcing materials used are glass (S-glass and E-glass), aramid, carbon fibers in the form of roving strands, mat (continuous filament mat, chopped strand mat), and fabrics. Particular properties of the fabric reinforcement can be altered by doing some treatments. Veils are also used sometimes to achieve a surface finish of high quality. These veils may be imprinted with some design or logos, which will appear on the surface of pultruded products. Most of the time, matrix materials are phenolic resins, epoxy, vinyl ester, unsaturated polyester, etc. Sometimes, filler materials are also used as per design requirements.

2.4.7.1 Applications

- Solid rods, long sheets, and tubes are easily manufactured.
- Products of constant cross-section are easily fabricated.
- Handles of high voltage components and rail covers.

2.4.7.2 Benefits of Pultrusion Process

- The cost of production is less because of automation and the quality of the product is high.
- Products fabricated using pultrusion are of better surface finish.
- The production rate is high because of its automation.
- Skilled labor is not required due to its simplicity.
- Handling and maintenance of tools are easy.

2.4.7.3 Limitations of Pultrusion

- This process is only used to produce constant cross-section products.
- Controlling the orientation of fiber is difficult.
- Products with thin walls cannot be produced using this method.

2.4.8 Filament Winding

Filament winding is a process in which unwound fibers are passed through a resin tank and the impregnation of the resin occurs. After the resin impregnation, these fibers are passed through a rotating mandrel for the purpose of collection. The rotation of mandrel is in a controlled manner such that during winding no extra tension is generated in the fibers. The detailed setup of filament winding can be seen in Figure 2.14. The exact amount of fiber tension is required for the curing of the composite. It also affects the volume fraction of fiber and porosity content in the composites. Fiber tension mostly depends on the fiber type, its geometry, winding pattern, and rotation of the mandrel. The fiber tension should be optimum such that there should be no fracture on the fiber surface taking place. For the removal of mandrel from composite parts, hydraulic rams are mostly used. In the case of complex products, mandrels are made of some soluble material that can collapse or dissolve or be melted after composite fabrication. The mandrel is the replica of the final product to be fabricated. The mandrel may be an integral part of the product in some cases. In this process, a high-volume fraction can be achieved. Recently,

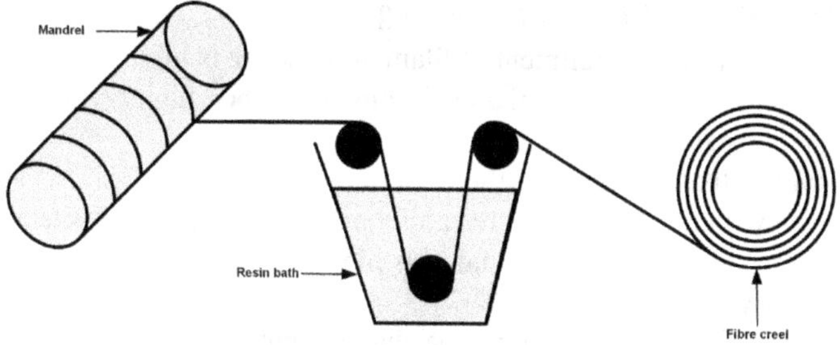

FIGURE 2.14 Schematic of filament winding.

each process of filament winding is controlled with a computer to enhance the productivity of the process. Matrix material used in this process may be phenolic resin, epoxy, polyester, polyvinyl ester, etc. Reinforcement materials used may be S and E-glass fibers, aramid fibers, boron fibers, carbon fibers, etc.

Filament winding comprises the following components:

(i) Fiber creel for unloading the fiber into resin tank
(ii) Resin tank for fiber impregnation
(iii) Carriage
(iv) Mandrel for the collection of composite

2.4.8.1 Applications

- Common products fabricated using filament winding are pipelines, gas cylinders, storage tank, vessels, missile cases, ducting, fishing rods, cement mixture, aircraft fuselages, sailboat mast, golf club shafts, etc.
- In the past few years, the range of application of filament winding has increased. Some of the advanced applications of filament winding are complex engineered non-spherical and non-cylindrical products.

2.4.8.2 Benefits of Filament Winding

- This process produces higher specified strength products.
- The composites produced have a high degree of uniformity, better fiber orientation, and distribution.
- Labor involvement is minimum because of the automatic process.
- This process is mainly used to fabricate the products which require high tolerance.
- It is possible to produce a product with fiber in a particular direction.
- Production cost is less than that of other processes because the materials it requires cost less.
- Variation in the design of part is easy by varying the materials, winding pattern, and curing option.
- There is no component size restriction.
- The products obtained from this technique are of low cost because of automation and high production volume.

2.4.8.3 Limitations of Filament Winding

- Cost related to the installment of filament winding is high.
- To achieve uniform dispersal and alignment of fiber, high precision control mechanisms are required.
- Removal of the mandrel is difficult in this process.
- Products with reverse curvature cannot be produced using this technique.
- Sometimes, the cost of the mandrel is high and the surface obtained is not satisfactory.
- The direction of winding cannot be changed within one layer of winding.

2.5 NOVEL FABRICATION TECHNIQUES OF POLYMER COMPOSITES: MICROWAVE-ASSISTED COMPRESSION MOLDING (MACM)

MACM is a recently developed technique for the fabrication of polymer composites. Over the last two decades, researchers have been using this technique and trying to develop a much better way to fabricate the polymer composites (Hawley et al., 1994). In the field of composites fabrication, MACM was established as a promising manufacturing route because of its numerous advantages compared to the conventional fabrication routes. The main advantage of the microwave heating is direct, selective, volumetric, instantaneous, controllable and reduced processing time and energy (Singh and Zafar, 2019a). Microwave consists of orthogonal magnetic and electric fields. The role of magnetic field is important for microwave heating of metallic materials. Whereas, in the case of polymeric materials or insulating materials, dielectric properties play a significant role in microwave heating. The heating of the polymer composites mainly occurs due to the resisting forces which occur due to frequent changes in orientations of diploes that increase molecular kinetic energy and result in volumetric heating. These resisting forces are inertial, frictional, elastic, and molecular interaction forces. As discussed above, heating of polymer composites is at molecular level in presence of microwave radiation. Therefore, in microwave-assisted heating, heating takes place from inside to outside, whereas in the case of conventional heating it is reversed i.e., it occurs from outside to inside. MACM claims developed polymer composites have better properties compared to composites fabricated using the conventional fabrication techniques (Arora et al., 2019).

2.5.1 Microwave Heating Mechanism of Polymer Composites

The transfer of heat energy in material takes place due to the interaction of the electromagnetic field at a molecular level, generated by the magnetron of the microwave applicator. This molecular interaction mainly depends on the dielectric properties of the materials. In the presence of the external electric field, the dipoles of the materials/ composites tried to align themselves with the field by rotation as shown in Figure 2.15. But due to inertial, frictional, elastic, and molecular interaction forces, the dipoles tried to achieve the initial orientation and thus due to this resistance, molecular heat generated which integrated into volumetric heat. The rapidness of

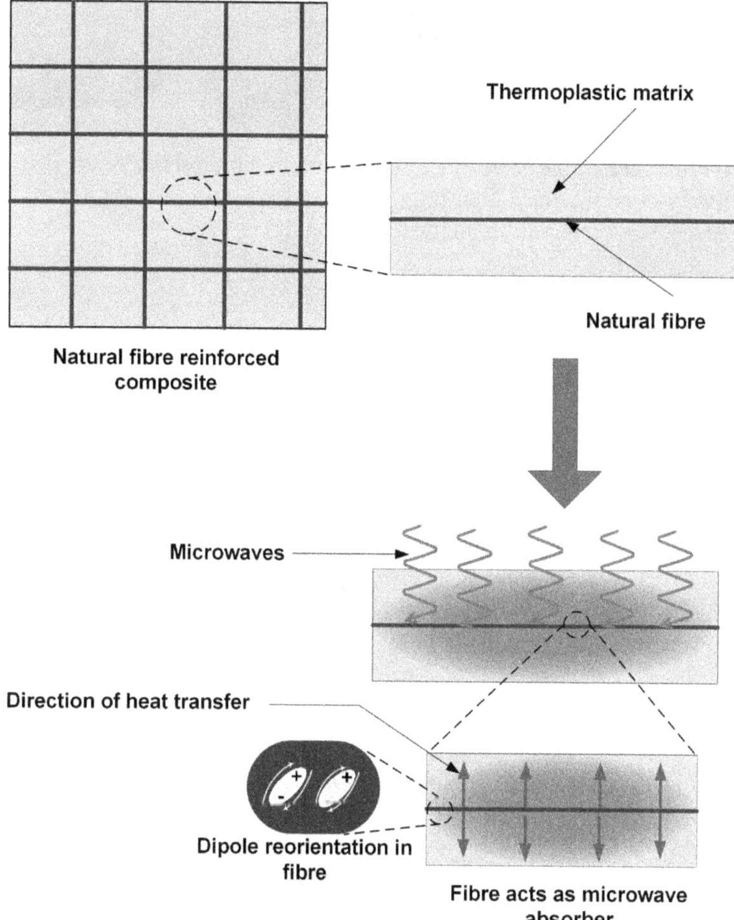

FIGURE 2.15 Illustration of microwave heating in polymer composites.

the dipoles achieving initial orientation depends on the phase lag due to high electric field frequency.

The generation of heat due to microwave interaction in a non-magnetic material mainly depends on its complex permittivity (ε^*) as shown in Equation 2.2 (Singh and Zafar, 2018).

$$\varepsilon^* = \varepsilon' - i\varepsilon'' \tag{2.2}$$

where ε' is electrical energy stored by the material/composites and ε'' is microwave energy dissipated by heated material/composites.

The above discussed complex permittivity can be separable in a real and complex term as Equations 2.3 and 2.4 display.

$$\varepsilon'(\omega) = \varepsilon_U + \frac{\varepsilon_R - \varepsilon_U}{1 + \omega^2\tau^2} \tag{2.3}$$

$$\varepsilon''(\omega) = \frac{(\varepsilon_R - \varepsilon_U)\omega\tau}{1+\omega^2\tau^2} \tag{2.4}$$

where ε_U is unrelaxed permittivity, ε_R is relaxed permittivity, ω is the angular frequency, and τ is relaxation time (s).

Local hindrance of the molecule reorientation in the presence of applied electric field is used to find the relaxation time. Arrhenius equation can be used to find out the dependency of temperature on the relaxation time.

$$\tau(T) = \tau_0 \exp\left(\frac{E_a}{kT}\right) \tag{2.5}$$

where, E_a is activation energy (J), k is the Boltzmann's constant (1.38×10^{-23} m^2 kg s^{-2} K^{-1}), and T is the temperature (K). In the case of some composites, fabrication surface treatment of the fibers is necessary and due to this, the dielectric behavior may get changed; as a result, additional interfacial, ionic, and dipole relation plays an important role in the process. Equation 2.5 shows that temperature distribution is exponential in nature, which reaffirms fast heating and thus energy and cost are saved.

Generation of heat within the non-magnetic materials or the heat dissipated inside the microwave applicator can be represented in Equation 2.6 as follows.

$$Q_e = \omega \cdot \varepsilon''_{\text{eff}} \cdot \varepsilon_0 \cdot E^2_{\text{rms}} \tag{2.6}$$

where $\omega = 2\pi f$ is the angular frequency (s^{-1}), $\varepsilon''_{\text{eff}}$ is effective dielectric loss (F/m), ε_o is the permittivity of free space (8.854×10^{-12} F/m), E_{rms} is the root mean square value of the electric field (V/m), and Q_e is the heat generated per unit volume (W/m^3).

The effective dielectric loss can be expressed in terms of dipolar polarization, ionic conduction, and Maxwell–Wagner polarization mechanisms. Therefore, Equation 2.6 can be expressed as:

$$\varepsilon''_{\text{eff}} = \varepsilon''_{\text{polarization}} + \varepsilon''_{\text{conduction}} = \varepsilon''_{\text{dipolar}} + \varepsilon''_{\text{interfacial}} + \sigma/\omega\varepsilon_0 \tag{2.7}$$

where σ is electrical conductivity (S/m).

Figure 2.15 shows the possible method of microwave heating in which natural fiber acts as a relatively higher microwave absorber compared to the polymer matrix. As discussed in Equation 2.2, the microwave heating of the materials mainly depends on the dielectric property of the material. Material with a high dielectric constant seems to interact better with being microwaved and heated rapidly than material with a lower value of dielectric constant. The microwave properties along with dielectric constants are shown in Table 2.5. From Table 2.5, it can be observed that the dielectric constant of natural fiber is in the range of 3–6, whereas for the PP and PE pellets, the dielectric constant is 2.26–2.420. Therefore, in this case, microwave interaction initially takes place with the fiber before occurring with the matrix. In some of the cases, the matrix may be a higher microwave absorber. In that case, initial heating takes place in the matrix and then it transfers into the fiber. Figure 2.16

TABLE 2.5
Dielectric Properties of Materials (Mishra and Sharma, 2016; Singh and Zafar, 2018)

Material properties	Materials			
	Ceramics	Metals (at 2.45 GHz)	Polymers	Natural fibers
ε'/μ_r	8.9–9.66	0.99–5000	2.2–6.0	3–6
ε''/ρ	0.009–0.029	1.6–13	0.0008–1.2	-
D_p/D_s	40–12563	0.05–3.39	79.9–77900	-

$\varepsilon'/\varepsilon''/D_p$ for ceramics, polymers, and natural fibers, $\mu_r/\rho/D_s$ for metals
where D_p is penetration depth(mm), $D_s$is skin depth (μm), μ_r is relative permeability (Hm^{-1}), and ρ is electrical resistivity (Ωm).

shows example of microwave heating of natural fiber reinforced polymer composites; it may be also used to fabricate synthetic fiber-reinforced polymer composites, HDPE/CNT composites, etc.

2.5.2 Fabrication of Composites Using MACM

Polymer composites of particular reinforcement percentage are fabricated by weighing the matrix and reinforcement. The matrix may be thermoplastic or thermosets and the reinforcement may be either natural fiber or synthetic, or some ceramic particulates. The matrix may either be in the form of mat or pellets and fibers may be in chopped or continuous form. An alumina mold of size 90 × 25 × 2.2 mm and 140 × 57 × 1.3 mm was used to fabricate different types of polymer composites i.e., HDPE/jute, HDPE/kenaf, PP/jute, PP/kenaf, HDPE/sisal, chopped coir/HDPE, HDPE/CNT, PP/CNT, Poly-L-lactide (PLLA) based foams and hybrid composites (Singh and Zafar, 2018; Arora et al., 2019; Singh et al., 2019b, a; Singh and Zafar, 2019a, b; Verma et al., 2019b, a; Verma and Zafar, 2019; Arora and Pathak, 2020; Verma et al., 2020). The size of the alumina mold was chosen according to the requirement of the sample size to test different properties of the composites as per the American Society for Testing and Materials (ASTM) standard. Alumina mold was used for fabrication because it is transparent to microwave up to 600°C temperature (Thostenson et al., 1999). Also, the surface finish of the alumina mold is good which is helpful during the removal of composites from the mold. No chemical reaction takes place in alumina mold in the presence of microwave heat. The microwave applicator used in the MACM is of industrial class (make: VB ceramics, Chennai; Model:700 DEG°C Premium) with a maximum output power of 1.1 kW and frequency of 2.45 GHz.

According to the types of composites, in some cases, stacked layers of fiber and matrix were put into the alumina mold. In chopped fiber or particulate reinforced composites, properly mixed charge was put into the mold and placed into the

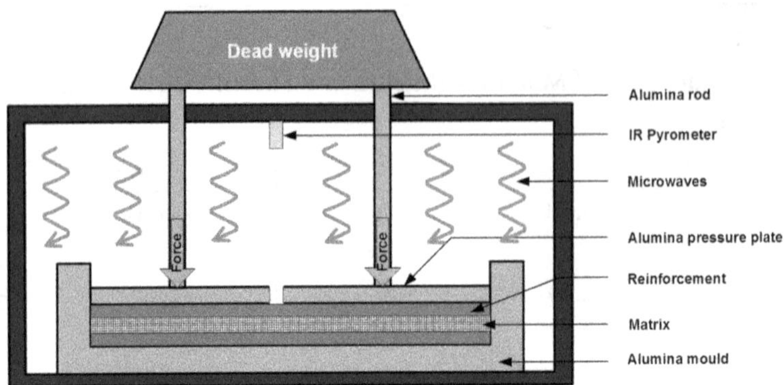

FIGURE 2.16 Setup of microwave-assisted compression molding.

microwave applicator. The requirement of heat for the fabrication was decided on the basis of microwave power and time of exposure. Prior to the fabrication of composites, some observations of the trial and error were taken and studied. The glass transition temperature (T_g) of any polymer is considered to be the main optimization parameter. Therefore, during trials, the whole concern was around attaining the glass transition temperature to be at particular microwave power and exposure time. The T_g for the PP, HDPE, and PLLA can be obtained using a standard differential scanning calorimeter (DSC) equipment as per ASTM D 3418 standard. To obtain optimum microwave power and exposure time, intensive trials were done.

To obtain good interfacial bonding between reinforcements and matrix an external pressure was applied on the alumina plate used to cover the stacked composite in the mold as shown in Figure 2.16. Because of uniform heating of microwave, matrix throughout the mold achieved the glass transition temperature and thus helps in proper wetting of the reinforcement. An infrared (IR) pyrometer (make: Raytek; Model: RAYXRTG5SFA) was mounted on the microwave applicator to monitor the real-time temperature of the polymer composite through a hole of 0.7 mm in the alumina pressure plate as shown in Figure 2.17. During the fabrication of the composites, it was observed that the power required was in the range of 720–900 W and the process cycle was in the range of 1140–1500 s (Singh and Zafar, 2018, 2019a). Figure 2.17 shows the optical image of some composites fabricated using MACM.

2.5.3 Advantages of MACM

- MACM is a novel, rapid, and efficient route to fabricating polymer composites.
- Uniform heating and pressure on the polymer composites are applied due to the good interfacial bonding created.
- The time and energy involved in this technique are less, thus cost is saved.
- Low-cost equipment is required.
- Volumetric and selective heating takes place.
- MACM is an environmentally friendly process.
- Precise heating control is achieved.
- The quality and properties of MACMed composites are improved.

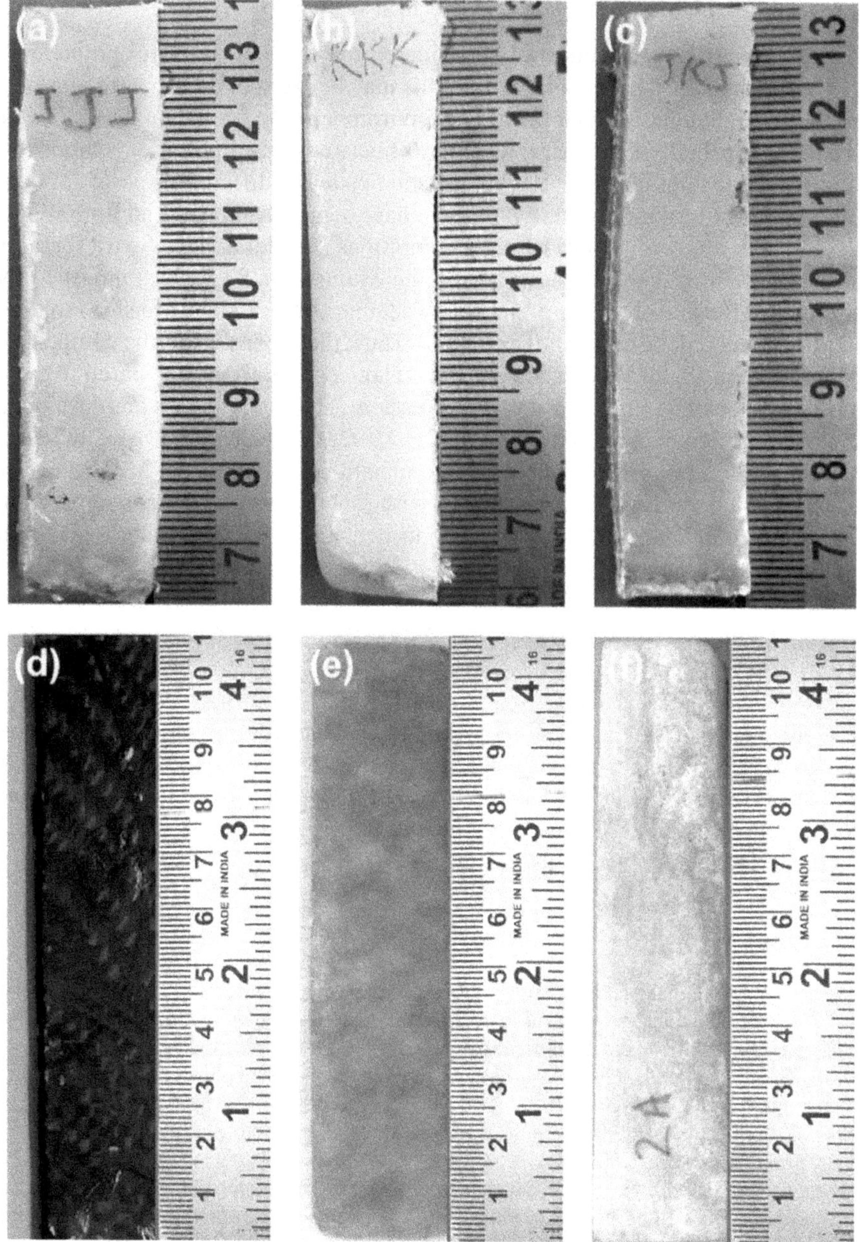

FIGURE 2.17 Composite samples fabricated using MACM (a) HDPE/jute (b) HDPE/kenaf (c) jute/kenaf/jute (d) epoxy/CNT (e) chopped coir/HDPE (f) PLLA foams.

2.6 CONCLUSION AND FUTURE PERSPECTIVE

Polymer composites are currently considered to be the most useful products in various industries. The reasons behind this may be the specific strength, durability, lightweightedness, economical and environmentally friendly nature of these products, especially in the case of natural fiber-reinforced polymer composites. For the fabrication of various types of composites, a different type of fabrication technique has been used. All of them have some advantages and limitations. Researchers are trying to reduce the limitations by adding some extra features to the conventional techniques. Taking the example of RTM, it is modified by adding the feature of vacuum and thus vacuum-assisted resin transfer molding (VARTM) technique came into existence. Thus, the properties of the composites obtained from VARTM have enhanced mechanical properties, compared to composite fabricated from RTM and compression. The latest technique developed in the field of composite manufacturing is MACM. MACM involves volumetric heating of the composites; thus, it is rapid in nature. The impact of MACM on the environment is less and cost associated with this technique is less, compared to other conventional techniques. Owing to its benefits, MACM could be a potential composite fabrication technique in the future for bulk production of the polymer composites.

ACKNOWLEDGMENT

The authors would like to thank the editor and publisher for giving us the opportunity to write this chapter. It has been a very educational experience. The authors are very thankful to the authors around the world who have published their studies and National Programme on Technology Enhanced Learning (NPTEL), which was very useful during the writing of this chapter.

REFERENCES

Arora, G. and Pathak, H. (2019a) Modeling of transversely isotropic properties of CNT-polymer composites using meso-scale FEM approach. *Compos Part B Eng* 166:588–597. doi: 10.1016/j.compositesb.2019.02.061.

Arora, G. and Pathak, H. (2019b) Numerical study on the thermal behavior of polymer nanocomposites. *J Phys Conf Ser* 1240:012050. doi: 10.1088/1742-6596/1240/1/012050.

Arora, G. and Pathak, H. (2019c) Multi-scale fracture analysis of fibre-reinforced composites. *Mater Today Proc*: 687–695.

Arora, G. and Pathak, H. (2020) Experimental and numerical approach to study mechanical and fracture properties of high-density polyethylene carbon nanotubes composite. *Mater Today Commun* 22:100829. doi: 10.1016/j.mtcomm.2019.100829.

Arora, G., Pathak, H. and Zafar, S. (2019) Fabrication and characterization of microwave cured high-density polyethylene/carbon nanotube and polypropylene/carbon nanotube composites. *J Compos Mater* 53(15):2091–2104. doi: 10.1177/0021998318822705.

Azom (2019) Viewed at https://www.azom.com. Accessed on 22 May 2019.

Hawley, M.C., Wei, J., and Adegbite, V. (1994) Microwave processing of polymer composites. *MRS Online Proc Libr Arch* 347:669–680.

long LIU, J., Zhu, Y., Wang, Q., and Ge, S. (2008) Biotribological behavior of ultra high molecular weight polyethylene composites containing bovine bone hydroxyapatite. *J China Univ Min Technol* 18(4):606–612. doi: 10.1016/S1006-1266(08)60303-X.

Mallick, P.K. (2007) *Fiber-Reinforced Composites: Materials, Manufacturing, and Design.* CRC Press.

Matweb (2019) Viewed at http://matweb.com. Accessed on 21 May 2019.

Mazumdar, S.K. (2002) *Composite Manufacturing Materials, Product, and Process Engineering.*

Mishra, R.R., Sharma, A.K. (2016) Microwave-material interaction phenomena: Heating mechanisms, challenges and opportunities in material processing. *Compos A* 81:78–97.

Mohanty, A.K., Misra, M., Drzal, L.T. (2005) *Natural Fibers, Biopolymers, and Biocomposites.* CRC Taylor and Francis, Page 46.

Noyes, J.V. (1983) Composites in the construction of the Lear Fan 2100 aircraft. *Composites* 14(2):129–139. doi: 10.1016/S0010-4361(83)80008-1.

Singh, M.K., Chauhan, D., Gupta, M.K., Diwedi, A. (2015) Optimization of process parameters of Aluminum Alloy (Al-6082 T-6) machined on CNC lathe machine for low surface roughness. *J Mater Sci Eng* 4. doi: 10.4172/2169-0022.1000202.

Singh, M.K., Verma, N., Zafar, S. (2019a) Optimization of process parameters of microwave processed PLLA/coir composites for enhanced mechanical behavior. *J Phys Conf Ser* 1240:012038. doi: 10.1088/1742-6596/1240/1/012038.

Singh, M.K., Zafar, S. (2018) Influence of microwave power on mechanical properties of microwave-cured polyethylene/coir composites. *J Nat Fibers*: 1–16. doi: 10.1080/15440478.2018.1534192.

Singh, M.K. and Zafar, S (2019a) Development and mechanical characterization of microwave-cured thermoplastic based natural fibre reinforced composites. *J Thermoplast Compos Mater* 32(10):1427–1442. doi: 10.1177/0892705718799832.

Singh, M.K. and Zafar, S. (2019b) Development and characterisation of poly-L-lactide-based foams fabricated through microwave-assisted compression moulding. *J Cell Plast* 55(5):523–541. doi: 10.1177/0021955X19850728.

Singh, M.K., Zafar, S., and Talha, M. (2019) Development of porous bio-composites through microwave curing for bone tissue engineering. *Mater Today Proc* 18:731–739. doi: 10.1016/j.matpr.2019.06.478.

Thostenson, E.T.T. and Chou, T-W (1999) Microwave processing: Fundamentals and applications. *Compos A* 30(9):1055–1071. doi: 10.1016/S1359-835X(99)00020-2.

Verma, N. and Vettivel, S.C. (2018) Characterization and experimental analysis of boron carbide and rice husk ash reinforced AA7075 aluminium alloy hybrid composite. *J Alloys Compd* 741:981–998. doi: 10.1016/j.jallcom.2018.01.185.

Verma, N. and Zafar, S. (2019) Investigations on mechanical performance of multi-layered microwave processed HDPE / sisal composites for automobile applications. *Appl Mech Mater* 895:64–69. doi: 10.4028/www.scientific.net/AMM.895.64.

Verma, N., Zafar, S., and Pathak, H. (2019a) Microwave-assisted composite fabrication of nano-hydroxyapatite reinforced ultra-highmolecular weight polyethylene composite. *Mater Res Express* 6(11):115333. doi: 10.1088/2053-1591/ab4b28.

Verma, N., Zafar, S., and Talha, M. (2019b) Influence of nano-hydroxyapatite on mechanical behavior of microwave processed polycaprolactone composite foams. *Mater Res Express* 6(8):085336. doi: 10.1088/2053-1591/ab260d.

Verma, N., Zafar, S., and Talha, M. (2020) Application of microwave energy for rapid fabrication of nano- hydroxyapatite reinforced polycaprolactone composite foam. *Manuf Lett* 23:9–13. doi: 10.1016/j.mfglet.2019.11.006.

Ville, J., Inceoglu, F., Ghamri, N., et al (2013) Influence of extrusion conditions on fiber breakage along the screw profile during twin screw compounding of glass fiber-reinforced PA. *Int Polym Process* 28(1):49–57. doi: 10.3139/217.2659.

Wang, X. et al. (2015) Composite components and manufacturing defects in autoclave molding technology. *J Reinf Plast Compos* 28. doi: 10.1177/0731684408093876.

Wulfsberg, J., Herrmann, A., Ziegmann, G., et al (2014) Combination of carbon fibre sheet moulding compound and prepreg compression moulding in aerospace industry. *Procedia Eng* 81:1601–1607. doi: 10.1016/j.proeng.2014.10.197.

Yao, S-S, Jin F-L, Rhee K.Y., et al (2018) Recent advances in carbon-fiber-reinforced thermoplastic composites: A review. *Compos B Eng* 142:241–250. doi: 10.1016/J.COMPOSITESB.2017.12.007.

Zeng, Z. and Grigg, R. (2006) A criterion for non-darcy flow in porous media. *Transp Porous Media* 63(1):57–69. doi: 10.1111/j.1439-0523.2006.01169.x.

3 Biodegradable and Biocompatible Polymer Composite

Biomedical Applications and Bioimplants

Naga Srilatha Cheekuramelli, Dattatraya Late, S. Kiran, and Baijayantimala Garnaik

CONTENTS

3.1 INTRODUCTION

There has been a rapid increase in polymeric materials and they have become a part of human life, which is probably due to their potential applications and their functional characteristics. However, the use of polymeric systems that are nondegradable induces serious problems to human life as well as to the environment and it affects the sustainable development of the ecosystem. Therefore, researchers look for an alternative in biodegradable polymeric systems [1]. These advanced functional materials are classed under biomaterials and it is postulated that the biological material induces interactions biologically to facilitate the medication or ecosystem to degrade it [2]. However, medication is only possible if the biomaterial can be compatible with the biological system [3]. In order to meet the ecofriendly as well as bio needs, researchers have extensively investigated various kinds of materials either experimentally and/or theoretically. They have experimented and quantified the responses by interacting with the biological systems and/or ecosystem and the materials of composed metals, ceramics, and polymers. Polymers possess significant properties such as flexibility in chemistry leading to a great diversity and versatility of chemical and physical functional properties attainable and even the probability of attaining other useful properties such as electronic, hydrophobic or hydrophilic, magnetic, and optical ones. The probabilities of degradable and compatible polymers are also possible via alteration of molecular chemistry and its chemical structure. Although natural biodegradable and compatible polymers are available and are used for bio medication, current demand needs to maintain sustainable ecosystems and the available resources. There is a need to develop new functional biodegradable and biocompatible polymers for biomedical applications. Since the last decade, the demand for a clean and green sustainable pollution free environment and reduction of fossil fuel utilization has been taken into consideration. Hence, the investigators are finding new avenues to develop and replace petroleum-based commodity plastics by biodegradable materials via renewable sources. Biomaterial-based polymeric systems can be realized from either biological synthesis from bioorganisms or chemically produced from natural resources. The researchers have suggested that these kinds of synthetically induced biopolymeric systems are also appropriate substitutes to address the problems which are comprised of biocompatibility, biodegradability, and lightweight. But these materials are mechanically unstable, have inadequate processing ability, lack long-term-durability and possess low chemical resistance. In order to overcome these drawbacks, new functional materials are being developed for the essential requirement, which can tolerate and withstand a wide range of functional applications. Biopolymeric materials are kinds of filler systems such as micro/nano sized fillers which can be blended or reinforced to produce biocomposites or nano biocomposites. Particularly, nano-level morphologies exhibit greater precise surface area, lower density, and optimum surface-energy than the conformed micro morphologies. This functionality renders the advanced properties via well-established material interactions for biomedical applications. Thus, bio-nanocomposites are advantageous to the following applications:, food packaging, nanomedicine, forestry, cosmetics, agriculture, medicinal applications, biomedical wearable electronic devices, bioimplants, and bioelectronics for biomedical implants, etc. Moreover, composites have attractive characteristics which include being lightweight and easy to manufacture at lower cost, which makes them a suitable replacement for metals and alloys either in industrial applications or biomedical implants. Polymers possess low tensile-strength and Young's modulus, compared to metals and alloys. These disadvantages can be overcome by the addition of nanofillers to improve thermal, optical, and magnetic properties, making it suitable for conventional material replacement [4–6]. Furthermore, carbon filler-based polymeric composite systems are the most common composite-based materials useful for functional applications. Carbon nanotubes and graphene are gaining importance in industrial and biomedical applications. Recently, researchers have developed bio-carbons like biochar that offer promising characteristics and have the potential to replace classical carbon fibers. These biobased carbons are produced from renewable sources that show high thermal properties and lower densities. These bio-carbons can be produced by much cleaner and economical ways which can exhibit high Young's modulus and tensile-strength [5, 6]. In engineering applications, bio-carbons are produced for improving tensile and flexural properties [7].

In this context, current advancement in bio-epoxies, their blends, and composite materials, along with the significance of bio epoxy-based materials has been discussed. Bioresins and curing agents realized via various synthetic routes from biobased resources are highlighted, and their electrical, magnetic, and optical properties are reviewed. Further, bio epoxy-based materials with broad functionalities and low-cost electrical and electronic devices are shown.

3.2 INTERESTING FACT ABOUT BIODEGRADABLE AND BIOCOMPATIBLE MATERIALS

Biobased-plastics are either biocompatible i.e., durable or biodegradable. For example, bio-PET called polyethylene terapthalate is a biocompatible nonbiodegradable biobased plastic that exists for more time than it is needed. The bio-PET could be replaced by fossil fuel-based PET, for example plant-based PET i.e., plant bottle is made by plants-based material and fully recyclable. Similarly, biobased polylactic acid (PLA) is biodegradable, which facilitates the production of many commercially available products including tea bags and drinking cups and can also produce various kinds of products through three-dimensional additive printing technique. PLA can be used for many commercial products and packaging systems. The World Food Development Authority also recognized and approved PLA as a biodegradable and compatible polymer for food-contact applications (food packaging). Also, polyhydroxyalkanoates (PHA) is polyester which is naturally produced by various microorganisms, that include lipids or bacteria's fermentation. That leads to biodegradability and can produce different biobased plastics with desired properties. These are all FDA-approved biodegradable or biocompatible polymeric systems or both. FDA-approved polymeric systems such as biodegradable, biocompatible, and other commercial nonapproved biodegradable polymer are not exposed directly to the environment. Biodegradability doesn't mean that a substance or a material can be automatically degraded under any natural atmospheric exposure. The right treatment is required at its end product. In most of the conditions, industrial composting with particular conditions was necessary to ensure biodegradation within a reasonable period, otherwise, it would take a longer time over years to degrade in open environment conditions; even biocompatible materials will start to degrade over 100 years. According to the European standards, the production of industrial compost must facilitate biodegradation within six months. Compostable means in accord with the European standards (EN14995 and EN13432), material needs to be approved/certified and then is composted in industrial plants. Industrial composting means that the composting of biodegradable plastics will only degrade under specific conditions that are implemented in industrial plants. Specific conditions include time, humidity, temperature, and being in the presence of bacteria and fungi. Besides biodegradability of plastic materials, the capacity for recycling and reusability of products are also important considerations to be implemented. This is because composting of biobased materials only makes sense for the particular applications and recycling of the biobased materials is more difficult. The tea bag prepared using PLA shows best compostabilty as compared to polypropylene (PP) tea bag when these bags are thrown into dustbin. Researchers have suggested that theoretically all biobased plastic materials can be recycled and reused. Most of the substances made from various kinds of fossil and biobased plastics which can be easily processed by recycling plants through tunable mechanical and physical proprieties. New products can be reproduced using different forms of granulating or remelting processes on separated plastic materials. During recycling, purity of feed materials is an essential parameter for the production of product. Purity of material is a deciding factor in the functionality for a particular application; if the material does not have a 100% yield, the application efficiency and reliability of that material may decrease. In some polymers, for example, biobased drop in plastics such as Bio-PE or Bio-PET is chemically identical to that of fossil fuel-based versions of PE and PET. They could be utterly assimilated during the establishment of a recycling stream in order to maintain purity. For example, biobased plastics need to be recycled in separate streams for each material type because the purity of the recycle stream is a major concern. In comparison to the fossil-based plastics, this is a small disadvantage. With the biobased plastics, the emission of CO_2 is increased, so in order to avoid this greenhouse gas concentration in the atmosphere, recycling of some plastics must be followed. Some countries tend to have more stringent regulations involving food and product environmental claims; when it comes to considerations of biodegradable claims, peoples are also following the regulations to limit the use of certain items. This is because warning labels are required to state items that have been found to cause cancer, defects in birth (teratogenicity), or other reproductive harm.

Biodegradation: In order to maintain an ecofriendly and sustainable environment, composting at industrial level is necessary through heating bioplastic at the optimum level of temperature that creates a perfect environment for microorganisms break it down. Without intense heat and critical conditions, bioplastics could not degrade within a time frame, the probability of degradation starts after 100 years or above, in either landfills or the home compost heap. In the marine environment, if they degrade by breaking down into microsized pieces, these pieces may last for decades, leading to abnormal, unpredicted conditions for marine life. This is dangerous for marine animals and the ecosystem. For example, PLA is not biodegradable on its own in the ocean. It can be composted in an industrial facility.

The compostable plastics are suitable to industrial composting which must meet specific requirements. Table 3.1 presents active standards for biodegradation of materials in various systems.

- Biodegradation of plastics in environment: chemical breakdown of material into CO_2, H_2O, biomass, and minerals. According to this standard, ≥70% degradation of reference material. The measured CO_2 or the biochemical oxygen demand (BOD) values from the blanks at the end of the test are within 20% of the mean.
- Quality of the final compost and ecotoxicity should be maintained under limits. The standard specifies that the ecotoxicity tests check the presence of toxic compounds. This involves examining to see if the germination and biomass production of plants is not adversely affected by the influence of composted packing.

TABLE 3.1
The Present Active Standards for Biodegradation of Materials in Various Systems

S.No	System	Version
1.	Soil	ASTM—D5988-12
		ISO 17556:2012
		NF U52-001
		UNI 11462:2012
		OECD-304A and 307
2.	Marine	ASTM—D7473-12, ASTM D7991-15
		EN 17294-2, EN 17033
		ISO 18830:2016
		OECD 208
3.	Biological	ASTM F1635-11
		ISO 10993-13
		ISO 10993-14-16

Miscellaneous: proper labeling; preparation of a test report by manufacturer—surface area should not decrease <10% during deployment (if best practices are used)—Annexes G and H of EN 17033

American Society for Testing and Materials International (ASTM)
International Standards of Organization (ISO)
NF and UNI: French and Italian Normalization Organizations (AFNOR, UNI)
EN: European Standard
Organization for Economic Co-operation and Development (OECD)

Biocompatibility: Understanding of biocompatibility is complicated because attributable problematic discretization existed in between investigators thinking and their experimental evaluations. However, according to the clinical investigation observations and benchtop animal experimental conclusions, researchers have pointed to conclusions, which are listed below.

- All of the majority of medical and bioimplant materials are standard and economically tolerable substances.
- Biocompatibility depends on various factors such as the fabrication method and its finish and/or shape, and/or material choice.
- Though impressive material characteristics are realized from the devices, it does not necessarily mean the material or material-enabled devices are useful for the medication; this should be validated clinically over some time.
- Reactivity in biologically active material or material enabled devices are time or site or species dependent. In humans, each individual varies from one another.

Table 3.2 illustrates the active standards for biocompatibility of materials.

TABLE 3.2
The Present Active Standards for Biocompatibility of Materials

S.No.	Feild	Version
1.	Biocompatibility	ISO 10993-1, 4–11
		(CFR) Title 21 Part 58
		Subchapter A–K, Parts 800–895
		OECD 423, 450–453, 471
		ICH- S1–S11, M3, M6–M7, Q3C, Q3D
		ASTM F04.16, F981-04.16, F1027-86.17

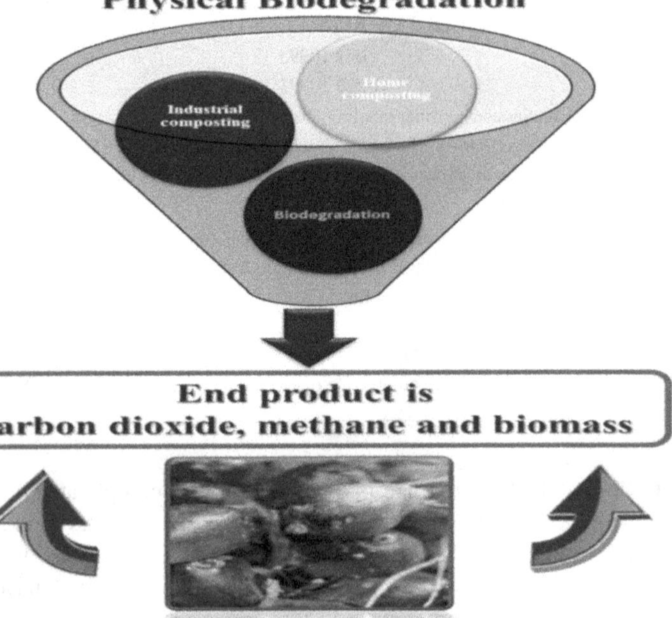

FIGURE 3.1 Biodegradation process.

All delineations suggested the various elemental dynamics subsidizing limitations of the usage of materials for biomedical applications. For example, in bioimplants, factors are, blood interaction intensified thrombus development, biological fluid induced protein gel accretion, chronic inflammatory scar development during the healing time, biofilm establishment, and its susceptibility to the infection. Figure 3.1 shows the biodegradation process.

Table 3.3 depicts the terms and description of biodegradation, the biodegradation process, and its end product realization.

TABLE 3.3
Terms and Description of Biodegradation, Biodegradation Process, and Its End Product Realization

Term	Process	Result
	Physical and mechanical process	
Home composting (Initial Stage)	Biodegradation and disintegration take place i.e., breakdown of material is caused by multiple factors, for example, water, wind, ultraviolet rays, and weather	Carbon dioxide, methane, and biomass (takes much longer time to degrade)
Industrial composting (Intermediate Stage)	Biodegradation and disintegration take place in an industrial facility to produce compost: controlled environment, high temperatures	Carbon dioxide, methane, and biomass (due to controlled environmental condition, less time is required to degrade)
Biodegradation (Final Stage Complete Degradation)	The chemical process of microorganisms breaks down the material to carbon dioxide (CO_2), methane, and biomass	Final end product is carbon dioxide, methane, and biomass
	Natural process	
Biodegradable in soil, fresh or sea water	Biodegradation and disintegration take place in an open environment: conditions such as low temperatures, the population of microorganisms like bacteria and fungi are needed to start the process	The final end product is carbon dioxide, methane, and biomass

3.3 NEED, NECESSITY, AND PREVENTION OF BIODEGRADABLE AND BIOCOMPATIBLE MATERIALS

Biodegradability: According to the World Ecofriendly and Sustainable Commission survey report, every year, landfill sites are filled with millions of tons of plastic waste which includes carrier bags, packaging, and refuse sacks. These waste items are buried and are increasing tremendously in amount which leads to unrecoverable damage to the Earth's ecosystem and marine system. According to the reports, China generates 16,000,000 tons of waste, India generates 4,500,000 tons, and the UK generates one million tons in which over 800,000 tons of this waste is polyethylene waste. A century is needed to degrade conventional polyethylene [8]. World survey reports describe that globally 3.2 million tons out of 26,000,000 tons of waste packages are produced by household waste in every year. 150 million tons of waste packages are from the industries and also increases each year. 2.5 million plastic bottles waste comes from the USA in every hour, out of which only 3% are recycled. Similarly, the UK produces 15,000,000 plastic bottles each day. 3% of these bottles are recycled and around 1% of plastic bags were recycled out of billions [9]. Incineration and recycling disposal procedures were available for these materials, but if in streams of industrial waste and in domestic waste, plastic waste is made with some other materials, that results in great expansivity in separation, especially for small carrier bags. In the present climate, scarcity of resources increases day by day. So, there is a high demand to focus on salvation of waste, post-consumer. Also, in order to avoid the buildup of these wastes in order to protect our ecosystem and marine system, researchers are finding permanent solutions by expanding the research and development of biodegradable materials [10]. According to the American Society for Testing and Materials (ASTM) and the International Organization for Standardization (ISO) standards, degradable polymers are those that undergo chemical changes in their chemical structure by various conditional environments. These chemical changes result in the loss of mechanical and physical properties [11]. The main pillars for biodegradable plastics are natural to the synthetic polymers. Renewable sources are producing large amounts of natural biopolymers as well as non-renewable sources are producing synthetic polymers such as resources of petroleum, etc. Besides the biocompatibility, they have other functionalities such as air penetrability, lower temperature scalability, economical compatibility, and availability. Such polymers are starch, chitosan, cellulose, PLA, PCL, polyhydroxybutyrate (PHB), etc., which are used for food and industrial packaging. The trend of blends of different biopolymers such as starch-PLA and starch-PCL blends are also useful for packaging and are semisynthetic.

Biocompatibility:

- Biocompatibility is life for biomaterials in biomedical applications in terms of quality in compatibility with the live atmosphere such as the human body that must not be toxical or deleterious and not lead to immunological ailments.
- In sense of regularity, biocompatibility means revealing of measurement toxicity quantification, which results from the contact with biocompatible components. This is an important activity because systemic toxicity impairs the entire biological system i.e., nervous or cell or immune system.
- Biocompatibility tests are proved by sensitization and allergic/irritation reaction. Sensitization is toxic and independent of dosage. It is delayed after an exposure and it may not be localized. Irritation reaction is an immediate activity after a single exposure. These kinds of tests are important because the presence of extractable chemical compounds or reagents during material and immune system interaction processes may influence biocompatibility. Therefore, cytotoxicity is an important and most sensitive characteristic of biocompatibility useful for evaluating the harmful reactions.
- According to the International Organization for Standardization and the United States of Pharmacopeia (USP), biocompatibility is an important part of complying with the standards. That ensures the quality of medicines and other health care technologies.

Figure 3.2 shows biomedical applications of biodegradable and biocompatible lightweight polymer composites.

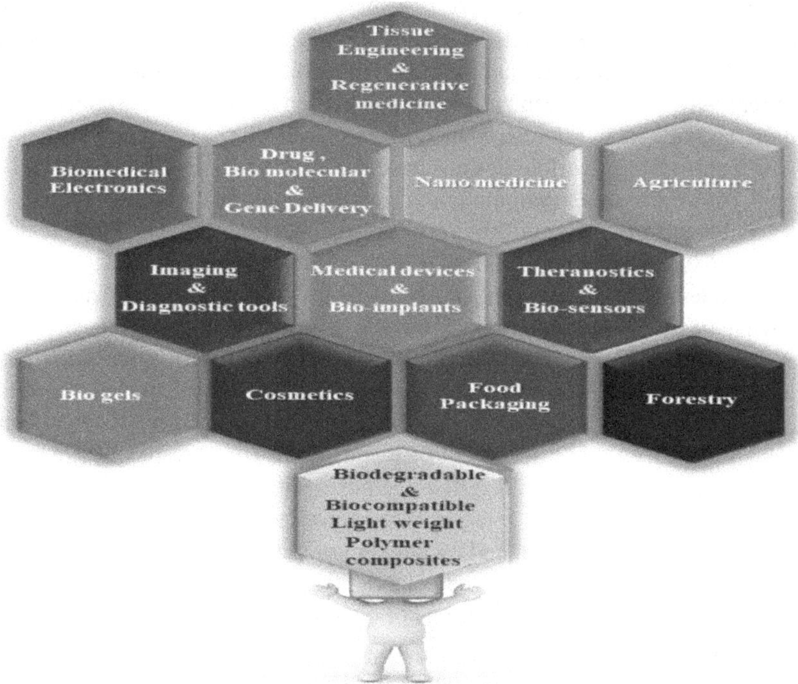

FIGURE 3.2 Biomedical applications of biodegradable and biocompatible lightweight polymer composites.

3.4 BIODEGRADABLE AND BIOCOMPATIBLE POLYMER COMPOSITES AND THEIR BIOMEDICAL APPLICATIONS

Researchers have explored various kinds of biodegradable and biocompatible resin materials for dental implants. However, dental material resin facilitated dental implants. One of the major challenges is the functional properties and mechanical performance of the material, which can be improved via nanotechnology, leading to the stabilization and improvement of dentistry. Recently, researchers have developed composites to improve dental implant performance via the micelle assisted hydrothermal method. The developed biodegradable and biocompatible nanoscaffolds of PLA/Al_2O_3 are generated and applicable as dental resin composites and are also applicable to other biomedical applications. They observed that the improved mechanical functionality, which is attributable to the interactions-induced intermolecular network. The flexural strength, modulus and compressive strength of PLA/Al_2O_3 nanocomposites have been observed as 88 MPa, 7.5 GPa and 157.2 MPa respectively [12]. Hydroxyapatite (HAp) is one of the remarkable minerals in the bone that gives bones their strength. The use of hydroxyapatite, with the biodegradable and bio-active qualities of polyxylitol sebacic-adipate (PXSA), resulted in a controlled release at target sites. Many research investigations also describe that vitamin K plays a vital role in bone metabolism. Recently, Zhipend et al. have reported that synthesized biopolymeric composite based on HAp/PXSA/K can be used for regeneration of tissue. Based on *in-vitro* studies, HAP/PXSA/K was reliable to stem cell culture or mesenchymal. They found that the HAP/PXSA/K biopolymeric material is more favorable than the pure HAP sample [13]. This biomaterial provides characteristic tissue regrowth and improvement in cell penetration of scaffold, favorable biodegradation, and favorable microstructures for the tissue engineering application [13]. In tissue engineering, scaffold designs have required a porosity in high level, large amounts of surface area, optimum pore size all over the matrix, that porous structures should be distributed uniformly with interconnection [14], where interaction of biological tissues can be improved with the incremental trend in surface area along with the biodegradability as necessary prerequisites. Recently, the fabrication of biodegradable scaffolds was the main aim of the creation of polyvinyl alcohol gelatin (PVA-GE) loaded with biphasic calcium phosphite (BCP)-electrospun blends for the engineering of bone tissue. After the application of biological tissues, to perform adhesion measurement and to access the bone scaffold potentiality, they performed their analysis of PVA-GE loaded with BCP cells of the osteoblastic of the nanofiber composite mat. Other results suggested that pores within the matrix and their interconnected structure provided a larger surface area which helped the residence of tissue cells, their proliferation, and exchanging of nutrients probably attributable to the scaffold environments. Thus, the potentiality suggests that the developed scaffolds could be utilized for the functions of bone tissue [15]. Prophylaxis-instigated infection of bones can be avoided via sustained drug release with the bioactivity of biodegradable materials, which has been demonstrated as an auspicious treatment for bone infection. Recently, poly (3-hydroxybutyrate-co-3-6% hydroxyvalerate) (PHBV), nanodiamond, and vancomycin (VC) loaded nanohydroxybutyrate was used to prepare nanocomposites via an injection moving technique. Two composites were fabricated; one with the help of a rotary evaporator and the second one with a spray dryer. The developed composite systems were able to present antibacterial activity and were compatible for sustained release of VC. The spray dryer-assisted method induces increasing crystallinity of PHBV from 57 to 73%; in addition to this, the glass transition temperature of these composites increases. Also, researchers have analyzed and observed the incremental PHBV flexural elastic modulus which is 34% and that was a good agreement with the results reported earlier. Moreover, *in-vitro* studies revealed good adhesion capability, cell growth, cytotoxicity, and non-cytotoxicity behavior. Interesting desirable observations were made where these composites could be a potential candidate in the material application of bone defect filling [16]. Self-healing materials with biodegradability and biocompatibility could be potentially applied in dentistry and other biomedical implant applications.

There are no potent remedies for disorderliness in neurodegenerative conditions like Parkinsons disease (PD). Moreover, most of the drugs wasted from the clinical trials due to low penetrability in the brain. In the models of PD, Hsp70 chaperone protein is reported as efficient, but it is a challenge to get it to target the brain. Recently, Tunesi et al. have developed a unique Hsp70 neural delivery system by using semi-interpenetratable collagen networks (COLL), which features biocompatibility, injectability with biodegradability, and also the structure of hyaluronic acid (low in molecular weight) along with particles of gelatin. Tat is a peptide which can be interpenetrated to the cell, thereafter fuses with Hsp70-1A which leads to the production of human recombinant that results in the protectiveness of neurons against 6-hydroxydopamine toxin in *in-vitro*. They assessed that in *in-vitro* from selected composites of COLL-LMW hyaluronic acid releases Hsp70 with Tat and also observed that it released 95% in four days. These experimental studies reveal that the COLL-LMW hyaluronic acid composites are available for at least 96 hrs in the brain by magnetic resonance and protecting of SH-SYSY neuron cells. The release studies of Hsp70 with Tat indicate that hydrogel is crucial for highly focused protectiveness to neurons. By injecting mice with Hsp70 with Tat composites against toxin 6-OHDA striatal injection, the immunostained CD11b and GFAP levels were shown to decrease after seven days. Thus, the developed drug system with biodegradable composites enables Hsp70 with Tat release efficiency [17]. Moreover, researchers have investigated nerve regeneration via biodegradable electroactive conduits acceleration, which is based on nanocomposite hydrogels. Homaeigohar et al. [18] have reported green functionalized graphite aided alginate nanocomposite biocompatible hydrogels. Mechanical stability is improved more than three orders compared to the pristine one which is probably due to well-dispersed nanofillers in the hydrogel. Also, stimulating biological activity has also been observed, which is a well-established phenomenon between electrical contact via reinforcement and intercellular signaling. According to *in-vitro* investigations, they observed the biocompatibility of a nanocomposite hydrogel via implantation within the biological cells. Thus, the significant hydrogels interaction presented a non-inflammable response and no adverse reaction after 14 days. These overall results suggest that these nanocomposite hydrogels could be potential candidates for stimulating nerve generation and nerve conduct material [18]. Though the repair of tissue by automatic curability of biodegradable hydrogels were attentive, its impact was unclear. Recently, Cheng et al. have developed a self-curable composite of chitosan-cellulose nanofiber (CS-CNF) which has stable-modulus 2 kPa and properties of automatic tunable curability. The self-curable pristine chitosan hydrogel has reversible dynamic-Schiff bonds, automatic curability of hydrogels with strain sensitiveness which is affected by the addition of a tiny supply of CNF filler. When hydrogel CS-CNF is fixed with neural stem cells, they result in the efficient healing with their own proliferation that leads to the augmentation of metabolism of oxygen, as well as neural differentiation taking place.

The metabolic change in the self-curing hydrogel ability was correlated with neural cell differentiation. Furthermore, in the brain injury of the zebra fish model, the result showed results were enhanced by 50% by using fixed hydrogel of CS-CNF—0.09 wt% inefficiency of regeneration of neurons compared to pristine CS hydrogel. They proposed a new contrivance, which has shown the moldable abilities of self-curing hydrogels with the maintenance of identical scope of stiffness. They clarified although that the main aspect of self-curable hydrogels performance in a biological system and the rationalization of the design of hydrogels results in an excellent potentiality of regeneration of tissue as well as injectability [19]. The protein-based system of polymeric blends was examined to figure out the reciprocal action and silk fibroin miscibility with polylactic acid in scales of a micron (μ) to μm that manage responses of cells and biodegradation of enzymes on the interfaces of biomaterials. Enhanced tunability of semicrystalline content of protein (silk fibroin) β sheets and biodegradability with noncrystallizable (PDLLA)-poly(D, L-lactic acid) favored providing thermal stability and hydrophobicity. This study concluded the essentiality of comprehension of various schemes for the composites of synthetic polymer fabrications that lead to great usage in applications of green chemistry and the biomedical field [20]. In some critical and complicated surgeries, critical conditions and well-predicted environment are required; these conditions enabled heart wall surgeries that needed coverage for enormous deformities for the subsequent evacuation of tumors in malignant stage. Generally, Gore-Tex is nondegradable and inert to tissue for the replacement of cardiac tissue. All innovative biodegradable biomaterials are united to stem cells such as the kind of nanocomposite hydrogels that stimulate the curing ability. The double-layered hybrid biodegradable nanocomposites have been developed using amorphous nanoparticles of calcium phosphate with pure poly-lactic-co-glycolic acid (PLGA). The adipose-based derived stem cells were cultured on electrospun disk and biochemical tests were carried out. The scaffolds were diffused in C5BL/LYS for the substitute to the heart wall. In *in-vivo*, mouse studies, the following were observed for one to two months after the surgery: the protrusion of scaffolds into biomaterials through bilayers, the duration of CD45 cells, the response in inflammation, and the content of extracellular matrix (ECM). The *in-vitro* study showed that the material got stable after two weeks. They concluded that these composite scaffolds were suitable for reconstruction of the heart wall as well as supporting regeneration and being compatible with host tissue [21].

Biodegradable materials have a premium passion for the progress of transitory implants. Most of the researches has been concentrated on the basis of Mg materials; recently, zinc and zinc-based materials, i.e., alloys have received a surge of attention in this field. Recently, huge surface areas of layered and flower-structured zinc oxide nanotubes were generated through anodic oxidation onto the surface of pure zinc. Electrochemical tests were performed on the solution of simulated body fluid (SBF) and it was shown that the layers of zinc oxide nanotube enhanced the electrochemical performance, compared to that of pure zinc. From the aspect of the degradation rate of bone, the zinc/zinc oxide nanotube composite is more suitable to fixation of bone, compared to pure zinc. Moreover, the immersion test showed that there is a high rate of deposition of Ca/P on the top of zinc oxide nanotube samples that indicates a superior biological compatibility leading to a number of biomedical applications due to semiconducting [22]. However, corrosion of biodegradable polymeric composites in bioimplants is a major drawback which predominantly depends on various conditions and a wide range of factors probably influence it. However, magnesium alloys have gained valid attention because of their exceptional characteristics of biocompatibility and degradability. The rate of the corrosion of Mg alloys leads to limited usage of materials in implantation. Sing et al. have investigated the corrosive behavior of AZ91Mg alloy and improved corrosion resistance by hydroxyapatite and iron oxide composite coating. Chitosan-based polymeric composite coatings were produced via the electro-phoric method. Different weight configurations of Fe_3O_4 were utilized (1, 3, and 5%). The results suggested that the composite (HA and 1wt% Fe_3O_4) coating exhibited an anticorrosive property and could be a potential candidate for temporary bioimplant applications [23]. As Mg0.8Ca alloy was developed with the addition of Si or F bioactive-coatings, biocompatibility investigations were performed via seeding premyoblastic-endothelial and preosteoblastic cell lines. The preosteoclastic, preosteoblastic, murine-endothelial, and premyoblastic cell cultures were also utilized in order to reveal biocompatibility. The experimental results showed the reduced degradation due to less oxidation, and it released P, MgCa, Si, and F from all the studied compositions; however, the optimal characteristics exhibited Si containing PEO coating utilization in cardiovascular-applications and bone-regeneration. The high fluorine content enabled coatings are attributed to negative endothelial-cell effects. Co-culture differentiation and *RAW*264.7, *MC*3*T*3 investigations using extracts of PEO-coated Mg0.8Ca demonstrated the improved osteoblastogenesis and osteoclastogenesis, compared to bare alloy [24].

Other than the biomedical and bioimplants application, biodegradable and biocompatible polymeric composites are dominantly required in order to maintain the ecological system, water purity; food-based customized goods, electrical and electronic applications. For example, non-biodegradability and toxicity-induced heavy-metal-ions are hazardous to the human body, hence long-term effects induce disorders to human health and the natural ecosystem [25, 26]. Removal of heavy metal-ions via the ultrahigh adsorption capability of biosorbent utilization is important to eradicating water pollution. Amino-functionalized nano-cellulose aerogels are important materials to adsorb Cu(II) ions, reversible-recycled Cu(II) ion capture enabled 2,2,6,6-tetramethylpiperidine-1-oxyl(TEMPO), oxidized cellulose nanofibrils (TO-CNF), cross-linked trimethylolpropane-tris-(2-methyl-1-aziridine) propionate(TMPTAP), and polymethyleneimine (PEI) has been reported by Liuting et al. The TO-CNF/TMPTAP/PEI aerogel showed an advanced cellular-structure which enabled in three-dimensional (3D) configuration with plentiful oxygen and amino-groups. This unique architecture is attributable to the efficient adsorption capability, which is 485.44 mg/g obtained via the Langmuir-isotherm model. The aerogel restores the 3D architecture, even in EDTA-2Na treatment, and can be used for adsorption without significant degradation. Moreover, reversibility, recyclability, ultrahigh adsorption performance capacity, and structural and chemical stability make aerogel the ideal choice for Cu(II) ions treatment applications [27]. The use of crude-glycerol, an industrial-biodiesel process-based byproduct mainly composed of fatty-acids glycerol and impurities leads to the development of biobased polycarbonates. Simultaneously, biobased-epoxide monomers were successively produced via biobased epichlorohydrin derived from refined crude-glycerol donated epoxy group. The biobased polycarbonates exhibited thermal properties such as decomposition at 230°C, good transparency and mechanical properties such as elongation (59.6%) and tensile-strength 1.69 Mpa; all these functionalities could lead to a new platform of biodegradable-biomaterials production via conversion of CO_2 and industrial wastes offering ecofriendly and environmental solutions [28]. Biodegradable and biosources-instigated nanocomposites are emerging materials, especially in the area of food packaging due to food-contact instigated issues with respect to consumer safety precautions. Recently, researchers have developed nanoclay systems as fatty food stimulants. They found that the presence of nanoclay leads to the nature of migrating-substance and it is influenced by the nature of the food stimulant. In other reports, researchers observed that the biopolymerbased material contact with the ethanol where the sorption by ethanol leads to the prevention of synthetic-polyester crystallization considered continuous homogeneous-phase [29]. Such a kind of bio-instigated compatibilized polymeric composites could be a potential food safety material in the future. Developed polypropylene/polylactide/nanoclay blend/composite films aerobically degraded to reveal CO_2 capture efficiency as per the ASTMS5338 method. Results elucidate ungraded mineralizable-intermediate carbon kinetics and degradation curves with two kinetic regimes. The first regime is comprised of slowly and moderately hydrolyzing carbon, which leads to the low degradation\and degradability rate. The second regime is comprised of hydrolyzable carbon inducing a relatively high-degradation rate and degradability. The absence of phase growth reveals the degradation rate profiles attributable to the hydrolyzable carbon. These approaches could be a method to use in the future to realize model and design-equipment for other waste-biodegradation systems [30].

Moreover, researchers have explored various kinds of composites that electronically stimulate the biological parts, for example, in the human body such as heartbeat monitoring systems, nanogenerators for heart rhythms and fibrillations, biobased electronic chips to stimulate brain frequencies, and electronic circuits-enabled eye lenses for stimulation monitoring systems. Recently, many research reports suggest synthetic nanostructures have functional advantages which could lead to many potential applications in biomedicine, bioelectronics, and wastewater treatment. Therefore, researchers are extensively investigating multifunctional composites; Aguilar et al. have investigated magnetic particle adsorption and developed smart bioplastic based on cellulose films. These composites were attained via the process of immersing cellulose-based biopolymers in a nanometre-sized $MnFe_2O_4$ ferrofluid. The magnetic measurements were conducted and revealed that the super paramagnetic shape is maintained in bioplastics. However, the optical transference did not show a drastic effect after the adsorption of magnetic particles [31]. This kind of biodegradable and biocompatible composites with functional characteristics either magnetic or electric characteristics enabled electronic devices could be a potential candidate as a bio implanted device to stimulate biological activities.

The complex biomineralized natural bone system is an orderly assembled apatite within a type one collagenous matrix [32, 33]. A number of research investigations carried out on bone tissue engineering found that additional surgical procedure is required to improve the functional quality of bone and tissue, which is probably due to their drawbacks during bone grafting [34]. This kind of bone grafting or repair is an engineered process with materials design which presents several difficulties. Though the mechanical, as well as porous morphological characteristics, are prevalent [35], bone graft must be osteoconductive and biocompatible with the host [36]. Although various techniques have been introduced by the researchers to realize proper bone graft substrate, fabrication of micro-architecture-enabled bone bases lead to incremental mass transport of nutrients and oxygen [37]. A recent research investigation revealed the native bone tissue morphology mimicking which exhibits incremental tissue regeneration and osteoblastic phenotype [36]. Among all these reports, porous architecture-based mixed composites were reported but the nonregular pore distribution and nonhomogeneous characteristics lead to an efficient bone replacement [38, 39]. Sharma and co-workers have reported that 10 wt% of ZnO in electrospun polycaprolactone/hydroxyapatite/ZnO scaffold exhibited appreciable balancing of mechanical properties with mineralization, antimicrobial activity with proliferation, and optimality in cell viability. Based on *in-vitro* results, they concluded that this grafted bone substrate acts as a potential bone graft substrate for the applications of bone tissue regeneration. They also concluded from *in-vitro* studies that Clopidogrel eluting electrospun polyurethane/polyethylene glycol nanofibrous scaffolds are assured to be hemocompatible biomaterial with thermoresistant qualities [40–41]. The chemical versatility induced chitosan provides extended nano-structures which enabled ordered pore sizes and distribution and makes them promising candidates for biomedical applications. Also, chitosan provided compatibility attributable to promoting the proliferation of osteoblasts without toxic and inflammatory reaction and antibacterial activity, which are important attributes for biomedical needs [42–44]. Moreover, S-3 glass fibers enabled special graded products to act as a fiber reinforcement for the biomaterial implants. The advanced functional biomaterial PHB glass fiber provides a longer life span and compatibility, with respect to thermoplastic materials such as polyphenylene sulfide, polyetheretherketone, and polyetherimide for biomedical applications over 30 days. These fibers exhibited important mechanical characteristics such as a tensile strength of 40% and a tensile modulus of 20%, as correlated to the standard E-glass fibers. AGY reports suggest that these fiber reinforcements could be a healing material used for dental implants in the future. Other medical implanted devices, such as those required in orthopedics may demand biocompatible reinforcement of structural composites.

Figure 3.3 depicts the types and requirement of criteria in the selection of bioimplants.

However, in interactions of materials with human blood, the materials should be capable of resisting blood cell adhesion and absorption of protein, which lead to triggering of organisms' defense system [45]. Researchers have introduced three kinds of polymeric surfaces [45, 46], such as 1. micro-phase separation surface domains, 2. biomembranes-based materials, 3. hydrophilicity-induced surfaces. These surfaces instigated physicochemical activities such as stiffness, surface-charge, wettability surface free energy, topography attributes to the induced chemical functionalities which are recommended to the implementation of these materials in biomedical devices [45, 47]. Blood-compatibility- and biocompatibility-induced poly(2-methoxyethyl acrylate) (PEMA) is also an excellent material for implanted applications and it has been approved in food-administration and medical drug applications [46, 48, 49]. PEMA-coated tubes were recently used in biomedical applications due to their reduced blood cell-activation when it is implemented in catheters for central veins and cardiopulmonary bypass. Moreover, PEMA's compatibility with platelet interaction elucidates coagulation, which is the most preeminent characteristics for the biomedical implanted applications [50].

Although biodegradable composites are preferred to the biomedical applications, device failures are due to their extracellular influences and molecular interactions at the interfaces [51, 52]. In general, series of tissue responses as well as the eliciting of non-specific protein adsorption are accompanied by denaturation and changes in protein conformation. Such kinds of protein changes participate during cell-signaling in the communication process and induce platelet adhesion, followed by blood clot and systemic inflammation [53]. Though biodegradable byproducts enabled toxicity effects, which induce complications in biomedical application. Overall, serve

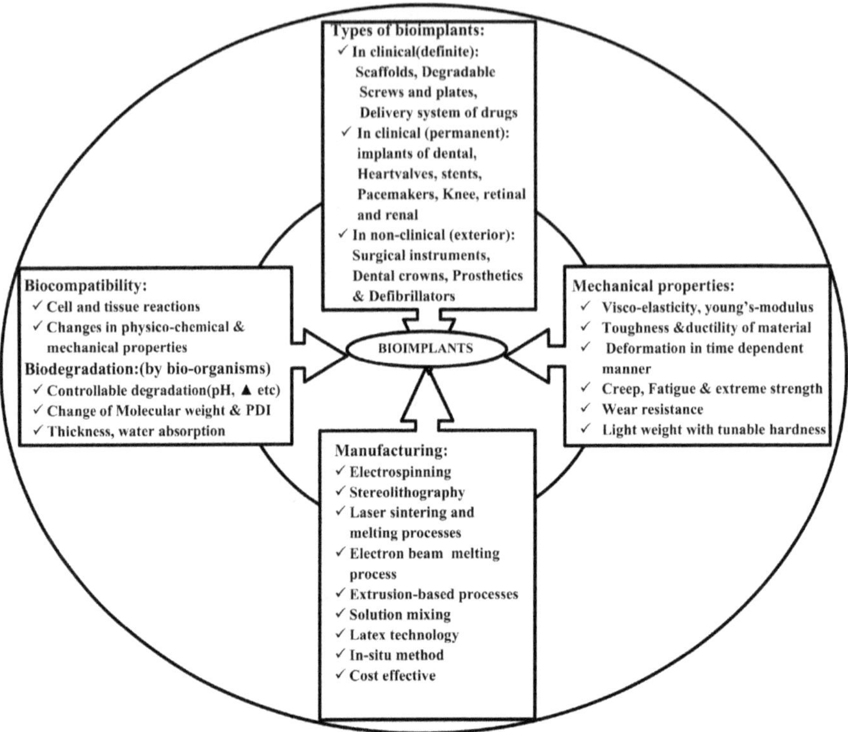

FIGURE 3.3 Types and requirement of criteria in the selection of bioimplants.

limitation alternatives experimented toward the end product application which could be a future solution for device failures, leading to efficient implantable applications [54].

The therapeutic outcomes in the field of dentistry may lead to unexpected side effects such as toxicological and allergic reactions, local or systemic. Toxicity on the local level is mainly due to the chemistry of interactions between the foreign toxiferous and host molecules and also is dependent on compatibility of tissue with biomaterials. Tissues of the stony tooth and gingival mucosa are involved in local reactions, including extreme wearing of restorative materials opposed by teeth. Kanca states that the altering of current testing methods of biocompatibility will discriminate between the effects on the pulp by bacteria and materials [55]. Studies realized on specific restorations of crucial dentin by the usage of cohesive bonds through hybridization will be effective in shedding light on controlling sensitivity issues postoperatively, along with how to avoid the contact of pulps with dental components indirectly as well as on how to heal the naked pulps. Eugenol zinc oxide cement prevents microleakage in bacteria in naked dental pulps with the healing capacity [56, 57]. The study showed that after the implantation of nine adhesive systems and composites of resin were tested, according to the directions of manufacturers, the results came back nontoxic and biocompatible to exposed and unexposed pulp tissues along with control of hemorrhage [58, 59]. The release of resin fluoride was considered as a secured adhesive-dental restorative resin, compared to the compatibility of non-fluoride resin [60]. Contrarily, based on results of other studies, KETAC-CEM showed less toxicity than experimental fluoride composite resin, compared with FUJI (type 2 of glass ionomer cement) [61]. After uptake of various amounts of fluoride into a storage medium, cement compomers of glass ionomer and giomers were compatible with multiple release times [62]. When compared to freshly mixed resins before the polymerization, it was found that restorative resins have slight toxicity toward cells even though the extract was exchanged more than once [63].

The enamel removal from dentin teeth was carried out with the cyclopolymerized system, where trimethylolpropane trimethacrylate is chemically intricate with the filler particle, experimental results elucidate no inflammation and no pulpal irritation [64]. Cement was assessed in a comparative biological study that revealed that Ketac-bond was a biologically accepted restorative material, attained via polymerized resin monomers in resin-modified glass ionomer cement and Cu^{2+} and Ag^{+} in metal-reinforced glass ionomer [65, 66]. Furthermore, investigation results elucidate that the glass ionomer and resin composite exhibit biocompatibility during an interconnecting tissue process [67]. Moreover, unbound resin leads to leaching into saliva at the initial-phase formation and may predispose both dental personnel and patients to allergic reactions [68]. This kind of systemic skin allergy induced side-effects may be classically introduced as acute or chronic. Researchers believe that dental biocompatible material assessability for *in-vitro* and clinical studies are required [69]. Human blood induced histamine via basophil is insensitive to the patients [70]. Various investigations were conducted during the last ten years to reveal information about composites and their leachability on cell growth and function, which is probably attributable to the regulated appropriate formulations of the chemistry of resin [71]. Cytotoxicity undergoing accelerated aging is relevant and reveals the improvement of dental restorative materials and have been advised for the treatment [72]. Moreover, improved biocompatibility due to lower or mocers cytotoxicity based on amine or amide-dimethacrylate-trialkoxysilane was reported [73]. Furthermore, thiolene formulations lead to improved water solubility and methacrylate-conversion is suggested for better biocompatibility when used with the dimethacrylate system [74]. Also, silicon-based investigations reveal that the methacrylate and silorate-based systems exhibit comparative biocompatibility, less cytotoxicity when used with the methacrylate systems [75, 76].

3.5 CONCLUSION

In this chapter, the advancement in biodegradable and biocompatible materials, due to their functionalities, has been described. Further, interesting facts about biodegradable and biocompatible materials, their significance, and the necessity for biodegradable and biocompatible materials and their applications such as biomedical and food packaging and other applications are illustrated. Further, this chapter describes biodegradable and biocompatible materials with widespread functionalities which could facilitate the delivery of efficient and low-cost biomedical applications and also protective materials for ecological system.

ACKNOWLEDGMENT

The present work was financially supported by the Council of Scientific and Industrial Research, Government of India, New Delhi (CSC-0302 and CSC-0134).

CONFLICT OF INTEREST

There is no conflict of interest to declare.

REFERENCES

1. Bogoeva-Gaceva, G., Dimeski, Dimko, and Srebrenkoska, Vineta. (2013) Undefined Biocomposites based on poly (lactic acid) and kenaf fibers: Effect of micro-fibrillated cellulose. mjcce.org.mk.
2. Biomaterials, D.W. (2009) *Undefined on the Nature of Biomaterials*. Elsevier.
3. Grainger, D.W. (1999) The Williams dictionary of biomaterials. *Mater. Today*, **2**(3), 29.
4. Mittal, G., Dhand, V., Rhee, K.Y., Park, S.-J., and Lee, W.R. (2015) A review on carbon nanotubes and graphene as fillers in reinforced polymer nanocomposites. *J. Ind. Eng. Chem.*, **21**, 11–25.
5. Reddy, M.M., Vivekanandhan, S., Misra, M., Bhatia, S.K., and Mohanty, A.K. (2013) Biobased plastics and bionanocomposites: Current status and future opportunities. *Prog. Polym. Sci.*, **38**(10–11), 1653–1689.
6. Rahman, A., Ali, I., Zahrani, A.L., Saeed, M., and Eleithy, R.H. (2011) A review of the applications of nanocarbon polymer composites. *Nano*, **06**(03), 185–203.
7. Ogunsona, E.O., Misra, M., and Mohanty, A.K. (2017) Impact of interfacial adhesion on the microstructure and property variations of biocarbons reinforced nylon 6 biocomposites. *Compos. Part A Appl. Sci. Manuf.*, **98**, 32–44.
8. Kumar, A., Karthick, K., and Arumugam, K.P. (2011) Properties of biodegradable polymers and degradation for sustainable development. cabdirect.org.
9. Meena, P.L., Goel, A., Rai, V., Rao, E., Singh Barwa, M., Manjeet, C., Barwa, S., Vinay, A., Goel, V., Rai, E., and Rao, S. (2017) Packaging material and need of biodegradable polymers: A review. *Int. J. Appl. Res.*, **3**(7), 886–896.
10. Unilever (2016) Unilever announces new global zero waste to landfill achievement, *Waste & Packaging*. Available at: https://www.unilever.com/news/press-releases/2016/Unilever-announces-new-global-zero-waste-to-landfill-achievement.html.
11. Kolybaba, M., Tabil, L.G., Panigrahi, S., Crerar, W.J., Powell, T., and Wang, B. (2003) Biodegradable polymers: Past, present, and future. ASABE/CSBE North Central Intersectional Meeting.
12. Ranjbar, M., Dehghan Noudeh, G., Hashemipour, M.-A., and Mohamadzadeh, I. (2019) A systematic study and effect of PLA/Al_2O_3 nanoscaffolds as dental resins: Mechanochemical properties. *Artif. Cells Nanomed., Biotechnol.*, **47**(1), 201–209.
13. Dai, Z., Dang, M., Zhang, W., Murugan, S., Teh, S.W., and Pan, H. (2019) Biomimetic hydroxyapatite/poly xylitol sebacic adibate/vitamin K nanocomposite for enhancing bone regeneration. *Artif. Cells Nanomed., Biotechnol.*, **47**(1), 1898–1907.
14. Patrick C.W., Jr., Mikos A.G., McIntire L.V. (1998) Prospectus of tissue engineering. *Front. Tissue Eng.*, **3**(14), 3–11.
15. Nguyen, L.T.B., Nguyen, T.-H., Huynh, C.-K., Lee, B.-T., and Ye, H. (2020) *Composite nano-fiber mats consisting of biphasic calcium phosphate loaded polyvinyl alcohol—gelatin for bone tissue engineering*. Springer, Singapore, 301–305.
16. Almeida Neto, G.R. de, Barcelos, M.V., Ribeiro, M.E.A., Folly, M.M., and Rodríguez, R.J.S. (2019) Formulation and characterization of a novel PHBV nanocomposite for bone defect filling and infection treatment. *Mater. Sci. Eng. C*, **104**, 110004.

17. Tunesi, M., Raimondi, I., Russo, T., Colombo, L., Micotti, E., Brandi, E., Cappelletti, P., Cigada, A., Negro, A., Ambrosio, L., Forloni, G., Pollegioni, L., Gloria, A., Giordano, C., and Albani, D. (2019) Hydrogel-based delivery of Tat-fused protein Hsp70 protects dopaminergic cells in vitro and in a mouse model of Parkinson's disease. *NPG Asia Mater.*, **11**(1), 28.
18. Homaeigohar, S., Tsai, T.-Y., Young, T.-H., Yang, H.J., and Ji, Y.-R. (2019) An electroactive alginate hydrogel nanocomposite reinforced by functionalized graphite nanofilaments for neural tissue engineering. *Carbohydr. Polym.*, **224**, 115112.
19. Cheng, K.-C., Huang, C.-F., Wei, Y., and Hsu, S. (2019) Novel chitosan–cellulose nanofiber self-healing hydrogels to correlate self-healing properties of hydrogels with neural regeneration effects. *NPG Asia Mater.*, **11**(1), 25.
20. Wang, F., Wu, H., Venkataraman, V., and Hu, X. (2019) Silk fibroin-poly(lactic acid) biocomposites: Effect of protein-synthetic polymer interactions and miscibility on material properties and biological responses. *Mater. Sci. Eng. C*, **104**, 109890.
21. Buschmann, J., Yamada, Y., Schulz-Schönhagen, K., Hess, S.C., Stark, W.J., Opelz, C., Bürgisser, G.M., Weder, W., and Jungraithmayr, W. (2019) Hybrid nanocomposite as a chest wall graft with improved integration by adipose-derived stem cells. *Sci. Rep.*, **9**(1), 10910.
22. Dong, H., Zhou, J., and Virtanen, S. (2019) Fabrication of ZnO nanotube layer on Zn and evaluation of corrosion behavior and bioactivity in view of biodegradable applications. *Appl. Surf. Sci.*, **494**, 259–265.
23. Singh, S., Singh, G., and Bala, N. (2019) Corrosion behavior and characterization of HA/Fe3O4/CS composite coatings on AZ91 Mg alloy by electrophoretic deposition. *Mater. Chem. Phys.*, **237**, 121884.
24. Santos-Coquillat, A., Esteban-Lucia, M., Martinez-Campos, E., Mohedano, M., Arrabal, R., Blawert, C., Zheludkevich, M.L., and Matykina, E. (2019) PEO coatings design for Mg-Ca alloy for cardiovascular stent and bone regeneration applications. *Mater. Sci. Eng. C*, **105**, 110026.
25. Zhang, J., Guo, W., Li, Q., Wang, Z., and Liu, S. (2018) The effects and the potential mechanism of environmental transformation of metal nanoparticles on their toxicity in organisms. *Environ. Sci. Nano*, **5**(11), 2482–2499.
26. Schutzendubel, A., and Polle, A. (2002) Plant responses to abiotic stresses: Heavy metal-induced oxidative stress and protection by mycorrhization. *J. Exp. Bot.*, **53**(372), 1351–1365.
27. Mo, L., Pang, H., Tan, Y., Zhang, S., and Li, J. (2019) 3D multi-wall perforated nanocellulose-based polyethylenimine aerogels for ultrahigh efficient and reversible removal of Cu(II) ions from water. *Chem. Eng. J.*, **378**, 122157.
28. Cui, S., Borgemenke, J., Liu, Z., Keener, H.M., and Li, Y. (2019) Innovative sustainable conversion from CO2 and biodiesel-based crude glycerol waste to bio-based polycarbonates. *J. CO2 Util.*, **34**, 198–206.
29. Lajarrige, A., Gontard, N., Gaucel, S., Samson, M.-F., and Peyron, S. (2019) The mixed impact of nanoclays on the apparent diffusion coefficient of additives in biodegradable polymers in contact with food. *Appl. Clay Sci.*, **180**, 105170.
30. Sable, S., Mandal, D.K., Ahuja, S., and Bhunia, H. (2019) Biodegradation kinetic modeling of oxo-biodegradable polypropylene/polylactide/nanoclay blends and composites under controlled composting conditions. *J. Environ. Manag.*, **249**, 109186.
31. Aguilar, N.M., Arteaga-Cardona, F., de Anda Reyes, M.E., Gervacio-Arciniega, J.J., and Salazar-Kuri, U. (2019) Magnetic Bioplastics based on isolated cellulose from cotton and sugarcane bagasse. *Mater. Chem. Phys.*, **238**, 121921.
32. Habibovic, P., Juhl, M.V., Clyens, S., Martinetti, R., Dolcini, L., Theilgaard, N., and van Blitterswijk, C.A. (2010) Comparison of two carbonated apatite ceramics in vivo. *Acta Biomater.*, **6**(6), 2219–2226.

33. Deng, Y., Sun, Y., Chen, X., Zhu, P., and Wei, S. (2013) Biomimetic synthesis and biocompatibility evaluation of carbonated apatites template-mediated by heparin. *Mater. Sci. Eng. C*, **33**(5), 2905–2913.
34. Zouhary, K.J. (2010) Bone graft harvesting From distant sites: Concepts and techniques. *Oral Maxillofac. Surg. Clin. North Am.*, **22**(3), 301–316.
35. Otsuka, M., and Hirano, R. (2011) Bone cell activity responsive drug release from biodegradable apatite/collagen nano-composite cements—In vitro dissolution medium responsive vitamin K2 release. *Colloids Surf. B Biointerfaces*, **85**(2), 338–342.
36. Chappuis, V., Gamer, L., Cox, K., Lowery, J.W., Bosshardt, D.D., and Rosen, V. (2012) Periosteal BMP2 activity drives bone graft healing. *Bone*, **51**(4), 800–809.
37. Bleek, K., and Taubert, A. (2013) New developments in polymer-controlled, bioinspired calcium phosphate mineralization from aqueous solution. *Acta Biomater.*, **9**(5), 6283–6321.
38. Schmitt, M., Weiss, P., Bourges, X., Amador del Valle, G., and Daculsi, G. (2002) Crystallization at the polymer/calcium-phosphate interface in a sterilized injectable bone substitute IBS. *Biomaterials*, **23**(13), 2789–2794.
39. Pérez, R.A., Won, J.-E., Knowles, J.C., and Kim, H.-W. (2013) Naturally and synthetic smart composite biomaterials for tissue regeneration. *Adv. Drug Deliv. Rev.*, **65**(4), 471–496.
40. Shitole, A.A., Raut, P.W., Sharma, N., Giram, P.S., Khandwekar, A.P., and Garnaik, B. (2019) Electrospun polycaprolactone/hydroxyapatite/ZnO nanofibers as potential biomaterials for bone tissue regeneration. *J. Mater. Sci. Mater. Med.*, **30**(51), 1–17.
41. Shitole, A.A., Giram, P.S., Raut, P.W., Rade, P.P., Khandwekar, A.P., Sharma, N., and Garnaik, B. (2019) Clopidogrel eluting electrospun polyurethane/polyethylene glycol thromboresistant, hemocompatible nanofibrous scaffolds. *J. Biomater. Appl.*, **33**(10), 1327–1347.
42. Thein-Han, W.W., and Misra, R.D.K. (2009) Biomimetic chitosan–nanohydroxyapatite composite scaffolds for bone tissue engineering. *Acta Biomater.*, **5**(4), 1182–1197.
43. Peter, M., Binulal, N.S., Nair, S.V., Selvamurugan, N., Tamura, H., and Jayakumar, R. (2010) Novel biodegradable chitosan–gelatin/nano-bioactive glass ceramic composite scaffolds for alveolar bone tissue engineering. *Chem. Eng. J.*, **158**(2), 353–361.
44. Zhang, Y., Venugopal, J.R., El-Turki, A., Ramakrishna, S., Su, B., and Lim, C.T. (2008) Electrospun biomimetic nanocomposite nanofibers of hydroxyapatite/chitosan for bone tissue engineering. *Biomaterials*, **29**(32), 4314–4322.
45. Dumitriu, S., and Popa, V.I. (2013) *Polymeric Biomaterials*. Boca Raton, FL: CRC Press
46. Tanaka, M., Hayashi, T., and Morita, S. (2013) The roles of water molecules at the biointerface of medical polymers. *Polym. J.*, **45**(7), 701–710.
47. Nel, A.E., Mädler, L., Velegol, D., Xia, T., Hoek, E.M.V., Somasundaran, P., Klaessig, F., Castranova, V., and Thompson, M. (2009) Understanding biophysicochemical interactions at the nano–bio interface. *Nat. Mater.*, **8**(7), 543–557.
48. Javakhishvili, I., Tanaka, M., Ogura, K., Jankova, K., and Hvilsted, S. (2012) Synthesis of graft copolymers based on poly(2-methoxyethyl acrylate) and investigation of the associated water structure. *Macromol. Rapid Commun.*, **33**(4), 319–325.
49. Miwa, Y., Ishida, H., Saitô, H., Tanaka, M., and Mochizuki, A. (2009) Network structures and dynamics of dry and swollen poly(acrylate)s. Characterization of high- and low-frequency motions as revealed by suppressed or recovered intensities (SRI) analysis of 13C NMR. *Polymer (Guildf)*, **50**(25), 6091–6099.
50. Tsuruta, T. (2010) On the role of water molecules in the interface between biological systems and polymers. *J. Biomater. Sci. Polym. Ed.*, **21**(14), 1831–1848.
51. Lendlein, A., Neffe, A.T., Pierce, B.F., and Vienken, J. (2011) Why are so few degradable polymeric biomaterials currently established in clinical applications? *Int. J. Artif. Organs*, **34**(2), 71–75.

52. Place, E.S., Evans, N.D., and Stevens, M.M. (2009) Complexity in biomaterials for tissue engineering. *Nat. Mater.*, **8**(6), 457–470.
53. Anderson, J.M. (2001) Biological responses to materials. *Annu. Rev. Mater. Res.*, **31**(1), 81–110.
54. Salthouse, T.N. (1984) Some aspects of macrophage behavior at the implant interface. *J. Biomed. Mater. Res.*, **18**(4), 395–401.
55. Kanca, J. (1990) Pulpal studies: Biocompatibility or effectiveness of marginal seal? *Quintessence Int.*, **21**(10), 775–779.
56. Cox, C.F., Suzuki, S., and Suzuki, S.H. (1995) Biocompatibility of dental adhesives. *J. Calif. Dent. Assoc.*, **23**(8), 35–41.
57. Cox, C.F., Sübay, R.K., Suzuki, S., Suzuki, S.H., and Ostro, E. (1996) Biocompatibility of various dental materials: Pulp healing with a surface seal. *Int. J. Periodontics Restorative Dent.*, **16**(3), 240–251.
58. Cox, C.F., Hafez, A.A., Akimoto, N., Otsuki, M., Suzuki, S., and Tarim, B. (1998) Biocompatibility of primer, adhesive and resin composite systems on non-exposed and exposed pulps of non-human primate teeth. *Am. J. Dent.*, **11 Spec No**, S55–63.
59. Akimoto, N., Momoi, Y., Kohno, A., Suzuki, S., Otsuki, M., Suzuki, S., and Cox, C.F. (1998) Biocompatibility of Clearfil Liner Bond 2 and Clearfil AP-X system on nonexposed and exposed primate teeth. *Quintessence Int.*, **29**(3), 177–188.
60. Benton, J.B., Zimmerman, B.F., Zimmerman, K.L., and Ralph Rawls, H. (1993) In vivo biocompatibility of an acrylic, fluoride-releasing, anion-exchange resin. *J. Appl. Biomater.*, **4**(1), 97–101.
61. Kasten, F.H., Pineda, L.F.R., Schneider, P.E., Ralph Rawls, H., and Foster, T.A. (1989) Biocompatibility testing of an experimental fluoride releasing resin using human gingival epithelial cells in vitro. *Vitro Cell. Dev. Biol.*, **25**(1), 57–62.
62. Mousavinasab, S.M., and Meyers, I. (2009) Fluoride release by glass ionomer cements, compomer and giomer. *Dent. Res. J. (Isfahan)*, **6**(2), 75–81.
63. Kato, M., Nishida, T., Kataoka, Y., Yokoyama, M., Ogitani, Y., Nakamura, M., and Kawahara, H. (1979) [Studies on the cytotoxic action of new restorative resins (in vitro) (author's transl)]. *Shika Rikogaku Zasshi*, **20**(49), 20–26.
64. EBSCOhost I 37375540 I A New Copolymerized Composite Resin System: A Multiphased Evaluation.
65. Stanislawski, L., Daniau, X., Lauti, A., and Goldberg, M. (1999) Factors responsible for pulp cell cytotoxicity induced by resin-modified glass ionomer cements. *J. Biomed. Mater. Res.*, **48**(3), 277–288.
66. Beer, R., Gängler, P., Wutzler, P., and Krehan, F. (1990) [Comparative biological testing of Ketac-Bond glass ionomer cement]. *Dtsch. Zahnarztl. Z.*, **45**(4), 202–208.
67. Martins, T.M., Bosco, A.F., Nóbrega, F.J.O., Nagata, M.J.H., Garcia, V.G., and Fucini, S.E. (2007) Periodontal tissue response to coverage of root cavities restored With resin materials: A histomorphometric study in dogs. *J. Periodontol.*, **78**(6), 1075–1082.
68. Geurtsen, W. (2000) Biocompatibility of resin-modified filling materials. *Crit. Rev. Oral Biol. Med.*, **11**(3), 333–355.
69. Goldberg, M., Lasfargues, J.J., and Legrand, J.M. (1994) Clinical testing of dental materials--Histological considerations. *J. Dent.*, **22**(Suppl 2), S25–28.
70. Babakhin, A.A., Volozhin, A.I., Zhuravleva, A.A., Kazarina, L.N., Babakhina, I.A., Dubova, L.V., and DuBuske, L.M. (2008) [Histamine releasing and immunomodulating activity of dental restorative materials]. *Stomatologiia (Mosk)*, **87**(4), 4–10.
71. Santerre, J.P., Shajii, L., and Tsang, H. (1999) Biodegradation of commercial dental composites by cholesterol esterase. *J. Dent. Res.*, **78**(8), 1459–1468.
72. Mattioli-Belmonte, M., Natali, D., Tosi, G., Torricelli, P., Totaro, I., Zizzi, A., Fini, M., Sabbatini, S., Giavaresi, G., and Biagini, G. (2006) Resin-based dentin restorative materials under accelerated ageing: Bio-functional behavior. *Int. J. Artif. Organs*, **29**(10), 1000–1011.

73. Moszner, N., Gianasmidis, A., Klapdohr, S., Fischer, U.K., and Rheinberger, V. (2008) Sol–gel materials: 2. Light-curing dental composites based on ormocers of cross-linking alkoxysilane methacrylates and further nano-components. *Dent. Mater.*, **24**(6), 851–856.
74. Boulden, J.E., Cramer, N.B., Schreck, K.M., Couch, C.L., Bracho-Troconis, C., Stansbury, J.W., and Bowman, C.N. (2011) Thiol–ene–methacrylate composites as dental restorative materials. *Dent. Mater.*, **27**(3), 267–272.
75. Krifka, S., Seidenader, C., Hiller, K.-A., Schmalz, G., and Schweikl, H. (2012) Oxidative stress and cytotoxicity generated by dental composites in human pulp cells. *Clin. Oral Investig.*, **16**(1), 215–224.
76. Castañeda, E.R., Silva, L.A.B., Gaton-Hernández, P., Consolaro, A., Rodriguez, E.G., Silva, R.A.B., Queiroz, A.M., and Nelson-Filho, P. (2011) FiltekTM Silorane and FiltekTM Supreme XT resins: Tissue reaction after subcutaneous implantation in isogenic mice. *Braz. Dent. J.*, **22**(2), 105–110.

4 Lightweight Polymer Composites from Wood Flour, Metals, Alloys, Metallic Fibers, Ceramics

E. Teke, M. Sütçü, V. Acar, and M. Ö. Seydibeyoğlu

CONTENTS

4.1 INTRODUCTION

In this part of the book, composites made from wood, metals, and ceramics will be explained. The need for composite materials is growing and new materials for the composite world are needed. The use of lightweight materials is one of the top topics of priority in many applications and using the right class of composites is a good solution to this issue. Moreover, the use of sustainable materials such as wood flour, novel metallic reinforcements, and the use of ceramics obtained from different natural resources present new way of composites. In this chapter, a brief introduction to composite materials is presented. Detailed investigation of three different classes of composites are explained in detail.

4.2 COMPOSITE MATERIALS

Composite materials, with two components, as reinforcement and matrix phase, are defined as a combination of at least two different materials. Classification of composites based on matrix type is given in Figure 4.1. The most important advantages of composite materials are their mechanical properties such as their high strength

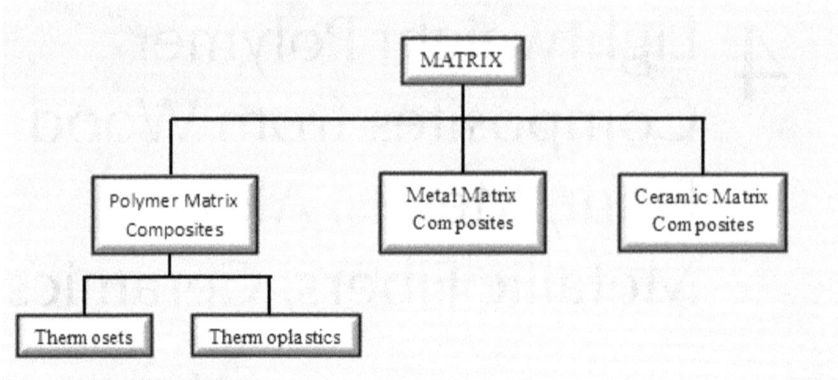

FIGURE 4.1 Classification of composites based on matrix type.

and stiffness and their low density, which provides for a weight reduction in the finished part, when compared with bulk materials (Campbell, 2010). The matrix phase may be metal, ceramic, or polymer material. Polymer matrix composite (PMC) is the material containing a resin matrix combined with a reinforcing dispersed phase Strong, 1999). Thermosets and thermoplastics are the two main kinds of polymers. Thermoplastic polymers that can be used more than once melt easily and are often solvent soluble. Thermoset polymers that cannot be used more than once are heavily cross linked, infusible, and also insoluble in solvents (Peng and Riedl, 1995).

A fiber or a particulate is usually used as the reinforcement. Fibers or particles reinforced in the matrix of another material would be the best example of modern-day composite materials. Large-particle and dispersion-strengthened composites are the two sub-classifications of particle reinforced composites, as shown in Figure 4.2. The distinction between these is based on the reinforcement or strengthening mechanism. For most of these composites, the particulate phase is harder and stiffer than the matrix. For dispersion-strengthened composites, particles are much smaller

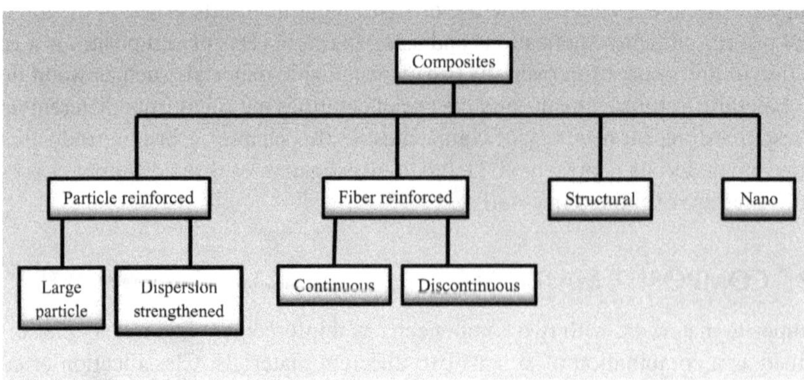

FIGURE 4.2 Classification of composites based on reinforced type.

(10 to 100 nm) whereas for large-particle reinforced composites, the particles are micro-sized (Callister and Rethwisch, 2014).

4.3 PARTICLE REINFORCED POLYMER COMPOSITES

As a result of modern technology, instead of using metals and alloys in different industries in the world, new materials are required that are lightweight, have a high modulus, and are tough and stable against the environment. With the advent of new technologies, there is a great need for materials with advanced properties (Mohammed, 2011).

Nowadays, due to the increasing interest in lightweight and high-performance materials, important research is being conducted in the area of composite materials. New focus areas are to achieve low density and to develop multifunctional composites that have more than one property tailored as per the design requirements. Automotive, aerospace, and wind energy sectors are considered to be the main users of composite materials (Gupta and Paramsothy, 2014).

Various polymer matrix composites are reinforced with metal particles as shown in Figure 4.3. The use of particulate fillers to change the physical and mechanical properties of polymers in many ways is preferred (Fu, 2008).

Reinforced materials like talc, mica, wood dust, sand, silica, alumina, metal flakes, and metal powder are normally used to modify the creep, impact, thermal, electrical, and magnetic properties in polymer composites (Pleşa et al., 2016). The particle shape, size, and surface area of such added particles affect mechanical properties of the composites (Fu, 2008). The reinforcements are typically hard, stiff materials usually of glass, ceramics, or metals whereas the matrix materials are generally ductile and tough like polymers, but brittle matrices are also used (Mohammed, 2011).

4.4 PARTICULATES

Particulates of various shapes and sizes are used as reinforcing particles. Various physical parameters such as particle size, shape, and chemical composition are used to characterize powder, however, these parameters also affect its flowability.

Mechanical properties of the materials are affected by not only chemical composition but also the particle shape, size, and surface area of such added particles (Nichols et al., 2002).

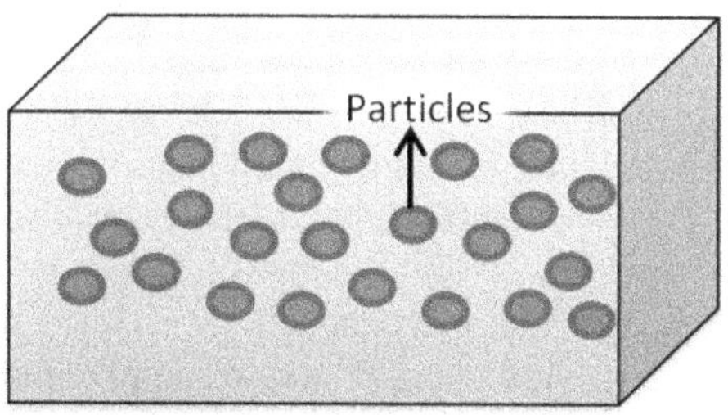

FIGURE 4.3 Particle-reinforced composite.

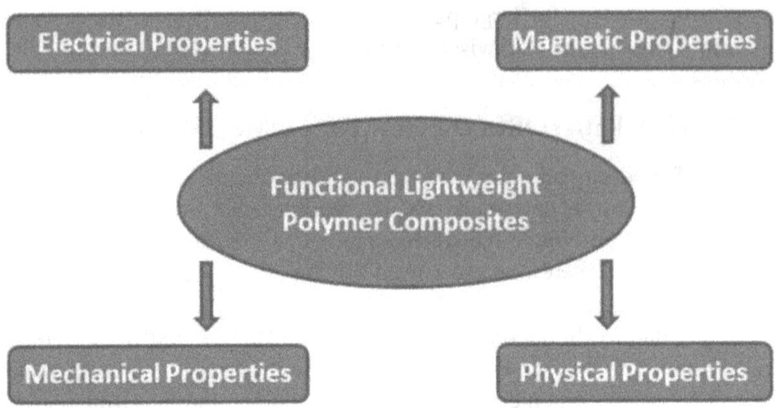

FIGURE 4.4 Various functionalities in the polymer composites.

Particle size distribution of powders plays a significant role in determining the critical chemical and physical properties of the particulate systems (Basim and Khalili, 2015). When the particle size decreases, the surface area increases and the reaction time is shortened. The surface area is increased as the particle size becomes small and also increased if the particle has pores (Dubois et al., 2010).

Particles have complex geometric features such as being dendritic, rounded, spherical, angular, platy, acicular, cylindrical, or cubic (Seville and Wu, 2016). The relations between measured sizes and particle volume or surface area are called shape coefficients (Allen, 1990).

Figure 4.4 lists properties that can be used to create various functionalities in the PMC. The improving of lightweight polymer composites to have specific functional properties and high performance is not so easy and studies are continuing in this area. The improving of functioning capabilities in the base material can reduce the requirements for additional systems on the application platform and can help in reducing the weight. As an example, an engine block is required to have high dimensional stability at low to high temperatures, fatigue life, wear resistance, and also low weight. Such complex requirements reveal the need for the development of multifunctional materials, which is a significant focus area of research (Gupta and Paramsothy, 2014).

Some metals have special electrical, thermal, and magnetic properties. Metal particles have a significant effect on improving the electrical, mechanical, magnetic, and thermal properties of composites (Nielsen and Landel, 1994; Peters, 1998). Copper particles, brass particles, steel particles, zinc particles, aluminum particles, and bronze particles are the most commonly used materials as fillers.

The reinforcing of a polymer with metallic particles results in an increase in both electrical and thermal conductivity of the composites. As an example, due to aluminum having high electrical and thermal conductivity, it can be used to impart electrical conductivity to the polymer composite and to improve thermal conductivity of the composite.

Polymer composites are playing a growing role in a wide range of applications including construction materials. In addition, they are becoming an important alternative for metals in applications related to the aerospace, automotive, marine, sporting goods, and electronics industries. Their light weight and superior mechanical and electrical properties are particularly suitable for some of these applications (Moses et al., 2008).

4.5 METAL REINFORCEMENTS

Metal reinforcements are used to improve the mechanical, electrical, magnetic, physical, and several other properties of both thermoset (epoxy, phenolic, polyester, silicone, bismaleimide, polyimide, polybenzimidazole, etc.) and thermoplastic (polyethylene, polystyrene, polyamides, nylon, polycarbonates, polysulfones, etc.) materials.

Tekçe et al. (2007) investigated thermal properties of copper-reinforced polyamide composites in the range of filler content 0–30% by volume for particle shape of short fibers and 0–60% by volume for particle shapes of plates and spheres. Different particle shapes (plates, fibers, spherical) of copper powders were used as conductive fillers. According to this, the composite's thermal conductivity is increased by the addition of copper fillers. According to the study by Tekçe et al., copper fiber is the most effective agent on the thermal conductivity of the composite.

The electrical conductivity of plastics can be improved by using metal filler particles. Two methods are available for providing electrical conductivity to plastics: (i) metal coating of the surface, (ii) adding conductive fillers into the polymer matrix.

Amoabenga and Velankar (2018) studied the electrical conductivity of polystyrene by simultaneously adding copper particles (30–80 microns), and a lead/tin solder alloy (15–25 microns).

In Bigg's study (1979), the electrical resistivity, thermal conductivity, and tensile strength of aluminum fiber-filled polypropylene were investigated. Electrical and thermal conductivity increased while tensile strength decreased.

Mohammed (2011) studied the mechanical behavior of copper powder reinforced epoxy. He found that increasing the weight ratio of copper powder to epoxy for different particle sizes leads to an increase in elasticity, rigidity, tensile yield stress, and ultimate tensile strength. In addition, it decreased the compression yield stress, fracture energy, and the impact strength due to increasing weight ratio of copper powder to epoxy with different particle size.

Short fibers are called discontinuous fibers and also can be used as particles because of their cylindrical shapes. Flakes are commonly used because they are less expensive than short fibers and can be aligned to have improved plane direction properties compared to short fibers. These composites, classified as continuous or discontinuous, generally obtain the highest strength and stiffness with continuous reinforcement. Discontinuous fibers are used only when the manufacturing economy requires the use of a process in which the fibers should be in this form.

Metal fibers have high stiffness, high strength, high tensile energy absorption, high electrical conductivity, and superior structural integrity following impact (Breuer et al., 2013).

However, fiber reinforced polymer (FRP) composites provide high corrosion resistance when using optimum fibers and resins, while the addition of steel fiber reinforcement can reduce this durability (Agarwal et al., 2010).

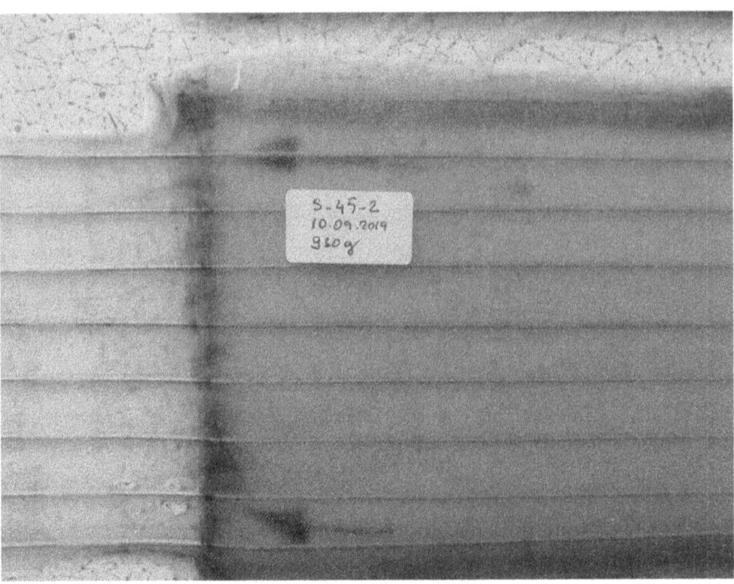

FIGURE 4.5 Copper-reinforced epoxy systems.

The results show the potential of steel fibers to improve the failure strain and energy dissipation performance of composites. However, there have been insufficient studies to investigate the corrosion resistance of FRP—stainless steel fiber reinforced polymer composites. (McBride et al., 2017; Triantafyllou et al., 2017).

In some of our studies we have been using metal structures in polymer matrices. In one of our studies, we have created a novel copper pipe reinforced epoxy composite to form a material that can be heated with hot water and heat-conducting fluids. Figure 4.5 shows the general form of the materials and Figure 4.6 shows many structures made with this material.

In another study, we have utilized Ti6Al4V alloy wastes in polyurethane foams to obtain new lightweight high-strength composites. This is a very new study conducted by our group. Figure 4.7 shows a general overview of these composites.

4.6 WOOD PLASTIC COMPOSITES

The use of woody materials is of great importance and is an abundant resources in nature. Wood flour is important in terms of ease of access, low price, good strength, and can be considered to be more uniform when compared with other natural fibers (Kumar et al., 2011). Wood flour is obtained by pulverizing the wood and the particle size can be considered as consistent but also might change from fine powder to the size of rice. There are various trees used in the wood flour. Some trees such as pine or fir have lower mechanical properties and some tress such as teak, oak, and poplar have high-quality flour due to hardwood present in these trees. The pulverization of the trees creates particles with an aspect ratio of 1–5. This aspect ratio is low but it still provides easy processing albeit with a sacrifice of some mechanical

FIGURE 4.6 A group of copper-reinforced composites.

FIGURE 4.7 Polyurethane foams (a) Pure and (b) Ti6Al4V reinforced.

properties. The wood fibers have been used in different applications but for polymer composites they have been used with plastics that are processed at less than 200°C (Panthapulakkal and Sain, 2007). Figure 4.8 shows the polymer compounded with wood particles.

At first, wood flour was mainly used as a filler for thermosetting resins including phenol formaldehyde, but these days it commonly used for many different polymers including polyvinyl chloride, polyethylene, polystyrene, polyamides, and even biopolymers. The wood flour can be used as a reinforcing element depending on the compatibilizer/coupling agent. The problem with mechanical properties with wood flour arises due to the mismatch of the thermoplastics being hydrophobic and

FIGURE 4.8 Polymer wood compounds.

wood fibers being hydrophilic. In order to increase the compatibility between the two phases, numerous coupling agents and compatibilizers have been used such as maleic anhydride-modified polypropylene (MAPP), silanes, titanate, and others. Also, humidity is one of the key factors in the success of composites. The particle size of the wood fiber is important and also the right aspect ratio is critical for the reinforcement (Kumar et al., 2011).

Compared to thermosetting resins, wood flour has been used more predominantly in the thermoplastic materials. The total tonnage of wood plastic composites is around 1.5 million ton/year where North America leads with two-thirds of this volume (Mehmood et al., 2010; Nova-Institute 2010). The main applications are door profiles, window profiles, decking materials, roofs, and railings. Besides construction materials, these materials have been used in automotive parts, musical instruments, and sound acoustic applications.

Furthermore, wood flour has been used in biodegradable polymers and it was shown that wood flour can increase the biodegradability of these biopolymers (Scaffaro et al., 2009; Finkenstadt and Tisserat, 2010). Wood flour can be used in various polymer processes including injection molding, extrusion, and/or compression molding (Kumar et al., 2011).

Figure 4.9 shows one of the commercial extruded wood polymer composites. Figure 4.10 is a closer look at the same product.

Figure 4.11 shows an example of injection-molded wood plastic composite developed by our group. These parts are intended to be used in automotive parts.

4.7 CERAMIC REINFORCED POLYMER COMPOSITES

Various types of fibers, whiskers, particles, and platelets have been added to overcome the disadvantages in terms of mechanical properties such as low modulus and shear strength of the polymer matrix. Ceramic particles are also widely used as fillers in the polymer matrix and improve the properties of the composite material produced. Ceramic particle-filled polymer composites form a class of materials that are strong, lightweight, and have remarkable toughness.

FIGURE 4.9 Wood plastic extruded sample.

FIGURE 4.10 Wood plastic composite enlarged.

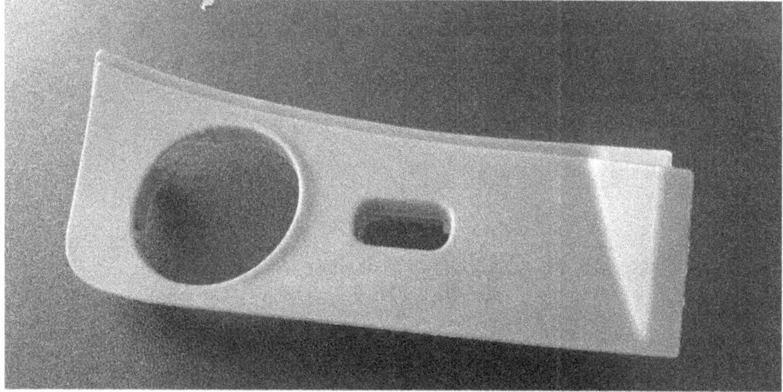

FIGURE 4.11 Wood plastic composites to be used in automotive parts.

Some natural or synthetic layered minerals like clays used in polymer composites increase properties such as mechanical strength, impact strength, thermal stability, and flame retardancy required for aerospace applications (Fischer, 2003).

Polymeric materials that exhibit superior properties such as high durability for harsh environments, high optical transmission, low solar absorptivity, and sufficient electrical conductivity for electrostatic charge dissipation are needed for ultra-lightweight spacecrafts in the future. For this purpose, thin film polymer composite materials from polyimide with metal oxide additives are prepared (Thompson et al., 2003).

Polymer matrix nano-composites (PMNCs) such as polymer/alumina (Al_2O_3) have improved their ballistic performance for which they have been used in rocket propellant preparation (Meda et al., 2005).

For thermal insulating applications, the highly porous polyurethane composites produced using mineral additives are quite remarkable. For instance, the polyurethane foam composites with basalt filler used up to 40wt% shows low thermal conductivity values up to 0.022 W/mK. In addition, basalt filler helps to improve the thermal stability of polyurethane, and polyurethane/basalt composite foams exhibit improved thermomechanical stability (Kurańska et al., 2019).

Lightweight polymer foam composites are also used in electromagnetic (EM)-wave- and microwave absorbing applications (Zhang et al., 2014); Wang et al., 2017; Yang et al., 2019). Because of the health and safety concerns associated with microwave use, high-performance microwave absorbent materials (MAMs) are used in civil and military fields (Wang et al., 2017). The high-performance and lightweight structure of these materials are essential requirements for effective and practical use. Polymer composite foams with ultralow density and an interconnected pore structure have attracted broad attention as MAMs. Their porous structure was provided with the emulsion method using a surfactant and Fe_3O_4 nanoparticles (5–15 wt%). Microwave absorbing and magnetic properties of the composites with increasing Fe_3O_4 nanoparticles have been enhanced (Phadtare et al., 2019). In another study, the ultralight three-dimensional polypyrrole/nano SiO_2 aerogel polymer composites were prepared by using an in situ gelation process (Xie et al., 2015). This study provides an easy method for producing a potential and excellent electromagnetic absorption material with low loading ratio and wide absorption bandwidth. Furthermore, following the cost-effective and environmental protection preparation procedure, lightweight and flexible silicone rubber/multi-walled carbon nanotubes (MWCNTs)/Fe_3O_4 nanocomposite foams were prepared by the supercritical carbon dioxide foaming process for effective EM protection performance (Yang et al., 2019).

In recent years, some representative applications related to polymer-based energy-harvesting systems have been quite remarkable. Polymer-based piezoelectric and pyroelectric energy harvesters have been developed with production of PVDF-HFP polymer films using ceramic nanoparticles like ZnO, $BaTiO_3$, La_2O_3, etc. (Costa et al., 2019). In the scope of biomedical devices, piezoelectric flexible and sponge-like nanogenerator hybrid implants based on zinc oxide (ZnO) nanowires and PVDF film in a conductive fiber embedded in polystyrene (PS) and polydimethylsiloxane (PDMS) were developed for harvesting energy from a human elbow.

FIGURE 4.12 Polyurethane alumina composites.

By utilizing the engineering properties of some ceramic particles, polymer-based functional composite materials can be developed as a result of research.

In a recent work carried out by our group, we have developed a new polyurethane foam with alumina particles to increase the heat insulating properties. One sample of this work is shown in Figure 4.12.

4.8 CONCLUSION

Compared to other material classes, composites are a quite new type of materials, dating back to the 1950s. The area of composite materials is a highly dynamic area with lots of inventions in the processing of composites, the use of new manufacturing techniques, and the use of new materials. The polymer matrix composites started with glass fiber at the end of the 1940s and still, glass fiber is the most predominant fiber used in the area of composite materials. However, the use of carbon fiber is increasing enormously and the use of aramid fibers has been spread in many applications. There are some other fibers arising as well like basalt fiber and ultrahigh-molecular-weight polyethylene.

Because of environmental concerns, there has been lots of research conducted on wood flour reinforced composites and this research resulted in many different commercial products. The use of ceramic fibers and ceramic powders are also important in terms of natural resources as they are also obtained from natural resources.

Moreover, there are very new studies on metal powder reinforced polymer composites besides battery research. The use of metallic materials does not only provide mechanical properties but also improvements in thermal and electrical conductivity of plastics.

Composites are the new generation materials opening many new dimensions and providing many new manufacturing methods to explore.

ACKNOWLEDGMENTS

TUBITAK 218M26 and IKCU ONP-MUM-002 project grants are greatly acknowledged for this chapter's figures and context.

REFERENCES

Agarwal, A., Garg, S., Rakesh, P., Singh, I., Mishra, B. (2010). Tensile behavior of glass fiber reinforced plastics subjected to different environmental conditions. *Indian Journal of Engineering and Materials Sciences*, 17, 471–476.

Allen, T. (1990). *Particle Size Measurement*, 4th edition, Chapman and Hall, New York.

Amoabenga, D., Velankar, S. (2018). Bulk Soldering: Conductive Polymer Composites filled with copper particles and solder. *Colloids and Surfaces. Part A: Physicochemical and Engineering Aspects*, 553, 624–632.

Basim, B., Khalili, M. (2015). Particle size analysis on wide size distribution powders; effect of sampling and characterization technique. *Advanced Powder Technology*, 26(1), 200–207.

Bigg, D. M. (1979). Mechanical, thermal, and electrical properties of metal fiber-filled polymer composites. *Polymer Engineering and Science*, 19(16), 1188–1192.

Breuer, U., Schmeer, S., Eberth, U. (2013). Carbon and metal fibre reinforced airframe structrues—A new approach to composite multifunctionality. *Proceedings Deutscher Luft-Und Raumfahrtkongress.*

Callister, W. D., Rethwisch, D. G. (2014). *Materials Science and Engineering*, 9th edition, John Wiley & Sons.

Campbell, F. C. (2010). *Introduction to Composite Materials: Structural Composite Materials*, ASM International, Materials Park, OH.

Costa, P., Nunes-Pereira, J., Pereira, N., Castro, N., Gonçalves, S., Lanceros-Mendez, S. (2019). Recent progress on piezoelectric, pyroelectric, and magnetoelectric polymer-based energy-harvesting devices. *Energy Technology*. http://www.ncbi.nlm.nih.gov/pubmed/1800852.

Dubois, I. E., Holgersson, S., Allard, S., Malmström, M. E. (2010). *Water-Rock Interaction III: Correlation between Particle Size and Surface Area for Chlorite and K-Feldspar,–Birkle & Torres-Alvarado* (eds). Taylor & Francis Group, London, ISBN 978-0-415-60426-0.

Finkenstadt, V. L., Tisserat, B. (2010). Poly(lactic acid) and Osage orange wood fiber composites for agricultural mulch films. *Industrial Crops and Products*, 31(2), 316–320. doi: 10.1016/j. indcrop.2009.11.012.

Fischer, H. (2003). Polymer nanocomposites: From fundamental research to specific applications. *Materials Science and Engineering C: Biomimetic Materials Sensors and Systems*, 23(6–8), 763–772.

Fu, S.-Y. (2008). Effects of particle size, particle/matrix interface adhesion and particle loading on mechanical properties of particulate–polymer composites. *Elsevier Composites: Part B*, 39, 933–961.

Gupta, N., Paramsothy, M. (2014). Metal- and polymer-matrix composites: Functional lightweight materials for high-performance structures, the Minerals, Metals & Materials Society, *JOM*, 66(6).

Kumar, V., Tyagi, L., Sinha, S. (2011). Wood flour reinforced plastic composites: A review. *Reviews in Chemical Engineering*, 27(5–6), 253–264.

Kurańska, M., Barczewski, M., Uram, K., Lewandowski, K., Prociak, A., Michałowski, S. (2019). Basalt waste management in the production of highly effective porous polyurethane composites for thermal insulating applications. *Polymer Testing*, 76, 90–100.

McBride, A. K., Turek, S. L., Zaghi, A. E., Burke, K. A. (2017). Mechanical behavior of hybrid glass/steel fiber reinforced epoxy composites. *Polymers*, 9(4), 151. doi: 10.3390/polym9040151.

Meda, L. U. I. S. A., Marra, G., Galfetti, L., Inchingalo, S., Severini, F., De Luca, L. (2005). Nano-composites for rocket solid propellants. *Composites Science and Technology*, 65(5), 769–773.

Mehmood, S., Khaliq, A., Ranjha, S. A. (2010). The use of post consumer plastic waste for the production of wood plastic composites: A review. *Proceedings Venice, Third International Symposium on Energy from Biomass and Waste Venice, Italy.*

Mohammed, M. A. (2011). Mechanical behavior for polymer matrix composite reinforced by copper powder. *College of Engineering Journal (NUCEJ)*, 14(2), 160–176.

Moses, Y., Simon, I., Maxwell, I. (2008). Mechanical properties of carbon fibre and metal particles filled epoxy composite. *International Journal of Emerging Technology and Advanced Engineering*, 3(11), 664–667.

Nichols, G., Byard, S., Bloxham, M. J., Botterill, J., Dawson, N. J., Dennis, A., Diart, V., North, N. C., Sherwood, J. D., et al. (2002). A review of the terms agglomerate and aggregate with a recommendation for nomenclature used in powder and particle characterization. *Journal of Pharmaceutical Sciences*, 91(10), 2103–2109.

Nielsen, L. E., Landel, R. F. (1994). *Mechanical Properties of Polymers and Composites*, 2nd edition, Marcel Deckker, New York, 377–459.

Nova-Institut GmbH. Chemiepark Knapsack. Accessed on July 24/2010, http://www.wpc-kongress.de/index.php?tpl=impressumlist&lng=en.

Panthapulakkal, S., Sain, M. (2007). Agro-residue reinforced high-density polyethylene composites: Fiber characterization and analysis of composite properties. *Composites—Part A: Applied Science and Manufacturing*, 38(6), 1445–1454.

Peng, W., Riedl, B. (1995). Thermosetting resins. *Journal of Chemical Education*, 72(7), 587–592.

Peters, S. T. (1998). *Handbook of Composites*, 2nd edition, Chapman and Hall, London, 242–243.

Phadtare, V. D., Parale, V. G., Lee, K. Y., Kim, T., Puri, V. R., Park, H. H. (2019). Flexible and lightweight Fe3O4/polymer foam composites for microwave-absorption applications. *Journal of Alloys and Compounds*, 805, 120–129.

Pleşa, I., Noţingher, P. V., Schlögl, S., Sumereder, C., Muhr, M., et (2016). Properties of polymer composites used in high-voltage applications, review. *Polymers*, 8(5), 173.

Scaffaro, R., Morreale, M., Lo Re, G., La Mantia, F. P. (2009). Effect of the processing techniques on the properties of ecocomposites based on vegetable oil-derived Mater-Bi® and wood flour. *Journal of Applied Polymer Science*, 114(5), 2855–2863.

Seville, J. P. K., Wu, C. (2016). *Particle Technology and Engineering: An Engineer's Guide to Particles and Powders: Fundamentals and Computational Approaches*, Chapter 2. Elsevier, Oxford, UK, ISBN: 978-0-08-098337-0.

Strong, A. B. (1999). *Plastics: Material & Processing*, 2nd edition, Prentice Hall, Upper Saddle River, NJ.

Tekce, H. S., Kumlutas, D., Tavman, İ. H. (2007). Effect of particle shape on thermal conductivity of copper reinforced polymer composites. *Journal of Reinforced Plastics and Composites*, 26(1), 113–121.

Thompson, C. M., Herring, H. M., Gates, T. S., Connell, J. W. (2003). Preparation and characterization of metal oxide/polyimide nanocomposites. *Composites Science and Technology*, 63(11), 1591–1598.

Triantafyllou, G. G., Rousakis, T. C., Karabinis, A. I. (2017). Corroded rc beams patch repaired and strengthened in flexure with fiber-reinforced polymer laminates. *Composites Part B Engineering*, 112, 125–136. doi: 10.1016/j.compositesb.2016.12.032.

Wang, Y., Du, Y., Xu, P., Qiang, R., Han, X. (2017). Recent advances in conjugated polymer-based microwave absorbing materials. *Polymers*, 9(1), 1–28.

Xie, A., Wu, F., Xu, Z., Wang, M. (2015). In situ preparation of ultralight three-dimensional polypyrrole/nano SiO2 composite aerogels with enhanced electromagnetic absorption. *Composites Science and Technology*, 117, 32–38.

Yang, J., Liao, X., Li, J., He, G., Zhang, Y., Tang, W.,Wang, G., Li, G. (2019). Light-weight and flexible silicone rubber/MWCNTs/Fe3O4 nanocomposite foams for efficient electromagnetic interference shielding and microwave absorption. *Composites Science and Technology.*

Zhang, L., Roy, S., Chen, Y., Chua, E. K., See, K. Y., Hu, X., Liu, M. (2014). Mussel-inspired polydopamine coated hollow carbon microspheres, a novel versatile filler for fabrication of high performance syntactic foams. *ACS Applied Materials and Interfaces*, 6(21), 18644–18652.

5 Lightweight Composite Materials in Transport Structures

M. Özgür Seydibeyoğlu, Alperen Doğru, M. Batıkan Kandemir, and Özay Aksoy

CONTENTS

5.1 INTRODUCTION

Nowadays, with the rapid development of new material types and production techniques, lightweight and durable composites are becoming more important for use in transportation vehicles. For instance, fiber-reinforced plastics (FRP) are frequently used in the marine industry due to their advantages such as being lightweight, having a high strength/weight ratio, high impact resistance, and high corrosion resistance against seawater compared to conventional metal materials like steel and aluminum. Besides mechanical properties, they are preferred for applications where stealthiness and sound isolation are important because of their good acoustic damping and radar reflectivity performance. FRP can also be optimized by modifying the reinforcement rate, weaving type, lay-up configuration, and laminate thickness for end-use applications, so they have the advantage of versatility of design (Eric Greene Associates, 1999). For fast and safe transportation in maritime transport, vessels are required to have high engine power and high strength. Since engine modification is a costly process, fiber-reinforced polymer composites (FRP) are used in boat hulls and decks because of their lightness and low fuel consumption (Selvaraju, 2011).

The aerospace industry is one of the fastest-growing sectors today. With developing technology, the materials used in this field began to change (Seydibeyoğlu, Mohanty, and Misra, 2017). The development of new materials and the better use of existing materials have played an important role in the aerospace industry. The structural performance, speed, safety, range, fuel economy, and service life of the aircraft are based on improvements in airframe and engine materials. Since the first strong flight of the Wright Brothers in 1903, the mechanical performance, durability, functionality, and quality of aerospace materials have changed drastically. In addition, the criteria used to select materials for aircraft have changed over the last 100 years (Brothers, 2012). Aircraft is the first major technology where the concept of weight is at the forefront. At the very beginning of aircraft's history, due to this irrevocable technical situation, wood was the only material suitable for building a flying machine (wood to metal). After the use of wood in aircraft structure, the aviation industry, which developed with the age of aluminum, had a different role in the commercial transportation and defense industries with the emergence of composite materials (Brothers, 2012). Composite materials are more widely used in structural components of aircraft due to their advantages. These materials are increasingly used in aircraft main structures (Boeing 787, Airbus A380) due to their superior strength properties compared to metallic materials (structural health monitoring). In particular, the demand for polymer materials, with its lightweight nature and advantages in fuel economy, has begun to increase (Seydibeyoğlu et al., 2017).

The use of composite materials is increasing due to the popularity of more lightweight vehicles with higher strength-to-weight ratio, which have as a result lower carbon emissions. The safer vehicles are obtained around the world using composite body vehicles rather than metallic structures. Although the use of composite materials makes better cars in terms of crashworthiness, durability, safety, environmental friendliness, and carbon emissions; metal body vehicles are still used due to low cost and the available supply of metals. In our research, it has been observed that 1 kg of vehicle weight reduction means 20 kg of lower carbon dioxide emissions (Ghassemieh, 2011).

5.2 HISTORY

Glass fiber was patented by Owens Corning in 1935, then polyester was patented by DuPont in 1936. In 1942, a fiberglass sailing dinghy was produced for the first time using these two materials by Ray Greene (Marsh, 2006). Since then, the use of FRP in marine transportation has been increasing. The use of composites in the marine industry from the past to the present is summarized in Figure 5.1.

The rapid growth of the aviation industry is driving momentum for the development of new aircraft materials (X. Zhang, Chen, and Hu, 2018). In order to keep up

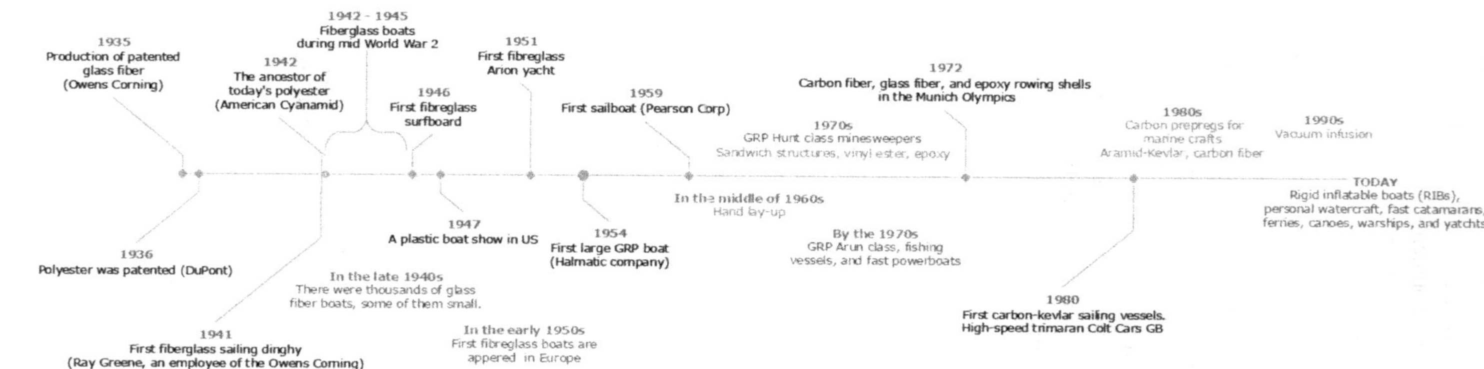

FIGURE 5.1 History of polymer composites in the marine industry (Marsh, 2006).

with the pace of change in the aerospace industry, material technology in aviation is developing very rapidly. The historical development of aerospace materials is shown in Figure 5.2. In this sense, developments in aircraft, spacecraft, engines, landing gear, etc. are triggering the development of new materials and manufacturing processes in the aerospace industry (Brothers, 2012). Aerospace materials are structural materials that carry the loads applied to the airframe from takeoff to flight and landing. Aircraft's structural materials are divided into primary and secondary structural parts. Advanced polymer has a role in the production of the following components: engine nacelles, engine cowls, horizontal and vertical stabilizers, center wing boxes, aircraft wings, pressure bulkheads, landing gear doors, floor beams, tall cones, flap track panels, and so on. All aerospace materials, called primary and secondary structural parts, have changed and evolved in the historical process. Figure 5.2 presents a timeline of changes in the new criteria applied in the selection of aerospace materials over the years (Starke and Staley, 1996). Lightness and strength were at the forefront when designing the first aircraft. Other design criteria, such as cost, toughness, and durability, have been given less importance than the search for lightness. Most of the critical criteria for material selection today are not considered important by first-generation aircraft designers. In the first aircraft, the designers' goal was to use materials that provided high strength with little weight (Brothers, 2012; Starke and Staley, 1996). At that time, the best primary structural material to reach the weight requirement was wood (Brothers, 2012). In order to provide the stiffness feature of aircraft, aluminum alloys have been used in the following years and aluminum was used for airframe after precipitation hardening was discovered by Alfred Wilm (Starke and Staley, 1996). Aluminum alloys have been the main material for structural parts of the aircraft for more than 80 years because of their well-known performance, being lightweight, strength, ease of manufacture, and reliability (X. Zhang et al., 2018). With the development of cladding and anodizing applications in manufacturing, the use of aluminum alloys as the main structural material in modern commercial aircraft has increased. Since about 1930, aluminum alloys have been the first-choice material for structural components of aircraft (Starke and Staley, 1996). In addition, with the discovery of jet engines, air transportation has been realized at high speeds and altitudes. As a result, research on aluminum alloys gained momentum. Because of all this, many researchers have been conducting research to develop materials with optimized properties to reduce weight, improve tolerance of damage, fatigue, and corrosion resistance (X. Zhang et al., 2018). In the following years, features such as fuel-saving, being lightweight, environmentally friendly, and long life with radar absorber for military aviation, have come to the fore. As a result of all of this, the concept of composite has occurred. Alloy materials are used extensively in primary structural parts. In secondary structures, composite materials were used to make improvements (Bai, 2013; Ranasinghe, Guan, Gardi, and Sabatini, 2019). Initially, composite materials were used to produce complex secondary structural parts that required high deformation in the forming process (Starke and Staley, 1996). Advanced composite materials have been developed primarily for the aerospace industry to improve the performance of commercial and military aircraft. They have a significant role in current and future aerospace components (Bai, 2013; Rabotnov et al., 1982). The reasons for the preference of composite materials in the

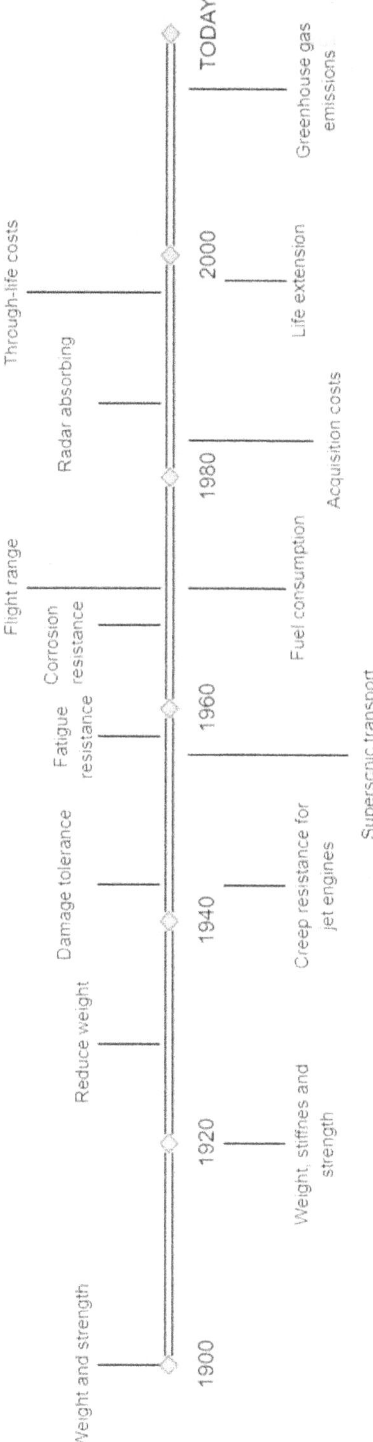

FIGURE 5.2 Historical timeline indicating when key criteria for materials selection were introduced into aircraft design (Brothers, 2012).

aerospace industry are their high specific strength and stiffness, high specific properties (property/density), damage tolerance, ability to form, corrosion resistance, and high-temperature resistance (Rabotnov et al., 1982; X. Zhang et al., 2018).

In 1945, the first composite prototype car was manufactured. In 1955, the first fiberglass-reinforced polymer composite body and composite leaf spring were produced in sports cars. Between 1980 and 1990, body panels from sheet molding compound (SMC) were used. After the 1990s carbon fiber-reinforced composite body begun to be used in high-performance vehicles; also, from the beginning of 2010, carbon fiber body vehicles manufactured at high speed begun to be sold at competitive prices. The chronology of the developments is presented in Figure 5.3.

5.3 POLYMER MATRIX COMPOSITES

Composites are engineering materials that are insoluble in each other, composed of multiple components, and have chemically distinguishing properties. They consist of two components: matrix (resin) and reinforcement phase (fiber). Composite materials are often referred to as reinforced plastics because they are produced with the polymer-based matrix. The matrix phase in composite material provides continuity by keeping the additive phase together. The additive phase gives strength and stiffness to the material. The reinforcements can be discontinuous fiber, continuous fiber, or particulate.

5.3.1 Resin Types

Polymer matrix composites are divided into two types: thermoset and thermoplastic, according to resin type. Thermoplastics which are in solid form at room temperature can be reshaped by heat. However, the thermoset, which is in liquid form at room temperature, will degrade if it is reheated after curing. Thermoset has higher strength due to its cross-linked structure. Thermoplastics such as polyamide (PA), polypropylene (PP), polyether ether ketone (PEEK), polyethersulphone (PES), and polycarbonate (PC) provide superior fracture toughness, high hardness, and impact resistance, long shelf life, easy recyclability, and repairability. In addition, they don't need organic solvents for curing. On the other hand, due to its disadvantages compared to thermosets such as high raw material cost, production difficulty, and low forming ability at room temperature, its use as a matrix in marine composite applications is limited (Arhant and Davies, 2019). It is only used for boatbuilding or fitting. Thermosets such as polyester, vinyl ester, and epoxy are usually used as the bonding matrix for composites used in vessel production. Polyester is more economical than others and does not require post-curing heating. However, it has lower mechanical strength and adhesion. In addition, its heat resistance is lower than other resins and absorbs more water. There are two types of polyester, orthophthalic and isophthalic, depending on the acid type it contains or the position of the acid groups around the benzene ring. Orthophthalic polyester is an economical standard resin. Isophthalic polyester has better chemical resistance and mechanical properties. Because of its high molecular weight, the isophthalic resin requires more styrene thinner for viscosity adjustment (Dagher, Iqbal, and Bogner, 2004). Epoxy has the best mechanical

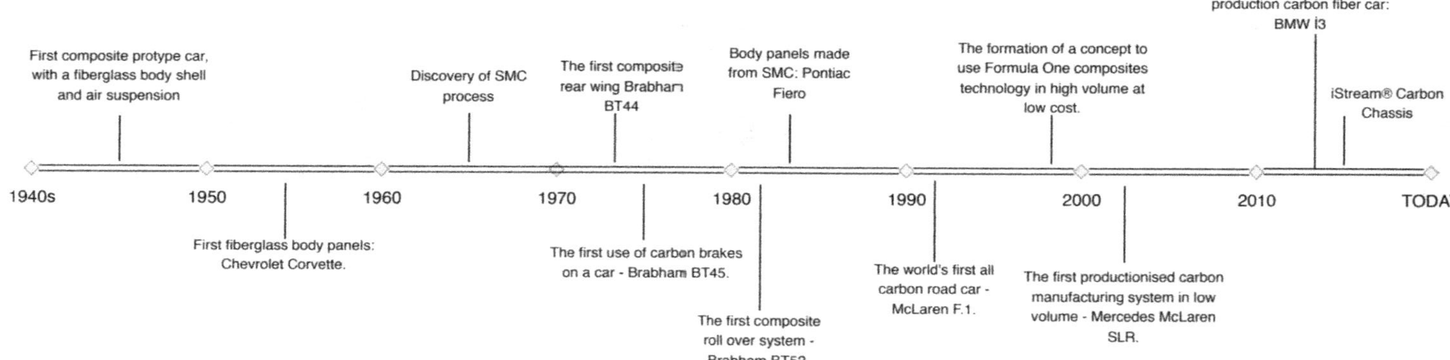

FIGURE 5.3 History of polymer composites in the automotive industry ("Gordon Murray Design announce 'iSTREAM® CARBON' a revolution in automotive manufacturing—Gordon Murray Design," n.d.).

TABLE 5.1
Typical Properties of Resins (Chalmers, 1994)

Resin type	Density (gm/cm3)	Young's modulus (GPa)	Poisson's ratio	Tensile strength/ yield (MPa)	Tensile failure strain (%)	Relative cost
Phenolic [s)]	1.15	3.0	—	50	2	0.8
Polyester (orthophthalic) [s)]	1.23	3.2	0.36	65	2	0.9
Polyester (isophthalic) [s)]	1.21	3.6	0.36	60	2.5	1
Vinyl ester [s)]	1.12	3.4	—	83	5	1.8
Epoxy [s)]	1.20	3.2	0.37	85	5	2.3
Polycarbonate (PC) [p)]	1.20	2.3	—	60	100	2.3
Polyethersulphone (PES) [p)]	1.35	2.8	—	84	60	6.0
Polyether-ether ketone (PEEK) [p)]	1.30	3.7	—	92	50	25

[s)] thermoset, [p)] thermoplastic

properties among these three resins as tabulated in Table 5.1. This two-component material has an excellent adhesion quality and shows minimum heat, shrinkage, and water permeability. In addition, it prevents delamination and provides a long service life. However, it is costly and requires a complex and precise post-curing step. Furthermore, polyester and epoxy thinners cause toxic gas release due to styrene which is a volatile organic compound. Vinyl ester is a formulation of epoxy resin and methacrylic acid, it has a mechanical performance between epoxy and polyester (Composite Resin Developments, 2014; Grenier, Dembsey, and Barnett, 1998). The resin types are tabulated in Table 5.1.

5.3.2 Fiber Types

Glass fiber, aramid fiber, and carbon fiber are mostly preferred as a reinforcement phase in composite vessel production. Glass fiber has a high strength-to-weight ratio and low production cost. It also exhibits good chemical resistance. Glass fibers are used for general purpose (E-glass) and special purpose (S-glass, D-glass, A-glass, electrical chemical resistance (ECR)-glass). Most of the glass fibers produced are E-glass. E-glass (lime aluminum borosilicate) has relatively good tensile strength, compressive strength, toughness, electrical property, and is low cost, but fatigue strength is poor. Therefore, S-glass (silicon dioxide, aluminum and magnesium oxides) is preferred for applications where higher tensile strength and fatigue strength are required (Wallenberger, Watson, and Hong Li, PPG Industries, 2001). Aramid fiber has a low weight, high tensile strength, high impact and fatigue resistance, and high weaving ability. However, its compressive strength is similar to glass fibers. It can be degraded in ultraviolet light. In addition, it may not combine well with the resin; in this case, the formation of microcracking and water absorption can be seen. Although there is a wide range of aramid fibers, the most commonly

TABLE 5.2
Typical Properties of Fibers (Chalmers, 1994)

Fiber type	Density (gm/cm³)	Young's modulus (GPa)	Poisson's ratio	Tensile strength (GPa)	Failure strain (%)	Relative cost
E-glass	2.55	72	0.2	2.4	3	1
S2, R-glass	2.5	88	0.2	3.4	3.5	8
Aramid (Kevlar 49)	1.45	124	—	2.8	2.5	15
High-strength carbon	1.74–1.81	248–345	—	3.1–4.5	0.9–1.8	45–50
High-modulus carbon	2.00–2.18	520–826	—	2.1–2.2	0.3–0.4	250–2700

used ones are Kevlar 29 and Kevlar 49 developed by DuPont. The main differences between these two fibers are the application areas, modulus of elasticity, and elongation at break. The elastic modulus of Kevlar 49 is approximately 30% higher than that of Kevlar 29. The elongation at break of Kevlar 29 is around 3.6%, while that of Kevlar 49 is around 2.5%. Kevlar 29 is used in armored military vehicle panels and bulletproof vests, while Kevlar 49 is preferred in boat hulls and the aviation industry (Ahmed et al., 2014). Carbon fiber is one of the most durable and stiff materials used commercially. It can maintain its mechanical properties even at high temperatures. Carbon fiber is a material resistant to corrosion and combustion. It represents prestige and luxury. The major disadvantages of its use are the high precursor cost and the complex production process. Electrical conductivity may also cause problems in applications where insulation is required. When combined with resin, it forms a brittle composite. While the percentage use of glass fibers in composites is 89%, natural fibers are around 10%, and carbon fiber is only 6% (Moreau, 2009). Although polyester and nylon thermoplastic fibers are widely used, they can be used in combination with glass fibers when necessary. Basalt fiber works are also still in progress (Davies and Verbouwe, 2018). The fiber types are listed in Table 5.2.

Glass fiber-reinforced thermoplastic composites up to 40% in mass can be used in non-structural parts in vehicle bodies such as radiator heads and body interior brackets. In structural parts, many composite manufacturing derivations can be processed. For example, radius pultrusion can be used in bumper beams and hot-pressed carbon fiber prepregs can also be used in bumper beams and other body side crash-absorbing components.

Natural fibers can be used in low-cost engineering applications instead of E-glass fibers, when high stiffness per unit weight is requested. Natural fibers, which are biodegradable and environmentally friendly, are traditionally used for ropes and chords. Natural fibers are sorted into animal, vegetable, and mineral groups. Vegetable natural fibers are grouped into seed, leaf, and fruit-based classifications (Nguong, Lee, and Sujan, 2013; Cheung, Ho, Lau, Cardona, and Hui, 2009; "Hybrid Cars I Explore The Best Hybrids from Toyota Toyota Motor Europe," n.d.). Because of their good mechanical properties, jute, hemp, flax, kenaf, and ramie are thought to be the best options for composite reinforcements. Bamboo, sisal, jute, wood, hemp, and flax fibers are also popular because of their good mechanical properties. Hemp, jute, flax, and sisal fibers are already mostly used in the automotive industry. Some disadvantages of natural fibers are their low thermal stability (possible degradation at 200–250°C) and poor adhesion of fibers because of their hydrophilic behavior resulting in swelling of fibers and high moisture content. Low adhesion with the polymeric matrix and highly moisturized natural fibers affect adhesion with hydrophobic matrix material negatively. These problems mostly result in degradation and a decrease in strength (Herrera-Franco and Valadez-González, 2005). The advantages of using natural fibers are low thermal expansion, high stiffness, low weight, good fatigue strength, high corrosion resistance, and good tensile-compressive-impact strength. The energy needed for the manufacturing of synthetic fibers is double that of natural fibers (Jauhari, Mishra, and Thakur, 2015). The usage of natural fiber composites in the automotive industry is summarized in Table 5.3. Figure 5.4 shows one example

TABLE 5.3
Natural Fiber Components Use in Vehicles by Automobile Companies (Kumar and Bharj, 2018)

Company name	Model name	Name of component
BMW	3, 5, and 7 series	Headliner panel, seatbacks, door panel, and noise insulation panels
FORD	Mondeo CD 162, Focus	Boot liner, pillar, and door panel
Toyota	Brevis, Harrier, Celsior, Raum	Tire cover, door panel, and seatback
Volkswagen	Passat, Bora, Fox, Polo, Golf	Boot liner, seat backs, boot lid finish panel, and door panel
Mercedes-Benz	Trucks	Internal engine cover, engine insulation, sun visor, interior insulation, bumper, wheel box, and roof cover
Audi	A2, A3, Avant, A6	Hat track, seatbacks, door panel, boot lining
BMW	3, 5, and 7 series	Headliner panel, seatbacks, door panel, and noise insulation panels
FORD	Mondeo CD 162, Focus	Boot liner, pillar, and door panel

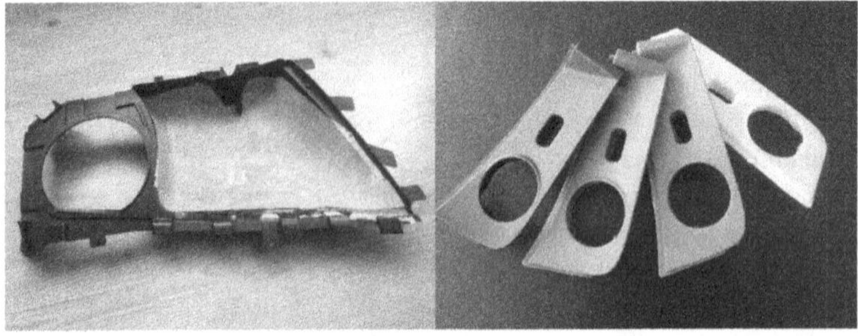

FIGURE 5.4 Overmolded and injection molded natural fiber-reinforced polypropylene components.

of flax-reinforced composite we developed with an over-molding technique and some other natural fiber-reinforced composites with injection molding.

5.4 PRODUCTION TECHNIQUES

Advanced composite technology is constantly changing and embracing new developments daily, yet the basics needed to successfully design, fabricate, and repair composite structures remain the same (Rana and Fangueiro, 2016). Within the basic framework, there are various methods employed in the manufacture of advanced composite parts. A key ingredient in the successful production application of a material or a component is a cost-effective and reliable manufacturing method. Cost-effectiveness depends largely on the rate of production, and reliability requires a uniform quality from part to part (Mallick, 2008). In accordance with these requirements, different manufacturing methods have been used from past to present.

In this chapter, the authors would like to focus on polymer matrix composites rather than ceramic and metal matrix composites due to their wider usage. Long fiber-reinforced thermoplastic composites, thermoforming, injection molding, resin transfer molding, high-pressure resin transfer molding (RTM), and sheet molding compound are widely used as automotive composite manufacturing techniques. Prepregs, sheet molding compound, bulk molding compound, glass mat thermoplastics, hand lay-up (wet and dry lay-up), spray-up, filament winding, pultrusion, vacuum infusion, seaman composites reaction injection molding proses (SCRIMP), autoclave, braiding, reaction injection molding, and thermoforming methods are also used. Each method provides different advantages due to its conventional methods. In tailgates, SMC-manufactured parts have more complex shapes and shorter manufacturing periods with respect to sheet metal-formed tailgates. Long fiber-reinforced thermoplastics (LFRT) parts have more stiff and cheaper parts with respect to injection molded parts on bumpers. Rapid manufacturing of structural composite components for automotive applications can be achieved with high pressure—RTM with highly reactive thermoset resins (Henning et al., 2019). Filament winding which is shown in Figure 5.9 is used in composite leaf spring applications (Robertson and Park, 1985).

5.4.1 Hand Lay-Up

Hand lay-up is the most basic and simplest open molding method. First, a gel coat (pigment-containing resin) is applied by spray gun or brush to achieve a high surface quality. When the gel coat becomes sticky, the reinforcing fiberglass fabric is placed on the mold and resin is poured onto it. Squeegees or rollers are used to ensure that the resin wets the fiber. A detailed drawing can be seen in Figure 5.5. At this stage, air gaps should be avoided. The process is repeated according to the desired thickness and number of layers. The hand lay-up method is low cost, is easy to repair and to assemble. It is suitable for epoxy, vinyl ester, and polyester resins. It also allows for high fiber content. However, it does not provide the same quality for all parts. This method, in which production volume is low to medium, is suitable for boats, tanks, housings, making building panels for prototypes and other large parts requiring high strength (Eric Greene Associates, 1999). Although it has been used in the

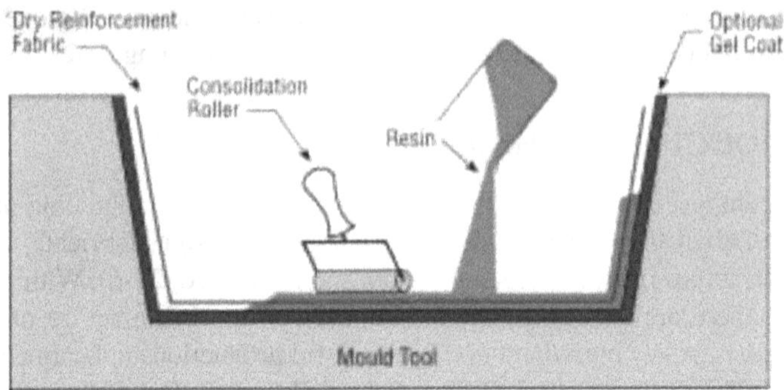

FIGURE 5.5 Illustration of the hand lay-up technique (Cripps, 2019d).

past, it is not preferred in the aerospace industry today. Nowadays, in the aerospace industry, the hand lay-up method is used for laying prepregs in molds. It can be used in the first stage of autoclave production. Prepregs can be placed in the molds by hand laying or by automatic laying machines (Mallick, 2008; Rana and Fangueiro, 2016). In the production of composites, laying of prepregs in molds is an important step. Nowadays, an automated tape lay-up process is used for the correct placement of prepregs. A number of techniques have been developed that require high capital investment in automatic laying (Rana and Fangueiro, 2016).

5.4.2 Spray-Up

Spray-up is the mechanized form of hand lay-up method. The mixture of chopped fibers and resin is sprayed into the mold with a spray gun as shown in Figure 5.6. Spraying should be perpendicular to the surface to prevent material waste. After spraying, a roller or squeegees are used to remove the air in the resin and smooth the surface. It is suitable for polyester resins. This method, in which production volume

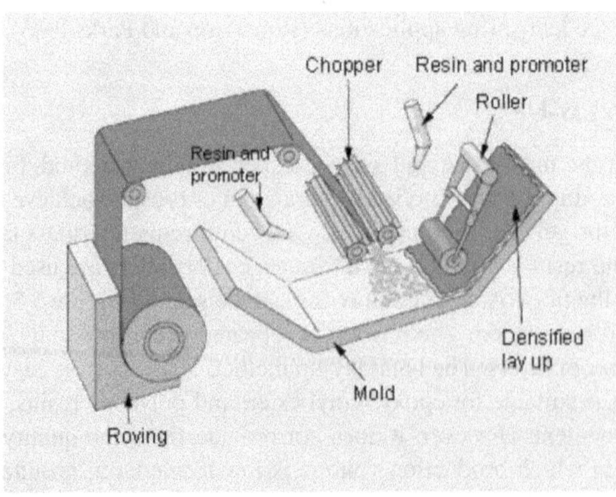

FIGURE 5.6 Illustration of the spray-up technique (Wacker Chemie AG, 2019).

is low to medium, is suitable for making boats, tanks, and tub/shower units (Eric Greene Associates, 1999).

5.4.3 Vacuum Bagging/Autoclave

Autoclave processes are the most widely used manufacturing methods in the production of advanced composites used in the aerospace industry. Autoclave molding is a preferred process for the aerospace industry. The aerospace parts are made using fully inserted carbon fiber epoxy prepreg layers (Rana and Fangueiro, 2016). With this method, high-performance parts that can be used in the aerospace industry can be produced (Balasubramanian, Sultan, and Rajeswari, 2018). The autoclave generally uses a pressure of between 50 and 100 psi. The prepregs are placed in a desired number of layers on the mold before being bagged under vacuum. The molds are then placed in the autoclave and heated until cured (Hodgkinson, 2000). Prepregs are laid in molds by automatic machines or by hand tilting. Prepreg hand laying is one of the most widely used methods in the production of aerospace composite parts. The production process takes place in a vacuum bag. The primary function of the vacuum bag is to create compression pressure and to ensure that the layers are joined. The laying process is carried out in an environmentally controlled clean room where temperature and humidity are regulated and monitored. The second function of the vacuum bag is to facilitate the removal of excess resin and gases between the layers during the process. The vacuum bag is still an essential part of the process in the autoclave because it acts as a membrane separating the tank pressure from the laminate (Rana and Fangueiro, 2016). Autoclaves operate by pressurizing gas (>10 atm) into the closed vessel. Air or inert gas is used to pressurize. This process provides a much more compressive force on a composite laminate during the curing process. Conventional epoxy aviation resins are generally cured in autoclave production at 120–135 or 180°C and up to 8 bars (Rabotnov et al., 1982). For advanced composite parts requiring high-temperature curing, an autoclave can be used (Rana and Fangueiro, 2016). This method has evolved over time, and nowadays, large autoclaves have been installed in the aerospace industry that can house complete wing or tail sections (Rabotnov et al., 1982). A schema is shown in Figure 5.7.

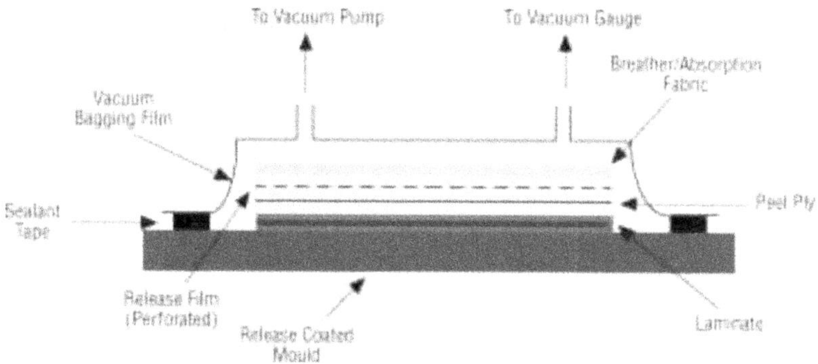

FIGURE 5.7 Illustration of the vacuum bagging technique (Cripps, 2019c).

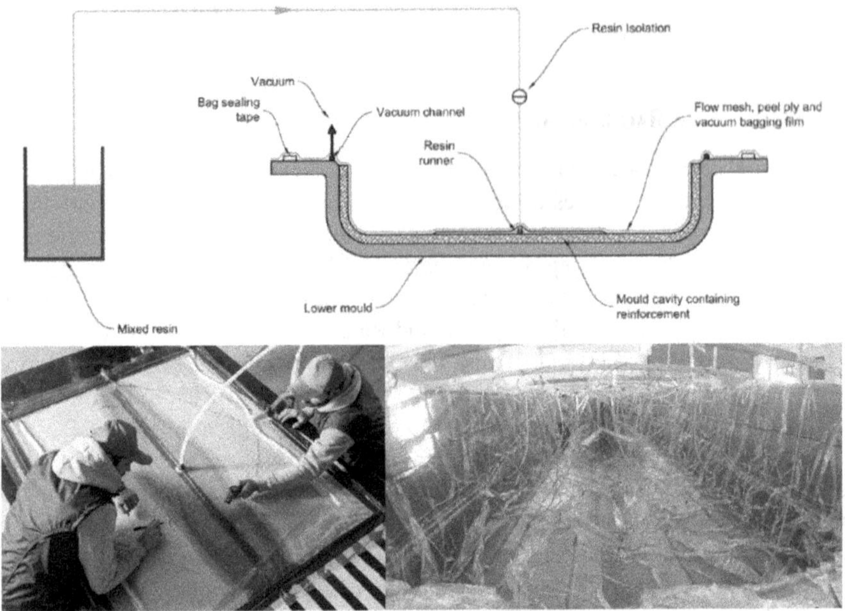

FIGURE 5.8 Illustration of the vacuum infusion technique (with kind permission of Trevor Osborne, 2013) and usage in the boat building process (with kind permission of Dromeas Yachts).

5.4.4 Vacuum Infusion

Vacuum infusion is a method of producing composite by impregnating resin on the vacuumed reinforcement material on the mold by means of pipes as shown in Figure 5.8. It is a method of producing high-quality moldings with minimal styrene emissions. Vacuum-infused laminates show better consolidation and strength. By removing compressed air and excess resin, higher fiber content can be achieved. The higher glass content means that for the same glass weight per square meter, the laminate will be thinner with infusion compared to with hand lay-up and this can lead to a reduction in stiffness. It is suitable for vinyl ester and polyester resin. The molds used in the hand lay-up method can be arranged and used for this method; however, steps and installation are complex. Parameters such as permeability of the laminate, the viscosity of the resin, and pressure differential in the cavity in relation to atmospheric pressure are important for product quality. Vacuum infusion is suitable to molding very large structures and is considered a low-volume molding process (Osborne, 2013).

5.4.5 Filament Winding

In the filament winding method, the fiber bundle is passed through the resin bath and wound onto the rotating mandrels seen in Figure 5.9. The simple helical path on the mandrel cylinder is defined as the geodetic path. The fiber bundles are continuously conveyed to the resin tank and the fiber bundles are saturated with resin in the resin

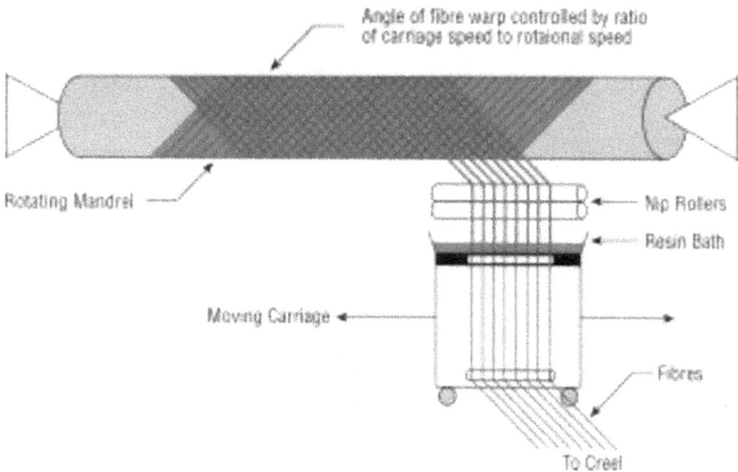

FIGURE 5.9 Illustration of the filament winding technique (Cripps, 2019a).

tank. The resin-saturated fibers are then wound onto a rotating mandrel. The mandrel acts as a mold and the profile of the mandrel is exactly the same as the desired end product. The fiber tension in the filament winding is critical because compression is achieved through the fiber tension. The tension affects the percentage of porosity in the final product. The fiber tension depends on the fiber type, geometry, and the winding arrangement required for the rotating mandrel. The fiber tension should be optimal because too high fiber tension can cause fiber breakage on the surface. These bundles of fibers are wound around the mandrel in a controlled manner and in a particular direction. The filament winding method is an open molded method capable of producing symmetrical shapes. The curing of the composite is usually done by heat in an oven and the final composite product is removed from the mandrel (Balasubramanian et al., 2018; Minsch, Herrmann, Gereke, Nocke, and Cherif, 2017). Circular shaped components can be produced using the filament winding process. This method is mainly used to produce pipes. The filament winding process is also used to make oxygen tanks, automotive drive shafts, spherical pressure vessels, and helicopter blades. One of the biggest advantages is the automation of the process so that labor times and costs can be kept low. In addition, high reliability and quality products can be obtained (Potter, 2000).

5.4.6 Resin Transfer Molding

Resin transfer molding is another manufacturing process that has received significant attention in both aerospace and automotive industries for its ability to produce composite parts with complex shapes at relatively high production rates (Mallick, 2008). RTM and resin film infusion (RFI) are curing methods that are being developed and have various variations. In conventional prepreg technology, the resin is saturated with fibers and processes are carried out mainly to remove air between the layers, to consolidate and cure the layers. In RTM and RFI, in the simplest sense, as shown in Figure 5.10, the saturate of the fabrics is placed in the closed mold under vacuum

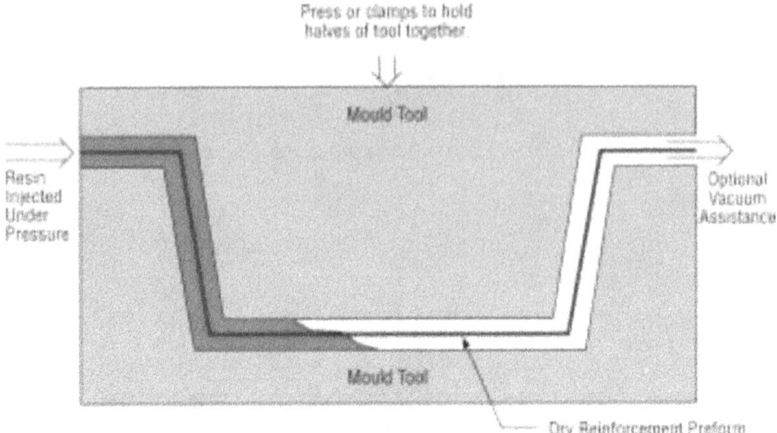

FIGURE 5.10 Illustration of the resin transfer technique (Cripps, 2019b).

resin. Closed mold is used in the RTM method and resin is transferred inside. In the RFI method, the mold is sealed with a vacuum bag and saturated with resin (Rana and Fangueiro, 2016). This is done by creating a pressure differential between the inside and the outside of the bag using a vacuum pump, a hose, and a sealed film membrane, and results in down-force on the bag equal to the ambient atmospheric pressure. The RFI method is used in aircraft radome manufacturing and the RTM method is used to produce small complex aircraft parts. This method is suitable for medium volume production where the process is not too slow but not too fast. Figure 5.11 shows the production of high-pressure resin transfer molding for the automotive part.

5.4.7 Prepreg

Some of the recent technological improvements enable manufacturers to manufacture much faster with the use of prepregs. Prepreg is the abbreviated word for resin pre-impregnated woven cloths that are prepared with a special resin. These pre-impregnated structures are stored in cold conditions to maintain the resin strength. This prepreg technology brings flexibility to manufacturers as it is easily shaped and also there is no need for the resin to flow again. Prepregs have been predominantly used in the aerospace industry since the 1970s. It has also been considered to be used in automotive applications as well. In recent years, there is a new technique to prepare unidirectional (UD) thermosetting resin-based prepreg composites. The reinforcing capability of the fibers increases with the orientation of the fibers and also reduces void problems occurring in the final composite structure. Furthermore, very recently, over the last five to six years, the interest in thermoplastic resins created a new class of thermoplastic UD prepregs.

5.4.8 Long Fiber Technology (LFT)

This is a new technology that is developed by Fraunhofer Institute Germany and whether it can be used for automotive parts is under intense research. It has been

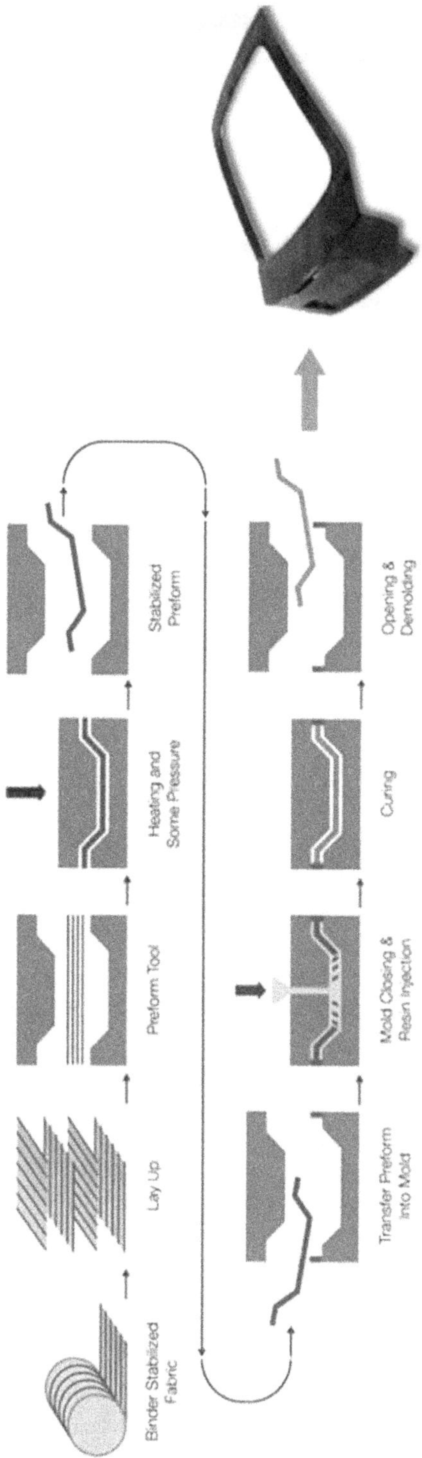

FIGURE 5.11 High-pressure resin transfer molding ("Advanced Materials | HP-RTM (High Pressure Resin Transfer Molding," n.d.).

FIGURE 5.12 An example of an LFT PP molding of underbody shields made by Polytec for the VW group (Automotive market, 2011).

developed to replace the thermoset-based composites that have many processing issues. The technology is based on the dispersion of the medium-long reinforcing fibers in thermoplastic resins. The technology is based on glass fiber and at Nagoya University, they have developed the technology with carbon fiber as well. Figure 5.12 shows some LFT products made in Japan with carbon fiber.

5.5 APPLICATIONS

The emergence of strong and stiff reinforcements like carbon fiber along with advances in polymer research to produce high-performance resins as matrix materials have helped meet the challenges posed by the complex designs of modern aircraft. Today, the use of advanced composites in the development of military aircraft, small and large civil aircraft, helicopters, satellites, and missiles plays an important role (Mangalgiri, 1999). The use of polymer composites in the aerospace industry in the production of lightweight and durable components has increased in the last 40 years. Starting with the production of relatively small parts, the process has progressed to structural parts such as ailerons and bodywork, and to large and critical elements such as wings and bodies. Until 1987, 350 composite components were put into service in various commercial aircraft (Mallick, 2008; McIlhagger, Archer, and McIlhagger, 2014). Airbus was the first commercial aircraft manufacturer to use composites extensively on its A310 aircraft, introduced in 1987. For large commercial aircraft, the development of primary structural applications began with the fin part of the Airbus A310 aircraft. The composite components reduced the weight of the aircraft by about 10%. Over the years, composites have replaced metal in many parts. For example, lower access panels and top panels of the wing leading edge, outer deflector doors, nose wheel doors, main wheel leg fairing doors, engine cowling panels, elevators, fin box, leading and trailing edges of fins, flap track fairings, flap access doors, rear and forward wing-body fairings, pylon fairings, nose radome, cooling air inlet fairings, tail leading edges, upper surface skin panels above the main wheel bay, glide slope antenna cover, and rudder (Cantor, Assender, and Grant, 2015; Mallick, 2008). The tail cone, vertical stabilizer, and horizontal stabilizer of the A320 aircraft of the Airbus company, launched in 1988, are made of composite material.

It was also the first commercial aircraft to use a fully composite tail (Mallick, 2008). Boeing 777 was introduced in 1995 and Boeing company first started to use composite materials in its tail cone. In the same year, the Airbus company began to use polymer stabilizers in the vertical stabilization and tailplanes of the A340 model. For example, the composite vertical stabilizer, 8.3 m high and 7.8 m wide, is about 400 kg lighter than the aluminum vertical stabilizer used previously (Cantor et al., 2015; Starke and Staley, 1996). The Airbus A380 was launched in 2006. 25% of the weight of the A380 is made of composite material. The main composite components in the A380 aircraft are the central torsion box, the aft-pressure bulkhead, the tail, and the flight control surfaces (Mallick, 2008). Carbon fiber-reinforced polymer composite material is used in the wings and fuselage of Boeing 787 and Airbus A400 M aircraft and makes up 50% of the weight of an Airbus A350 aircraft (McIlhagger et al., 2014). The Boeing B-787 Dreamliner is designed and manufactured from highly advanced materials. It is referred to as 'breakthrough' in the aviation industry. Approximately 50% of the mass of the whole aircraft is designed from composite materials as shown in Figure 5.15. The wing, tail, and fuselage are made of composite materials (Koniuszewska and Kaczmar, 2016). In the A350, designed by Airbus, more than 50% of the mass of the aircraft is made of composite materials as shown in Figure 5.16. Polymer composite applications are used in a wide range of applications from military aviation to civil aviation platforms, as significant weight savings are achieved compared to metallic designs (Cantor et al., 2015).

There are many uses of FRP, from sailboats, kayaks, and canoes to speed boats, luxury yachts, navy ships, and submarines (Figures 5.13 and 5.14). More than 95% of marine vehicles are made of glass fiber-reinforced composite (GRP) because of their low cost. In other applications requiring high performance, carbon fiber is used (Selvaraju, 2011). Boat hulls, masts, bulkheads, deckhouses, seatings, modular tanks, cable ladders-trays, rudders, ventilation ducts, radomes, sonar domes, railings, watertight doors, hatch covers, low-pressure pipes, and propellers are made of FRP. It is also used in submarine external structure (fin and casing), submarine control

FIGURE 5.13 'Galaxy of Happiness' is all-composite 53-meter multihull trimaran (Milberg, 2016).

FIGURE 5.14 The deckhouse of the 'USS Michael Monsoor' ship is polygonal composite and it is covered with a material that can absorb radar waves and increase the destroyer's stealth (Tribune News Service, 2019).

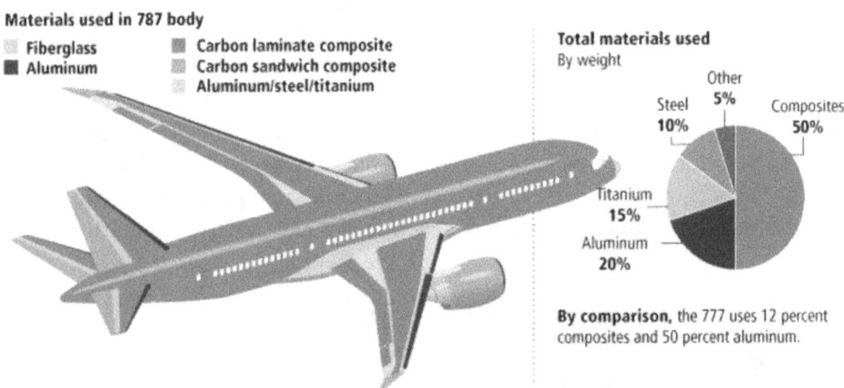

FIGURE 5.15 Boeing 787 body materials (Airways, 2011).

surfaces, and components for diesel engines and heat exchanger on large warships. On the other hand, the use of GRP in the construction of large ships is limited partly due to problems with hull deflections and flammability of plastics (Chalmers, 1994; Eric Greene Associates, 1999; Ertuğ, 2013; Neşer, 2017).

Carbon fiber-reinforced polymer applications are widely used in German cars. In this technology, a high-pressure resin transfer molding (hp-RTM) technique is used with carbon fiber textiles and fast cross-linking of epoxy chemicals. In this technology there are some disadvantages, which are difficulties in high-speed mass production, However, there are several other drawbacks to this technology, including difficulties

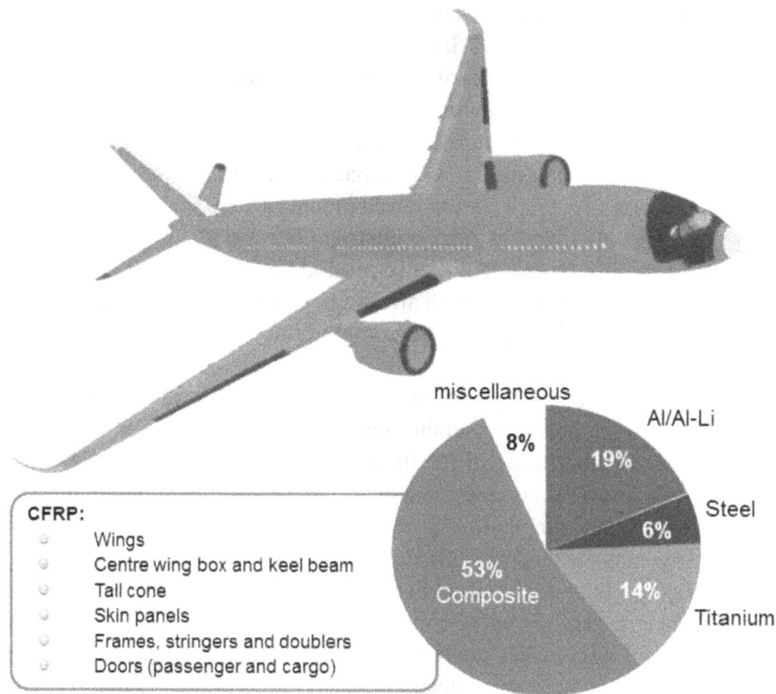

FIGURE 5.16 Airbus A350 materials (Airbus, 2013).

in a much higher rate of production and slow and hard joining of composite materials to other regions in the vehicle body. The use of carbon fiber-based thermoplastic matrix composites seem to be one option to solve the above disadvantages.

5.6 FUTURE TRENDS

Today's most important goals are to recycle FRP by separating fibers and thermoset resins [19], to reduce the need for petroleum-based synthetic materials, and to produce more environmentally friendly sustainable composites (Asmatulu, Twomey, and Overcash, 2014). In this context, natural fibers such as flax, hemp, and cotton began to be used as additives. Since their mechanical properties are poor, their application is limited currently. Low styrene rate resins are developed to reduce emissions. Bio-based resins are also being developed. The biodiesel product, bio glycerin, can be used as the raw material of polyester resin. Vegetable oils from soya beans, castor trees, or linseed can also be used as a base material for polyester (Moreau, 2009). In addition, reducing fuel consumption, weight, and greenhouse gas emissions in vehicles are other critical issues in the transportation sector (Spero and Raval, 2019). Therefore, the joining of hybrid structures like metal and FRP by adhesives is increasing (D. Zhang, Zhang, Fan, and Zhao, 2019). The use of 3D-printed thermoplastics parts is increasing in boat components. For example, the production time of carbon fiber catamaran has been shortened thanks to parts such as bearing cages printed with the 3D printer. Studies on anticorrosive polymer nanocomposite coatings that are resistant to the marine environment are being carried out (Mardare and Benea, 2017). The use of composite materials in the maritime sector is expected to grow to $1.5 million over the next five years (Stratview Research, 2019).

Fuel economy, increasing the carrying capacity of aircraft and improving maneuverability encourages research on the development of new materials characterized by low weight and good mechanical properties. For example, American Airlines operates a fleet of approximately 600 aircraft and has saved 11,000 gallons of fuel per year by reducing the weight of each aircraft by just 1 kilo (Moniruzzaman, 2012). Today, in the aerospace industry where competition is at the forefront, efforts are being made to use composite materials in structural components of aircraft more. Parts integration reduces the number of parts, the number of manufacturing operations, and also, the number of assembly operations. In this context, it is aimed to produce integrated parts that are difficult to manufacture with additive manufacturing. The manufacturing of complex shapes has been made possible by advances in additive manufacturing. Another issue that limits the use of composites in aircraft structural components is electrical conductivity. In this context, special conductive coatings or conductive layers are used in composite production. In order to eliminate these constraints, research is being conducted on nano-enhanced polymers materials. Environmental approaches come to the forefront as the lightening of aircraft reduces emissions. Topology optimization of aircraft components and new material issues are always at the forefront for reducing emissions and increasing the payload ratio. In addition, recycling is an issue that cannot be ignored within the aviation industry as it is in every field. Current and future waste management and environmental legislation, as with all engineering materials, require that aircraft be properly recycled and recycled from end-of-life products. Recycling will ultimately save resources and energy. Although the more conventional matrix materials used today are thermoset epoxies, thermoplastic materials are becoming increasingly important

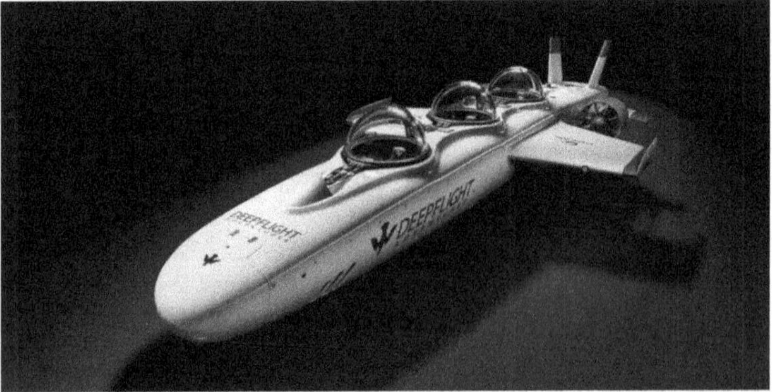

FIGURE 5.17 'Super Falcon 3S' is the first composite submarine certified by Lloyds Register (DeepFlight Inc., 2019).

FIGURE 5.18 'Solar Impulse 2' is a long-range solar-powered aircraft which is made of carbon fiber and epoxy composite (Ryan, 2016).

and research is being carried out in this field. Within this scope, research is carried out for the development of new production methods. Figure 5.17 shows one of the recent examples of a transportation vehicle, a submarine, made with composites. Figure 5.18 shows another impressive example of an airplane flying with solar energy made with carbon composites.

The mass production speed of the mentioned composite part is not as high as the rate of extremely high production vehicle sub-systems of 100,000 units in one year. That's why it is requested that carbon fiber-reinforced thermoplastic (CFRTP) parts be more convenient for mass production technology by automotive industries. One possible solution could be long glass fiber thermoplastic direct consolidation (LFT-D), created in Fraunhofer Institute/Germany. On the other hand, the carbon

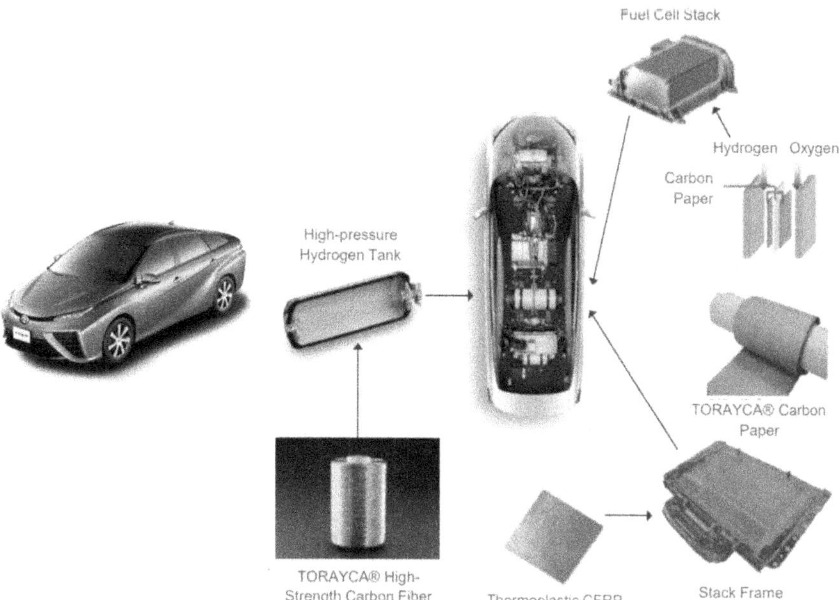

FIGURE 5.19 'Toyota Mirai' is the first mass-production hydrogen fuel cell electric vehicle that has carbon fiber parts (TORAY, n.d.).

fiber-reinforced LFT-D technique is still being developed in Germany (Rohan, Mcdonough, Ugresic, Potyra, and Henning, 2015). In Japan National Composites Center at Nagoya University, carbon fiber-based LFT-D technology (C-LFT-D) has been studied within the last five years and some chassis components were manufactured. An alternative technology of discontinuous CFRTP's alternative technology is being developed by Professor Takahashi's group at the University of Tokyo (Ishikawa et al., 2018). The first fuel-cell car in the world, the 'Toyota Mirai' used carbon fiber-reinforced thermoplastic polymers ("Hybrid Cars I Explore The Best Hybrids from Toyota Toyota Motor Europe," n.d.) with paper type carbon fiber mat in a fuel cell stack base (see Figure 5.19).

5.7 CONCLUSION

The composite industry is growing with a very fast pace and its use in transport industries is one of the most important uses of composite materials. Although marine applications are older than many other uses, aerospace and defense industry is leading the way for composites with many new advanced techniques. In the last two to three decades, with more and more concerns around sustainability, automotive applications are very critical as well. These innovations in the composite materials used in transportation bring many novel solutions for other industries including energy, medical, and sporting goods. From the materials perspective, composite materials are the ultimate material class that will replace all other types of materials.

REFERENCES

Advanced Materials I HP-RTM (High Pressure Resin Transfer Molding). (n.d.). Retrieved September 5, 2019 from http://www.katcon.com/advancedmaterials/hp-rtm-high-pressure-resin-transfer-molding/.

Ahmed, D., Hongpeng, Z., Haijuan, K., Jing, L., Yu, M., & Muhuo, Y. (2014). Microstructural developments of poly (p-phenylene terephthalamide) fibers during heat treatment process: A review. *Materials Research*, *17*(5), 1180–1200. doi:10.1590/1516-1439.250313.

Airbus. (2013). A350XWB special edition. *Airbus Technical Magazine*, June.

Airways, J. (2011). Boeing 787 Dreamliner flight deck. Retrieved August 22, 2019, from Modern Airliners website: http://www.modernairliners.com/boeing-787-dreamliner/boeing-787-dreamliner-specs/.

Arhant, M., & Davies, P. (2019). Thermoplastic matrix composites for marine applications. In: *Marine Composites* (pp. 31–53). doi:10.1016/b978-0-08-102264-1.00002-9.

Asmatulu, E., Twomey, J., & Overcash, M. (2014). Recycling of fiber-reinforced composites and direct structural composite recycling concept. *Journal of Composite Materials*, *48*(5), 593–608. doi:10.1177/0021998313476325.

Automotive market: LFT PP demand growing in most applications / Increasing role in automotive design / AMI report. Plasteurope.com. (2011). Retrieved September 5, 2019 from https://www.plasteurope.com/news/detail.asp?id=220163.

Bai, J. (2013). Introduction. In: *Advanced Fibre-Reinforced Polymer (FRP) Composites for Structural Applications* (pp. 1–4). doi:10.1533/9780857098641.1.

Balasubramanian, K., Sultan, M. T. H., & Rajeswari, N. (2018). Manufacturing techniques of composites for aerospace applications. In: *Sustainable Composites for Aerospace Applications*. doi:10.1016/b978-0-08-102131-6.00004-9.

Brothers, W. (2012). Aerospace materials: Past, present and future. *Introduction to Aerospace Materials*, 15–38. doi:10.1533/9780857095152.15.

Cantor, B., Assender, H., & Grant, P. (2015). Aerospace materials. In: *Aerospace Materials*. doi:10.1201/9781420034721.

Chalmers, D. W. (1994). The Potential for the use of composite materials in marine structures. *Marine Structures*, *7*(2–5), 441–456. doi:10.1016/0951-8339(94)90034-5.

Cheung, H.Y, Ho, M. P., Lau, K. T., Cardona, F., & Hui, D. (2009). Natural fibre-reinforced composites for bioengineering and environmental engineering applications. *Composites Part B: Engineering*, *40*(7), 655–663. doi:10.1016/j.compositesb.2009.04.014.

Composite Resin Developments. (2014). *Epoxy Resin vs Vinylester vs Polyester Use and Application Overview*. Retrieved from https://www.compresdev.co.uk/CRD The Use and Application of Epoxy Resin vs Vinylester vs Polyester.pdf.

Dagher, H. J., Iqbal, A., & Bogner, B. (2004). Durability of isophthalic polyester composites used in civil engineering applications. *Polymers and Polymer Composites*, *12*(3), 169–182. doi:10.1177/096739110401200302.

David Cripps, G. (2019a). Filament winding. Retrieved August 22, 2019 from https://netcomposites.com/guide/manufacturing/filament-winding/

David Cripps, G. (2019b). Resin transfer. Retrieved August 22, 2019 from https://netcomposites.com/guide/manufacturing/resin-transfer/

David Cripps, G. (2019c). Vacuum bagging. Retrieved August 22, 2019 from https://netcomposites.com/guide/manufacturing/vacuum-bagging/

David Cripps, G. (2019d). Wet/hand lay-up. Retrieved August 22, 2019 from https://netcomposites.com/guide/manufacturing/wet-hand-lay-up/

Davies, P., & Verbouwe, W. (2018). Evaluation of basalt fibre composites for marine applications. *Applied Composite Materials*, *25*(2), 299–308. doi:10.1007/s10443-017-9619-3.

DeepFlight Inc. (2019). DeepFlight super falcon 3S - DeepFlight Inc. Retrieved August 6, 2019 from http://www.deepflight.com/deepflight-super-falcon-3s/

Eric Greene Associates. (1999). *Composite Materials* (2nd ed.). Eric Greene Associates.

Ertuğ, B. (2013). Advanced fiber-reinforced composite materials for marine applications. *Advanced in Materials Research*, *772*, 173–177. doi:10.4028/www.scientific.net/amr.772.173.

Ghassemieh, E. (2011). Materials in automotive application, state of the art and prospects. *New Trends and Developments in Automotive Industry*. doi:10.5772/13286.

Gordon Murray Design announce "iSTREAM® CARBON" a revolution in automotive manufacturing—Gordon Murray Design. (n.d.). Retrieved September 24, 2019 from https://www.gordonmurraydesign.com/news-articles/gordon-murray-design-announce-istream®-carbon-a-revolution-in-automotive-manufacturing.html.

Grenier, A. T., Dembsey, N. A., & Barnett, J. R. (1998). Fire characteristics of cored composite materials for marine use. *Fire Safety Journal*, *30*(2), 137–159. doi:10.1016/S0379-7112(97)00059-3.

Henning, F., Kärger, L., Dörr, D., Schirmaier, F. J., Seuffert, J., & Bernath, A. (2019). Fast processing and continuous simulation of automotive structural composite components. *Composites Science and Technology*, *171*, 261–279. doi:10.1016/j.compscitech.2018.12.007.

Herrera-Franco, P. J., & Valadez-González, A. (2005). A study of the mechanical properties of short natural-fiber reinforced composites. *Composites Part B: Engineering*, *36*(8), 597–608. doi:10.1016/j.compositesb.2005.04.001.

Hodgkinson, J. M. (2000). Mechanical testing of advanced fibre composites. In: *Mechanical Testing of Advanced Fibre Composites*. doi:10.1533/9781855738911.

Hybrid cars | Explore the best hybrids from Toyota Toyota Motor Europe. (n.d.). Retrieved August 21, 2019 from https://www.toyota-europe.com/hybrid/

Ishikawa, T., Amaoka, K., Masubuchi, Y., Yamamoto, T., Yamanaka, A., Arai, M., & Takahashi, J. (2018). Overview of automotive structural composites technology developments in Japan. *Composites Science and Technology*, *155*, 221–246. doi:10.1016/J.COMPSCITECH.2017.09.015.

Jauhari, N., Mishra, R., & Thakur, H. (2015). Natural fibre reinforced composite laminates—A review. *Materials Today: Proceedings*, *2*(4–5), 2868–2877. doi:10.1016/j.matpr.2015.07.304.

Koniuszewska, A. G., & Kaczmar, J. W. (2016). Application of polymer based composite materials in transportation. *Progress in Rubber, Plastics and Recycling Technology*, *32*(1), 1–23.

Kumar, S., & Bharj, R. S. (2018). Emerging composite material use in current electric vehicle: A review. *Materials Today: Proceedings*, *5*(14), 27946–27954. doi:10.1016/j.matpr.2018.10.034.

Mallick, P. K. (2008). *Fiber-reinforced composites: Materials, manufacturing, and design*. Retrieved from https://www.crcpress.com/Fiber-Reinforced-Composites-Materials-Manufacturing-and-Design-Third/Mallick/p/book/9780849342059.

Mangalgiri, P. D. (1999). Composite materials for aerospace applications. *Bulletin of Materials Science*, *22*(3), 657–664. doi:10.1007/BF02749982.

Mardare, L., & Benea, L. (2017). Development of anticorrosive polymer nanocomposite coating for corrosion protection in marine environment. *IOP Conference Series: Materials Science and Engineering*, *209*(1). doi:10.1088/1757-899X/209/1/012056.

Marsh, G. (2006). 50 years of reinforced plastic boats. *Reinforced Plastics*, *50*(9), 16–19. doi:10.1016/S0034-3617(06)71125-0.

McIlhagger, A., Archer, E., & McIlhagger, R. (2014). Manufacturing processes for composite materials and components for aerospace applications. *Polymer Composites in the Aerospace Industry*, 53–75. doi:10.1016/B978-0-85709-523-7.00003-7.

Milberg, E. (2016). All-composite yacht "galaxy of happiness" debuts in Monaco. Retrieved August 21, 2019 from http://compositesmanufacturingmagazine.com/2016/10/composite-yacht-galaxy-happiness-debuts-monaco/

Minsch, N., Herrmann, F. H., Gereke, T., Nocke, A., & Cherif, C. (2017). Analysis of filament winding processes and potential equipment technologies. *Procedia CIRP*, *66*, 125–130. doi:10.1016/j.procir.2017.03.284.

Moniruzzaman, M. (2012). Materials innovation—Aerospace manufacturing and design. Retrieved August 22, 2019, from JEC Composites Magazine website: https://www.aerospacemanufacturinganddesign.com/article/amd0312-metal-aircraft-food-tray-arms/

Moreau, R. (2009). *Nautical activities: What impact on the environment? A life cycle approach for "clear blue" boating—Commissioned by the European Confederation of Nautical Industries—ECNI*. Retrieved from www.ecni.org.

Neşer, G. (2017). Polymer based composites in marine use: History and future trends. *Procedia Engineering*, *194*, 19–24. doi:10.1016/j.proeng.2017.08.111.

Nguong, C. W., Lee, S. N. B., & Sujan, D. (2013). *A review on natural fibre reinforced polymer composites*. Retrieved from https://espace.curtin.edu.au/bitstream/handle/20.500.11937/31977/196455_106598_Waset2013_polymer.pdf?sequence=2.

Potter, K. (2000). An introduction to composite products: Design, development and manufacture. *Industrial Robot: An International Journal*, *27*(2), 276. doi:10.1108/ir.2000.04927bae.001.

Rabotnov, Y. N., Tupolev, A. A., Kut'inov, V. F., Kogaev, V. P., Berezin, A. V., & Šulimenkov, V. V. (1982). Use of carbon fiber-reinforced plastics in aircraft construction. *Mechanics of Composite Materials*, *17*(4), 455–465. doi:10.1007/BF00605914.

Rana, S., & Fangueiro, R. (2016). *Advanced composite materials for aerospace engineering: Processing, properties and applications*. Retrieved from https://books.google.com.tr/books/about/Advanced_Composite_Materials_for_Aerospa.html?id=QauSBgAAQBAJ&source=kp_cover&redir_esc=y.

Ranasinghe, K., Guan, K., Gardi, A., & Sabatini, R. (2019). Review of advanced low-emission technologies for sustainable aviation. *Energy*. doi:10.1016/j.energy.2019.115945.

Robertson, Richard E., & Allen Park, M. (1985). *Process of making a filament wound composite material leaf spring*. Retrieved from https://patents.google.com/patent/US4749534.

Rohan, K., Mcdonough, T. J., Ugresic, V., Potyra, E., & Henning, F. (2015). Mechanical study of direct long fiber thermoplastic carbon/polyamide 6 and its relations to processing parameters. *Society of Plastics Engineers Automotive Composites Conference & Exhibition (ACCE), c*(4), 1–24. Retrieved from http://www.temp.speautomotive.com/SPEA_CD/SPEA2014/pdf/ET/ET6.pdf.

Ryan, Fletcher (2016). Solar Impulse 2 makes triumphant landing to NYC. Retrieved September 24, 2019 from https://www.aerotime.aero/aerotime.extra/23179-solar-impulse-2-makes-triumphant-landing.

Selvaraju, S. (2011). Applications of composites in marine industry. *Engineering Research and Studies*, *2*(2), 89–91. Retrieved from http://citeseerx.ist.psu.edu/viewdoc/download?doi=10.1.1.464.963&rep=rep1&type=pdf.

Seydibeyoğlu, M. Ö., Mohanty, A. K., & Misra, M. (2017). Fiber technology for fiber-reinforced composites. In: *Fiber Technology for Fiber-Reinforced Composites*, Elsevier.

Spero, J., & Raval, Anjli. (2019). Pollution: The race to clean up the shipping industry. *Financial Times*. Retrieved July 28, 2019 from https://www.ft.com/content/642b6b62-70ab-11e9-bf5c-6eeb837566c5.

Starke, E. A., & Staley, J. T. (1996). Application of modern aluminum alloys to aircraft. *Progress in Aerospace Sciences*, *32*(2–3), 131–172. doi:10.1016/0376-0421(95)00004-6.

Stratview Research. (2019). *Marine Composites Market | Industry Analysis 2019–2024*. Retrieved from https://www.stratviewresearch.com/505/marine-composites-market.html.

TORAY. (n.d.). Toray's carbon fiber materials adopted by Toyota's fuel cell vehicle MIRAI. Retrieved September 24, 2019 from https://cs2.toray.co.jp/news/toray/en/newsrrs02.nsf/0/d144e85f1c996c5149257d950007222b.

Trevor Osborne. (2013). An introduction to vacuum infusion—Materials Today. Retrieved August 5, 2019, from. *Materials Today* website: https://www.materialstoday.com/composite-processing/features/an-introduction-to-vacuum-infusion/

Tribune News Service. (2019). US Navy's stealthy new 'ship killer' Michael Monsoor has nothing to shoot from its hi-tech guns. *South China Morning Post*. Retrieved August 6, 2019 from https://www.scmp.com/news/world/united-states-canada/article/2183789/us-navys-stealthy-new-ship-killer-michael-monsoor.

Wacker Chemie AG. (2019). Hand lay-up. doi:10.1007/978-0-387-30160-0_5685.

Wallenberger, Frederick T., Watson, James C., Li, Hong, & PPG Industries, I. (2001). Glass fiber. In: *ASM Handbook: Composites* (vol. 21). Retrieved from https://www.asminternational.org/news/-/journal_content/56/10192/06781G/PUBLICATION.

Zhang, D., Zhang, Q., Fan, X., & Zhao, S. (2019, December). Review on joining process of carbon fiber-reinforced polymer and metal: Applications and outlook. *Xiyou Jinshu Cailiao Yu Gongcheng/Rare Metal Materials and Engineering*, *48*, 44–54. doi:10.1016/s1875-5372(19)30018-9.

Zhang, X., Chen, Y., & Hu, J. (2018). Recent advances in the development of aerospace materials. *Progress in Aerospace Sciences*, *97*(August 2017), 35–60. doi:10.1016/j.paerosci.2018.01.001.

6 Hybrid Thermoplastic and Thermosetting Composites

N. R. Singha, P. K. Chattopadhyay, M. Karmakar, and H. Mondal

CONTENTS

ABBREVIATIONS

MOF: metal-organic frameworks
TS: tensile strength
TM: tensile modulus
FS: flexural strength
FM: flexural modulus
IS: impact strength
BS: bending strength
BM: bending modulus

YS: Young's modulus
EM: elastic modulus
EAB: elongation at break
SS: shear strength
PFR: phenol-formaldehyde resin
VAE: vinyl acetate-ethylene
LDPE: low density polyethylene
RTM: resin transfer molding
CNT: carbon nanotube
MWCNT: multi-walled carbon nanotube
PVA: polyvinyl alcohol
GO: graphene oxide

6.1 INTRODUCTION

In recent times, polymer has been considered as one of the basic components of various traditional and smart materials including organic, inorganic, organic+inorganic-hybrid composites, MOFs, and nanocomposites [1]. The term "polymer" originated from the Greek words "πολύς" (i.e., *polus*, meaning "many") and "μέρος" (i.e., *meros*, meaning "part") that denotes a high molecular-weight molecule comprising multiple repetitions of lower molecular mass. Though the term "polymer" was initialized by Jöns Jacob Berzelius in 1833, the modern idea of polymers was first suggested by Hermann Staudinger in 1920. Because of the never-ending benefits of polymeric materials, modern civilization utilizes polymers extensively for diversified prospective applications in medical, aeronautics, waste management, and engineering fields, along with fulfilling the basic needs in our day-to-day lives [2–5]. Among the various polymeric materials, continuously increasing attention is being paid by researchers in developing high-strength, lightweight polymer composites and organic+inorganic-hybrid composites possessing superior thermal and electrical properties [6–11].

The term "hybrid" is used in numerous scientific fields to express something of mixed origin or composition. According to Asby [12], a hybrid material is a combination of two or more materials in a predetermined geometry and scale to serve the specific engineering purpose. In the polymer field, hybrid composites are fabricated by adding more than one reinforcing fillers to a single matrix [13] and incorporating a particular type of reinforcing material into a mixture of different matrices [14], as illustrated in Figure 6.1a and b, respectively. The main strategy behind fabricating hybrid materials is to introduce new properties, which are completely different from individual constituents. Such new properties in hybrid materials are accomplished through interactions of orbitals of the constituents at the molecular or nanometer level. In fact, a hybrid material must demonstrate superior functions or properties compared to traditional composites. In this regard, a polymer nanocomposite can be considered a hybrid material, if discrete nanoparticle phases are uniformly distributed throughout the continuous phase. Moreover, substantial chemical interaction between the individual components is essential to promote composite into a hybrid material. Indeed, there is no clear demarcation to identify hybrid materials from nanocomposites, as both nanocomposite and hybrid material possess properties which are not existent in parent components.

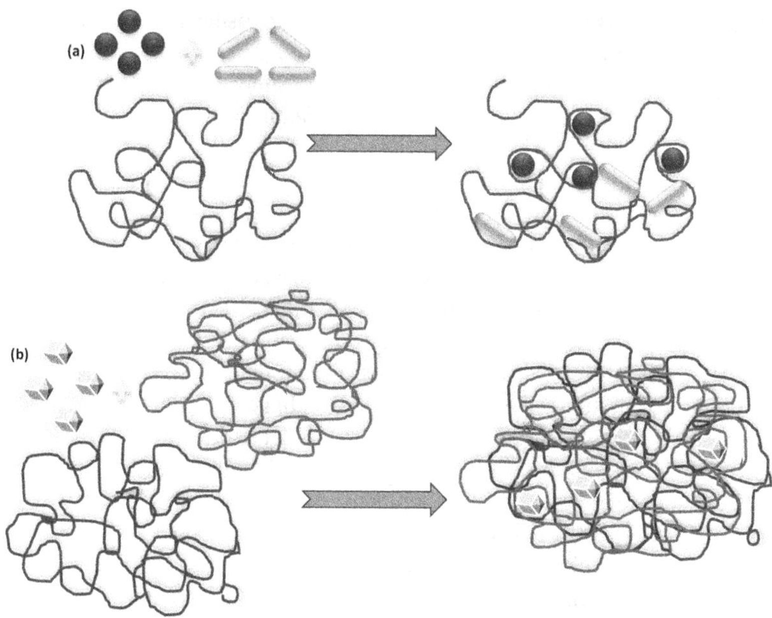

FIGURE 6.1 Hybrid particulate composites (a) containing two different fillers in a single matrix and (b) having single filler in a mixture of matrices.

The design of lightweight materials is attracting interests from different consumers and from aerospace, wind energy, and automotive industries. Notably, the fiber-reinforced composites are attracting elevated interest in weight-sensitive applications because of low density, excellent stiffness, and strength [15]. In this regard, the fiber-reinforced polymer composites include a group of materials, in which a polymer matrix is reinforced by fibers of nano-to-micro meter diameters. In fact, reinforcement is particularly essential to achieve the additional stiffness and strength from the reinforcing fibers. The extent of reinforcement in fibrous composites depends on the orientation of fibers compared to the principal loading direction. In this context, the design matrix of fibrous composites varies largely with the fiber orientation, stacking sequence, and preform types. The mostly studied design matrices are: (a) unidirectional composites possessing significantly stiffer and stronger reinforcement in the fiber direction compared to the similar composites comprising fibers in multiple directions, (b) placement of 0° plies outside the 0°-/90°-laminate to increase the longitudinal flexural stiffness keeping the tensile stiffness unchanged, and (c) sandwiching textiles/random mats between the unidirectional plies to increase the impact. Such elevation of mechanical strength can be monitored from the improvement of the various mechanical properties, such as TS, TM, FS, FM, IS, BS, BM, YM, EM, EAB, and SS of the composites compared to component(s). Continuous investigations are going on in search of lightweight fiber-reinforced composites possessing excellent stiffness, toughness, and temperature resistance. In this regard, numerous attempts are in process to develop suitable hybrid composite materials, as metal fibers have higher density, and polymeric fibers suffer from stiffness and low temperature resistance. Additionally, hybrid composites feature very high corrosion resistivity and excellent fatigue life. Therefore, hybrid composite materials containing at least two different types of fibers can show enough potential to manufacture ductile, lightweight composite materials featuring toughness, stiffness, and temperature resistance. In general, coexistence of two types of fibers combines the beneficial effects of individual fibers in a single composite, along with circumventing the disadvantages of individual fibers.

Notably, natural fiber-reinforced polymer composites are attracting much attention because of uncomplicated eco-/cost-friendly synthesis, robust structural integrity, better mechanical properties, recyclability, and potential applicability. In contrast to the generally used composites, natural fiber-reinforced polymer composites contain natural fibers as the reinforcing materials, which elevate the volume fraction of loading and the possibility of tuning mechanical properties. In general, these biopolymers find extensive applications as surgical lubricant and tablet disintegrants, and in targeted drug delivery, tissue engineering, bandage for wound healing, and as a medium for bacterial culture because of biodegradability, biocompatibility, non-toxicity, and antibacterial activity. Therefore, natural fiber-reinforced composites are a better option for replacing petrochemical-based composites. In this context, most of the bio-renewable polymers can be employed as basic components for fabricating hybrid natural fiber-reinforced composites. Indeed, bio-renewable polymers can be classified as (a) biomass-derived polymers, i.e., polysaccharides, proteins, and lipids, (b) bio-monomer-derived polymers, i.e., polyacrylate and polyethylene (PE), and (c) polymers fabricated from natural or genetically-modified organisms, i.e., bacterial cellulose and chitin. Again, natural fibers can broadly be classified into plant and animal fibers, of which plant fibers are mainly non-wood, i.e., bast, leaf, straw, seed, and glass fibers, and wood, i.e., soft and hard wood fibers. Recently, biocomposites combining properties of at least two different bio-fibers are attracting the utmost attention.

Mostly, thermoset polymers, such as epoxy, vinyl ester, phenolic, and polyimide, are used to fabricate composites. In fact, the journey of thermosetting hybrid polymer composites started after the discovery of Bakelite resin, a thermosetting PFR, by the Belgian-American chemist Leo Baekeland in 1907. Such composites are attracting the utmost attention as the impendent alternative to metal and ceramic-based materials in different industries, such as automotive, marine, aeronautics, sportsgoods, pollution control, electrical, and electronic. In this regard, the high melt viscosity of thermoplastic polymers is the major drawback toward fabrication of fiber-reinforced thermoplastic materials. In fact, high melt viscosity often resists dispersion and movement of fibrous fillers of high aspect ratios. Indeed, thermoplastic composites feature recyclability, toughness, and superior impact resistance over thermosets [16]. Notably, a thermoplastic polymer can be molded at a certain elevated temperature, and once the material becomes cooled, it usually comes back to solid state. In other words, thermoplastics can be reshaped by various molding techniques, such as injection molding, compression molding, calendering, and extrusion. On heating, pre-existing reversible intermolecular forces in thermoplastic materials weaken, and the material converts into a viscoelastic fluid. However, upon removal of external heat, the intermolecular forces can be regained, and the material regains original or near original state. In contrast, thermosetting polymers are composed of irreversible chemical bonds generated mostly at the time of curing. Since thermosets contain highly stable irreversible bonds, these bonds cannot be easily ruptured upon heating. In fact, excessive heating is essential to cleave huge numbers of bonds. Since this kind of bond rupture is mostly irreversible, the material does not reform upon cooling. Importantly, if the temperature of thermoplastic material is higher than the glass transition temperature but lesser than the melting point, physical properties of thermoplastic material alter significantly without phase transition. Moreover, if the temperature of the material is less than the glass transition temperature, the material becomes partly or fully amorphous depending on their inherent crystallizing tendencies.

Importantly, thermoplastic hybrid composites are flexible, elastic, and recyclable, and thus, can be reprocessed unlike thermosetting materials. Therefore, this chapter has been subdivided into two major sections dealing with the fabrication, properties, and applications of thermoplastic and thermosetting type hybrid composites, emphasizing the structure-property-performance relationships of thermoplastic hybrid composites.

6.2 HYBRID COMPOSITES BASED ON THERMOPLASTIC MATRICES

Different thermoplastic materials, such as polyetherimide [15], poly(vinylidene fluoride) [17], poly(ether sulfone) [18], nylons [19], epoxy [20], polyurethane [21, 22], polypropylene [23–25], low-density polyethylene (LDPE) [26], polyester [27, 28], natural rubber [29], styrene-butadiene rubber [30], and ethylene propylene diene monomer (EPDM) [31], have been employed as matrices for the fabrication of preferably lightweight hybrid polymer composites. However, all the polymers have several advantages, along with some inherent limitations. For instance, polyetherimide, a high-performance thermoplastic, exhibits high modulus, strength, and superior high temperature stability. This lightweight polymer has wide applications in aerospace, electronics, and other fields under extreme conditions. However, extremely high viscosity of polyetherimide severely impedes the uniform dispersion of nanoparticles, making the fabrication of nanocomposites more difficult and challenging [15]. In contrast, poly(vinylidene fluoride), a semi-crystalline thermoplastic engineering polymer, possesses high dielectric permittivity, chemical resistance, and excellent thermal stability [17]. Moreover, the elevated thermal conductivity of poly(vinylidene fluoride)-composite facilitates the fabrication of conductive parts in microelectronics materials. In contrast, poly(vinylidene fluoride) membranes are susceptible to bio-fouling [18]. Additionally, being a flexible hydrophobic aliphatic polymer, polypropylene is not compatible with hydrophilic natural fillers. Therefore, ex situ additions of coupling and compatibilizing agents are required to overcome the difficulty in dispersion of hydrophilic fillers in polypropylene. Though polyester is less hydrophobic compared to polypropylene, pretreatment of either filler or matrix is essential for encouraging the compatibility between continuous and discrete phases [27, 28].

6.2.1 Preparation

The thermoplastic hybrid polymer composites are fabricated by different methods. However, prior to the preparation of composite(s), functionalization of fillers is an important step to ensuring sufficient polymer-filler interfacial interactions.

6.2.1.1 Functionalization and Preparation of Fillers

Both graphitic nanoparticles and carbon nanotubes (CNTs) usually undergo chemical treatment to introduce functional groups to the microstructure for elevating performances of composites prepared from these modified fillers. Moreover, such functionalization elevates the pre-existing van der Waals forces and strong π–π interactions within graphene platelets and CNTs preventing the individual aggregation of fillers within polymer matrix [32]. A group of workers treated CNTs and expanded graphites with strong nitric acid for 4 h at 140°C to introduce oxygen-containing functional groups onto surfaces [17]. Thereafter, the oxidized fillers are filtered and washed with adequate distilled water until neutral condition is achieved, followed by drying in an oven. Moreover, the acid treatment follows to prepare functionalized CNTs [32]. Herein, the pristine CNTs are treated with 3:1 (v/v) sulfuric acid/nitric acid at room temperature for 8 h. Thereafter, the mixture is diluted and washed by deionized water, followed by drying at 60°C for 96 h in vacuum. Sometimes, modified graphene is prepared by initial preparation of GO via modified Hummers method [18], followed by the dispersion of GO in water by sonication. Afterward, GO is converted into hydrazine-reduced graphene sheets via reaction between hydrazine hydrate and oxidized graphene. Finally, functionalized graphene sheets are prepared by the reaction of residual epoxide and carboxyl groups on the hydrazine reduced with hydroquinone [32].

For preparing the nanofibers and mats, such as carbon and glass nanofibers, sometimes an electrospinning process is implemented, by which electrospun carbon nanofibers are prepared via electrospinning of a spin dope containing 10 wt. % polyacrylonitrile in dimethyl formamide solvent, followed by stabilization in air at 280°C and carbonization in argon at 1200°C. Similarly, electrospun glass nanofiber mats are fabricated by electrospinning of a spin dope mixture comprising 13.3 and 9.0 wt. % of tetraethyl orthosilicate and polyvinyl pyrrolidone, respectively, in the solvent mixture of dimethyl formamide and dimethyl sulfoxide, followed by pyrolysis in air at 800°C [16].

In contrast, for the preparation of the natural fiber-based fillers, size reduction is carried out by initial grinding of dry natural fibers, followed by pulverization in a pulverizing machine and sieving to separate micro fillers [21, 22]. Notably, various pretreatments, such as cyanoethylation [33], acetylation [33], alkali treatment [27], and UV irradiation [28], as illustrated in Figure 6.2a, b, c, and d, respectively, are often pursued for improving the reinforcing abilities of natural fillers, including bamboo [27], jute [28], and sisal [33], in producing PE-based hybrid composites. In this regard, cyanoethylation is basically incorporation of cyanoethyl groups in sisal or any natural fiber by means of initial dewaxing of sisal fibers, followed by refluxing with acrylonitrile, acetone, and pyridine-catalyst at 60°C for 2 h. Thereafter, the treated fibers are washed with acetic acid and

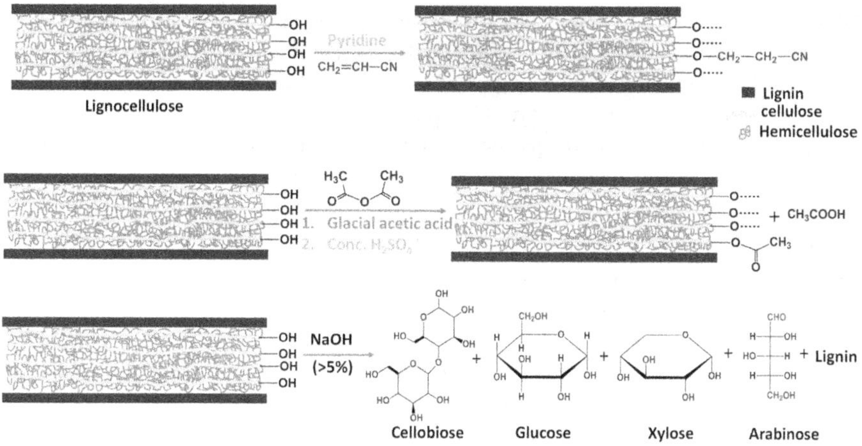

FIGURE 6.2 Various modifications of lignocellulosic fibers: (a) cyanoethylation, (b) acetylation, and (c) excess alkali pretreatment.

acetone, followed by washing with deionized water and vacuum drying. In contrast, acetylation of natural fiber is achieved via treatment of alkali-treated fibers with glacial acetic acid, acetic anhydride, and concentrated H_2SO_4, followed by the usual washing and drying.

6.2.1.2 Solution Mixing and Casting/Molding

Solution mixing generally involves initial addition of polymer matrix and particulate phases to an appropriate solvent, followed by mixing by ultrasonication, to obtain homogeneous dispersion for casting on suitable support, as shown in Figure 6.3. For instance, while preparing hybrid polyethylene imide and graphitic nanoplatelet functionalized MWCNT-based hybrid composites, a certain amount of polyethylene imide is dissolved in dichloromethane, followed by the addition of graphitic nanoplatelets and functionalized MWCNTs. Thereafter, the mixture undergoes 2 h bath-type sonication, followed by 1 h horn-type ultrasonication, to achieve homogeneous suspension through breakdown of fillers. Finally, the homogeneous suspension is

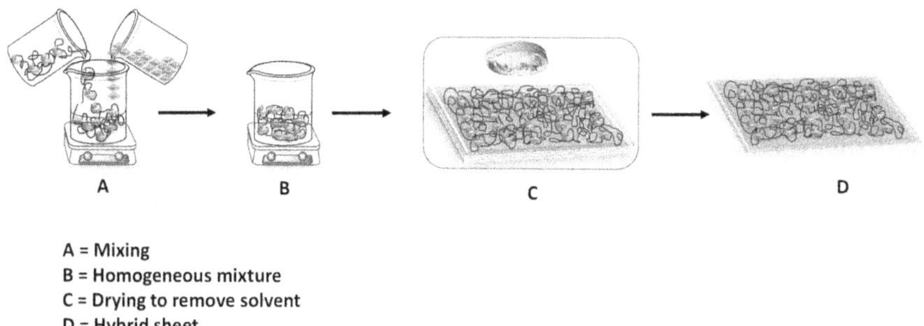

FIGURE 6.3 Solution casting and moulding.

casted on glass slide, and solvent is subsequently evaporated [15]. Moreover, the mixing by sonication method is adopted for preparing poly(vinylidene fluoride)-based hybrid composites containing expanded graphite and oxidized CNTs/graphitic nanoparticles, followed by compression molding [17].

Similarly, functionalized graphene and CNT and poly(ether sulfone)-based composites are fabricated by solution casting method [32]. Herein, *N*-methyl-2-pyrrolidone is used as solvents to prepare ultrasonication-driven uniform dispersion of graphene and CNT. Finally, the dispersion is sprayed on the glass plate, and the coated glass plate is dried as usual at the desired temperature under vacuum in drying chamber, as shown in Figure 6.3. Sometimes, ball milling technique is followed for size reduction of dispersed phases. Thereafter, the as-obtained particles of desired particle size are dispersed in the colloidal dispersion of polymer, followed by drying and film formation on suitable support [34]. Layer-by-layer assembly and casting method have been followed by various researchers to produce laminated hybrid composites comprising multiple layers arranged in an alternating manner [34, 35]. For instance, Zhao et al. have fabricated hybrid multi-layered composite matrices comprising seven layers of electrospun nanofiber mats interleaved by eight layers of conventional carbon fiber fabrics [16]. Indeed, multi-step assembly process is actuated at 200°C in a vacuum oven. Initially, the sandwiched electrospun nanofiber mats, carbon fiber fabrics, and cyclic butylene terephthalate pellets are placed in a vacuum oven, followed by the addition of slow-acting tin-based catalyst. In the next step, the in situ polymerization of cyclic butylene terephthalate is executed under pressure. Finally, all the layers become assembled under pressure and vacuum to produce multilayer-reinforced hybrid composites [35].

6.2.1.3 Overmolding Injection

This integrated manufacturing process is employed to design hybrid polymer composites, as shown in Figure 6.4. For this, composite based on short glass fiber-reinforced nylon 66 is injected directly on the backside of thermoformed organic sheets comprising nylon6 matrix reinforced by unidirectional continuous carbon fibers. At the time of injection molding, 3800 kN clamping force and 135 bar injection pressure are maintained. Thus, this method is slightly different from the conventional injection molding process carried out for preparing hybrid thermoplastic composites containing natural and glass fibers. Notably, prior to injection molding, the ingredients are usually admixed at the desired temperature in a twin screw extruder containing two counter-rotating intermeshed rollers.

6.2.1.4 Mechanical Mixing and Compression Molding

This traditional method of composite preparation is followed in fabricating hybrid thermoplastic polyurethane-based hybrid composites containing glass and palm sugar fibers in an internal mixer [21, 22], followed by molding the pulp at a pre-designated temperature under compressive force.

6.2.2 Composites Containing Natural Fibers

Nowadays, natural fiber-filled composites are becoming increasingly popular. For instance, the engine and transmission covers of Mercedes-Benz transit buses now

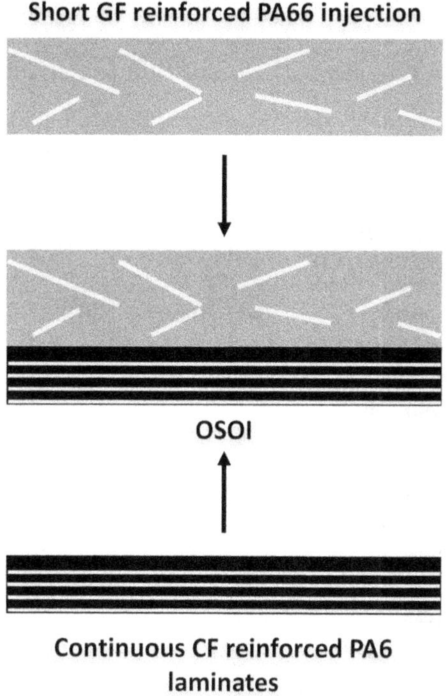

FIGURE 6.4 Fibrous hybrid composites fabricated by overmolding injection process.

contain natural fiber-reinforced PE-resin [33]. In order to fabricate cheap hybrid fibrous composites, thermoplastic polyurethane is chosen as the matrix to which glass fiber and biodegradable sugar palm (*Arenga pinnata* or *Arenga saccharifera*) fibers are incorporated as reinforcing fillers [21, 22]. These hybrid bio-composites are prepared by melt-mixing and compounding, followed by pressing at high temperatures. Notable enhancements in thermal resistance, dynamic mechanical properties, impact resistance, flexural resistance, and tensile properties are created in all the hybrid composites containing variable amounts of glass fiber and sugar palm [21, 22]. In this regard, the lowest damping factor is recorded in the case of hybrid composites carrying 30 wt. % glass fiber compared to 10 wt. % palm sugar fiber. Additionally, a relatively higher extent of glass fiber deteriorates the stretchiness of the thermoplastic polyurethane (PU) matrix, whereas sugar palm fiber increases the elongation at break. Such findings can be correlated with higher rigidity of the glass fiber compared to the added natural fibers. In this regard, several hybrid composites comprising glass and natural fibers, such as bamboo [23, 24, 36, 37], kenaf [38], banana [39], hemp [25], jute [28], kapok [40], sisal [33], and date palm wood flour [41], have been reported, and almost all of these materials have advantages in terms of thermal and mechanical properties, as tabulated in Table 6.1. In this regard, hemp is one of the important, readily available, lignocellulosic bast fibers. Moreover, if two natural fibers are admixed within a thermoplastic matrix, such as polypropylene [23, 24, 36, 39, 41], polyester [27, 33, 40], and LDPE [26], the resultant hybrid composites feature

TABLE 6.1
Comparative Tensile Properties of Various Hybrid Composites Containing Natural Fillers

Composites	Matrices	Fillers (wt. %)	TS (MPa)	TM (MPa)	Ref.
Polypropylene	Isotactic polypropylene	—	32.03	585.96	24
Polypropylene-bamboo fiber	Isotactic polypropylene	Bamboo fiber (10)	36.32	710.41	24
Polypropylene-bamboo fiber	Isotactic polypropylene	Bamboo fiber (20)	38.21	812.57	24
Polypropylene-bamboo fiber	Isotactic polypropylene	Bamboo fiber (30)	43.96	1240.2	24
Polypropylene-bamboo fiber	Isotactic polypropylene	Bamboo fiber (40)	40.25	1290.82	24
Polypropylene-glass fiber-bamboo fiber	Isotactic polypropylene	Bamboo fiber (25) and glass fiber (5)	45.47	1311.06	24
Polypropylene-glass fiber-bamboo fiber	Isotactic polypropylene	Bamboo fiber (20) and glass fiber (10)	47.22	1380.80	24
Polypropylene-glass fiber-bamboo fiber	Isotactic polypropylene	Bamboo fiber (15) and glass fiber (15)	51.00	1426.14	24
Polypropylene-glass fiber-bamboo fiber	Isotactic polypropylene	Bamboo fiber (15), glass fiber (15), and maleic anhydride-grafted polypropylene (1)	54.50	1491.49	24
Polypropylene-glass fiber-bamboo fiber	Isotactic polypropylene	Bamboo fiber (15), glass fiber (15), and maleic anhydride-grafted polypropylene (3)	58.25	1520.50	24
Polypropylene-banana fiber	Isotactic polypropylene	Banana fiber (10)	37.50 ± 1.10	640.00 ± 3.20	39
Polypropylene-banana fiber	Isotactic polypropylene	Banana fiber (20)	41.00 ± 0.93	815.00 ± 4.02	39
Polypropylene-banana fiber	Isotactic polypropylene	Banana fiber (30)	45.25 ± 0.86	985 ± 4.10	39
Polypropylene-banana fiber	Isotactic polypropylene	Banana fiber (40)	39.00 ± 1.20	1045.00 ± 3.22	39
Polypropylene-glass fiber- banana fiber	Isotactic polypropylene	Banana fiber (25) and glass fiber (5)	44.50 ± 0.82	1070.00 ± 2.23	39
Polypropylene-glass fiber- banana fiber	Isotactic polypropylene	Banana fiber (20) and glass fiber (10)	49.00 ± 0.74	1155.00 ± 3.53	39
Polypropylene-glass fiber- banana fiber	Isotactic polypropylene	Banana fiber (15) and glass fiber (15)	57.00 ± 0.91	1350.00 ± 5.56	39

(*Continued*)

TABLE 6.1 (CONTINUED)
Comparative Tensile Properties of Various Hybrid Composites Containing Natural Fillers

Composites	Matrices	Fillers (wt. %)	TS (MPa)	TM (MPa)	Ref.
Polypropylene-glass fiber- banana fiber	Isotactic polypropylene	Banana fiber (15), glass fiber (15) and maleic anhydride-grafted polypropylene (1)	59.00 ± 1.06	1440.00 ± 2.10	39
Polypropylene-glass fiber- banana fiber	Isotactic polypropylene	Banana fiber (15), glass fiber (15), and maleic anhydride-grafted polypropylene (2)	64.00 ± 1.20	1692.00 ± 1.23	39
Polypropylene-hemp fiber	Polypropylene	Hemp fiber (40) and maleic anhydride-grafted polypropylene (5)	52.50 ± 0.60	3770.00	25
Polypropylene-glass fiber- hemp fiber	Polypropylene	Glass fiber (5), hemp fiber (40), and maleic anhydride-grafted polypropylene (5)	53.70 ± 1.60	4070.00 ± 50.00	25
Polypropylene-glass fiber- hemp fiber	Polypropylene	Glass fiber (10), hemp fiber (30), and maleic anhydride-grafted polypropylene (5)	57.90 ± 0.70	4250.00 ± 40.00	25
Polypropylene-glass fiber- hemp fiber	Polypropylene	Glass fiber (15), hemp fiber (25), and maleic anhydride-grafted polypropylene (5)	59.50 ± 0.90	4400.00 ± 10.00	25
Polypropylene-wood flour	Polypropylene	Wood flour (40)	41.54	2671.00	42
Polypropylene-kenaf fiber	Polypropylene	Kenaf fiber (40)	46.03	3229.00	42
Polypropylene-kenaf fiber-wood flour	Polypropylene	Wood flour (10) and kenaf fiber (30)	40.26	2771.00	42
Polypropylene-kenaf fiber-wood flour	Polypropylene	Wood flour (20) and kenaf fiber (20)	40.51	2891.00	42
Polypropylene-kenaf fiber-wood flour	Polypropylene	Wood flour (10) and kenaf fiber (35)	43.33	3008.00	42
Polyester-glass fiber-bamboo fiber	Polyester	Glass fiber (10) and bamboo fiber (30)	—	7912.00	27

(Continued)

TABLE 6.1 (CONTINUED)
Comparative Tensile Properties of Various Hybrid Composites Containing Natural Fillers

Composites	Matrices	Fillers (wt. %)	TS (MPa)	TM (MPa)	Ref.
Polyester-glass fiber-bamboo fiber	Polyester	Glass fiber (10) and alkali treated bamboo fiber (30)	—	11478.00	27
Polyester-kapok fiber	Polyester	Untreated kapok fiber (9 vol%)	67.34	975.40	40
Polyester-kapok fiber	Polyester	NaOH-kapok fiber (9 vol%)	79.10	1425.70	40
Polyester-glass fiber-kapok fiber	Polyester	Glass fiber (6.75 vol%) and untreated kapok fiber (2.25 vol%)	102.55	1228.70	40
Polyester-glass fiber-kapok fiber	Polyester	Glass fiber (6.75 vol%) and alkali-treated kapok fiber (2.25 vol%)	107.60	2362.60	40
Polyester-glass fiber-kenaf fiber	Polyester	Glass fiber (15 vol%) and alkali-treated kenaf fiber (15 vol%)	39.28	23000.00	38
Polyester-banana fiber-sisal fiber	Polyester	Banana fiber (20 vol%)	—	1010.00	43
Polyester-banana fiber-sisal fiber	Polyester	Sisal fiber (20 vol%)	—	1069.00	43
Polyester-banana fiber-sisal fiber	Polyester	Banana fiber (15 vol%) and sisal fiber (5 vol%)	—	1090.00	43
Polyester-banana fiber-sisal fiber	Polyester	Banana fiber (30 vol%)	—	1312.00	43
Polyester-banana fiber-sisal fiber	Polyester	Sisal fiber (30 vol%)	—	1185.00	43
Polyester-banana fiber-sisal fiber	Polyester	Banana fiber (22.5 vol%) and sisal fiber (7.5 vol%)	—	1469.00	43
Polyester-banana fiber-sisal fiber	Polyester	Banana fiber (40 vol%)	—	1352.00	43
Polyester banana fiber-sisal fiber	Polyester	Sisal fiber (40 vol%)	—	1079.00	43
Polyester-banana fiber-sisal fiber	Polyester	Banana fiber (30 vol%) and sisal fiber (10 vol%)	—	1536.00	43

(*Continued*)

TABLE 6.1 (CONTINUED)
Comparative Tensile Properties of Various Hybrid Composites Containing Natural Fillers

Composites	Matrices	Fillers (wt. %)	TS (MPa)	TM (MPa)	Ref.
Polyester-banana fiber-sisal fiber	Polyester	Banana fiber (50 vol%)	—	1412.00	43
Polyester-banana fiber-sisal fiber	Polyester	Sisal fiber (50 vol%)	—	1110.00	43
Polyester-banana fiber-sisal fiber	Polyester	Banana fiber (37.5 vol%) and sisal fiber (12.5 vol%)	—	1647.00	43

interesting property enhancement, as presented in Table 6.1. The property development is associated with better dispersion of individual fillers in thermoplastic matrix in the presence of another. In this regard, several coupling agents, such as maleic anhydride-grafted polypropylene [23–25, 36, 39], have been incorporated to optimize compatibility among the discrete hydrophilic natural fillers and hydrophobic continuous phases, resulting in better tensile properties in the polypropylene-based hybrid composites. Herein, coupling effect is attributed mostly to the hydrogen bonding between hydroxyl groups of natural fibers, Si–O functionality of glass, and carbonyl groups of the maleic anhydride-grafted polypropylene [24, 36], as illustrated in Figure 6.5. Accordingly, for preparing the date palm wood flour and glass fiber-filled hybrid polypropylene composites, coupling or compatibilizing agents are not applied [41].

Arbelaiz and co-workers have modified both matrix and fillers by maleic anhydride, vinyl trimethoxy silane, maleic anhydride-polypropylene copolymer, and alkalization to examine the changes in interfacial bonding of modified matrices with chemically treated fillers [44]. The matrix modification led to better mechanical performance compared to fiber surface modification. In fact, silane or maleic anhydride grafted polypropylene matrix features better mechanical properties compared to composites based on polypropylene matrix modified by maleic anhydride-polypropylene. Notably, maleic anhydride-polypropylene modified composites show higher mechanical properties, as maleic anhydride-polypropylene is capable of binding with both flax and glass fibers, resulting in enriched interfacial adhesion via hydrogen bonding among matrix and both types of fibers, as shown in Figure 6.5. In case of glass fiber/natural fiber based hybrid composites, the tensile properties of the hybrid composites are enhanced with the increased glass fiber content [27], as glass fibers are significantly stiffer and stronger compared to natural fibers [25]. Similarly, the overall mechanical properties of hybrid composites are usually contributed to by the mechanical characteristics, i.e., tensile modulus, strength, and stiffness, of the individual fiber and fiber contents. For instance, in the case of polypropylene and kenaf fiber-wood flour hybrid composite, higher tensile properties are obtained for composites comprising higher extents of kenaf fiber, attributed to greater inherent

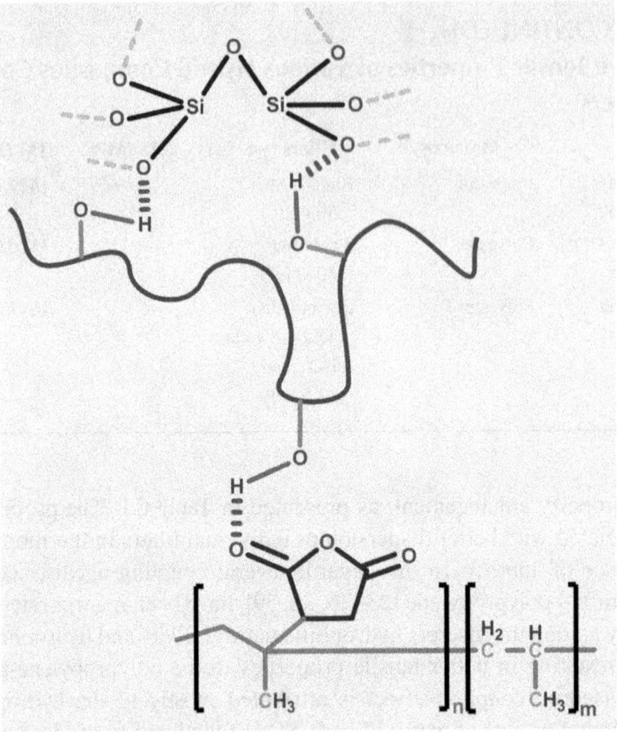

FIGURE 6.5 Hydrogen-bonding interaction among Si–O of glass, O–H of natural fiber, and >C=O of maleic anhydride grafted polypropylene.

mechanical stiffness of kenaf fiber over wood flour. In contrast, although banana and sisal fibers possess similar mechanical strengths, synergistic mechanical property enhancement is noted in all the hybrid composites containing higher percentage of banana fiber, as shown in Table 6.1. Sometimes, the reinforcing capabilities of natural fibers are improved by pretreatments, such as alkali pretreatment [27, 33, 40], mercerization [38], cyanoethylation [33], acetylation [33], and UV irradiation [28], resulting in the better tensile properties of PE-based composites, as presented in Table 6.1. Herein, alkali pretreatment is done to remove unwanted greasy materials, discouraging the bonding between natural fibers and matrices. More importantly, by means of such pretreatment of lignocellulose materials, delignification up to a desired level can be carried out, so that cellulosic portion can be exposed for interaction with polymer matrix, as illustrated in Figure 6.6. In this regard, the impact of various pretreatment techniques, such as cyanoethylation, acetylation, and alkali pretreatment, as shown in Figure 6.2, are compared in fabricating sisal/glass fiber based hybrid composites. Notably, the optimum tensile property is registered if sisal fibers are chemically modified with 5% alkali [33]. However, excessive alkali pretreatment is harmful against the reinforcing effect. In fact, if the concentration of alkali is very high (>5%), an excess delignification of natural fiber makes the fiber weaker, as shown in Figure 6.2c.

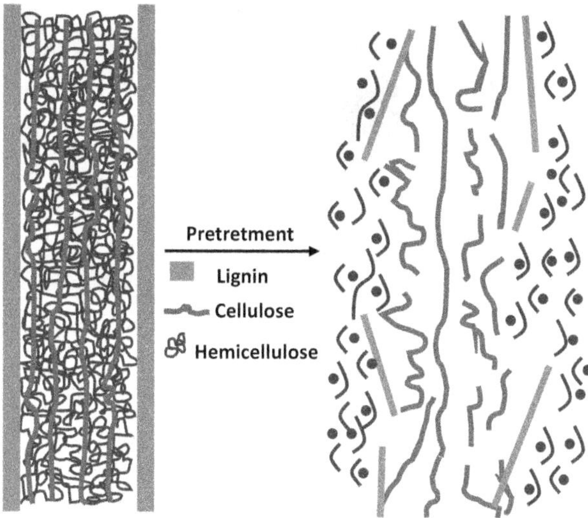

FIGURE 6.6 Pretreatment of lignocellulose materials resulted breakdown of lignin and exposure of cellulose for interacting with the polymer matrix of hybrid composite.

6.2.3 Composites Containing Synthetic Fibers

The demand for lighter weight fiber-filled polymer composites is increasing day by day in the automotive industry [19]. However, because of the excellent specific mechanical properties, carbon fiber-reinforced plastics are becoming increasingly popular in aeronautic and aerospace industries [34]. Recently, short glass fiber and carbon fiber filled hybrid nylon-based composites have been prepared by overmolding injection process. Herein, composite is composed of two types of nylon matrix layers, i.e., nylon 6 and nylon 66, filled with carbon fiber and short glass fiber, respectively [19]. In this regard, physical interactions are the primary driving force behind attachment of two composite layers. The bonding mechanisms between the two nylon-based composite layers are governed by micro mechanical interlocks, such as microwave and hook-like structures, diffusion reaction, and non-bond interactions. Herein, the diffusion reaction mostly involves cicatrization process, by which one nylon matrix interpenetrates into another, followed by a certain degree of co-crystallization. Moreover, such interactions simulated by molecular dynamics method show that the interfacial bonding between two layers is dominated by van der Waals forces and electrostatic interaction, whereas chemical forces contribute slightly to the overall interfacial bonding strength. In impact-resistant multi-layered composites, toughening layers are incorporated in between two carbon fiber/bismaleimide unidirectional prepregs, as shown in Figure 6.7. In this regard, the toughening layer is the nanocomposite of MWCNT dispersed in thermoplastic polyetherketone cardo, i.e., phenolphthalein containing polyetherketone. For preparing the toughening layer, MWCNTs undergo extensive size reduction by means of ball milling [34], resulting in better dispersion of particulate fillers. Though such size reduction increases impact resistance, the drop in the aspect ratio of nanotubes

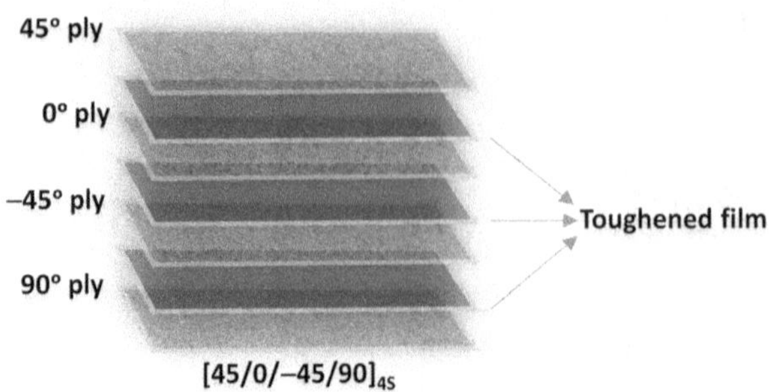

FIGURE 6.7 Impact resistant multi-layered hybrid composites comprising of toughening film layers in between carbon fiber/ bismaleimide unidirectional prepregs.

is not suitable for electrically conductive films. However, impact resistance of the multi-layered assembly is improved substantially, as realized from the appreciable reduction in damage area after the execution of impact test. Moreover, as a result of incorporation of toughening film, relatively ductile fracture behavior and receded crack propagation are created in the assembly, suggesting elevated impact resistance in the laminated nanocomposites. Since high melt viscosity often resists the dispersion of particulate phases within thermoplastic matrices, recent measures have been taken to reduce the viscosity of the matrix adding cyclic butylene terephthalate oligomer [35]. In this work, the hybrid multi-layered composite matrices are prepared by seven layers of electrospun nanofiber mats interleaved by eight layers of conventional fabrics, as illustrated in Figure 6.8. These mats are prepared via addition of either electrospun carbon or electrospun glass nanofibers, whereas conventional fabric layers are composed of carbon fibers. The manifold increase

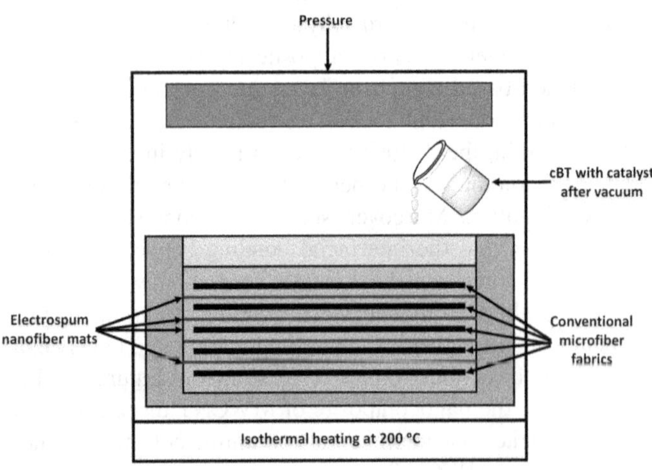

FIGURE 6.8 Multi-laminated hybrid composites comprising of nanofiber containing mats and conventional carbon microfiber layers.

in overall mechanical properties, including flexural and tensile properties, is noted in the assembly as a result of better dispersion of either nanofibers in the laminated composites, as presented in Table 6.1. Both the carbon fiber-filled composites show better tensile properties compared to glass fiber-filled reinforced composites. Thus, addition of either electrospun carbon or electrospun glass nanofibers in polybutylene terephthalate matrix results in a larger interfacial bonding area, leading to enhanced effective interfacial adhesion and energy consumption for debonding at the polybutylene terephthalate-nanofiber interface.

Hybrid fibrous composites have tremendous potential to be used as structural components of building, bridges, and other civil infrastructures. Recently, lightweight but reinforced polymer composites have effectively been utilized as substitutes for heavy building materials to construct lightweight strong structures or foundations. In fact, traditionally used building materials are composed of heavy metallic components, mostly manufactured from steel, and these materials are susceptible to corrosion and fatigue. In this regard, epoxy matrix-based hybrid thermoplastic composite rods comprising carbon and glass fibers have been developed to serve as composite tendons suitable as construction materials. Notably, epoxy compounds are frequently used as thermosetting resin, and thermoplastic epoxy compounds are pretty much rare. In fact, these hybrid rods contain a carbon fiber bundle core surrounded by a glass fiber tubular membrane sheath, and the thermoplastic epoxy is evenly impregnated as a matrix. Herein, the thermoplastic epoxy matrix of the hybrid rods is fabricated by compounding difunctional epoxy resin with difunctional phenolic compound in 100:6.5 ratio. Eventually, tensile properties of all the composite rods become at par with the conventionally used metallic components, as presented in Table 6.2. Moreover, both tensile modulus and strength increase with the volume fraction of carbon fiber, suggesting carbon-fiber impregnated enhancement of tensile properties in the longitudinal direction of the fiber reinforcements.

6.2.4 Composites Containing Two Different Nanocarbons

An extraordinary synergy is obtained if two different nanocarbon-based reinforcing fillers are simultaneously incorporated in polyvinyl acetate (PVA)-based thermoplastic matrix [48]. In this regard, various PVA-based hybrid composites containing different binary combinations of nanocarbon fillers, such as nanodiamond, single-walled nanotube, and few-layer graphene, have been fabricated. Almost all of these hybrid composites demonstrate synergistic enhancement in tensile modulus, as shown in Table 6.2. In recent times, researchers have been strongly inclined to develop advanced hybrid composites comprising potential fillers, such as graphene and CNTs, as these fillers possess superior intrinsic electrical and mechanical properties. In this regard, appreciably higher mechanical properties of graphene are realized from the huge Young's modulus (YM) and intrinsic strength of single graphene sheet at 1 TPa and 130 GPa, respectively [49]. Moreover, graphene has higher electrical conductivity, as its current density exceeds 108 A cm^{-2} [50]. In addition, thermal conductivity of graphene is nearly 5000 W m^{-1} K^{-1}, which is pretty much similar to that of MWCNT. However, the shapes of graphene and MWCNT are completely different. The shape of graphene is similar to a plate, i.e., two-dimensional (2D),

TABLE 6.2
Comparative Thermal, Electrical, and Mechanical Properties of Various Hybrid Composites Containing Synthetic and Nanocarbon Fillers

Composites	Matrices	Fillers (wt.%)	Resistivity (ohm sq^{-1})/ percolation threshold (%)	Thermal (Wm^{-1} K^{-1})/electrical conductivity (S m^{-1})	TS (MPa)	TM (MPa)	Ref.
Poly (vinylidene fluoride)	Poly (vinylidene fluoride)	—	–/–	2.19/–	—	—	17
Poly (vinylidene fluoride)-expanded graphite	Poly (vinylidene fluoride)	Expanded graphite (3.1)	–/–	2.52/–	—	—	17
Poly (vinylidene fluoride)-CNT[1)]	Poly (vinylidene fluoride)	CNT[1)] (6.1)	–/–	2.66/–	—	—	17
Poly (vinylidene fluoride)-expanded graphite-CNT[1)]	Poly (vinylidene fluoride)	Expanded graphite and CNT[1)] (5.2)	–/–	3.20/–	—	—	17
Polyethylene imide	Polyethylene imide	—	1.68×10^{16}/–	–/–	—	—	15
Polyethylene imide-graphene	Polyethylene imide	Graphene (0.5.0)	7.63×10^{14}/–	–/–	—	—	15
Polyethylene imide-graphene	Polyethylene imide	Graphene (5.0)	2.25×10^{5}/–	–/–	—	—	15
Polyimide-MWCNT[2)]	Polyimide	Functionalized MWCNT[2)] (7.0)	6.60×10^{10} ohm cm/–	–/–	—	—	45
Polyethylene imide- MWCNT[2)]	Polyethylene imide	Functionalized MWCNT[2)]	–/1–2	–/–	—	—	46

(Continued)

TABLE 6.2 (CONTINUED)
Comparative Thermal, Electrical, and Mechanical Properties of Various Hybrid Composites Containing Synthetic and Nanocarbon Fillers

Composites	Matrices	Fillers (wt.%)	Resistivity (ΩΩ sq^{-1})/ percolation threshold (%)	Thermal (Wm^{-1} K^{-1})/electrical conductivity (S m^{-1})	TS (MPa)	TM (MPa)	Ref.
Polyethylene imide- MWCNT[2]	Polyethylene imide	MWCNT[2]	–/1.5	–/–	—	—	47
Polyethylene imide- MWCNT[2]-graphene	Polyethylene imide	COOH functionalized MWCNT[2] (0.25) and graphene (0.25)	5.056 × 10^{6}/–	–/–	—	—	15
Poly(ether sulfone)	Poly(ether sulfone)	—	–/–	–/–	70.20 ± 0.90	1824.00 ± 36.20	32
Poly(ether sulfone)-CNT[1]	Poly(ether sulfone)	CNT (5.0)	–/–	–/–	74.60 ± 1.04	2381.12 ± 50.10	32
Poly(ether sulfone)-fCNT[1]	Poly(ether sulfone)	Functionalized CNT (5.0)	–/–	–/1.43 ± 0.028 × 10^{-4}	92.70 ± 1.60	2394.51 ± 48.54	32
Poly(ether sulfone)-fGraphene	Poly(ether sulfone)	Functionalized graphene (5.0)	–/–	–/5.82 ± 0.11 × 10^{-4}	39.00 ± 0.86	3098.10 ± 60.01	32
Poly(ether sulfone)-fCNT[1]-fGraphene	Poly(ether sulfone)	Functionalized CNT (2.5) and functionalized graphene (2.5)	–/0.22	–/1.27 ± 0.021 × 10^{-3}	78.45 ± 1.14	3607.98 ± 66.21	32
Poly(ether sulfone)-CNT[1]-fGraphene	Poly(ether sulfone)	CNT (2.5) and functionalized Graphene (2.5)	–/–	–/2.55 ± 0.050 × 10^{-3}	54.50 ± 0.96	2692.00 ± 54.52	32

(Continued)

TABLE 6.2 (CONTINUED)
Comparative Thermal, Electrical, and Mechanical Properties of Various Hybrid Composites Containing Synthetic and Nanocarbon Fillers

Composites	Matrices	Fillers (wt.%)	Resistivity (ΩΩ sq^{-1})/ percolation threshold (%)	Thermal (Wm^{-1} K^{-1})/electrical conductivity (S m^{-1})	TS (MPa)	TM (MPa)	Ref.
Polybutylene terephthalate-carbon nanofiber-conventional carbon fiber	Polybutylene terephthalate	Carbon nanofiber (1.0)	–/–	–/–	694.50 ± 41.70	69200 ± 2800	35
Polybutylene terephthalate- glass nanofiber- conventional carbon fiber	Polybutylene terephthalate	Glass nanofiber (1.0)	–/–	–/–	215.00 ± 4.20	18300 ± 2400	35
Polybutylene terephthalate-conventional carbon fiber	Polybutylene terephthalate	—	–/–	–/–	606.40 ± 24.70	67500 ± 3900	35
Polybutylene terephthalate-conventional glass fiber	Polybutylene terephthalate	—	–/–	–/–	203.20 ± 12.9	16900 ± 700	35
Epoxy-carbon fiber-glass fiber	Epoxy	—	–/–	–/–	1420–1840	65000–91000	20
Polyvinyl alcohol	Polyvinyl alcohol	—	–/–	–/–	—	660 ± 30	48
Polyvinyl alcohol-few-layer graphene	Polyvinyl alcohol	Few-layer graphene (0.6)	–/–	–/–	—	890 ± 80	48
Polyvinyl alcohol- nanodiamond	Polyvinyl alcohol	Nanodiamond (0.6)	–/–	–/–	—	1330 ± 350	48
Polyvinyl alcohol- SWCNT[3)]	Polyvinyl alcohol	SWCNT[c] (0.6)	–/–	–/–	—	7800 ± 340	48

(Continued)

TABLE 6.2 (CONTINUED)
Comparative Thermal, Electrical, and Mechanical Properties of Various Hybrid Composites Containing Synthetic and Nanocarbon Fillers

Composites	Matrices	Fillers (wt.%)	Resistivity (ΩΩ sq^{-1})/ percolation threshold (%)	Thermal (Wm^{-1} K^{-1})/electrical conductivity (S m^{-1})	TS (MPa)	TM (MPa)	Ref.
Polyvinyl alcohol-SWCNT[3)]- few-layer graphene	Polyvinyl alcohol	Single-walled nanotube (0.4) and few-layer graphene (0.2)	–/–	–/–	—	9300 ± 430	48
Polyvinyl alcohol-SWCNT[3)]- few-layer graphene	Polyvinyl alcohol	single-walled nanotube (0.2) and few-layer graphene (0.4)	–/–	–/–	—	8600 ± 340	48
Polyvinyl alcohol-nanodiamond - few-layer graphene	Polyvinyl alcohol	Nanodiamond (0.4) and few-layer graphene (0.2)	–/–	–/–	—	1300 ± 70	48
Polyvinyl alcohol-nanodiamond - few-layer graphene	Polyvinyl alcohol	Nanodiamond (0.2) and few-layer graphene (0.4)	–/–	–/–	—	1600 ± 100	48
Polyvinyl alcohol-nanodiamond –SWCNT[3)]	Polyvinyl alcohol	Nanodiamond (0.4) and SWCNT[3)] (0.2)	–/–	–/–	—	9300 ± 360	48
Polyvinyl alcohol-nanodiamond –SWCNT[3)]	Polyvinyl alcohol	Nanodiamond (0.2) and SWCNT[3)] (0.4)	–/–	–/–	—	7500 ± 50	48

[1)] Carbon nanotube, [2)] multi-walled carbon nanotube, and [3)] single-walled carbon nanotube

whereas MWCNTs are rod-shaped and unidimensional. Most importantly, graphene has significantly higher specific surface area, i.e., 2600 $m^2 g^{-1}$, compared to MWCNT, which has a specific surface area, i.e., ~100 $m^2 g^{-1}$ [51]. All of these advantages of both graphene and CNTs can be fruitfully utilized in fabricating high-performance polymer composites, if both of these fillers are homogeneously dispersed to circumvent the agglomeration in a continuous polymer matrix, increasing largely the specific surface area offered by individual graphene platelet or MWCNT-rod. In order to separate individual particulate matters within a composite or nanocomposite, exfoliation of the graphene layers in primary graphene nanoplatelets or expanded graphite agglomerate is an effective and established technique, leading to elevated mechanical, thermal, and electrical properties of the composites. Previously, attempts had been made to fabricate thermoplastic exfoliated nanocomposites by simple mechanical mixing, latex compounding, and in situ reduction process [52, 53]. In this regard, exfoliation of graphene is really difficult because of the strong intrinsic van der Waals forces of attraction, i.e., >2 eV nm^{-2}, between adjacent sheets. In this regard, hybrid polyimide-based composites containing 0.5% graphitic nanoplatelets and functionalized MWCNTs demonstrate greater electrical conductivity, thermal conductivity, and dynamic mechanical properties compared to the same proportion of either individual graphitic nanoplatelet or MWCNT-filled composites. Herein, a protective action of mechanically robust graphitic nanoplatelets prevents unwanted fragmentation of the MWCNTs during high power sonication [15]. Accordingly, the high aspect ratio of MWCNT is little affected, as breakdown of MWCNTs is effectively shielded by mechanically strong graphitic nanoplatelets. Thus, MWCNTs of high aspect ratio and graphitic nanoplatelets can jointly produce an interconnected hybrid network structure, which facilitates the electron transport throughout the composite matrix. In this context, MWCNTs have the ability to produce conducting channels between the adjacent graphitic nanoplatelets, offering the facilitation of the electrical transport [15]. Such easier electrical transport is attributed to the percolation network structure being constituted of multiple triangular-shaped junctions among MWCNTs and nanoplatelets, as shown in Figure 6.9. The elevation of thermal conductivity can be explained by the increasingly available conducting pathways and bridges for transport of heat energy within the hybrid matrix containing unbroken MWCNTs of high aspect ratio interconnecting the graphitic nanoplatelets through frequent junction points. Similarly, the improvement in dynamic mechanical property is attributed to increasingly uniform dispersion of both the particulate phases, along with the unaltered aspect ratio of CNTs. Another objective for preparing graphene-CNT composites is to reduce the overall cost, as graphene is appreciably cheaper compared to CNTs. The hybrid nanocomposites comprising expanded graphite and CNTs result in a similar synergistic improvement in thermal conductivity compared to the nanocomposites filled with individual filler [17]. Herein, the expansion of graphite is mediated via interpenetration of polymer chains in between the adjacent graphite layers, followed by exfoliation of intercalated graphite compounds. Such enhancement of thermal conductivity is contributed to via formations of more conductive pathways and associated minimization of the interfacial phonon scattering between matrix and fillers. Additionally, the overall improved dispersion of fillers in the matrix, along with the in situ formed CNTs bridges connecting expanded graphite platelets, facilitate the flow of heat. In fact, homogeneous

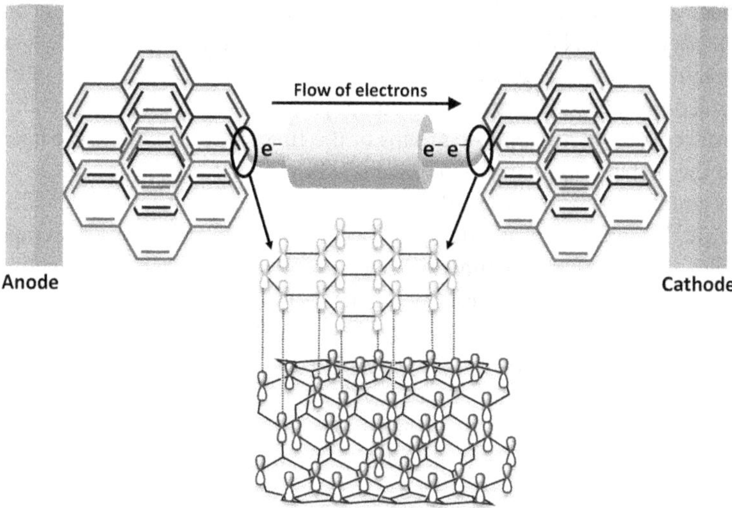

FIGURE 6.9 π–π interaction between expanded graphite platelets and MWCNT.

dispersion of expanded graphite is actuated through the breakdown of expanded graphite agglomerates via preventing the π–π interaction between expanded graphite platelets, as shown in Figure 6.9. In fact, the interfering tendency of CNTs against formation of graphite agglomerates is exploited to produce more uniform combined dispersion of CNTs and graphitic platelets in solution phase [17]. Similarly, relatively elevated thermal and electrical conductivities are recorded in hybrid composites comprising both functionalized graphene and functionalized CNTs [32]. In this regard, π–π interaction between the functionalized graphene nanosheets and functionalized CNTs is evidenced from the morphological changes in the respective transmission electron microscopy (TEM) photomicrographs, as shown in Figure 6.10 [18]. In fact, the surfaces of the individual functionalized CNTs wrap with functionalized graphene nanosheets via π–π stacking interactions between functionalized graphene nanosheets and functionalized CNTs, as shown in Figure 6.10. Accordingly,

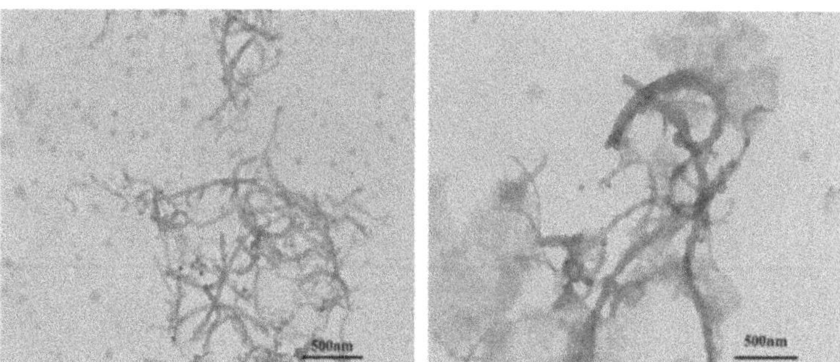

FIGURE 6.10 π–π interaction between the functionalized graphene nanosheets and functionalized CNTs resulted roughening of the functionalized CNT-surface. (Reproduced from Zhang et al. 2013 [18] with permission.)

the surface of functionalized CNTs becomes rough because of the attachment of graphene sheets on CNTs, as illustrated in Figure 6.10. It appears that the π–π stacking interactions among graphene nanosheets is intervened and superseded by stacking interactions between functionalized graphene nanosheets and functionalized CNTs. Moreover, the hydrophilic groups of the functionalized graphene nanosheets increase the dispersion of functionalized CNTs via facilitated interactions with the polar solvent, resulting in lesser extents of aggregations. As observed from Table 6.2, the electrical conductivity of the hybrid composites is much higher compared to composites filled with either functionalized graphene nanosheets or functionalized CNTs because of the higher density of the percolating network via bridging of CNTs connecting two individual graphene nanosheets. Accordingly, the alterations of contact geometry of mixed fillers in polymer matrix of hybrid composites lead to low volume percolation concentration, i.e., 0.22%, in hybrid composites filled with both functionalized graphene nanosheets and functionalized CNTs. Since the nature of dispersion has a huge impact on mechanical properties of composites, both tensile modulus and strength are increased considerably in hybrid composites compared to composites filled with one type of filler. In this regard, the functionalized graphene nanosheets are more efficient compared to either pristine or modified CNTs in improving the tensile moduli of hybrid composites because of the huge YM and intrinsic strength of single graphene sheet at 1 TPa and 130 GPa, respectively [49]. Moreover, the integration of one-dimensional, oxidized CNTs with 2D-GO results in a strong synergism between individual particulate matters in producing superior ultrafiltration hybrid membrane of higher antifouling capability [18]. Herein, the interaction between one-dimensional, oxidized CNTs with 2D-GO was originated from π–π attractions between the surface of MWCNT and the π-conjugated aromatic domains of GO. In this regard, the adhesive forces operating between hybrid composite membrane and foulants are analyzed via atomic force microscopy. The fouling potential of the hybrid membrane depends on the magnitude of interfacial forces between the model foulants, such as bovine serum albumin and the membrane, as illustrated in Figure 6.11. In fact, superior antifouling property is ascribed to the

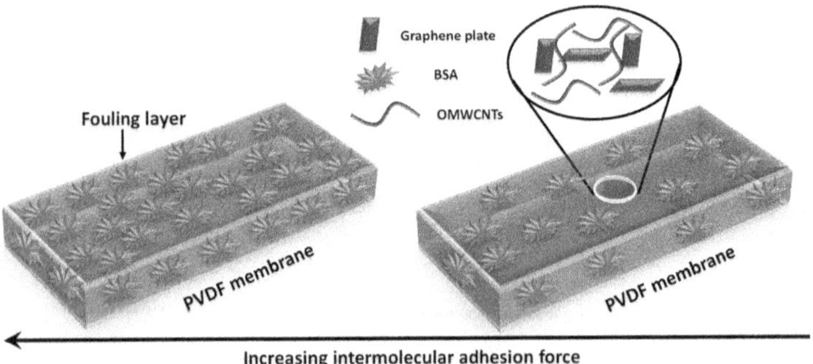

FIGURE 6.11 Lower adhesion force between hybrid composite membrane and foulant cake layer resulted greater water flux through the less compact cake layer.

decrease in adhesive forces between membrane and foulants. Lower adhesive force between foulants and membrane produces loose cake layer facilitating the water permeation, along with poor binding between cake layer and the hybrid composite membrane.

6.3 HYBRID COMPOSITES BASED ON THERMOSETTING MATRICES

Various polymer matrices including different resins, such as epoxy [54–66], VAE [67], Rooflite [68], and PFRs [69], along with polyester [70–75], LDPE [76, 77], and polypropylene [78–80], as shown in Figure 6.12, have been selected to prepare multi-component hybrid composites containing various fibrous and non-fibrous discrete fillers. In this context, with the variation of either discrete or continuous phases or both, interesting property developments have been noticed in almost all composites, as presented in Table 6.3. Herein, the fabrication of composites has been carried out mostly by hand layup, compression molding, casting method, and resin transfer molding techniques.

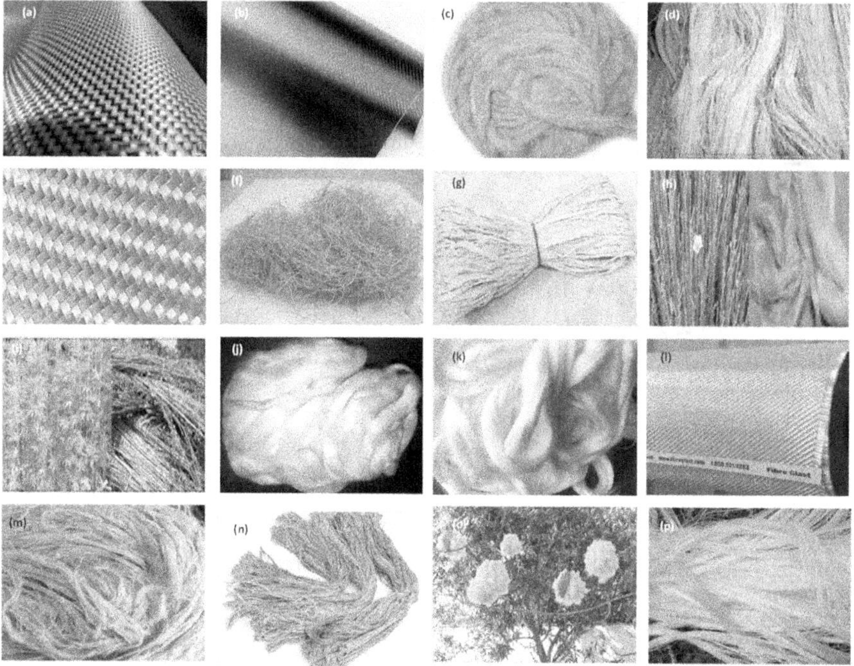

FIGURE 6.12 Different fiber used for synthesizing thermosetting polymers: (a) kevlar, (b) 3D carbon, (c) bamboo, (d) banana, (e) basalt, (f) OPEFB, (g) raffia, (h) flax, (i) kenaf, (j) ramie, (k) silk, (l) E-glass, (m) hemp, (n) jute, (o) kapok, and (p) sisal.

TABLE 6.3
Synthesis and Important Properties of the Thermosetting Hybrid Polymers

Synthesis techniques	Component 1	Component 2	Matrix	Configuration/Composition	Important properties	Ref
Hand layup	Sisal	Glass	Polyester	G:S = 16:40	FS[1)] = 280.70 MPa, FM[2)] = 12200 MPa, and IS[3)] = 135 kJ m^{-2}	70
	Sisal	Glass	Polyester	G:S = 12:40 (G-facings = 4 vol.%)	FS = 295.00 MPa, FM = 12930 MPa, and IS = 208 kJ m^{-2}	70
	Sisal	Glass	Polyester	G:S = 8:40 (G-facings = 8 vol.%)	FS = 294.60 MPa, FM = 16850 MPa, and IS = 154 kJ m^{-2}	70
	Sisal	Glass	Polyester	G:S = 16:40 (asymmetric with G-tensile side)	FS = 299.20 MPa, FM = 22790 MPa, and IS = 238 kJ m^{-2}	70
	Sisal	Glass	Polyester	G:S = 15:46	TS[4)] = 65.20 MPa, FS = 89.20 MPa, and EAB[5)] = 11.30%	71
	Sisal	Glass	Polyester	G:S = 8:34 + dry red mud (20 wt.%)	TS = 45.20 MPa, TM[6)] = 5950 MPa, FS = 98100 MPa, and EAB = 8.10%	71
	Sisal	Glass	Polyester	G:S = 8:34 + aqueous red mud (20 wt.%)	TS = 38.20 MPa	71
	Sisal	Glass	Polyester	G:S = 25:75 (fiber content = 5 vol.%)	IS = 2.36 J m^{-2}	72
	NaOH-sisal	Glass	Polyester	G:S = 50:50 (fiber content = 5 vol.%)	IS = 3.00 J m^{-2}	72
	Sisal	Glass	Polyester	G:S = 50:50 (fiber content = 5 vol.%)	IS = 2.90 J m^{-2}	72
	Silane-sisal	Glass	Polyester	G:S = 50:50 (fiber content = 5 vol.%)	IS = 2.90 J m^{-2}	72
	Sisal	Glass	Polyester	G:S = 75:25 (fiber content = 5 vol.%)	IS = 3.86 J m^{-2}	72
	Sisal	Glass	Polyester	G:S = 25:75 (fiber content = 8 vol.%)	IS = 4.00 J m^{-2}	72
	NaOH-sisal	Glass	Polyester	G:S = 50:50 (fiber content = 8 vol.%)	IS = 5.46 J m^{-2}	72
	Sisal	Glass	Polyester	G:S = 50:50 (fiber content = 8 vol.%)	IS = 5.20 J m^{-2}	72

(*Continued*)

TABLE 6.3 (CONTINUED)
Synthesis and Important Properties of the Thermosetting Hybrid Polymers

Synthesis techniques	Component 1	Component 2	Matrix	Configuration/Composition	Important properties	Ref
	Silane-sisal	Glass	Polyester	G:S = 50:50 (fiber content = 8 vol.%)	IS = 5.30 J m^{-2}	72
	Sisal	Glass	Polyester	G:S = 75:25 (fiber content = 8 vol.%)	IS = 5.76 J m^{-2}	72
	Sisal	Glass	Polyester	Composite of fiber length = 30 mm	TS = 176.20 MPa and IS = 18.00 J m^{-2}	73
	Flax	—	Epoxy resin	Flax thread diameter = 0.2 mm	IPE[7)] = 16 J, SFE[8)] = 15.9 J, LS[9)] = 113.4 MPa, and maximum load = 1514 N	54
	Flax	—	Epoxy resin	Flax thread diameter = 0.9 mm	IPE = 17.8 J, SFE = 16.6 J, LS = 162.8 MPa, and maximum load = 1954 N	54
	Flax	—	Epoxy resin	Flax thread diameter = 2.3 mm	IPE = 16 J, SFE = 15.0 J, LS = 115.5 MPa, and maximum load = 1583 N	54
	—	E-glass	Epoxy resin	—	IPE = 78 J and SFE = 69.1 J	54
	Sisal	Glass	Polyester	Dry 8G laminate (fiber content = 20 wt.%)	TS = 80.28 MPa, TM = 1186 MPa, FS = 123.54 MPa, FM = 4270 MPa, and IS = 37.80 kJ m^{-2}	74
	Sisal	Glass	Polyester	Dry 2G/4S/2G laminate (fiber content = 20 wt.%)	TS = 83.46 MPa, TM = 1235 MPa, FS = 136.13 MPa, FM = 5230 MPa, and IS = 38.14 kJ m^{-2}	74
	Sisal	Glass	Polyester	Dry G/2S/2G/2S/G laminate (fiber content = 20 wt.%)	TS = 88.51 MPa, TM = 1428 MPa, FS = 149.59 MPa, FM = 5490 MPa, and IS = 41.89 kJ m^{-2}	74

(Continued)

TABLE 6.3 (CONTINUED)
Synthesis and Important Properties of the Thermosetting Hybrid Polymers

Synthesis techniques	Component 1	Component 2	Matrix	Configuration/Composition	Important properties	Ref
	Sisal	Glass	Polyester	Wet 8G laminate (fiber content = 20 wt.%)	TS = 76.37 MPa, TM = 1157 MPa, FS = 108.70 MPa, FM = 4110 MPa, and IS = 32.98 kJ m^{-2}	74
	Sisal	Glass	Polyester	Wet 2G/4S/2G laminate (fiber content = 20 wt.%)	TS = 80.70 MPa, TM = 1205 MPa, FS = 124.36 MPa, FM = 4990 MPa, and IS = 34.46 kJ m^{-2}	74
	Sisal	Glass	Polyester	Wet G/2S/2G/2S/G laminate (fiber content = 20 wt.%)	TS = 86.41 MPa, TM = 1397 MPa, FS = 138.70 MPa, FM = 5510 MPa, and IS = 38.74 kJ m^{-2}	74
	—	—	Epoxy resin	Glass fiber = 0 wt.%	TS = 18.6 MPa, EAB = 3.89%, TM = 525 MPa, FS = 45.86 MPa, and FM = 640 MPa	55
	Silk fiber	—	Epoxy resin	Glass fiber = 0 wt.%	TS = 58.35 MPa, EAB = 7.49%, TM = 844 MPa, FS = 60.81 MPa, and FM = 1503 MPa	55
	Silk fiber	Glass	Epoxy resin	Glass fiber = 5 wt.%	TS = 60.99 MPa, EAB = 10.74%, TM = 891 MPa, FS = 94.31 MPa, and FM = 1847 MPa	55
	Silk fiber	Glass	Epoxy resin	Glass fiber = 10 wt.%	TS = 64.87 MPa, EAB = 11.66%, TM = 922 MPa, FS = 97.31 MPa, and FM = 3015 MPa	55

(*Continued*)

TABLE 6.3 (CONTINUED)
Synthesis and Important Properties of the Thermosetting Hybrid Polymers

Synthesis techniques	Component 1	Component 2	Matrix	Configuration/Composition	Important properties	Ref
	Silk fiber	Glass	Epoxy resin	Glass fiber = 15 wt.%	TS = 70.12 MPa, EAB = 12.06%, TM = 944 MPa, FS = 106.5 MPa, and FM = 4221 MPa	55
	Silk fiber	Glass	Epoxy resin	Glass fiber = 20 wt.%	TS = 77.81 MPa, EAB = 13.26%, TM = 992 MPa, FS = 108.2 MPa, and FM = 5251 MPa	55
	Silk fiber	Glass	Epoxy resin	Glass fiber = 25 wt.%	TS = 84.04 MPa, EAB = 14.33%, TM = 1008 MPa, FS = 114.5 MPa, and FM = 5440 MPa	55
	Silk fiber	Glass	Epoxy resin	Volumetric density = 1510 kg m^{-3}	TS = 61.15 MPa, TM = 1300 MPa, FS = 96.32 MPa, FM = 5320 MPa, and EAB = 6.59%	81
	Ramie fiber	—	Epoxy resin	20 wt.% reinforcement	TS = 42 MPa, FS = 50 MPa, and IS = 79 J m^{-2}	56
	—	Glass fiber	Epoxy resin	20 wt.% reinforcement	TS = 81 MPa, FS = 62 MPa, and IS = 141 J m^{-2}	56
	Ramie fiber	Glass fiber	Epoxy resin	20 wt.% reinforcement	TS = 85 MPa, FS = 69 MPa, and IS = 143 J m^{-2}	56
	Ramie fiber	—	Epoxy resin	30 wt.% reinforcement	TS = 55 MPa, FS = 65 MPa, and IS = 96 J m^{-2}	56
	—	Glass fiber	Epoxy resin	30 wt.% reinforcement	TS = 88 MPa, FS = 69 MPa, and IS = 180 J m^{-2}	56

(Continued)

TABLE 6.3 (CONTINUED)
Synthesis and Important Properties of the Thermosetting Hybrid Polymers

Synthesis techniques	Component 1	Component 2	Matrix	Configuration/Composition	Important properties	Ref
	Ramie fiber	Glass fiber	Epoxy resin	30 wt.% reinforcement	TS = 90 MPa, FS = 71 MPa, and IS = 190 J m^{-2}	56
	Sisal	Glass	Epoxy resin	G:S = 2:2	TS = 37.40 MPa, FS = 87.34 MPa, and IS = 14.69 kJ m^{-2}	82
	Sisal	Glass	Epoxy resin	G:S = 2:4	TS = 48.00 MPa, FS = 159.00 MPa, and IS = 44.92 kJ m^{-2}	82
	Sisal	Glass	Epoxy resin	G:S = 2:6	TS = 25.20 MPa, FS = 19.76 MPa, and IS = 36.54 kJ m^{-2}	82
	Woven glass fiber	Kevlar fiber	Epoxy	100 wt.% epoxy	TM = 10500 MPa, FS = 156 MPa, FM = 9000 MPa, and YM[10)] = 627 MPa	58
	Woven glass fiber	Kevlar fiber	Epoxy + MR[23)]	91 wt.% epoxy + 9 wt.% MR	TM = 14970 MPa, FS = 136 MPa, FM = 10350 MPa, and YM = 699.2 MPa	58
	Woven glass fiber	Kevlar fiber	Epoxy + MR+NS[24)]	80 wt.% epoxy + 8 wt.% MR + 11 wt.% NS	TM = 9880 MPa, FS = 128 MPa, FM = 9700 MPa, and YM = 615.9 MPa	58
	Sisal	Glass	Epoxy resin	G:S = 19.62:12.26	TS = 108.22 MPa, TM = 532.20 MPa, FS = 205.64 MPa, FM = 12720 MPa, and IS = 0.65 kJ m^{-2}	83
	Sisal	Glass	Epoxy resin	G:S = 26.15:7.26	TS = 168.82 MPa, TM = 721.60 MPa, FS = 241.22 MPa, FM = 14960 MPa, and IS = 0.98 kJ m^{-2}	83

(Continued)

TABLE 6.3 (CONTINUED)
Synthesis and Important Properties of the Thermosetting Hybrid Polymers

Synthesis techniques	Component 1	Component 2	Matrix	Configuration/Composition	Important properties	Ref
	Sisal	Glass	Polyester	G:S = 60:40, fiber orientation = 0 degree	TS = 168.40 MPa, FS = 228.76 MPa, EAB = 10.8%, and IS = 16.00 J m^{-2}	75
	Curaua fiber	Glass fiber	Polyester	Alternative stacking laminate	TS = 92.2 MPa, YM = 2340 MPa, and PE = 3.7%	84
	Sisal	Glass	Epoxy resin	G:S = 90:10 (longitudinal)	TS = 168.82 MPa, FS = 269.32 MPa, and IS = 17.20 J m^{-2}	66
	Sisal	Glass	Epoxy resin	G:S = 80:20 (longitudinal)	TS = 171.25 MPa, FS = 212.32 MPa, and IS = 18.40 J m^{-2}	66
	Sisal	Glass	Epoxy resin	G:S = 70:30 (longitudinal)	TS = 158.69 MPa, FS = 234.12 MPa, and IS = 15.10 J m^{-2}	66
	Sisal	Glass	Epoxy resin	G:S = 60:40 (longitudinal)	TS = 142.32 MPa, FS = 208.45 MPa, and IS = 13.20 J m^{-2}	66
	Sisal	Glass	Epoxy resin	G:S = 90:10 (transverse)	TS = 173.96 MPa, FS = 258.58 MPa, and IS = 16.90 J m^{-2}	66
	Sisal	Glass	Epoxy resin	G:S = 80:20 (transverse)	TS = 185.25 MPa, FS = 232.65 MPa, and IS = 15.70 J m^{-2}	66
	Sisal	Glass	Epoxy resin	G:S = 70:30 (transverse)	TS = 135.23 MPa, FS = 217.38 MPa, and IS = 12.40 J m^{-2}	66
	Sisal	Glass	Epoxy resin	G:S = 60:40 (transverse)	TS = 138.28 MPa, FS = 205.15 MPa, and IS = 11.70 J m^{-2}	66
	Sisal	Glass	Polyester	G:S = 25:75 (fiber content = 5 vol.%)	FS = 46.77 MPa and FM = 2450 MPa	85

(Continued)

TABLE 6.3 (CONTINUED)
Synthesis and Important Properties of the Thermosetting Hybrid Polymers

Synthesis techniques	Component 1	Component 2	Matrix	Configuration/Composition	Important properties	Ref
	NaOH-sisal	Glass	Polyester	G:S = 50:50 (fiber content = 5 vol.%)	FS = 54.82 MPa and FM = 2630 MPa	85
	Sisal	Glass	Polyester	G:S = 50:50 (fiber content = 5 vol.%)	FS = 48.44 MPa and FM = 2400 MPa	85
	Silane-sisal	Glass	Polyester	G:S = 50:50 (fiber content = 5 vol.%)	FS = 47.07 MPa and FM = 2350 MPa	85
	Sisal	Glass	Polyester	G:S = 75:25 (fiber content = 5 vol.%)	FS = 68.84 MPa and FM = 2830 MPa	85
	Sisal	Glass	Polyester	G:S = 25:75 (fiber content = 8 vol.%)	FS = 58.64 MPa and FM = 3260 MPa	85
	NaOH-sisal	Glass	Polyester	G:S = 50:50 (fiber content = 8 vol.%)	FS = 66.19 MPa and FM = 3570 MPa	85
	Sisal	Glass	Polyester	G:S = 50:50 (fiber content = 8 vol.%)	FS = 52.86 MPa and FM = 3200 MPa	85
	Silane-sisal	Glass	Polyester	G:S = 50:50 (fiber content = 8 vol.%)	FS = 48.54 MPa and FM = 3140 MPa	85
	Sisal	Glass	Polyester	G:S = 75:25 (fiber content = 8 vol.%)	FS = 76.78 MPa and FM = 3780 MPa	85
	Jute	E-glass fiber	Polyester		TS = 125%, TM = 49%, BS = 162%, and BM[11]): 235%	28
	Sisal	Glass	Polyester	Red mud = 2 wt.%, particle size = 4 μm	TS = 142.69 MPa, FS = 334.53 MPa, and IS = 11.10 J m^{-2}	86

(Continued)

TABLE 6.3 (CONTINUED)
Synthesis and Important Properties of the Thermosetting Hybrid Polymers

Synthesis techniques	Component 1	Component 2	Matrix	Configuration/Composition	Important properties	Ref
	Sisal	Glass	Polyester	Red mud = 4 wt.%, particle size = 4 μm	TS = 140.78 MPa, FS = 358.50 MPa, and IS = 17.50 J m^{-2}	86
	Sisal	Glass	Polyester	Red mud = 6 wt.%, particle size = 4 μm	TS = 153.39 MPa, FS = 360.67 MPa, and IS = 20.60 J m^{-2}	86
	Sisal	Glass	Polyester	Red mud = 8 wt.%, particle size = 4 μm	TS = 162.51 MPa, FS = 417.17 MPa, and IS = 22.60 J m^{-2}	86
	Sisal	Glass	Polyester	Red mud = 2 wt.%, particle size = 13 μm	TS = 126.43 MPa, FS = 360.67 MPa, and IS = 13.80 J m^{-2}	86
	Sisal	Glass	Polyester	Red mud = 4 wt.%, particle size = 13 μm	TS = 131.66 MPa, FS = 468.59 MPa, and IS = 18.90 J m^{-2}	86
	Sisal	Glass	Polyester	Red mud = 6 wt.%, particle size = 13 μm	TS = 138.42 MPa, FS = 487.04 MPa, and IS = 21.50 J m^{-2}	86
	Sisal	Glass	Polyester	Red mud = 8 wt.%, particle size = 13 μm	TS = 153.42 MPa, FS = 501.77 MPa, and IS = 22.70 J m^{-2}	86
	Sisal	Glass	Polyester	G:S = 25:75 (fiber content = 5 vol%)	TS = 22.20 MPa and EAB = 6.00%	87
	NaOH-sisal	Glass	Polyester	G:S = 50:50 (fiber content = 5 vol%)	TS = 23.94 MPa and EAB = 6.50%	87
	Sisal	Glass	Polyester	G:S = 50:50 (fiber content = 5 vol%)	TS = 18.97 MPa	87
	Silane-sisal	Glass	Polyester	G:S = 50:50 (fiber content = 5 vol%)	TS = 28.91 MPa	87
	Sisal	Glass	Polyester	G:S = 75:25 (fiber content = 5 vol%)	TS = 25.93 MPa and EAB = 7.20%	87
	Sisal	Glass	Polyester	G:S = 25:75 (fiber content = 8 vol%)	TS = 23.90 MPa and EAB = 6.50%	87
	NaOH-sisal	Glass	Polyester	G:S = 50:50 (fiber content = 8 vol%)	TS = 27.00 MPa and EAB = 7.50%	87
	Sisal	Glass	Polyester	G:S = 50:50 (fiber content = 8 vol%)	TS = 20.95 MPa	87
	Silane-sisal	Glass	Polyester	G:S = 50:50 (fiber content = 8 vol%)	TS = 34.00 MPa	87

(*Continued*)

TABLE 6.3 (CONTINUED)
Synthesis and Important Properties of the Thermosetting Hybrid Polymers

Synthesis techniques	Component 1	Component 2	Matrix	Configuration/Composition	Important properties	Ref
	Sisal	Glass	Polyester	G:S = 75:25 (fiber content = 8 vol%)	TS = 30.26 MPa and EAB = 8.20%	87
	Banana fiber	—	Polyester	Composite orientation = plain	TS = 72 MPa	88
	—	Kenaf	Polyester	Composite orientation = plain	TS = 115 MPa	88
	Banana fiber	Kenaf	Polyester	Composite orientation = plain	TS = 140 MPa	88
	NaOH-banana fiber	—	Polyester	Composite orientation = plain	TS = 80 MPa	88
	SLS[$]-banana fiber	—	Polyester	Composite orientation = plain	TS = 85 MPa	88
	—	NaOH-kenaf	Polyester	Composite orientation = plain	TS = 90 MPa	88
	—	SLS-kenaf	Polyester	Composite orientation = plain	TS = 98 MPa	88
	NaOH-banana fiber	NaOH-kenaf	Polyester	Composite orientation = plain	TS = 99 MPa	88
	SLS[14)]-banana fiber	SLS-kenaf	Polyester	Composite orientation = plain	TS = 110 MPa	88
	Banana fiber	—	Polyester	Composite orientation = twill	TS = 61 MPa	88
	—	Kenaf fiber	Polyester	Composite orientation = twill	TS = 98 MPa	88
	Banana fiber	Kenaf fiber	Polyester	Composite orientation = twill	TS = 120 MPa	88
	Banana fiber	—	Polyester	Composite orientation = random	TS = 58 MPa	88
	—	Kenaf fiber	Polyester	Composite orientation = random	TS = 70 MPa	88
	Banana fiber	Kenaf fiber	Polyester	Composite orientation = random	TS = 73 MPa	88
	—	OPEFB[19)]	Epoxy resin	OPEFB: jute fiber = 100:0	TS = 22.6 MPa and TM = 2230 MPa	59
	Jute fiber	OPEFB	Epoxy resin	OPEFB: jute fiber = 4:1	TS = 25.3 MPa and TM = 2620 MPa	59

(Continued)

TABLE 6.3 (CONTINUED)
Synthesis and Important Properties of the Thermosetting Hybrid Polymers

Synthesis techniques	Component 1	Component 2	Matrix	Configuration/Composition	Important properties	Ref
	Jute fiber	OPEFB	Epoxy resin	OPΞFB: jute fiber = 1:1	TS = 28.3 MPa and TM = 2900 MPa	59
	Jute fiber	OPEFB	Epoxy resin	OPΞFB: jute fiber = 1:4	TS = 37.9 MPa and TM = 3310 MPa	59
	Jute fiber	—	Epoxy resin	OPEFB: jute fiber = 0:100	TS = 45.5 MPa and TM = 3890 MPa	59
	Jute fiber	Carbon	Epoxy resin	Resin was derived from bisphenol-C and bisphenol-C-formaldehyde	TS = 10 MPa and FS = 17 MPa	60
	NaOH-jute fiber	Carbon	Epoxy resin		TS = 14.64 MPa and FS = 19.33 MPa	60
	Glass	Carbon	Epoxy resin		TS = 21.4 MPa and FS = 24.53 MPa	60
Solution mixing	Sisal	Glass	LDPE[25)]	G:S = 2:27, longitudinal	TS = 15.97 MPa, TM = 160.6 MPa, and EAB = 7.64%	76
	Sisal	Glass	LDPE	G:S = 4:24, longitudinal	TS = 16.68 MPa, TM = 171.4 MPa, and EAB = 7.56%	76
	Sisal	Glass	LDPE	G:S = 6:21, longitudinal	TS = 16.92 MPa, TM = 187.5 MPa, and EAB = 7.16%	76
	Sisal	Glass	LDPE	G:S = 8:18, longitudinal	TS = 17.60 MPa, TM = 192.3 MPa, and EAB = 6.90%	76
	Sisal	Glass	LDPE	G:S = 10:15, longitudinal	TS = 17.76 MPa, TM = 196.6 MPa, and EAB = 6.60%	76
	Sisal	Glass	LDPE	G:S = 14:9, longitudinal	TS = 19.98 MPa, TM = 200 MPa, and EAB = 5.88%	76
	Sisal	Glass	LDPE	G:S = 16:6, longitudinal	TS = 20.98 MPa, TM = 210 MPa, and EAB = 5.62%	76
	NaOH-sisal	Glass	LDPE	G:S = 10:15, longitudinal	TS = 19.66 MPa, TM = 210 MPa, and EAB = 5.20%	76

(Continued)

TABLE 6.3 (CONTINUED)
Synthesis and Important Properties of the Thermosetting Hybrid Polymers

Synthesis techniques	Component 1	Component 2	Matrix	Configuration/Composition	Important properties	Ref
	Sisal	Glass	LDPE	G:S = 2:27, random	TS = 7.72 MPa, TM = 147.3 MPa, and EAB = 12.52%	76
	Sisal	Glass	LDPE	G:S = 4:24, random	TS = 7.83 MPa, TM = 152 MPa, and EAB = 11.79%	76
	Sisal	Glass	LDPE	G:S = 6:21, random	TS = 7.97 MPa, TM = 160 MPa, and EAB = 10.89%	76
	Sisal	Glass	LDPE	G:S = 8:18, random	TS = 8.11 MPa, TM = 175.23 MPa, and EAB = 10.39%	76
	Sisal	Glass	LDPE	G:S = 10:15, random	TS = 8.26 MPa, TM = 189.6 MPa, and EAB = 10.12%	76
	Sisal	Glass	LDPE	G:S = 14:9, random	TS = 8.70 MPa, TM = 190.3 MPa, and EAB = 9.64%	76
	Sisal	Glass	LDPE	G:S = 16:6, random	TS = 9.00 MPa, TM = 194.6 MPa, and EAB = 9.56%	76
	NaOH-sisal	Glass	LDPE	G:S = 10:15, random	TS = 8.89 MPa, TM = 200 MPa, an EAB = 9.90%	76
	Sisal	Glass	LDPE	G:S = 30:70, fiber length = 6 mm	TS = 27.86 MPa, TM = 800.33 MPa, and EAB = 5.00%	77
	Sisal	Glass	LDPE	G:S = 50:50, fiber length = 6 mm	TS = 29.75 MPa, TM = 1000.13 MPa, and EAB = 5.00%	77
	Sisal	Glass	LDPE	G:S = 70:30, fiber length = 6 mm	TS = 31.23 MPa, TM = 1136.36 MPa, and EAB = 4.00%	77
	NaOH-sisal	Glass	LDPE	G:S = 30:70, fiber length = 6 mm	TS = 31.26 MPa, TM = 831.27 MPa, and EAB = 5.00%	77

(*Continued*)

TABLE 6.3 (CONTINUED)
Synthesis and Important Properties of the Thermosetting Hybrid Polymers

Synthesis techniques	Component 1	Component 2	Matrix	Configuration/Composition	Important properties	Ref
	NaOH-sisal	Glass	LDPE	G:S = 50:50, fiber length = 6 mm	TS = 31.83 MPa, TM = 1081.09 MPa, and EAB = 4.00%	77
	NaOH-sisal	Glass	LDPE	G:S = 70:30, fiber length = 6 mm	TS = 31.45 MPa, TM = 1139.53 MPa, and EAB = 3.00%	77
	Acetylated sisal	Glass	LDPE	G:S = 30:70, fiber length = 6 mm	TS = 32.23 MPa, TM = 919.53 MPa, and EAB = 5.00%	77
	Acetylated sisal	Glass	LDPE	G:S = 50:50, fiber length = 6 mm	TS = 32.86 MPa, TM = 1110.09 MPa, and EAB = 4.00%	77
	Acetylated sisal	Glass	LDPE	G:S = 70:30, fiber length = 6 mm	TS = 31.51 MPa, TM = 1140.33 MPa, and EAB = 3.00%	77
	SA-sisal	Glass	LDPE	G:S = 30:70, fiber length = 6 mm	TS = 32.79 MPa, TM = 1000.78 MPa, and EAB = 5.00%	77
	SA-sisal	Glass	LDPE	G:S = 50:50, fiber length = 6 mm	TS = 33.93 MPa, TM = 1120.78 MPa, and EAB = 4.00%	77
	SA-sisal	Glass	LDPE	G:S = 70:30, fiber length = 6 mm	TS = 31.93 MPa, TM = 1141.64 MPa, and EAB = 4.00%	77
	$KMnO_4$-sisal	Glass	LDPE	G:S = 30:70, fiber length = 6 mm	TS = 32.97 MPa, TM = 1081.82 MPa, and EAB = 5.00%	77
	$KMnO_4$-sisal	Glass	LDPE	G:S = 50:50, fiber length = 6 mm	TS = 34.63 MPa, TM = 1182.24 MPa, and EAB = 4.00%	77
	$KMnO_4$-sisal	Glass	LDPE	G:S = 70:30, fiber length = 6 mm	TS = 32.92 MPa, TM = 112.56 MPa, and EAB = 3.00%	77
	MAPE-sisal	Glass	LDPE	G:S = 30:70, fiber length = 6 mm	TS = 34.97 MPa, TM = 1139.32 MPa, and EAB = 5.00%	77

(Continued)

TABLE 6.3 (CONTINUED)
Synthesis and Important Properties of the Thermosetting Hybrid Polymers

Synthesis techniques	Component 1	Component 2	Matrix	Configuration/Composition	Important properties	Ref
	MAPE-sisal	Glass	LDPE	G:S = 50:50, fiber length = 6 mm	TS = 36.23 MPa, TM = 1228.28 MPa, and EAB = 4.00%	77
	MAPE-sisal	Glass	LDPE	G:S = 70:30, fiber length = 6 mm	TS = 35.35 MPa, TM = 1209.51 MPa, and EAB = 3.00%	77
	Silane-sisal	Glass	LDPE	G:S = 30:70, fiber length = 6 mm	TS = 34.34 MPa, TM = 1185.00 MPa, and EAB = 4.00%	77
	Silane-sisal	Glass	LDPE	G:S = 50:50, fiber length = 6 mm	TS = 37.04 MPa, TM = 1458.35 MPa, and EAB = 4.00%	77
	Silane-sisal	Glass	LDPE	G:S = 70:30, fiber length = 6 mm	TS = 38.98 MPa, TM = 1606.77 MPa, and EAB = 3.00%	77
	DCP-sisal	Glass	LDPE	G:S = 30:70, fiber length = 6 mm	TS = 35.03 MPa, TM = 1303.70 MPa, and EAB = 4.70%	77
	DCP-sisal	Glass	LDPE	G:S = 50:50, fiber length = 6 mm	TS = 38.84 MPa, TM = 1583.28 MPa, and EAB = 3.90%	77
	DCP-sisal	Glass	LDPE	G:S = 70:30, fiber length = 6 mm	TS = 40.73 MPa, TM = 1726.98 MPa, and EAB = 2.20%	77
	BPO-sisal	Glass	LDPE	G:S = 30:70, fiber length = 6 mm	TS = 35.77 MPa, TM = 1401.75 MPa, and EAB = 5.00%	77
	BPO-sisal	Glass	LDPE	G:S = 50:50, fiber length = 6 mm	TS = 39.08 MPa, TM = 1626.00 MPa, and EAB = 4.00%	77
	BPO-sisal	Glass	LDPE	G:S = 70:30, fiber length = 6 mm	TS = 41.92 MPa, TM = 1776.18 MPa, and EAB = 2.00%	77
Compression molding	GFRP[15)]	FFRP[20)]	Phenol	Stacking sequence = GF	TS = 450.1 MPa and TM = 40100 MPa	69

(Continued)

TABLE 6.3 (CONTINUED)
Synthesis and Important Properties of the Thermosetting Hybrid Polymers

Synthesis techniques	Component 1	Component 2	Matrix	Configuration/Composition	Important properties	Ref
	GFRP	FFRP	Phenol	Stacking sequence = GGFF	TS = 412.5 MPa and TM = 40800 MPa	69
	GFRP	FFRP	Phenol	Stacking sequence = GGGGFFFF	TS = 392.5 MPa and TM = 39700 MPa	69
	Curaua fiber	—	Polyester	G:C = 0:100 (fiber content = 20 wt.%)	IS = 31.20 kJ m^{-2}	89
	Curaua fiber	Glass fiber	Polyester	G:C = 70:30 (fiber content = 20 wt.%)	IS = 89.70 kJ m^{-2}	89
	—	Glass fiber	Polyester	G:C = 100:0 (fiber content = 20 wt.%)	IS = 88.50 kJ m^{-2}	89
	Curaua fiber	—	Polyester	G:C = 0:100 (fiber content = 30 wt.%)	IS = 27.90 kJ m^{-2}	89
	Curaua fiber	Glass fiber	Polyester	G:C = 30:70 (fiber content = 30 wt.%)	IS = 76.90 kJ m^{-2}	89
	Curaua fiber	Glass fiber	Polyester	G:C = 50:50 (fiber content = 30 wt.%)	IS = 106.90 kJ m^{-2}	89
	Curaua fiber	Glass fiber	Polyester	G:C = 70:30 (fiber content = 30 wt.%)	IS = 110.00 kJ m^{-2}	89
	—	Glass fiber	Polyester	G:C = 100:0 (fiber content = 30 wt.%)	IS = 123.70 kJ m^{-2}	89
	Curaua fiber	—	Polyester	G:C = 0:100 (fiber content = 40 wt.%)	IS = 32.60 kJ m^{-2}	89
	Curaua fiber	Glass fiber	Polyester	G:C = 70:30 (fiber content = 40 wt.%)	IS = 149.00 kJ m^{-2}	89
	—	Glass fiber	Polyester	G:C = 100:0 (fiber content = 40 wt.%)	IS = 159.30 kJ m^{-2}	89
	Sisal	Glass	Polyester	G:S = 15.8:13.9 (thickness = 3 mm)	TS = 32.3 MPa, FS = 47.0 MPa, FM = 2285 MPa, and IS = 37.5 kJ m^{-2}	90
	SSF[16)]	CNT[21)]	Polypropylene	SSF: CNT = 1.75:1.75	EMI SE[12)] = 57.5 dB, YM = 2.1%, and EM[13)] = 1800 MPa	78
	Sisal fibril	Kenaf fiber	Polyester	K:S = 5:5 (fiber content = 10 wt.%)	TS = 49.14 MPa, TM = 2920 MPa, FS = 96.61 MPa, FM = 4420 MPa, and IS = 9.15 kJ m^{-2}	91

(Continued)

TABLE 6.3 (CONTINUED)
Synthesis and Important Properties of the Thermosetting Hybrid Polymers

Synthesis techniques	Component 1	Component 2	Matrix	Configuration/Composition	Important properties	Ref
	Sisal fibril	Kenaf fiber	Polyester	K:S = 10:10 (fiber content = 20 wt.%)	TS = 61.29 MPa, TM = 3670 MPa, FS = 111.04 MPa, FM = 5570 MPa, and IS = 12.47 kJ m^{-2}	91
	Sisal fibril	Kenaf fiber	Polyester	K:S = 15:15 (fiber content = 30 wt.%)	TS = 77.36 MPa, TM = 4890 MPa, FS = 136.57 MPa, FM = 5890 MPa, and IS = 19.58 kJ m^{-2}	91
	Sisal fibril	Kenaf fiber	Polyester	K:S = 20:20 (fiber content = 40 wt.%)	TS = 91.33 MPa, TM = 6580 MPa, FS = 150.35 MPa, FM = 6310 MPa, and IS = 22.31 kJ m^{-2}	91
	Sisal fibril	—	Polyester	K:S = 0:40 (fiber content = 40 wt.%)	TS = 73.77 MPa, TM = 4540 MPa, FS = 115.94 MPa, FM = 4500 MPa, and IS = 7.53 kJ m^{-2}	91
	—	Kenaf fiber	Polyester	K:S = 40:0 (fiber content = 40 wt.%)	TS = 77.33 MPa, TM = 5500 MPa, FS = 100.24 MPa, FM = 5790 MPa, and IS = 7.54 kJ m^{-2}	91
	Rafia fiber	Glass fiber	—	G:R = 2:1 (Sandwich composite)	FM = 3400 MPa	92
	NaOH-rafia fiber	Glass fiber	—	G:R = 2:1 (Sandwich composite)	FM = 3600 Mpa	92
	Composite of nylon 6 and GNP[17)]	h-BN[22)]	VAE[26)] resin	GNP = 1.97 vol.% and h-BN = 16.85 vol.%	Thermal conductor (2.69 $Wm^{-1} k^{-1}$) and electrical insulator (4.13×10^{-9} S m^{-2})	67
	Jute fiber	—	Epoxy	J:B = 100:0	TS = 16.62 MPa, TM = 664 MPa, FS = 57.22 MPa, FM = 8956 MPa, and IS = 13.44 kJ m^{-2}	61

(Continued)

TABLE 6.3 (CONTINUED)
Synthesis and Important Properties of the Thermosetting Hybrid Polymers

Synthesis techniques	Component 1	Component 2	Matrix	Configuration/Composition	Important properties	Ref
	Jute fiber	Banana fiber	Epoxy	J:B = 75:25	TS = 17.89 MPa, TM = 682 MPa, FS = 58.60 MPa, FM = 9065 MPa, and IS = 15.81 kJ m^{-2}	61
	Jute fiber	Banana fiber	Epoxy	J:B = 50:50	TS = 18.96 MPa, TM = 724 MPa, FS = 59.84 MPa, FM = 9170 MPa, and IS = 18.23 KJ m^{-2}	61
	Jute fiber	Banana fiber	Epoxy	J:B = 25:75	TS = 18.25 MPa, TM = 720 MPa, FS = 59.30 MPa, FM = 9056 MPa, and IS = 17.89 KJ m^{-2}	61
	—	Banana fiber	Epoxy	J:B = 0:100	TS = 17.92 MPa, TM = 718 MPa, FS = 58.06 MPa, FM = 9048 MPa, and IS = 16.92 KJ m^{-2}	61
	Sisal	Glass	Epoxy resin	G:S = 17:22	TS = 52.00 MPa, TM = 1570.00 MPa, FS = 159.00 MPa, and IS = 11.70 J m^{-2}	93
	Sisal	Glass	Epoxy resin	G:S = 23:15	TS = 93.00 MPa, TM = 2210.00 MPa, FS = 184.00 MPa, and IS = 13.30 J m^{-2}	93
	Sisal	—	Epoxy resin	B:S = 1.00:0.00	TS = 18 MPa and YM = 260 MPa	94
	Sisal	Banana	Epoxy resin	B:S = 0.75/0.25	TS = 22 MPa and YM = 300 MPa	94
	Sisal	Banana	Epoxy resin	B:S = 0.50/0.50	TS = 8 MPa and YM = 370 MPa	94
	Sisal	Banana	Epoxy resin	B:S = 0.25:0.75	TS = 22 MPa and YM = 360 MPa	94
	—	Banana	Epoxy resin	B:S = 0.00:1.00	TS = 23 MPa and YM = 350 MPa	94

(Continued)

TABLE 6.3 (CONTINUED)
Synthesis and Important Properties of the Thermosetting Hybrid Polymers

Synthesis techniques	Component 1	Component 2	Matrix	Configuration/Composition	Important properties	Ref
	Woven flax fiber	Glass fiber	MSO[27)]	Stacking sequence = W(0°)/W(0°)/W(0°)/G/W(0°)/W(0°)/W(0°)	TS = 119 MPa, YM = 14000 MPa, FS = 201 MPa, and FM = 24000 MPa	95
	Sisal	Glass	Epoxy resin	Sisal fiber length = 35 cm	FS = 347.00 MPa, EAB = 5.20%, and IS = 8.00 J m^{-2}	96
	Glass fiber	Sugar-palm fiber	Polyurethane	G:SP = 10:30	TS = 21 MPa, TM = 710 MPa, EAB = 3.9 mm, FS = 17 MPa, and FM = 250 MPa	22
	Glass fiber	Sugar-palm fiber	Polyurethane	G:SP = 20:20	TS = 17 MPa, TM = 650 MPa, EAB = 2.8 mm, FS = 17.5 MPa, and FM = 275 MPa	22
	Glass fiber	Sugar-palm fiber	Polyurethane	G:SP = 30:10	TS = 16 MPa, TM = 620 MPa, EAB = 1.6 mm, FS = 25 MPa, and FM = 410 MPa	22
	Glass fiber	—	Polyurethane	G:SP = 40:0	TS = 18 MPa, TM = 700 MPa, EAB = 1.0 mm, FS = 34 MPa, and FM = 450 MPa	22
	Glass fiber	Palmyra fiber	Rooflite resin	Fiber length = 40 mm	TS = 39.30 MPa, TM = 1433.75 MPa, FS = 56.24 MPa, FM = 2600 MPa, IS = 5.63 J cm^{-2}, and SS = 6.96 MPa	68

(Continued)

TABLE 6.3 (CONTINUED)
Synthesis and Important Properties of the Thermosetting Hybrid Polymers

Synthesis techniques	Component 1	Component 2	Matrix	Configuration/Composition	Important properties	Ref
	Banana fiber	Sisal fiber	Polyester	Volume fraction = 0.19	TM = 1347 MPa, EAB = 4%, and FM = 2247 MPa	97
	Banana fiber	Sisal fiber	Polyester	Volume fraction = 0.32	TM = 1443 MPa, EAB = 5%, and FM = 2376 MPa	97
	Banana fiber	Sisal fiber	Polyester	Volume fraction = 0.40	TM = 1601 MPa, EAB = 7%, and FM = 2842 MPa	97
Combination of compression molding and hand layup	—	Glass fiber	PFR28)	O/G = 0	TS = 82 MPa, EAB = 6.5%, and FS = 84 MPa	98
	Oil palm	Glass fiber	PFR	O/G = 0.2	TS = 56 MPa, EAB = 4.7%, and FS = 50 MPa	98
	Oil palm	Glass fiber	PFR	O/G = 0.4	TS = 40 MPa, EAB = 3.0%, and FS = 58 MPa	98
	Oil palm	Glass fiber	PFR	O/G = 0.6	TS = 64 MPa, EAB = 4.0%, and FS = 60 MPa	98
	Oil palm	Glass fiber	PFR	O/G = 0.8	TS = 30 MPa, EAB = 3.7%, and FS = 35 MPa	98
	Oil palm	—	PFR	O/G = 1.0	TS = 42 MPa, EAB = 4.0%, and FS = 55 MPa	98
	Hemp	—	Polyester	H:G = 51.4:0 (volume fraction ratio)	TS = 46.4 MPa and TM = 7200 MPa	99
	Hemp skin	Glass core	Polyester	H:G = 35.8:11.1 (volume fraction ratio)	TS = 70.1 MPa and TM = 8300 MPa	99
	Hemp core	Glass skin	Polyester	H:G = 36.6:11.3 (volume fraction ratio)	TS = 81.6 MPa and TM = 8300 MPa	99

(Continued)

TABLE 6.3 (CONTINUED)
Synthesis and Important Properties of the Thermosetting Hybrid Polymers

Synthesis techniques	Component 1	Component 2	Matrix	Configuration/Composition	Important properties	Ref
	Sisal	—	Epoxy resin	S:C = 100:0	TS = 24.16 MPa and TM = 1370 MPa	100
	Sisal	Carbon fiber	Epoxy resin	S:C = 75:25	TS = 31.35 MPa and TM = 1680 MPa	100
	Sisal	Carbon fiber	Epoxy resin	S:C = 50:50	TS = 38.30 MPa and TM = 1970 MPa	100
	Sisal	Carbon fiber	Epoxy resin	S:C = 25:75	TS = 50.85 MPa and TM = 2370 MPa	100
	—	Carbon fiber	Epoxy resin	S:C = 0:100	TS = 122.11 MPa and TM = 2980 MPa	100
	NaOH-sisal	—	Epoxy resin	S:C = 100:0	TS = 78.22 MPa and TM = 1960 MPa	100
	NaOH-sisal	Carbon fiber	Epoxy resin	S:C = 75:25	TS = 84.74 MPa and TM = 1990 MPa	100
	NaOH-sisal	Carbon fiber	Epoxy resin	S:C = 50:50	TS = 93.97 MPa and TM = 2170 MPa	100
	NaOH-sisal	Carbon fiber	Epoxy resin	S:C = 25:75	TS = 107.51 MPa and TM = 2780 MPa	100
	NaOH-sisal	Carbon fiber	Epoxy resin	S:C = 0:100	TS = 122.11 MPa and TM = 2980 MPa	100
Injection molding	Sisal	Glass	Polypropylene	G:S = 10:20 (+PP-g-MA: 3 phr)	TS = 29.62 MPa, TM = 2330.00 MPa, FS = 66.74 MPa, FM = 4030 MPa, and IS = 16.67 kJ m^{-2}	79

(Continued)

TABLE 6.3 (CONTINUED)
Synthesis and Important Properties of the Thermosetting Hybrid Polymers

Synthesis techniques	Component 1	Component 2	Matrix	Configuration/Composition	Important properties	Ref
	Sisal	Glass	Polypropylene	G:S = 15:15 (+PP-g-MA: 3 phr)	TS = 31.48 MPa, TM = 2420.00 MPa, FS = 68.49 MPa, FS = 4040 MPa, and IS = 18.35 kJ m^{-2}	79
	Sisal	Glass	Polypropylene	G:S = 20:10 (+PP-g-MA: 3 phr)	TS = 31.59 MPa, TM = 2430.00 MPa, FS = 68.84 MPa, FS = 4130 MPa, and IS = 20.01 kJ m^{-2}	79
	Sisal	Glass	Polypropylene	G:S = 5:25	TS = 41.75 MPa, TM = 970.01 MPa, FS = 47.38 MPa, FS = 1900 MPa, and IS = 59.30 J m^{-2}	80
	Sisal	Glass	Polypropylene	G:S = 10:20	TS = 42.32 MPa, TM = 990.56 MPa, FS = 52.51 MPa, FS = 2060 MPa, and IS = 60.16 J m^{-2}	80
	Sisal	Glass	Polypropylene	G:S = 15:15	TS = 45.44 MPa, TM = 1095.26 MPa, FS = 53.86 MPa, FS = 2260 MPa, and IS = 63.27 J m^{-2}	80
	Sisal	Glass	Polypropylene	G:S = 20:10	TS = 41.77 MPa, TM = 101.19 MPa, FS = 51.49 MPa, FS = 2040 MPa, and IS = 62.16 J m^{-2}	80
	Sisal	Glass	Polypropylene	G:S = 25:5	TS = 40.96 MPa, TM = 980.13 MPa, FS = 50.50 MPa, FS = 1970 MPa, and IS = 59.85 J m^{-2}	80

(Continued)

TABLE 6.3 (CONTINUED)
Synthesis and Important Properties of the Thermosetting Hybrid Polymers

Synthesis techniques	Component 1	Component 2	Matrix	Configuration/Composition	Important properties	Ref
	Sisal	Glass	Polypropylene	G:S = 15:15 (+1 wt.% MAPP)	TS = 50.50 MPa, TM = 1491.49 MPa, FS = 64.50 MPa, FS = 2660 MPa, and IS = 76.50 J m^{-2}	80
	Sisal	Glass	Polypropylene	G:S = 15:15 (+2 wt.% MAPP)	TS = 55.10 MPa, TM = 1685.90 MPa, FS = 67.49 MPa, FS = 2800 MPa, and IS = 81.57 J m^{-2}	80
	Sisal	Glass	Polypropylene	G:S = 15:15 (+3 wt.% MAPP)	TS = 51.21 MPa, TM = 1520.50 MPa, FS = 62.6 MPa, FS = 2740 MPa, and IS = 78.00 J m^{-2}	80
Casting	Bamboo fiber	—	Polyester	G:B = 0:40	YM = 7912 MPa	27
	Bamboo fiber	Glass	Polyester	G:B = 10:30	YM = 7912 MPa	27
	Bamboo fiber	Glass	Polyester	G:B = 20:20	YM = 4960 MPa	27
	Bamboo fiber	Glass	Polyester	G:B = 30:10	YM = 8343 MPa	27
	—	Glass	Polyester	G:B = 40:0	YM = 4921 MPa	27
	NaOH-bamboo	—	Polyester	G:B = 0:40	YM = 9244 MPa	27
	NaOH-bamboo	Glass	Polyester	G:B = 10:30	YM = 11478 MPa	27
	NaOH-bamboo	Glass	Polyester	G:B = 20:20	YM = 10755 MPa	27
	NaOH-bamboo	Glass	Polyester	G:B = 30:10	YM = 8527 MPa	27
	—	Glass	Polyester	G:B = 40:0	—	27
	Kapok	—	Polyester	G:K = 0:100 (Fabric content = 9 vol.%)	TS = 67.34 MPa and TM = 975.4 MPa	40
	Kapok	Glass	Polyester	G:K = 25:75 (Fabric content = 9 vol.%)	TS = 78.05 MPa and TM = 1133.2 MPa	40

(Continued)

TABLE 6.3 (CONTINUED)
Synthesis and Important Properties of the Thermosetting Hybrid Polymers

Synthesis techniques	Component 1	Component 2	Matrix	Configuration/Composition	Important properties	Ref
	Kapok	Glass	Polyester	G:K = 50:50 (Fabric content = 9 vol.%)	TS = 82.11 MPa and TM = 1182.9 MPa	40
	Kapok	Glass	Polyester	G:K = 75:25 (Fabric content = 9 vol.%)	TS = 102.55 MPa and TM = 1228.7 MPa	40
	—	Glass	Polyester	G:K = 100:0 (Fabric content = 9 vol.%)	TS = 112.87 MPa and TM = 2469.2 MPa	40
	NaOH-kapok	—	Polyester	G:K = 0:100 (Fabric content = 9 vol.%)	TS = 79.1 MPa and TM = 1425.7 MPa	40
	NaOH-kapok	Glass	Polyester	G:K = 25:75 (Fabric content = 9 vol.%)	TS = 94.1 MPa and TM = 1605.4 MPa	40
	NaOH-kapok	Glass	Polyester	G:K = 50:50 (Fabric content = 9 vol.%)	TS = 98.6 MPa and TM = 1645.4 MPa	40
	NaOH-kapok	Glass	Polyester	G:K = 75:25 (Fabric content = 9 vol.%)	TS = 107.6 MPa and TM = 2362.6 MPa	40
	NaOH-kapok	Glass	Polyester	G:K = 100:0 (Fabric content = 9 vol.%)	TS = 112.8 MPa and TM = 2469.2 MPa	40
	Kapok	—	Polyester	G:K = 0:100 (Fabric content = 9 vol.%)	FS = 136.1 MPa and FM = 8269.9 MPa	100
	Kapok	Glass	Polyester	G:K = 25:75 (Fabric content = 9 vol.%)	FS = 170.1 MPa and FM = 10345.3 MPa	100
	Kapok	Glass	Polyester	G:K = 50:50 (Fabric content = 9 vol.%)	FS = 192.8 MPa and FM = 14285.4 MPa	100

(Continued)

TABLE 6.3 (CONTINUED)
Synthesis and Important Properties of the Thermosetting Hybrid Polymers

Synthesis techniques	Component 1	Component 2	Matrix	Configuration/Composition	Important properties	Ref
	Kapok	Glass	Polyester	G:K = 75:25 (Fabric content = 9 vol.%)	FS = 213.9 MPa and FM = 16649.4 MPa	100
	—	Glass	Polyester	G:K = 100:0 (Fabric content = 9 vol.%)	FS = 230.7 MPa and FM = 17355.5 MPa	100
	NaOH-kapok	—	Polyester	G:K = 0:100 (Fabric content = 9 vol.%)	FS = 157.0 MPa and FM = 8349.9 MPa	100
	NaOH-kapok	Glass	Polyester	G:K = 25:75 (Fabric content = 9 vol.%)	FS = 183.8 MPa and FM = 10562.1 MPa	100
	NaOH-kapok	Glass	Polyester	G:K = 50:50 (Fabric content = 9 vol.%)	FS = 209.5 MPa and FM = 11500.1 MPa	100
	NaOH-kapok	Glass	Polyester	G:K = 75:25 (Fabric content = 9 vol.%)	FS = 222.5 MPa and FM = 12557.6 MPa	100
	NaOH-kapok	Glass	Polyester	G:K = 100:0 (Fabric content = 9 vol.%)	FS = 230.7 MPa and FM = 17355.5 MPa	100
	Bamboo	—	Epoxy	G:B = 0:40	YM = 7879 MPa	101
	Bamboo	Glass	Epoxy	G:B = 30:10	YM = 9283 MPa	101
	Bamboo	Glass	Epoxy	G:B = 20:20	YM = 8183 MPa	101
	Bamboo	Glass	Epoxy	G:B = 10:30	YM = 8111 MPa	101
	—	Glass	Epoxy	G:B = 40:0	YM = 5790 MPa	101
	NaOH-bamboo	—	Epoxy	G:B = 0:40	YM = 9133 MPa	101
	NaOH-bamboo	Glass	Epoxy	G:B = 30:10	YM = 11542 MPa	101
	NaOH-bamboo	Glass	Epoxy	G:B = 20:20	YM = 9205 MPa	101

(*Continued*)

TABLE 6.3 (CONTINUED)
Synthesis and Important Properties of the Thermosetting Hybrid Polymers

Synthesis techniques	Component 1	Component 2	Matrix	Configuration/Composition	Important properties	Ref
	NaOH-bamboo	Glass	Epoxy	G:B = 10:30	YM = 8949 MPa	101
	Sisal	Glass	Epoxy	Fiber length = 1 cm	IS = 10.89 J cm^{-2}	102
	Sisal	Glass	Epoxy	Fiber length = 2 cm	IS = 11.46 J cm^{-2}	102
	Sisal	Glass	Epoxy	Fiber length = 3 cm	IS = 10.03 J cm^{-2}	102
	NaOH-sisal	Glass	Epoxy	Fiber length = 1 cm	IS = 11.19 J cm^{-2}	102
	NaOH-sisal	Glass	Epoxy	Fiber length = 2 cm	IS = 12.87 J cm^{-2}	102
	NaOH-sisal	Glass	Epoxy	Fiber length = 3 cm	IS = 10.81 J cm^{-2}	102
RTM	Carbon fiber	—	Epoxy resin	Bisphenol A-type epoxy resin: amine = 100:45	TS = 768.56 MPa, TM = 3010 MPa, BS = 767.64 MPa, BM = 3070 MPa, and IS = 16.87 KJ m^{-2}	103
	Sisal	Glass	Polypropylene	G:S = 13.5:6.5	TS = 62.60 MPa, TM = 1327.00 MPa, FS = 181.10 MPa, FM = 8610 MPa, EAB = 4.20%, and IS = 68.60 kJ m^{-2}	104
	Neat	Polyamide 6	Epoxy resin	Resin infusion time = 4.03 sec	EM = 19800 MPa, TS = 268 MPa	105
	CNT	Polyamide 6	Epoxy resin	Resin infusion time = 4.78 sec	EM = 19600 MPa, TS = 252 MPa	105
	RGO	Polyamide 6	Epoxy resin	Resin infusion time = 4.64 sec	EM = 20900 MPa, TS = 267 MPa	105
	Clay	Polyamide 6	Epoxy resin	Resin infusion time = 4.23 sec	EM = 21300 MPa, TS = 280 MPa	105
	GO	Polyamide 6	Epoxy resin	Resin infusion time = 4.35 sec	EM = 25200 MPa, TS = 282 MPa	105
	GNP	Polyamide 6	Epoxy resin	Resin infusion time = 4.19 sec	EM = 20100 MPA, TS = 302 MPa	105

(Continued)

TABLE 6.3 (CONTINUED)
Synthesis and Important Properties of the Thermosetting Hybrid Polymers

Synthesis techniques	Component 1	Component 2	Matrix	Configuration/Composition	Important properties	Ref
	OPEFB[18)]	—	Polyester	G:OPEFB = 100:0 (wt. fraction = 45%)	TS = 30 MPa, TM = 3250 MPa, EAB = 3.9 mm, FS = 55 MPa, and FM = 1800 MPa	106
	OPEFB	Glass	Polyester	G:OPEFB = 70:30 (wt. fraction = 45%)	TS = 62 MPa, TM = 3500 MPa, EAB = 3.0 mm, FS = 98 MPa, and FM = 7000 MPa	106
	OPEFB	Glass	Polyester	G:OPEFB = 30:70 (wt. fraction = 45%)	TS = 50 MPa, TM = 3750 MPa, EAB = 3.7 mm, FS = 68 MPa, and FM = 5500 MPa	106
	OPEFB	Glass	Polyester	G:OPEFB = 50:50 (wt. fraction = 45%)	TS = 55 MPa, TM = 4750 MPa, EAB = 3.25 mm, FS = 80 MPa, and FM = 6000 MPa	106
	OPEFB	Glass	Polyester	G:OPEFB = 90:10 (wt. fraction = 45%)	TS = 80 MPa, TM = 5500 MPa, EAB = 2.8 mm, FS = 85 MPa, and FM = 8200 MPa	106
	Sisal (3 cm)	Glass	Polyester	G:S = 25:75 (fiber content = 10 vo.%)	FS = 133.70 MPa and EAB = 4.30%	107
	Sisal (3 cm)	Glass	Polyester	G:S = 50:50 (fiber content = 10 vo.%)	FS = 166.50 MPa and EAB = 4.50%	107
	Sisal (3 cm)	Glass	Polyester	G:S = 75:25 (fiber content = 10 vo.%)	FS = 264.10 MPa and EAB = 5.10%	107
	Sisal (4 cm)	Glass	Polyester	G:S = 25:75 (fiber content = 10 vo.%)	FS = 141.80 MPa and EAB = 4.30%	107
	Sisal (4 cm)	Glass	Polyester	G:S = 50:50 (fiber content = 10 vo.%)	FS = 184.60 MPa and EAB = 5.00%	107
	Sisal (4 cm)	Glass	Polyester	G:S = 75:25 (fiber content = 10 vo.%)	FS = 269.10 MPa and EAB = 5.20%	107
	Sisal (3 cm)	Glass	Polyester	G:S = 25:75 (fiber content = 20 vo.%)	FS = 266.00 MPa and EAB = 8.30%	107
	Sisal (3 cm)	Glass	Polyester	G:S = 50:50 (fiber content = 20 vo.%)	FS = 276.60 MPa and EAB = 6.30%	107

(*Continued*)

TABLE 6.3 (CONTINUED)
Synthesis and Important Properties of the Thermosetting Hybrid Polymers

Synthesis techniques	Component 1	Component 2	Matrix	Configuration/Composition	Important properties	Ref
	Sisal (3 cm)	Glass	Polyester	G:S = 75:25 (fiber content = 20 vo.%)	FS = 361.10 MPa and EAB = 6.90%	107
	Sisal (4 cm)	Glass	Polyester	G:S = 25:75 (fiber content = 20 vo.%)	FS = 218.10 MPa and EAB = 7.50%	107
	Sisal (4 cm)	Glass	Polyester	G:S = 50:50 (fiber content = 20 vo.%)	FS = 282.10 MPa and EAB = 6.40%	107
	Sisal (4 cm)	Glass	Polyester	G:S = 75:25 (fiber content = 20 vo.%)	FS = 372.20 MPa and EAB = 6.50%	107
	OPEFB	Glass	Vinyl ester	G:OPEFB = 50:50	TM = 850 MPa, TS = 95 MPa, EAB = 0.4 mm, FM = 8000 MPa, and FS = 240 MPa	108
	Arctic flax	—	Epoxy resin	Volume fraction = 0.42	TS = 280 MPa, TM = 35 GPa, and EAB = 0.9%	109
	Banana fiber	Sisal fiber	Polyester	Volume fraction = 0.19	TM = 1621 MPa, EAB = 4%, and FM = 2276 MPa	110
	Banana fiber	Sisal fiber	Polyester	Volume fraction = 0.32	TM = 1874 MPa, EAB = 6%, and FM = 2512 MPa	110
	Banana fiber	Sisal fiber	Polyester	Volume fraction = 0.40	TM = 1941 MPa, EAB = 6%, and FM = 3010 MPa	110
Vacuum infusion process	Flax, basalt	Glass	Epoxy resin	G:F:B = 2.30:11.72:7.16	TS = 153.16 MPa and FS = 137.95 MPa	110
	Hemp, basalt	Glass	Epoxy resin	G:H:B = 2.59:8.56:11.38	TS = 128.84 MPa and FS = 126.22 MPa	
	Hemp, basalt	Flax	Epoxy resin	F:H:B = 9.11:7.5:5.57	TS = 115.97 MPa and FS = 128.46 MPa	

(Continued)

TABLE 6.3 (CONTINUED)
Synthesis and Important Properties of the Thermosetting Hybrid Polymers

Synthesis techniques	Component 1	Component 2	Matrix	Configuration/Composition	Important properties	Ref
Others	3D carbon fiber	—	Epoxy resin	Impregnation of carbon fiber within the thermosetting matrix at 130°C	TS = 1476.11 MPa, TM = 100280 MPa, FS = 858.05 MPa, and FM = 71950 MPa	111
	OPEFB	Glass	Epoxy resin	Relative volume fraction = 0.8	TS = 100 MPa, EAB = 4.8% and YM = 4000 MPa	112
	OPEFB	Jute fiber	Epoxy resin	OPEFB:J = 4:1	TS = 32 MPa, TM = 2600 MPa, FS = 49 MPa, and FM = 3100 MPa	113
	OPEFB	Jute fiber	Epoxy resin	OPEFB:J = 4:1	TS = 27.41 MPa and TM = 2590 MPa	114
	Fineapple leaf fiber	glass	polyester	G:S = 0:30	TS = 68 MPa and FS = 100 MPa	33
				G:S = 3:27	TS = 82 MPa and FS = 125 MPa	33
	sisal fiber	glass	polyester	G:S = 5:25	TS = 96 MPa and FS = 128 MPa	33
				G:S = 6:24	TS = 100 MPa and FS = 140 MPa	33
				G:S = 8:22	TS = 102 MPa and 142 MPa	33
	kenaf fiber	E-glass fiber	polyester	G:K:polyester = 7.5:22.5:70	TS = 21 MPa, FS = 12500 MPa, TM = 1250 MPa, and IS = 80 J m^{-2}	38
	NaOH-kenaf fiber	E-glass fiber	polyester	G:K:polyester = 7.5:22.5:70	TS = 26 MPa, FS = 37500 MPa, TM = 2000 MPa, and IS = 120 J m^{-2}	38
	Sisal	Glass	Epoxy resin	G:S = 3.46:18.88 (strain: 0–0.02)	TS = 43.00 MPa, TM = 531.00 MPa, and IS = 20.29 kJ m^{-2}	115

(Continued)

TABLE 6.3 (CONTINUED)
Synthesis and Important Properties of the Thermosetting Hybrid Polymers

Synthesis techniques	Component 1	Component 2	Matrix	Configuration/Composition	Important properties	Ref
	Sisal	Glass	Epoxy resin	G:S = 3.46:18.88 (strain: 0.02–0.1)	TM = 1046.00 MPa	115
	Sisal	Glass	Epoxy resin	G:S = 14.00:8.45 (strain: 0–0.02)	TS = 71.22 MPa, TM = 634.00 MPa, and IS = 35.25 kJ m^{-2}	115
	Sisal	Glass	Epoxy resin	G:S = 14.00:8.45 (strain: 0.02–0.1)	TM = 1048.00 MPa	115
	Sisal	Glass	Polypropylene	G:S = 17:10 (porosity = 5 vol.%)	TS = 22.40 MPa, TM = 3649.00 MPa, FS = 52.60 MPa, FM = 4510 MPa, and EAB = 1.60%	116
	Sisal	Glass	Polypropylene	G:S = 12:19 (porosity = 5 vol.%)	TS = 18.30 MPa, TM = 3423.00 MPa, FS = 46.20 MPa, FM = 4210 MPa, and EAB = 1.67%	116
	Sisal	Glass	Polypropylene	G:S = 6:26 (porosity = 3 vol.%)	TS = 21.80 MPa, TM = 3136.00 MPa, FS = 47.90 MPa, FM = 3800 MPa, and EAB = 2.15%	116

1) Flexural strength, 2) flexural modulus, 3) impact strength, 4) tensile strength, 5) elongation at break, 6) tensile modulus, 7) impact permeation energy, 8) static flexural energy, and 9) linear stiffness, 10) Young's modulus, 11) bending modulus, 12) electromagnetic interference shielding effect, 13) elastic modulus, 14) sodium lauryl sulfate, 15) glass fiber-reinforced polymer composite, 16) stainless steel fiber, 17) graphite nanoplatelet, 18) oil palm empty fruit bunch, 19) oil palm empty fruit bunch, 20) flax fiber-reinforced polymer composite, 21) carbon nano tube, 22) hexagonal boron nitride, 23) micro rubber, 24) nano silica, 25) low density polyethylene, 26) vinyl acetate-ethylene, 27) methacrylated soybean oil, and 28) phenol-formaldehyde resin

6.3.1 Synthesis Technique

6.3.1.1 Hand Lay-Up

The simplest technique for fabricating thermoset composites is the hand layup method. Herein, the reinforcing fibers are placed manually, followed by the wetting through resin, as shown in Figure 6.13.

6.3.1.2 Compression Molding

This technique is widely employed for the production of thermosetting hybrid composites. Herein, the preheated molding material is initially positioned in an open-heated mold cavity, followed by closing the top with a plug member, and subsequent application of pressure and heat until the thermosetting hybrid composite is cured, as illustrated in Figure 6.14. This is better compared to the hand layup method because of the uniform resin coverage, better bonding, and compaction. In compression molding, high volume and pressure can be applied, making it suitable for synthesizing complex and high-strength fiber-glass reinforcements. However, although low amounts of wastes are generated, the compression molding technique has several disadvantages, such as poor product consistency, struggle to regulate flashing, production of fewer knit lines, and smaller fiber length compared to injection molding.

6.3.1.3 Casting Method

One of the oldest processes for manufacturing hybrid thermosetting composites is the casting method, in which the homogeneous liquid mixture of the components and resin is poured into a mold containing a hollow cavity. Thereafter, the material is converted into the solid state. The solid part, known as casting, is then isolated from the mold to obtain the thermosetting composite sheet, as shown in Figure 6.15.

6.3.1.4 Resin Transfer Molding (RTM)

Osborne Industries Inc. implemented the closed-mold molding process in 1976, which was referred to as RTM. RTM is commonly employed for the production of thermosetting composites because of high surface quality, use of diversified reinforcements, high dimensional tolerance, low cost, lesser wastage, and lesser labor requirements. For executing RTM, the preform matrix is created to inject the resin, followed by packing into a mold cavity that has a similar shape as the composite. Thereafter, the mold cavity is closed and clamped to regulate the thickness of composite precisely, and permits the smooth finish on both A and B sides of the part. In this context, gel coatings are sometimes employed in the inner side of the mold to achieve the high-quality durable composites. Thereafter, the resin is propelled into the heated mold under pressure eliminating air through vents until the mold is filled. The injection phase should assure the complete impregnation of the preform to circumvent the formation of dry spot areas between the layers. Finally, curing is initiated to make the resin into rigid thermosetting composites.

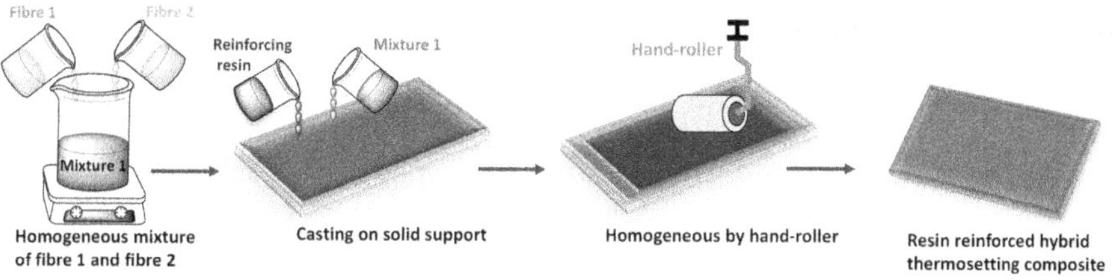

FIGURE 6.13 Hand layup synthesis of thermosetting composites.

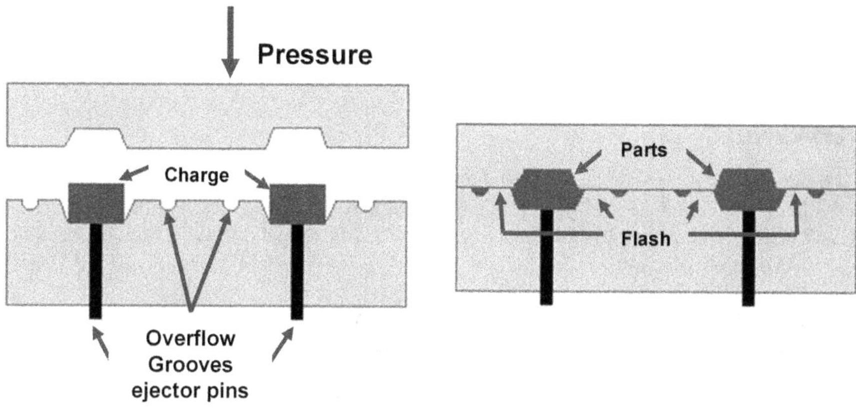

FIGURE 6.14 Compression molding synthesis of thermosetting.

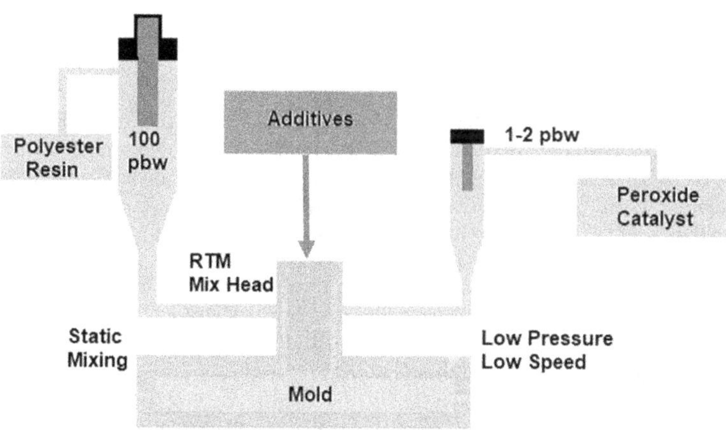

FIGURE 6.15 Synthesis of thermosetting composites through casting.

6.4 CONCLUSIONS

Hybrid thermoplastic/thermosetting composites have advantageous mechanical, thermal, and electrical properties, attributed mainly to stronger interfacial adhesion and receded cohesive forces. In some cases, one particulate phase facilitates breaking of aggregates, and thus, homogenous distributions of the other in a common matrix. Sometimes, interfacial adhesion is improved via additions of suitable coupling agents. Otherwise, either particulate or continuous phases are chemically treated to introduce functionalities, so as to ensure sufficient bonding between the constituents. Importantly, thermoplastic hybrid polymer composites are flexible and recyclable unlike their thermosetting counterpart. Researchers are in search of smart, ultra-lightweight, multifunctional, hybrid polymer composites for suitable high-performance sensor(s) and in vitro and/or in vivo device application(s) of the future.

REFERENCES

1. Singha, N. R.; Karmakar, M.; Mahapatra, M.; Mondal, H.; Dutta, A.; Deb, M.; Mitra, M.; Roy, C.; Chattopadhyay, P. K. An In Situ Approach for the Synthesis of a Gum Ghatti-g-Interpenetrating Terpolymer Network Hydrogel for the High-Performance Adsorption Mechanism Evaluation of Cd(II), Pb(II), Bi(III) and Sb(III). *J. Mater. Chem. A* 2018, *6*(17), 8078–8100.
2. Karmakar, M.; Mondal, H.; Mahapatra, M.; Chattopadhyay, P. K.; Chatterjee, S.; Singha, N. R. Pectin-Grafted Terpolymer Superadsorbent via N–H Activated Strategic Protrusion of Monomer for Removals of Cd(II), Hg(II), and Pb(II). *Carbohydr. Polym.* 2019, *206*, 778–791.
3. Mondal, H.; Karmakar, M.; Chattopadhyay, P. K.; Singha, N. R. Starch-g-Tetrapolymer Hydrogel via In Situ Attached Monomers for Removals of Bi(III) and/or Hg(II) and dye(s): RSM-Based Optimization. *Carbohydr. Polym.* 2019, *213*, 428–440.
4. Karmakar, M.; Mondal, H.; Ghosh, T.; Chattopadhyay, P. K.; Maiti, D. K.; Singha, N. R. Chitosan-grafted tetrapolymer Using Two Monomers: pH-Responsive High-Performance Removals of Cu(II), Cd(II), Pb(II), Dichromate, and Biphosphate and Analyses of Adsorbed Microstructures. *Environ. Res.* 2019, *179*(B), 108839.
5. Singha, N. R.; Karmakar, M.; Mahapatra, M.; Mondal, H.; Dutta, A.; Roy, C.; Chattopadhyay, P. K. Systematic Synthesis of Pectin-g-(Sodium Acrylate-Co-N-Isopropylacrylamide) Interpenetrating Polymer Network for Superadsorption of Dyes/M(II): Determination of Physicochemical Changes in Loaded Hydrogels. *Polym. Chem.* 2017, *8*(20), 3211–3237.
6. Singha, N. R.; Parya, T. K.; Ray, S. K. Dehydration of 1,4-Dioxane by Pervaporation Using Filled and Crosslinked Polyvinyl Alcohol Membrane. *J. Membr. Sci.* 2009, *340*(1–2), 35–44.
7. Singha, N. R.; Karmakar, M.; Chattopadhyay, P. K.; Roy, S.; Deb, M.; Mondal, H.; Mahapatra, M.; Dutta, A.; Mitra, M.; Roy, J. S. D. Structures, Properties, and Performances-Relationships of Polymeric Membranes for Pervaporative Desalination. *Membranes* 2019, *9*(5), 58.
8. Roy, S.; Singha, N. R. Polymeric Nanocomposite Membranes for Next Generation Pervaporation Process: Strategies, Challenges and Future Prospects. *Membranes* 2017, *7*(3), 53.
9. Singha, N. R.; Ray, S.; Ray, S. K.; Konar, B. B. Removal of Pyridine from Water by Pervaporation Using Filled SBR Membranes. *J. Appl. Polym. Sci.* 2011, *121*(3), 1330–1334.

10. Singha, N. R.; Ray, S. K. Removal of Pyridine from Water by Pervaporation Using Crosslinked and Filled Natural Rubber Membranes. *J. Appl. Polym. Sci.* 2012, *124*(S1), E99–E107.
11. Singha, N. R.; Das, P.; Ray, S. K. Recovery of Pyridine from Water by Pervaporation Using Filled and Crosslinked EPDM Membranes. *J. Ind. Eng. Chem.* 2013, *19*(6), 2034–2045.
12. Ashby, M. F.; Bréchet, Y. J. M. Designing Hybrid Materials. *Acta Mater.* 2003, *51*(19), 5801–5821.
13. Fu, S. Y.; Xu, G.; Mai, Y. W. On the Elastic Modulus of Hybrid Particle/Short-Fiber/Polymer Composites. *Compos. Part B Eng.* 2002, *33*(4), 291–299.
14. Thwe, M. M.; Liao, K. Durability of Bamboo-Glass Fiber Reinforced Polymer Matrix Hybrid Composites. *Compos. Sci. Technol.* 2003, *63*(3–4), 375–387.
15. Swolfs, Y.; Gorbatikh, L.; Verpoest, I. Fibre Hybridisation in Polymer Composites: A Review. *Compos. Part A* 2014, *67*, 181–200.
16. Zhao, Y.; Ma, X.; Xua, T.; Salem, D. R.; Fong, H. Hybrid Multi-Scale Thermoplastic Composites Reinforced with Interleaved Nanofiber Mats Using in-Situ Polymerization of Cyclic Butylene Terephthalate. *Compos. Commun.* 2019, *12*, 91–97.
17. Shou, Q.; Cheng, J.; Fang, J.; Lu, F.; Zhao, J.; Tao, X.; Liu, F.; Zhang, X. Thermal Conductivity of Poly Vinylidene Fluoride Composites Filled with Expanded Graphite and Carbon Nanotubes. *J. Appl. Polym. Sci.* 2013, *127*(3), 1697–1702.
18. Zhang, J.; Xu, Z.; Shan, M.; Zhou, B.; Li, Y.; Li, B.; Niu, J.; Qian, X. Synergetic Effects of Oxidized Carbon Nanotubes and Graphene Oxide on Fouling Control and Anti-Fouling Mechanism of Polyvinylidene Fluoride Ultrafiltration Membranes. *J. Membr. Sci.* 2013, *448*, 81–92.
19. Wang, Q.; Sun, L.; Li, L.; Yang, W.; Zhang, Y.; Dai, Z.; Xiong, Z. Experimental and Numerical Investigations on Microstructures and Mechanical Properties of Hybrid Fiber Reinforced Thermoplastic Polymer. *Polym. Test.* 2018, *70*, 215–225.
20. Naito, K.; Oguma, H. Tensile Properties of Novel Carbon/Glass Hybrid Thermoplastic Composite Rods. *Compos. Struct.* 2017, *161*, 23–31.
21. Atiqah, A.; Jawaid, M.; Sapuan, S. M.; Ishak, M. R.; Alothman, O. Y. Thermal Properties of Sugar Palm/Glass Fiber Reinforced Thermoplastic Polyurethane Hybrid Composites. *Compos. Struct.* 2018, *202*, 954–958.
22. Afzaluddin, A.; Jawaid, M.; Salit, M. S.; Ishak, M. R. Physical and Mechanical Properties of Sugar Palm/Glass Fiber Reinforced Thermoplastic Polyurethane Hybrid Composites. *J. Mater. Res. Technol.* 2019, *8*(1), 950–959.
23. Nayak, S. K.; Mohanty, S.; Samal, S. K. Influence of Short Bamboo/Glass Fiber on the Thermal, Dynamic Mechanical and Rheological Properties of Polypropylene Hybrid Composites. *Mater. Sci. Eng.* 2009, *523*(1–2), 32–38.
24. Samal, S. K.; Mohanty, S.; Nayak, S. K. Polypropylene-Bamboo/Glass Fiber Hybrid Composites: Fabrication and Analysis of Mechanical, Morphological, Thermal, and Dynamic Mechanical Behavior. *J. Reinf. Plast. Compos.* 2009, *28*(22), 2729–2747.
25. Panthapulakkal, S.; Sain, M. Injection-Molded Short Hemp Fiber/Glass Fiber-Reinforced Polypropylene Hybrid Composites—Mechanical, Water Absorption and Thermal Properties. *J. Appl. Polym. Sci.*, 2007, *103*, 2432–2441.
26. Valente, M.; Sarasini, F.; Marra, F.; Tirillò, J.; Pulci, G. Hybrid Recycled Glass Fiber/Wood Flour Thermoplastic Composites: Manufacturing and Mechanical Characterization. *Compos. Part A* 2011, *42*(6), 649–657.
27. Reddy, E. V. S.; Rajulu, A. V.; Reddy, K. H.; Reddy, G. R. Chemical Resistance and Tensile Properties of Glass and Bamboo Fibers Reinforced Polyester Hybrid Composites. *J. Reinf. Plast. Compos.* 2010, *29*(14), 2119–2123.
28. Al-Kafi, A.; Abedin, M. Z.; Beg, M. D. H.; Pickering, K. L.; Khan, M. A. Study on the Mechanical Properties of Jute/Glass Fiber-Reinforced Unsaturated Polyester Hybrid

Composites: Effect of Surface Modification by Ultraviolet Radiation. *J. Reinf. Plast. Compos.* 2006, *25*(6), 575–588.

29. Mahapatra, M.; Karmakar, M.; Mondal, B.; Singha, N. R. Role of ZDC/S Ratio for Pervaporative Separation of Organic Liquids through Modified EPDM Membranes: Rational Mechanistic Study of Vulcanization. *RSC Adv.* 2016, *6*(73), 69387–69403.
30. Karmakar, M.; Mahapatra, M.; Singha, N. R. Separation of Tetrahydrofuran Using RSM Optimized Accelerator-Sulfur-Filler of Rubber Membranes: Systematic Optimization and Comprehensive Mechanistic Study. *Korean J. Chem. Eng.* 2017, *34*(5), 1416–1434.
31. Mahapatra, M.; Karmakar, M.; Dutta, A.; Singha, N. R. Fabrication of Composite Membranes for Pervaporation of Tetrahydrofuran-Water: Optimization of Intrinsic Property by Response Surface Methodology and Studies on Vulcanization Mechanism by Density Functional Theory. *Korean J. Chem. Eng.* 2018, *35*(9), 1889–1910.
32. Zhang, S.; Yin, S.; Rong, C.; Huo, P.; Jiang, Z.; Wang, G. Synergistic Effects of Functionalized Graphene and Functionalized Multi-Walled Carbon Nanotubes on the Electrical and Mechanical Properties of Poly(Ether Sulfone) Composites. *Eur. Polym. J.* 2013, *49*(10), 3125–3134.
33. Mishra, S.; Mohanty, A. K.; Drzal, L. T.; Misra, M.; Parija, S.; Nayak, S. K.; Tripathy, S. S. Studies on Mechanical Performance of Biofibre/Glass Reinforced Polyester Hybrid Composites. *Compos. Sci. Technol.* 2003, *63*(10), 1377–1385.
34. Xu, X.; Zhou, Z.; Hei, Y.; Zhang, B.; Bao, J.; Chen, X. Improving Compression-After-Impact Performance of Carbon–Fiber Composites by CNTs/Thermoplastic Hybrid Film Interlayer. *Compos. Sci. Technol.* 2014, *95*, 75–81.
35. Chee, S. S.; Jawaid, M.; Sultan, M. T. H.; Alothman, O. Y.; Abdullah, L. C. Thermomechanical and Dynamic Mechanical Properties of Bamboo/Woven Kenaf Mat Reinforced Epoxy Hybrid Composites. *Compos. Part B Eng.* 2019, *163*, 165–174.
36. Thwe, M. M.; Liao, K. Effects of Environmental Aging on the Mechanical Properties of Bamboo-Glass Fiber Reinforced Polymer Matrix Hybrid Composites. *Compos. Part A* 2002, *33*(1), 43–52.
37. Shuib, S.; Ismail, N. F.; Nazri, M. N.; Romli, A. Z. Bamboo and Glass Fibre Hybrid Laminated Composites as Locking Compression Plate (LCP) for Tibia Fracture Treatment. *J. Phys. Conf. Ser.* 2019, *1150*, 012028.
38. Atiqah, A.; Maleque, M. A.; Jawaid, M.; Iqbal, M. Development of Kenaf-Glass Reinforced Unsaturated Polyester Hybrid Composite for Structural Applications. *Compos. Part B Eng.* 2014, *56*, 68–73.
39. Samal, S. K.; Mohanty, S.; Nayak, S. K. Banana/Glass Fiber-Reinforced Polypropylene Hybrid Composites: Fabrication and Performance Evaluation. *Polym. Plast. Technol. Eng.* 2009, *48*(4), 397–414.
40. Reddy, G. V.; Naidu, S. V.; Rani, T. S. Kapok/Glass Polyester Hybrid Composites: Tensile and Hardness Properties. *J. Reinf. Plast. Compos.* 2008, *27*(16–17), 1775–1787.
41. AlMaadeed, M. A.; Kahraman, R.; Khanam, P. N.; Madi, N. Date Palm Wood Flour/Glass Fibre Reinforced Hybrid Composites of Recycled Polypropylene: Mechanical and Thermal Properties. *Mater. Des.* 2012, *42*, 289–294.
42. Mirbagheri, J.; Tajvidi, M.; Hermanson, J. C.; Ghasemi, I. Tensile Properties of Wood Flour/Kenaf Fiber Polypropylene Hybrid Composites. *J. Appl. Polym. Sci.* 2007, *105*(5), 3054–3059.
43. Idicula, M.; Joseph, K.; Thomas, S. Mechanical Performance of Short Banana/Sisal Hybrid Fiber Reinforced Polyester Composites. *J. Reinf. Plast. Compos.* 2010, *29*(1), 12–29.
44. Arbelaiz, A.; Fernández, B.; Cantero, G.; Llano-Ponte, R.; Valea, A.; Mondragon, I. Mechanical Properties of Flax Fibre/Polypropylene Composites. Influence of Fibre/Matrix Modification and Glass Fibre Hybridization. *Compos. Part A* 2005, *36*(12), 1637–1644.

45. Yuen, S. M.; Ma, C. C. M.; Lin, Y. Y.; Kuan, H. C. Preparation, Morphology, and Properties of Acid and Amine Modified Multiwalled Carbon Nanotube/Polyimide Composite. *Compos. Sci. Technol.* 2007, *67*(11–12), 2564–2573.
46. Isayev, A. I.; Kumar, R.; Lewis, T. M. Ultrasound Assisted Twin Screw Extrusion of Polymer Nanocomposites Containing Carbon Nanotubes. *Polymer* 2009, *50*(1), 250–260.
47. Kumar, S.; Rath, T.; Mukherjee, M.; Khatua, B. B.; Das, C. K. Multi-Walled Carbon Nanotube/Polymer Composites in the Presence of Acrylic Elastomer. *J. Nanosci. Technol.* 2009, *9*, 2981–2990.
48. Prasad, K. E.; Das, B.; Maitra, U.; Ramamurty, U.; Rao, C. N. R. Extraordinary Synergy in the Mechanical Properties of Polymer Matrix Composites Reinforced with 2 Nanocarbons. *Proc. Natl. Acad. Sci. U.S.A.* 2009, *106*(32), 13186–13189.
49. Lee, C.; Wei, X.; Kaysar, J.; Hone, J. Measurement of the Elastic Properties and Intrinsic Strength of Monolayer Graphene. *Science* 2008, *32*(5887), 385–388.
50. Novoselov, K. S.; Geim, A. K.; Morozov, S. V.; Jiang, D.; Zhang, Y.; Dubonos, S. V.; Grigorieva, I. V.; Firsov, A. A. Electric Field Effect in Atomically Thin Carbon Films. *Science* 2004, *306*(5696), 666–669.
51. Rao, C. N. R.; Sood, A. K.; Voggu, R.; Subrahmanyam, K. S. Some Novel Attributes of Graphene. *J. Phys. Chem. Lett.* 2010, *1*(2), 572–580.
52. Zhan, Y.; Wu, J.; Xia, H.; Yan, N.; Fei, G.; Yuan, G. Dispersion and Exfoliation of Graphene in Rubber by an Ultrasonically-Assisted Latex Mixing and In Situ Reduction Process. *Macromol. Mater. Eng.* 2011, *296*(7), 590–602.
53. Wang, L. L.; Zhang, L. Q.; Tian, M. Effect of Expanded Graphite (EG) Dispersion on the Mechanical and Tribological Properties of Nitrile Rubber/EG Composites. *Wear* 2012, *276*, 85–93.
54. Santulli, C.; Janssen, M.; Jeronimidis, G. Partial Replacement of E-Glass Fibers with Flax Fibers in Composites and Effect on Falling Weight Impact Performance. *J. Mater. Sci.* 2005, *40*(13), 3581–3585.
55. Padma, P. S.; Rai, S. K. Mechanical Performance of Biofiber/Glass-Reinforced Epoxy Hybrid Composites. *J. Ind. Text.* 2006, *35*(3), 217–226.
56. Srinivasan, V. S.; Boopathy, S. R.; Sangeetha, D.; Ramnath, B. V. Evaluation of Mechanical and Thermal Properties of Banana-Flax Based Natural Fibre Composite. *Mater. Des.* 2014, *60*, 620–627.
57. Giridharan, R. Preparation and Property Evaluation of Glass/Ramie Fibers Reinforced Epoxy Hybrid Composites. *Compos. Part B Eng.* 2019, *167*, 342–345.
58. Gokuldass, R.; Ramesh, R. Mechanical Strength Behavior of Hybrid Composites Tailored by Glass/Kevlar Fibre-Reinforced in Nano-Silica and Micro-Rubber Blended Epoxy. *Silicon* 2019. DOI: 10.1007/s12633-018-0064-1.
59. Jawaid, M.; Khalil, H. P. S. A.; Hassan, A.; Dungani, R.; Hadiyane, A. Effect of Jute Fibre Loading on Tensile and Dynamic Mechanical Properties of Oil Palm Epoxy Composites. *Compos. Part B* 2013, *45*, 619–624.
60. Patel, V. A.; Bhuva, B. D.; Parsania, P. H. Performance Evaluation of Treated-Untreated Jute-Carbon and Glasscarbon Hybrid Composites of Bisphenol-C Based Mixed Epoxy-Phenolic Resins. *J. Reinf. Plast. Compos.* 2009, *28*(20), 2549–2556.
61. Boopalan, M.; Niranjanaa, M.; Umapathy, M. J. Study on the Mechanical Properties and Thermal Properties of Jute and Banana Fiber Reinforced Epoxy Hybrid Composites. *Compos. Part B Eng.* 2013, *51*, 54–57.
62. Kim, G.; Qin, H.; Fang, X.; Sun, F. C.; Mather, P. T. Hybrid Epoxy-Based Thermosets Based on Polyhedral Oligosilsesquioxane: Cure Behavior and Toughening Mechanisms. *J. Polym. Sci. Pol. Phys.* 2003, *41*(24), 3299–3313.
63. Thakur, V. K.; Thakur, M. K. Processing and Characterization of Natural Cellulose Fibers/Thermoset Polymer Composites. *Carbohyd. Polym.* 2014, *109*, 102–117.

64. Zeng, K.; Zheng, S. Nanostructures and Surface Dewettability of Epoxy Thermosets Containing Hepta(3,3,3-Trifluoropropyl) Polyhedral Oligomeric Silsesquioxane-Capped Poly(Ethylene Oxide). *J. Phys. Chem. B* 2007, *111*(50), 13919–13928.
65. Ramesh, M.; Palanikumar, K.; Reddy, K. H. Influence of Fiber Orientation and Fiber Content on Properties of Sisal-Jute-Glass Fiber-Reinforced Polyester Composites. *J. Appl. Polym. Sci.* 2016, *133*(6), 1–9.
66. Palanikumar, K.; Ramesh, M.; Reddy, K. H. Experimental Investigation on the Mechanical Properties of Green Hybrid Sisal and Glass Fiber Reinforced Polymer Composites. *J. Nat. Fibers* 2016, *13*(3), 321–331.
67. Zhang, X.; Wu, K.; Liu, Y.; Yu, B.; Zhang, Q.; Chen, F.; Fu, Q. Preparation of Highly Thermally Conductive but Electrically Insulating Composites by Constructing a Segregated Double Network in Polymer Composites. *Compos. Sci. Technol.* 2019, *175*, 135–142.
68. Velmurugan, R.; Manikandan, V. Mechanical Properties of Palmyra/Glass Fiber Hybrid Composites. *Compos. Part A—Appl. S.* 2007, *38*(10), 2216–2226.
69. Zhang, Y.; Li, Y.; Ma, H.; Yu, T. Tensile and Interfacial Properties of Unidirectional Flax/Glass Fiber Reinforced Hybrid Composites. *Compos. Sci. Technol.* 2013, *88*, 172–177.
70. Pavithran, C.; Mukherjee, P. S.; Brahmakumar, M.; Damodaran, A. D. Impact Properties of Sisal-Glass Hybrid Laminates. *J. Mater. Sci.* 1991, *26*(2), 455–459.
71. Singh, B.; Gupta, M.; Verma, A. Mechanical Behaviour of Particulate Hybrid Composite Laminates as Potential Building Materials. *Constr. Build. Mater.* 1995, *9*(1), 39–44.
72. John, K.; Naidu, S. V. Sisal Fiber/Glass Fiber Hybrid Composites: The Impact and Compressive Properties. *J. Reinf. Plast. Compos.* 2004, *23*(12), 1253–1258.
73. Ramesh, M.; Palanikumar, K.; Reddy, K. H. Mechanical Property Evaluation of Sisal-Jute-Glass Fiber Reinforced Polyester Composites. *Compos. Part B Eng.* 2013, *48*, 1–9.
74. Gupta, M. K.; Deep, V. Effect of Water Absorption and Stacking Sequences on the Properties of Hybrid Sisal/Glass Fibre Reinforced Polyester Composite. *Proc. Inst. Mech. Eng.* 2018. DOI: 10.1177/1464420718811867.
75. Ahmed, K. S.; Vijayarangan, S.; Rajput, C. Mechanical Behavior of Isothalic Polyester-Based Untreated Woven Jute and Glass Fabric Hybrid Composites. *J. Reinf. Plast. Compos.* 2006, *25*(15), 1549–1569.
76. Kalaprasad, G.; Thomas, S.; Pavithran, C.; Neelakantan, N. R.; Balakrishnan, S. Hybrid Effect in the Mechanical Properties of Short Sisal/Glass Hybrid Fiber Reinforced Low Density Polyethylene Composites. *J. Reinf. Plast. Compos.* 1996, *15*(1), 48–73.
77. Kalaprasad, G.; Francis, B.; Thomas, S.; Kumar, C. R.; Pavithran, C.; Groeninckx, G.; Thomas, S. Effect of Fibre Length and Chemical Modifications on the Tensile Properties of Intimately Mixed Short Sisal/Glass Hybrid Fibre Reinforced Low Density Polyethylene Composites. *Polym. Int.* 2004, *53*(11), 1624–1638.
78. Shajari, S.; Arjmand, M.; Pawar, S. P.; Sundararaj, U.; Sudak, L. J. Synergistic Effect of Hybrid Stainless Steel Fiber and Carbon Nanotube on Mechanical Properties and Electromagnetic Interference Shielding of Polypropylene Nanocomposites. *Compos. Part B Eng.* 2019, *165*, 662–670.
79. Jarukumjorn, K.; Suppakarn, N. Effect of Glass Fiber Hybridization on Properties of Sisal Fiber-Polypropylene Composites. *Compos. Part B Eng.* 2009, *40*(7), 623–627.
80. Nayak, S. K.; Mohanty, S. Sisal Glass Fiber Reinforced PP Hybrid Composites: Effect of MAPP on the Dynamic Mechanical and Thermal Properties. *J. Reinf. Plast. Compos.* 2010, *29*(10), 1551–1568.
81. Da Silva, C. C.; Freire Júnior, R. C. S.; Ford, E. T. L. C.; dos Santos, J. K. T.; de Aquino, E. M. F.; Aquino, E. M. Fd Mechanical Behavior and Water Absorption in Sisal/Glass Hybrid Composites. *Matéria* 2018, *23*(4), 4. DOI: 10.1590/S1517707620180004.0580.

82. Rana, R. S.; Kumre, A.; Rana, S.; Purohit, R. Characterization of Properties of Epoxy Sisal/Glass Fiber Reinforced Hybrid Composite. *Mater. Today Proc.* 2017, *4*(4), 5445–5451.
83. Arpitha, G. R.; Sanjay, M. R.; Senthamaraikannan, P.; Barile, C.; Yogesha, B. Hybridization Effect of Sisal/Glass/Epoxy/Filler Based Woven Fabric Reinforced Composites. *Exp. Tech.* 2017, *41*(6), 577–584.
84. Silva, R. V.; Aquino, E. M. F.; Rodrigues, L. P. S.; Barros, A. R. F. Curaua/Glass Hybrid Composite: The Effect of Water Aging on the Mechanical Properties. *J. Reinf. Plast. Compos.* 2009, *28*(15), 1857–1868.
85. John, K.; Naidu, S. V. Effect of Fiber Content and Fiber Treatment on Flexural Properties of Sisal Fiber/Glass Fiber Hybrid Composites. *J. Reinf. Plast. Compos.* 2004, *23*(15), 1601–1605.
86. Prabu, V. A.; Kumaran, S. T.; Uthayakumar, M.; Manikandan, V. Influence of Red Mud Particle Hybridization in Banana/Sisal and Sisal/Glass Composites. *Part. Sci. Technol.* 2018, *36*(4), 402–407.
87. John, K.; Naidu, S. V. Tensile Properties of Unsaturated Polyester-Based Sisal Fiber-Glass Fiber Hybrid Composites. *J. Reinf. Plast. Compos.* 2004, *23*(17), 1815–1819.
88. Alavudeen, A.; Rajini, N.; Karthikeyan, S.; Thiruchitrambalam, M.; Venkateshwaren, N. Mechanical Properties of Banana/Kenaf Fiber-Reinforced Hybrid Polyester Composites: Effect of Woven Fabric and Random Orientation. *Mater. Des.* 2015, *66*, 246–257.
89. Júnior, J. H. S. A.; Júnior, H. L. O.; Amico, S. C.; Amado, F. D. R. Study of Hybrid Intralaminate Curaua/Glass Composites. *Mater. Des.* 2012, *42*, 111–117.
90. Amico, S. C.; Angrizani, C. C.; Drummond, M. L. Influence of the Stacking Sequence on the Mechanical Properties of Glass/Sisal Hybrid Composites. *J. Reinf. Plast. Compos.* 2010, *29*(2), 179–189.
91. Nimanpure, S.; Hashmi, S. A. R.; Kumar, R.; Bhargaw, H. N.; Kumar, R.; Nair, P.; Naik, A. Mechanical, Electrical, and Thermal Analysis of Sisal Fibril/Kenaf Fiber Hybrid Polyester Composites. *Polym. Compos.* 2019, *40*(2), 664–676.
92. Ouarhim, W.; Essabir, H.; Bensalah, M.; Zari, N.; Bouhfid, R.; Qaiss, A. K. Structural Laminated Hybrid Composites Based on Raffia and Glass Fibers: Effect of Alkali Treatment, Mechanical and Thermal Properties. *Compos. Part B Eng.* 2018, *154*, 128–137.
93. Arthanarieswaran, V. P.; Kumaravel, A.; Kathirselvam, M. Evaluation of Mechanical Properties of Banana and Sisal Fiber Reinforced Epoxy Composites: Influence of Glass Fiber Hybridization. *Mater. Des.* 2014, *64*, 194–202.
94. Venkateshwaran, N.; Elayaperumal, A.; Sathiya, G. K. Prediction of Tensile Properties of Hybrid-Natural Fiber Composites. *Compos. Part B Eng.* 2012, *43*(2), 793–796.
95. Adekunle, K.; Cho, S.; Ketzscher, R.; Skrifvars, M. Mechanical Properties of Natural Fiber Hybrid Composites Based on Renewable Thermoset Resins Derived from Soybean Oil, for Use in Technical Applications. *J. Appl. Polym. Sci.* 2012, *124*, 4530–4541.
96. Ramesh, M.; Palanikumar, K.; Reddy, K. H. Evaluation of Mechanical and Interfacial Properties of Sisal/Jute/Glass Hybrid Fiber Reinforced Polymer Composites. *Trans. Indian Inst. Met.* 2016, *69*(10), 1851–1859.
97. Idicula, M.; Sreekumar, P. A.; Joseph, K.; Thomas, S. Natural Fiber Hybrid Composites – A Comparison between Compression Molding and Resin Transfer Molding. *Polym. Compos.* 2009, *30*(10), 1417–1425.
98. Sreekala, M. S.; George, J.; Kumaran, M. G.; Thomas, S. The Mechanical Performance of Hybrid Phenol-Formaldehyde-Based Composites Reinforced with Glass and Oil Palm Fibres. *Compos. Sci. Technol.* 2002, *62*(3), 339–353.
99. Shahzad, A. Impact and Fatigue Properties of Hemp-Glass Fiber Hybrid Biocomposites. *J. Reinf. Plast. Compos.* 2011, *30*(16), 1389–1398.

100. Khanam, P. N.; Khalil, H. P. S. A.; Jawaid, M.; Reddy, R.; Narayana, C. S.; Naidu, S. V. Sisal/Carbon Fibre Reinforced Hybrid Composites: Tensile, Flexural and Chemical Resistance Properties. *J. Polym. Environ.* 2010, *18*(4), 727–733.
101. Rao, H. R.; Rajulu, A. V.; Reddy, G. R.; Reddy, K. H. Flexural and Compressive Properties of Bamboo and Glass Fiber-Reinforced Epoxy Hybrid Composites. *J. Reinf. Plast. Compos.* 2010, *29*(10), 1446–1450.
102. Kumar, M. A.; Reddy, G. R.; Bharathi, Y. S.; Naidu, S. V.; Naidu, V. N. P. Frictional Coefficient, Hardness, Impact Strength, and Chemical Resistance of Reinforced Sisal-Glass Fiber Epoxy Hybrid Composites. *J. Compos. Mater.* 2010, *44*(26), 3195–3202.
103. Sun, Z.; Xiao, J.; Tao, L.; Wei, Y.; Wang, S.; Zhang, H.; Zhu, S.; Yu, M. Preparation of High-Performance Carbon Fiber-Reinforced Epoxy Composites by Compression Resin Transfer Molding. *Materials* 2019, *12*(1), 13.
104. Schmidt, T. M.; Goss, T. M.; Amico, S. C.; Lekakou, C. Permeability of Hybrid Reinforcements and Mechanical Properties of Their Composites Molded by Resin Transfer Molding. *J. Reinf. Plast. Compos.* 2009, *28*(23), 2839–2850.
105. Kim, B.; Cha, S.; Park, Y. Ultra-High-Speed Processing of Nanomaterial-Reinforced Woven Carbon Fiber/Polyamide 6 Composites Using Reactive Thermoplastic Resin Transfer Molding. *Compos. Part B Eng.* 2018, *143*, 36–46.
106. Khalil, H. P. S. A.; Hanida, S.; Kang, C. W.; Fuaad, N. A. N. Agro-Hybrid Composite: The Effects on Mechanical and Physical Properties of Oil Palm Fiber (EFB)/Glass Hybrid Reinforced Polyester Composites. *J. Reinf. Plast. Compos.* 2007, *26*(2), 203–218.
107. Ornaghi Jr., H. L.; Bolner, A. S.; Fiorio, R.; Zattera, A. J.; Amico, S. C. Dynamic Mechanical Analysis of Hybrid Composite Molded by Resin Transfer Molding. *J. Appl. Polym. Sci.* 2010, *118*, 887–896.
108. Khalil, H. P. S. A.; Kang, C. W.; Khairul, A.; Ridzuan, R.; Adawi, T. O. The Effect of Different Laminations on Mechanical and Physical Properties of Hybrid Composites. *J. Reinf. Plast. Compos.* 2009, *28*(9), 1123–1137.
109. Oksman, K. High Quality Flax Fibre Composites Manufactured by the Resin Transfer Moulding Process. *J. Reinf. Plast. Compos.* 2001, *20*(7), 621–627.
110. Petrucci, R.; Santulli, C.; Puglia, D.; Sarasini, F.; Torre, L.; Kenny, J. M. Mechanical Characterisation of Hybrid Composite Laminates Based on Basalt Fibres in Combination with Flax, Hemp and Glass Fibres Manufactured by Vacuum Infusion. *Mater. Des.* 2013, *49*, 728–735.
111. Ming, Y.; Duan, Y.; Wang, B.; Xiao, H.; Zhang, X. A Novel Route to Fabricate High-Performance 3D Printed Continuous Fiber-Reinforced Thermosetting Polymer Composites. *Materials* 2019, *12*(9), 1369.
112. Hariharan, A. B. A.; Khalil, H. P. S. A. Lignocellulose Based Hybrid Bilayer Laminate Composite: Part I-Studies on Tensile and Impact Behavior of Oil Palm Fiber-Glass Fiber-Reinforced Epoxy Resin. *J. Compos. Mater.* 2005, *39*(8), 663–684.
113. Jawaid, M.; Khalil, H. P. S. A.; Bakar, A. A. Woven Hybrid Composites: Tensile and Flexural Properties of Oil Palm-Woven Jute Fibres Based Epoxy Composites. *Mater. Sci. Eng.* 2011, *528*(15), 5190–5195.
114. Jawaid, M.; Khalil, H. P. S. A.; Bakar, A. A.; Khanam, P. N. Chemical Resistance, Void Content and Tensile Properties of Oil Palm/Jute Fibre Reinforced Polymer Hybrid Composites. *Mater. Des.* 2011, *32*(2), 1014–1019.
115. Hashmi, S. A. R.; Naik, A.; Chand, N.; Sharma, J.; Sharma, P. Development of Environment Friendly Hybrid Layered Sisal-Glass-Epoxy Composites. *Compos. Interfaces* 2011, *18*(8), 671–683.
116. Aslan, M.; Tufan, M.; Küçükömeroğlu, T. Tribological and Mechanical Performance of Sisal-Filled Waste Carbon and Glass Fiber Hybrid Composites. *Compos. B Eng.* 2018, *140*, 241–249.

7 Design and Modeling of Lightweight Polymer Composite Structures

Akarsh Verma, Naman Jain, Avinash Parashar, Vinay K. Singh, M. R. Sanjay, and Suchart Siengchin

CONTENTS

7.1 INTRODUCTION

Today's demand cannot be fulfilled by the monolithic metals and alloys. With the dynamic changes in technology, the properties of a single material cannot meet the present requirements. Therefore, the combination of several materials is required to overcome day-to-day life problems. For example, material used in satellites should have high dimensional stability as temperature in space varies from −160°C to 94°C [1], therefore, the material should have a coefficient of thermal expansion of about 10^{-7} m/m/°C that cannot be achieved with metals. improving efficiency, fuel saving, and reduction in weight without decrease in the stiffness and strength of the product are the major challenges for the automobile industries. These requirements also cannot be met with metals and ceramics; therefore, material with the combined properties of more than one material is required and these are known as composites [2–7]. A composite is a structural material, obtained when two or more materials are combined together without being soluble in each other at the macroscopic level. Composites are further classified on the basis of matrix such as the metal matrix composites (MMC), ceramic matrix composites (CMC), and polymer matrix composites (PMC) and according to the reinforcement used such as the particulate composites, fibrous composites, and laminate composites [8–12]. Composites play an essential role in today's world due to the many advantages they have over metals and alloys such as their low weight, corrosion resistance, low manufacturing cost, high creep resistance, etc. The major advantages of composites are their high specific strength (ratio of strength to density of a material) and specific modulus [13, 14]. Specific strength and specific modulus of some well-known composites are compared with those of metals as shown in Table 7.1. Furthermore, the automobile industries are looking for materials that are cheap, corrosion-resistant, low weight, easily manufactured, and available. These demands can only be fulfilled by the polymer matrix composites [15–24]. Taking an example of an aircraft, if weight of the aircraft is reduced by 1 lbm it can save about 1360 liters of fuel per year [25]. As per the survey of MIT in 2008, a 10 kg reduction of weight in a car and a truck can save $104 and $130, respectively [25].

7.2 MACRO-MECHANICAL BEHAVIOR OF LAMINA

Lamina is a thin single layer of composite in which fibers are oriented in unidirectional or woven form. When numbers of such layers are stack together, it is known as laminate (as shown in Figure 7.1). Laminate composites can be considered as leaf spring of the suspension system in automobiles. Unlike metals which are isotropic in nature, the stiffness of composites varies from point to point throughout the material, whether the point is considered on the fiber, matrix, or at the fiber-matrix interface. Therefore, the study of macro-mechanical properties and modeling of lamina become essential to understanding the fiber reinforced composites.

The stress-strain relationship for a linear isotropic material [26–29] subjected to three-dimensional (3D) state of stress can be written in the form of Equations 7.1 and 7.2:

TABLE 7.1
Comparison between Specific Weight and Specific Strength of Polymer-Based Composites with Commonly Used Material

	Specific gravity	Young's modulus (GPa)	Ultimate strength (MPa)	Specific modulus (GPa-m³/kg)	Specific strength (MPa-m³/kg)
Unidirectional glass/epoxy	1.8	38.60	1062	0.02144	0.5900
Cross-ply graphite/epoxy	1.6	95.98	373.0	0.0600	0.2331
Cross-ply glass/epoxy	1.8	23.58	88.25	0.01310	0.0490
Quasi-isotropic graphite/epoxy	1.6	69.64	276.48	0.04353	0.1728
Quasi-isotropic glass/epoxy	1.8	18.96	73.08	0.01053	0.0406
Steel	7.8	206.84	648.1	0.02652	0.08309
Aluminum	2.6	68.95	275.8	0.02652	0.1061

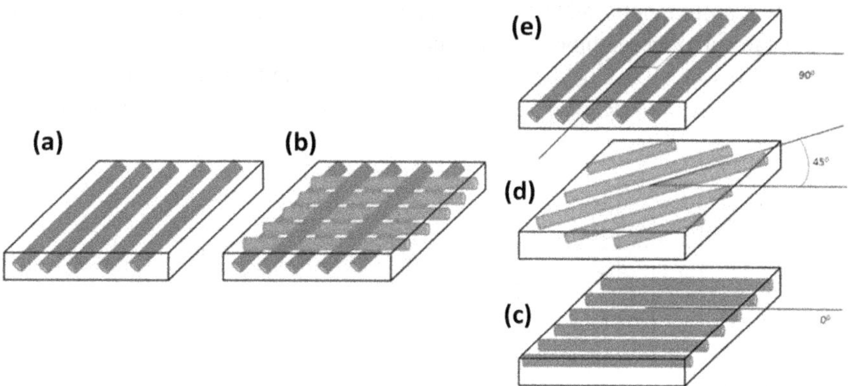

FIGURE 7.1 Types of lamina composites (a) Unidirectional, (b) Bi-directional/woven, (c) Unidirectional in the direction of stress applied, (d) Unidirectional at an angle of 45° of stress applied, and (e) Unidirectional perpendicular to the stress applied.

$$\begin{Bmatrix} \varepsilon_{xx} \\ \varepsilon_{yy} \\ \varepsilon_{zz} \\ \gamma_{yz} \\ \gamma_{xz} \\ \gamma_{xy} \end{Bmatrix} = \begin{bmatrix} \frac{1}{E} & \frac{-\mu}{E} & \frac{-\mu}{E} & 0 & 0 & 0 \\ \frac{-\mu}{E} & \frac{1}{E} & \frac{-\mu}{E} & 0 & 0 & 0 \\ \frac{-\mu}{E} & \frac{-\mu}{E} & \frac{1}{E} & 0 & 0 & 0 \\ 0 & 0 & 0 & \frac{1}{G} & 0 & 0 \\ 0 & 0 & 0 & 0 & \frac{1}{G} & 0 \\ 0 & 0 & 0 & 0 & 0 & \frac{1}{G} \end{bmatrix} \begin{Bmatrix} \sigma_{xx} \\ \sigma_{yy} \\ \sigma_{zz} \\ \tau_{yz} \\ \tau_{xz} \\ \tau_{xy} \end{Bmatrix} \tag{7.1}$$

$$\begin{Bmatrix} \sigma_{xx} \\ \sigma_{yy} \\ \sigma_{zz} \\ \tau_{xy} \\ \tau_{xz} \\ \tau_{yz} \end{Bmatrix} = \begin{bmatrix} \frac{E(1-\mu)}{(1-2\mu)(1+\mu)} & \frac{\mu E}{(1-2\mu)(1+\mu)} & \frac{\mu E}{(1-2\mu)(1+\mu)} & 0 & 0 & 0 \\ \frac{\mu E}{(1-2\mu)(1+\mu)} & \frac{E(1-\mu)}{(1-2\mu)(1+\mu)} & \frac{\mu E}{(1-2\mu)(1+\mu)} & 0 & 0 & 0 \\ \frac{\mu E}{(1-2\mu)(1+\mu)} & \frac{\mu E}{(1-2\mu)(1+\mu)} & \frac{E(1-\mu)}{(1-2\mu)(1+\mu)} & 0 & 0 & 0 \\ 0 & 0 & 0 & G & 0 & 0 \\ 0 & 0 & 0 & 0 & G & 0 \\ 0 & 0 & 0 & 0 & 0 & G \end{bmatrix} \begin{Bmatrix} \varepsilon_{xx} \\ \varepsilon_{yy} \\ \varepsilon_{zz} \\ \gamma_{yz} \\ \gamma_{xz} \\ \gamma_{xy} \end{Bmatrix} \tag{7.2}$$

where E is the Young's modulus/modulus of elasticity of the material, μ is the Poisson's ratio, and G is the shearing modulus of material. In general, composite materials are not linear and isotropic in nature, therefore, the stress-strain relationship contains a higher number of constants compared to Equation 7.2 and are represented in Equation 7.3. Due to the symmetry, $C_{ij} = C_{ji}$ and, therefore, only 21 elastic constants are there to determine such a material and the material is then known as anisotropic [30–35].

$$\begin{Bmatrix} \sigma_{xx} \\ \sigma_{yy} \\ \sigma_{zz} \\ \tau_{xy} \\ \tau_{xz} \\ \tau_{yz} \end{Bmatrix} = \begin{bmatrix} C_{11} & C_{12} & C_{13} & C_{14} & C_{15} & C_{16} \\ C_{12} & C_{22} & C_{23} & C_{24} & C_{25} & C_{26} \\ C_{13} & C_{23} & C_{33} & C_{43} & C_{53} & C_{63} \\ C_{14} & C_{24} & C_{34} & C_{44} & C_{54} & C_{64} \\ C_{15} & C_{25} & C_{35} & C_{45} & C_{55} & C_{65} \\ C_{16} & C_{26} & C_{36} & C_{46} & C_{56} & C_{66} \end{bmatrix} \begin{Bmatrix} \varepsilon_{xx} \\ \varepsilon_{yy} \\ \varepsilon_{zz} \\ \gamma_{yz} \\ \gamma_{xz} \\ \gamma_{xy} \end{Bmatrix} \tag{7.3}$$

(a) *Monoclinic material*

When the anisotropic material has one plane of symmetry, it is known as monoclinic material. Let us assume that the plane of symmetry is x-z plane, as shown in Figure 7.2. The change in result is x' = x, y' = y and z' = -z, and therefore, $\sigma'_{xx} = \sigma_{xx}$, $\sigma'_{yy} = \sigma_{yy}$, $\sigma'_{zz} = \sigma_{zz}$, $\tau'_{yz} = -\tau_{yz}$, $\tau'_{xz} = -\tau_{xz}$ and $\tau'_{xy} = \tau_{xy}$.

Considering the first condition, i.e., $\sigma'_{xx} = \sigma_{xx}$

$$C'_{11}\varepsilon'_{xx} + C'_{12}\varepsilon'_{yy} + C'_{13}\varepsilon'_{zz} + C'_{14}\gamma'_{yz} + C'_{15}\varepsilon'_{xz} + C'_{16}\varepsilon'_{xy} = C_{11}\varepsilon_{xx} + C_{12}\varepsilon_{yy}$$

$$+ C_{13}\varepsilon_{zz} + C_{14}\gamma_{yz} + C_{15}\varepsilon_{xz} + C_{16}\varepsilon_{xy}$$

this condition can only be true when $C_{14} = C_{15} = 0$.

Similarly,

$$\sigma'_{yy} = \sigma_{yy}, \text{ then } C_{24} = C_{25} = 0 \quad \sigma'_{yz} = -\sigma_{yz}, \text{ then } C_{46} = 0$$

$$\sigma'_{zz} = \sigma_{zz}, \text{ then } C_{34} = C_{35} = 0 \quad \sigma'_{xz} = -\sigma_{xz}, \text{ then } C_{56} = 0$$

Therefore, for the monoclinic material, there are only 13 independent constants, as shown in Equation 7.4.

$$\begin{Bmatrix} \sigma_{xx} \\ \sigma_{yy} \\ \sigma_{zz} \\ \tau_{xy} \\ \tau_{xz} \\ \tau_{yz} \end{Bmatrix} = \begin{bmatrix} C_{11} & C_{12} & C_{13} & 0 & 0 & C_{16} \\ C_{12} & C_{22} & C_{23} & 0 & 0 & C_{26} \\ C_{13} & C_{23} & C_{33} & 0 & 0 & C_{63} \\ 0 & 0 & 0 & C_{44} & C_{54} & 0 \\ 0 & 0 & 0 & C_{45} & C_{55} & 0 \\ 0 & 0 & 0 & 0 & 0 & C_{66} \end{bmatrix} \begin{Bmatrix} \varepsilon_{xx} \\ \varepsilon_{yy} \\ \varepsilon_{zz} \\ \gamma_{yz} \\ \gamma_{xz} \\ \gamma_{xy} \end{Bmatrix} \tag{7.4}$$

(b) *Symmetry with respect to two orthogonal planes*

Let us consider one more symmetric plane i.e., x-y other than the x-z (as in the case of monoclinic material). Both the planes are orthogonal to each other, as shown in Figure 7.2. The change in result is x' = x, y' = −y

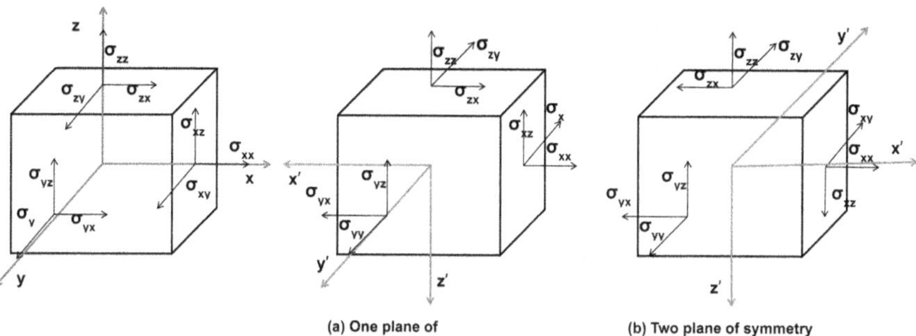

FIGURE 7.2 Stress acting inside the body (a) when one plane of symmetry exists, and (b) when two planes of symmetry exists.

and z' =-z, and therefore, the $\sigma'_{xx} = \sigma_{xx}$, $\sigma'_{yy} = \sigma_{yy}$, $\sigma'_{zz} = \sigma_{zz}$, $\tau'_{yz} = \tau_{yz}$, $\tau'_{xz} = -\tau_{xz}$ and $\tau'_{xy} = -\tau_{xy}$. This can only be true when $C_{16} = C_{26} = C_{36} = C_{45} = 0$. Therefore, the number of constant reduces to 9 [36–41], as shown in Equation 7.5.

$$\begin{Bmatrix} \sigma_{xx} \\ \sigma_{yy} \\ \sigma_{zz} \\ \tau_{xy} \\ \tau_{xz} \\ \tau_{yz} \end{Bmatrix} = \begin{bmatrix} C_{11} & C_{12} & C_{13} & 0 & 0 & 0 \\ C_{12} & C_{22} & C_{23} & 0 & 0 & 0 \\ C_{13} & C_{23} & C_{33} & 0 & 0 & 0 \\ 0 & 0 & 0 & C_{44} & 0 & 0 \\ 0 & 0 & 0 & 0 & C_{55} & 0 \\ 0 & 0 & 0 & 0 & 0 & C_{66} \end{bmatrix} \begin{Bmatrix} \varepsilon_{xx} \\ \varepsilon_{yy} \\ \varepsilon_{zz} \\ \gamma_{yz} \\ \gamma_{xz} \\ \gamma_{xy} \end{Bmatrix} \tag{7.5}$$

(c) *Transversal isotropic and isotropic material*

In orthotropic material, if one plane of material is isotropy the material is known as transversal isotropic [42, 43] i.e., when x-direction is normal to y-z plane, then the stiffness matrix is given as shown in Equation 7.6. In transversally isotropic material, $E_y = E_z$, $\mu_{xy} = \mu_{xz}$, $G_{xy} = G_{xz}$ and $G_{yz} = \dfrac{E_y}{2(1+\mu_{yz})}$. If all the plane of orthotropic material is isotropy, then material is known as isotropic material and the stiffness matrix is shown in Equation 7.7. In isotropic material, $E_x=E_y=E_z=E$, $\mu_{xy}=\mu_{yz}=\mu_{xz}=\mu$, $G_{xy}=G_{xz}=G_{yz}=G$ and $G=\dfrac{E}{2(1+\mu)}$.

$$\begin{Bmatrix} \sigma_{xx} \\ \sigma_{yy} \\ \sigma_{zz} \\ \tau_{xy} \\ \tau_{xz} \\ \tau_{yz} \end{Bmatrix} = \begin{bmatrix} C_{11} & C_{12} & C_{13} & 0 & 0 & 0 \\ C_{12} & C_{22} & C_{23} & 0 & 0 & 0 \\ C_{13} & C_{23} & C_{33} & 0 & 0 & 0 \\ 0 & 0 & 0 & \dfrac{C_{22}-C_{23}}{2} & 0 & 0 \\ 0 & 0 & 0 & 0 & C_{55} & 0 \\ 0 & 0 & 0 & 0 & 0 & C_{66} \end{bmatrix} \begin{bmatrix} \varepsilon_{xx} \\ \varepsilon_{yy} \\ \varepsilon_{zz} \\ \gamma_{yz} \\ \gamma_{xz} \\ \gamma_{xy} \end{bmatrix} \tag{7.6}$$

$$\begin{Bmatrix} \sigma_{xx} \\ \sigma_{yy} \\ \sigma_{zz} \\ \tau_{xy} \\ \tau_{xz} \\ \tau_{yz} \end{Bmatrix} = \begin{bmatrix} C_{11} & C_{12} & C_{13} & 0 & 0 & 0 \\ C_{12} & C_{22} & C_{23} & 0 & 0 & 0 \\ C_{13} & C_{23} & C_{33} & 0 & 0 & 0 \\ 0 & 0 & 0 & \dfrac{C_{11}-C_{22}}{2} & 0 & 0 \\ 0 & 0 & 0 & 0 & \dfrac{C_{11}-C_{22}}{2} & 0 \\ 0 & 0 & 0 & 0 & 0 & \dfrac{C_{11}-C_{22}}{2} \end{bmatrix} \begin{Bmatrix} \varepsilon_{xx} \\ \varepsilon_{yy} \\ \varepsilon_{zz} \\ \gamma_{yz} \\ \gamma_{xz} \\ \gamma_{xy} \end{Bmatrix} \quad (7.7)$$

A stress-strain relationship matrix can also be defined in terms of compliance matrix form as shown in Equation 7.8 for the orthotropic material.

Where $[C]=[S]^{-1}$

$$\begin{Bmatrix} \varepsilon_{xx} \\ \varepsilon_{yy} \\ \varepsilon_{zz} \\ \gamma_{yz} \\ \gamma_{xz} \\ \gamma_{xy} \end{Bmatrix} = \begin{bmatrix} S_{11} & S_{12} & S_{13} & 0 & 0 & 0 \\ S_{12} & S_{22} & S_{23} & 0 & 0 & 0 \\ S_{13} & S_{23} & S_{33} & 0 & 0 & 0 \\ 0 & 0 & 0 & S_{44} & 0 & 0 \\ 0 & 0 & 0 & 0 & S_{55} & 0 \\ 0 & 0 & 0 & 0 & 0 & S_{66} \end{bmatrix} \begin{Bmatrix} \sigma_{xx} \\ \sigma_{yy} \\ \sigma_{zz} \\ \tau_{xy} \\ \tau_{xz} \\ \tau_{yz} \end{Bmatrix} \quad (7.8)$$

To determine the relation between engineering constant and elements of compliance matrix, let us consider the following conditions:

Condition I: $\sigma_{xx} \neq 0$, $\sigma_{yy} = 0$, $\sigma_{zz} = 0$, $\tau_{yz} = 0$, $\tau_{xz} = 0$, and $\tau_{xy} = 0$

$$\left. \begin{aligned} \varepsilon_{xx} &= S_{11}\sigma_{xx} \\ \varepsilon_{yy} &= S_{12}\sigma_{xx} \\ \varepsilon_{zz} &= S_{13}\sigma_{xx} \end{aligned} \right\} \quad (7.9)$$

$$\gamma_{yz} = 0 \qquad \gamma_{xz} = 0 \qquad \gamma_{xy} = 0$$

Therefore,

$$E_x = \frac{1}{S_{11}} \quad (7.10)$$

$$\mu_{xy} = -\frac{S_{12}}{S_{11}} \quad \& \quad \mu_{xz} = -\frac{S_{13}}{S_{11}} \quad (7.11)$$

Condition II: $\sigma_{xx} = 0$, $\sigma_{yy} \neq 0$, $\sigma_{zz} = 0$, $\tau_{yz} = 0$, $\tau_{xz} = 0$, and $\tau_{xy} = 0$

$$E_y = \frac{1}{S_{22}} \quad (7.12)$$

$$\mu_{yx} = -\frac{S_{12}}{S_{22}} \quad \& \quad \mu_{yz} = -\frac{S_{23}}{S_{22}} \tag{7.13}$$

Condition III: $\sigma_{xx} = 0$, $\sigma_{yy} = 0$, $\sigma_{zz} \neq 0$, $\tau_{yz} = 0$, $\tau_{xz} = 0$, and $\tau_{xy} = 0$

$$E_z = \frac{1}{S_{33}} \tag{7.14}$$

$$\mu_{zx} = -\frac{S_{13}}{S_{33}} \quad \& \quad \mu_{zy} = -\frac{S_{23}}{S_{33}} \tag{7.15}$$

Condition IV: $\sigma_{xx} = 0$, $\sigma_{yy} = 0$, $\sigma_{zz} = 0$, $\tau_{yz} \neq 0$, $\tau_{xz} = 0$, and $\tau_{xy} = 0$
Condition V: $\sigma_{xx} = 0$, $\sigma_{yy} = 0$, $\sigma_{zz} = 0$, $\tau_{yz} = 0$, $\tau_{xz} \neq 0$, and $\tau_{xy} = 0$
Condition VI: $\sigma_{xx} = 0$, $\sigma_{yy} = 0$, $\sigma_{zz} = 0$, $\tau_{yz} = 0$, $\tau_{xz} = 0$, and $\tau_{xy} \neq 0$

$$\left.\begin{aligned} G_{yz} &= \frac{\tau_{yz}}{\gamma_{yz}} = \frac{1}{S_{44}} \\ G_{xz} &= \frac{\tau_{xz}}{\gamma_{xz}} = \frac{1}{S_{55}} \\ G_{xy} &= \frac{\tau_{xy}}{\gamma_{xy}} = \frac{1}{S_{66}} \end{aligned}\right\} \tag{7.16}$$

Hence, we need to define 12 engineering material constants that are three Young's modulus, one for each axis; six Poisson's ratio, two for each plane; and three shear moduli, one for each plane. Now, the compliance matrix can be rewritten in terms of material constants as shown in Equation 7.17. Moreover, six Poisson's ratios are not independent, as from Equations 7.11, 7.13, and 7.15 we get the following relations as shown in Equation 7.18.

$$[S] = \begin{bmatrix} \frac{1}{E_x} & \frac{-\mu_{12}}{E_x} & \frac{-\mu_{13}}{E_x} & 0 & 0 & 0 \\ \frac{-\mu_{21}}{E_y} & \frac{1}{E_y} & \frac{-\mu_{23}}{E_y} & 0 & 0 & 0 \\ \frac{-\mu_{31}}{E_z} & \frac{-\mu_{32}}{E_z} & \frac{1}{E_z} & 0 & 0 & 0 \\ 0 & 0 & 0 & \frac{1}{G_{23}} & 0 & 0 \\ 0 & 0 & 0 & 0 & \frac{1}{G_{31}} & 0 \\ 0 & 0 & 0 & 0 & 0 & \frac{1}{G_{12}} \end{bmatrix} \tag{7.17}$$

$$\frac{\mu_{xy}}{\mu_{yx}}=\frac{\dfrac{S_{12}}{S_{11}}}{\dfrac{S_{12}}{S_{22}}}=\frac{S_{22}}{S_{11}}=\frac{E_x}{E_y}.\ \text{Similarly,}\ \ \frac{\mu_{xz}}{\mu_{zx}}=\frac{E_x}{E_z}\ \text{and}\ \frac{\mu_{yz}}{\mu_{zy}}=\frac{E_y}{E_z} \tag{7.18}$$

The inverse of the compliance matrix represents the stiffness matrix as shown in Equation 7.19.

$$[S]=\begin{bmatrix} \dfrac{1-\mu_{yz}\mu_{zy}}{E_yE_z\Delta} & \dfrac{\mu_{yx}+\mu_{yz}\mu_{zx}}{E_yE_z\Delta} & \dfrac{\mu_{zx}+\mu_{yx}\mu_{zy}}{E_yE_z\Delta} & 0 & 0 & 0 \\ \dfrac{\mu_{yx}+\mu_{yz}\mu_{zx}}{E_yE_z\Delta} & \dfrac{1-\mu_{xz}\mu_{zx}}{E_xE_z\Delta} & \dfrac{\mu_{zy}+\mu_{xy}\mu_{zx}}{E_xE_z\Delta} & 0 & 0 & 0 \\ \dfrac{\mu_{zx}+\mu_{yx}\mu_{zy}}{E_yE_z\Delta} & \dfrac{\mu_{zy}+\mu_{xy}\mu_{zx}}{E_xE_z\Delta} & \dfrac{1-\mu_{xy}\mu_{yx}}{E_xE_y\Delta} & 0 & 0 & 0 \\ 0 & 0 & 0 & G_{23} & 0 & 0 \\ 0 & 0 & 0 & 0 & G_{31} & 0 \\ 0 & 0 & 0 & 0 & 0 & G_{12} \end{bmatrix} \tag{7.19}$$

$$\text{Where } \Delta=\left(1-\mu_{xy}\mu_{yx}-\mu_{yz}\mu_{zy}-\mu_{xz}\mu_{zx}-2\mu_{yx}\mu_{zy}\mu_{xz}\right)\Big/E_xE_yE_z \tag{7.20}$$

The diagonal element of stiffness matrix should be positive as per the first law of thermodynamics. Therefore, it results in Equation 7.21.

$$\begin{gathered} 1-\mu_{xy}\mu_{yx}>0,\ 1-\mu_{yz}\mu_{zy}>0,\ 1-\mu_{zx}\mu_{xz}>0 \\ \&\ 1-\mu_{xy}\mu_{yx}-\mu_{yz}\mu_{zy}-\mu_{zx}\mu_{xz}-2\mu_{xz}\mu_{yx}\mu_{zy}>0 \end{gathered} \tag{7.21}$$

From the above equations, we can conclude through the equations 7.22 and 7.23.

$$\left.\begin{aligned} &1-\mu_{xy}\ \mu_{yx}>0 \\ &\mu_{xy}<\frac{1}{\mu_{yx}}=\frac{E_x}{E_y\mu_{xy}} \\ &\left|\mu_{xy}\right|<\left|\sqrt{\frac{E_x}{E_y}}\right| \end{aligned}\right\} \tag{7.22}$$

Similarly,

$$\left|\mu_{yx}\right|<\left|\sqrt{\frac{E_y}{E_x}}\right|,\ \left|\mu_{zy}\right|<\left|\sqrt{\frac{E_z}{E_y}}\right|,\ \left|\mu_{yz}\right|<\left|\sqrt{\frac{E_y}{E_z}}\right|,\ \left|\mu_{xz}\right|<\left|\sqrt{\frac{E_x}{E_z}}\right|\ \&\ \left|\mu_{zx}\right|<\left|\sqrt{\frac{E_z}{E_x}}\right| \tag{7.23}$$

7.3 PLANE-STRESS CONDITION

If the state of stress only acts at one plane, then it is considered under plane stress criteria, i.e., stress on the third direction that are $\sigma_{zz} = 0$, $\tau_{yz} = 0$, and $\tau_{xz} = 0$. This type of condition will occur when the thickness of the material is much less in one direction, compared to the other two directions. If the thickness of lamina is very small and does not carry out of plane loads, then the condition of plane stress occurs. The following conditions apply: $\sigma_{zz} = 0$, $\tau_{yz} = 0$, and $\tau_{xz} = 0$, which results in $\varepsilon_{zz} = S_{12}\varepsilon_{xx} + S_{23}\varepsilon_{yy}$ and $\gamma_{yz} = \gamma_{xz} = 0$.

Therefore, ε_{zz}, γ_{yz}, and γ_{xz} can be omitted from stress-strain relationship and the stress problem for orthogonal material is written in the form of Equation 7.24:

$$\begin{Bmatrix} \sigma_{xx} \\ \sigma_{yy} \\ \sigma_{xy} \end{Bmatrix} = \begin{bmatrix} C_{11} & C_{12} & 0 \\ C_{21} & C_{22} & 0 \\ 0 & 0 & C_{66} \end{bmatrix} \begin{Bmatrix} \varepsilon_{xx} \\ \varepsilon_{yy} \\ \varepsilon_{xy} \end{Bmatrix} \tag{7.24}$$

Similarly, applying the condition as applied for the orthotropic material in a 3D case, we get Equation 7.25.

$$\begin{Bmatrix} \sigma_{xx} \\ \sigma_{yy} \\ \sigma_{xy} \end{Bmatrix} = \begin{bmatrix} \frac{E_x}{1-\mu_{yx}\mu_{xy}} & \frac{\mu_{xy}E_y}{1-\mu_{yx}\mu_{xy}} & 0 \\ \frac{\mu_{xy}E_y}{1-\mu_{yx}\mu_{xy}} & \frac{E_y}{1-\mu_{yx}\mu_{xy}} & 0 \\ 0 & 0 & G_{12} \end{bmatrix} \begin{Bmatrix} \varepsilon_{xx} \\ \varepsilon_{yy} \\ \varepsilon_{xy} \end{Bmatrix}. \tag{7.25}$$

7.4 HOOK'S LAW FOR ANGLE LAMINA IN TWO DIMENSIONS (2D)

In general, the orientation of fibers in lamina does not occur only unidirectionally because of low strength and stiffness of materials in other directions. In laminate composites, in some lamina the orientation of fiber is at a particular angle as shown in Figure 7.3.Thus, it is necessary to develop a stress-strain relationship for lamina composites also. Assuming, axis in x-y coordinate as global axes, while 1–2 are local axes, Direction 1 is parallel to the fiber orientation, while Direction 2 is perpendicular to the fiber.

Stress along the coordinates 1–2 can be related to the global coordinate system [44–46] by the following Expressions 26 and 27:

$$\sigma_1 ac = \sigma_x AB\cos\theta + \sigma_y BC\sin\theta + \tau_{xy} AB\cos\theta + \tau_{xy} BC\cos\theta \tag{7.26}$$

$$\sigma_1 = \sigma_x \cos\theta^2 + \sigma_y \sin\theta^2 + 2\tau_{xy}\sin\theta\cos\theta \tag{7.27}$$

Similarly,

$$\sigma_2 = \sigma_x \sin\theta^2 + \sigma_y \cos\theta^2 - 2\tau_{xy}\sin\theta\cos\theta \tag{7.28}$$

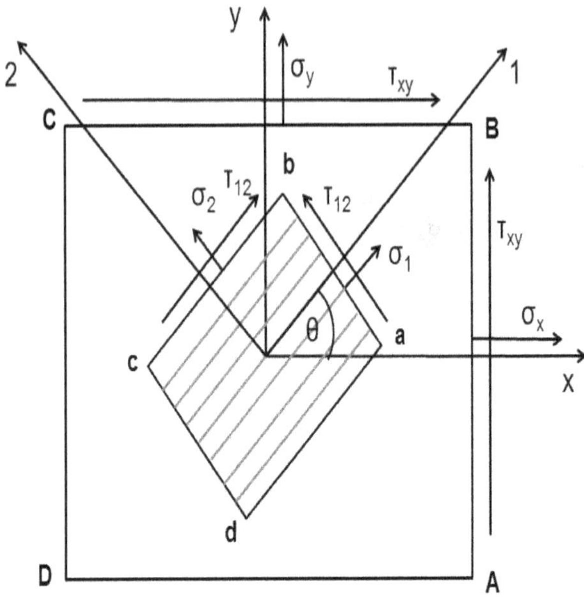

FIGURE 7.3 Stress distribution in an angle Lamina.

$$\tau_{12} = -\sigma_x \sin\theta\cos\theta + \sigma_y \sin\theta\cos\theta + 2\tau_{xy}\left(\cos\theta^2 - \sin\theta^2\right) \tag{7.29}$$

$$\begin{Bmatrix} \sigma_1 \\ \sigma_2 \\ \tau_{12} \end{Bmatrix} = \begin{bmatrix} \cos\theta^2 & \sin\theta^2 & 2\sin\theta\cos\theta \\ \sin\theta^2 & \cos\theta^2 & -2\sin\theta\cos\theta \\ -\sin\theta\cos\theta & \sin\theta\cos\theta & \cos\theta^2 - \sin\theta^2 \end{bmatrix} \begin{Bmatrix} \sigma_x \\ \sigma_y \\ \tau_{xy} \end{Bmatrix} \tag{7.30}$$

$$\begin{Bmatrix} \sigma_1 \\ \sigma_2 \\ \tau_{12} \end{Bmatrix} = \left[T\right] \begin{Bmatrix} \sigma_x \\ \sigma_y \\ \tau_{xy} \end{Bmatrix} \tag{7.31}$$

$$\begin{Bmatrix} \sigma_x \\ \sigma_y \\ \tau_{xy} \end{Bmatrix} = \begin{bmatrix} \cos\theta^2 & \sin\theta^2 & 2\sin\theta\cos\theta \\ \sin\theta^2 & \cos\theta^2 & -2\sin\theta\cos\theta \\ -\sin\theta\cos\theta & \sin\theta\cos\theta & \cos\theta^2 - \sin\theta^2 \end{bmatrix}^{-1} \begin{Bmatrix} \sigma_1 \\ \sigma_2 \\ \tau_{12} \end{Bmatrix} \tag{7.32}$$

$$\begin{Bmatrix} \sigma_x \\ \sigma_y \\ \tau_{xy} \end{Bmatrix} = \begin{bmatrix} \cos\theta^2 & \sin\theta^2 & -2\sin\theta\cos\theta \\ \sin\theta^2 & \cos\theta^2 & \sin\theta\cos\theta \\ \sin\theta\cos\theta & -\sin\theta\cos\theta & \cos\theta^2 - \sin\theta^2 \end{bmatrix} \begin{Bmatrix} \sigma_1 \\ \sigma_2 \\ \tau_{12} \end{Bmatrix} \tag{7.33}$$

Similarly, we can determine the transformation matrix for strain as well.

$$\begin{Bmatrix} \varepsilon_1 \\ \varepsilon_2 \\ \dfrac{\varepsilon_{12}}{2} \end{Bmatrix} = \begin{bmatrix} \cos\theta^2 & \sin\theta^2 & -2\sin\theta\cos\theta \\ \sin\theta^2 & \cos\theta^2 & \sin\theta\cos\theta \\ \sin\theta\cos\theta & -\sin\theta\cos\theta & \cos\theta^2-\sin\theta^2 \end{bmatrix} \begin{Bmatrix} \varepsilon_{xx} \\ \varepsilon_{yy} \\ \dfrac{\varepsilon_{xy}}{2} \end{Bmatrix} \tag{7.34}$$

$$\begin{Bmatrix} \varepsilon_1 \\ \varepsilon_2 \\ \varepsilon_{12} \end{Bmatrix} = [R][T][R]^{-1} \begin{Bmatrix} \varepsilon_{xx} \\ \varepsilon_{yy} \\ \varepsilon_{xy} \end{Bmatrix} \tag{7.35}$$

where

$$[T] = \begin{bmatrix} \cos\theta^2 & \sin\theta^2 & 2\sin\theta\cos\theta \\ \sin\theta^2 & \cos\theta^2 & -2\sin\theta\cos\theta \\ -\sin\theta\cos\theta & \sin\theta\cos\theta & \cos\theta^2-\sin\theta^2 \end{bmatrix} \quad \& \quad [T] = \begin{bmatrix} 1 & 0 & 0 \\ 0 & 1 & 0 \\ 0 & 0 & 2 \end{bmatrix}$$

For local coordinates, using the stress-strain relationship in Equation 7.34.

$$\begin{Bmatrix} \sigma_x \\ \sigma_y \\ \tau_{xy} \end{Bmatrix} = [T]^{-1}[C] \begin{Bmatrix} \varepsilon_1 \\ \varepsilon_2 \\ \varepsilon_{12} \end{Bmatrix} \tag{7.36}$$

After putting the value of local strain in term of global strain in Equation 7.36 we get,

$$\begin{Bmatrix} \sigma_x \\ \sigma_y \\ \tau_{xy} \end{Bmatrix} = [T]^{-1}[C][R][T][R]^{-1} \begin{Bmatrix} \varepsilon_{xx} \\ \varepsilon_{yy} \\ \varepsilon_{xy} \end{Bmatrix} = [Q] \begin{Bmatrix} \varepsilon_{xx} \\ \varepsilon_{yy} \\ \varepsilon_{xy} \end{Bmatrix} \tag{7.37}$$

After evaluating the term $[T]^{-1}[C][R][T][R]^{-1}$, we get,

$$\left.\begin{aligned} Q_{11} &= C_{11}\cos\theta^4 + C_{22}\sin\theta^4 + 2(C_{12}+2C_{66})\sin\theta^2\cos\theta^2 \\ Q_{11} &= C_{12}\left(\cos\theta^4+\sin\theta^2\right) + 2(C_{11}+C_{22}-4C_{66})\sin\theta^2\cos\theta^2 \\ Q_{22} &= C_{11}\sin\theta^4 + C_{22}\cos\theta^4 + 2(C_{12}+2C_{66})\sin\theta^2\cos\theta^2 \end{aligned}\right\} \tag{7.38}$$

$$Q_{16} = (C_{11}-C_{22}-4C_{66})\cos\theta^3\sin\theta^2 - (C_{22}-C_{12}-2C_{66})\sin\theta^3\cos\theta^2$$

$$Q_{26} = (C_{11}+C_{22}-4C_{66})\cos\theta\sin\theta^3 - (C_{22}-C_{12}-2C_{66})\cos\theta^3\sin\theta$$

$$Q_{66} = (C_{11}+C_{22}-2C_{12}-4C_{66})\sin\theta^2\cos\theta^2 + 2C_{66}\left(\sin\theta^4+\cos\theta^4\right)$$

$$\begin{Bmatrix} \sigma_x \\ \sigma_y \\ \tau_{xy} \end{Bmatrix} = \begin{bmatrix} Q_{11} & Q_{12} & Q_{16} \\ Q_{22} & Q_{21} & Q_{26} \\ Q_{61} & Q_{62} & Q_{66} \end{bmatrix} \begin{Bmatrix} \varepsilon_{xx} \\ \varepsilon_{yy} \\ \varepsilon_{xy} \end{Bmatrix} \quad \& \quad \begin{Bmatrix} \varepsilon_{xx} \\ \varepsilon_{yy} \\ \varepsilon_{xy} \end{Bmatrix} = \begin{bmatrix} \bar{S}_{11} & \bar{S}_{12} & \bar{S}_{16} \\ \bar{S}_{22} & \bar{S}_{21} & \bar{S}_{26} \\ \bar{S}_{61} & \bar{S}_{62} & \bar{S}_{66} \end{bmatrix} \begin{Bmatrix} \sigma_x \\ \sigma_y \\ \tau_{xy} \end{Bmatrix} \tag{7.39}$$

where $\left[Q\right]=\left[\bar{S}\right]^{-1}$

If only axial stress is applied in the unidirectional lamina, no coupling will occur in the material as shown in Equation 7.24. In an angle lamina when only normal stress is applied to the lamina, the shear stresses are non-zero. Therefore, coupling takes place between the normal and shearing terms. To determine the material constants for angle lamina, similar conditions are applied as for the unidirectional lamina one at a time, i.e., $\sigma_{xx} \neq 0$, $\sigma_{yy} = 0$, and $\tau_{xy} = 0$; $\sigma_{xx} = 0$, $\sigma_{yy} \neq 0$, and $\tau_{xy} = 0$ and $\sigma_{xx} = 0$, $\sigma_{yy} = 0$, and $\tau_{xy} \neq 0$.

$$\begin{aligned} &\text{Case I:} && \varepsilon_{xx} = \bar{S}_{11}\sigma_{xx}, && \varepsilon_{yy} = \bar{S}_{12}\sigma_{xx} && \& && \gamma_{yz} = \bar{S}_{16}\sigma_{xx} \\ &\text{Case II:} && \varepsilon_{xx} = \bar{S}_{12}\sigma_{yy}, && \varepsilon_{yy} = \bar{S}_{22}\sigma_{yy} && \& && \gamma_{yz} = \bar{S}_{26}\sigma_{yy} \\ &\text{Case III:} && \varepsilon_{xx} = \bar{S}_{16}\tau_{xy}, && \varepsilon_{yy} = \bar{S}_{26}\tau_{xy} && \& && \gamma_{yz} = \bar{S}_{66}\tau_{xy} \end{aligned}$$

Now, the material constants are defined as the following (Equations 7.40 and 7.41):

$$\overline{E_x} = \frac{1}{\bar{S}_{11}} \quad \& \quad E_x = \frac{1}{\bar{S}_{22}} \tag{7.40}$$

$$\overline{\mu_{xy}} = -\frac{\bar{S}_{12}}{\bar{S}_{11}}, \quad \overline{\mu_{yx}} = -\frac{\bar{S}_{12}}{\bar{S}_{22}} \quad \& \quad \overline{G_{xy}} = \frac{1}{\bar{S}_{66}} \tag{7.41}$$

As discussed earlier, an angle lamina coupling occurs between the normal stress and shear strain unlike the unidirectional lamina, which is known as the shear coupling. It is a non-dimensional parameter similar to the Poisson's ratio. It also relates the shear stress in x-y plane to the normal stress in x-direction. It is denoted by 'm_x' and the relation between normal stresses and shear strain in x-direction is mathematically defined as

$$\frac{1}{m_x} = -\frac{\sigma_{xx}}{\varepsilon_{xy}E_1} = -\frac{1}{\bar{S}_{16}E_1} \quad \& \quad \frac{1}{m_y} = -\frac{\sigma_{xx}}{\varepsilon_{xy}E_1} = -\frac{1}{\bar{S}_{26}E_1} \tag{7.42}$$

The stress-strain relationship for an angle lamina can be rewritten in the terms of material constant of angle lamina such as

$$\begin{Bmatrix} \varepsilon_{xx} \\ \varepsilon_{yy} \\ \varepsilon_{xy} \end{Bmatrix} = \begin{bmatrix} \dfrac{1}{\overline{\overline{E_x}}} & -\dfrac{\mu_{xy}}{\overline{\overline{E_x}}} & -\dfrac{m_x}{\overline{\overline{E_x}}} \\ -\dfrac{\overline{\mu_{xy}}}{\overline{\overline{E_x}}} & \dfrac{1}{\overline{\overline{E_y}}} & -\dfrac{m_y}{\overline{\overline{E_x}}} \\ -\dfrac{m_x}{\overline{\overline{E_x}}} & -\dfrac{m_y}{\overline{\overline{E_x}}} & \dfrac{1}{\overline{\overline{G_{xy}}}} \end{bmatrix} \begin{Bmatrix} \sigma_x \\ \sigma_y \\ \tau_{xy} \end{Bmatrix} \tag{7.43}$$

Values of material constants for angle lamina can also be written in term of material constant of unidirectional lamina using Equations 7.38 and 7.39:

$$\frac{1}{\overline{E_x}} = \overline{S_{11}} = S_{11}\cos\theta^4 + (2S_{12} + S_{66})\sin\theta^2\cos\theta^2 + S_{22}\sin\theta^4$$

$$= \frac{1}{E_x}\cos\theta^4 + \left(\frac{1}{G_{xy}} - \frac{2\mu_{12}}{E_x}\right)\sin\theta^2\cos\theta^2 + \frac{1}{E_y}\sin\theta^4 \quad (7.44)$$

$$\overline{\mu_{xy}} = -\overline{E_x}\overline{S_{12}}$$

$$= \overline{E_x}\left[S_{12}\left(\cos\theta^4 + \sin\theta^2\right) + (S_{11} + S_{22} - S_{66})\sin\theta^2\cos\theta^2\right]$$

$$= \overline{E_x}\left[\frac{\mu_{xy}}{E_x}\left(\cos\theta^4 + \sin\theta^2\right) - \left(\frac{1}{E_x} + \frac{1}{E_y} - \frac{1}{G_{xy}}\right)\sin\theta^2\cos\theta^2\right] \quad (7.45)$$

$$\frac{1}{\overline{E_y}} = \overline{S_{11}} = S_{11}\sin\theta^4 + (2S_{12} + S_{66})\sin\theta^2\cos\theta^2 + S_{22}\cos\theta^4$$

$$= \frac{1}{E_x}\sin\theta^4 + \left(\frac{1}{G_{xy}} - \frac{2\mu_{12}}{E_x}\right)\sin\theta^2\cos\theta^2 + \frac{1}{E_y}\cos\theta^4 \quad (7.46)$$

$$\frac{1}{\overline{G_{xy}}} = \overline{S_{66}}$$

$$= 2(2S_{11} + 2S_{22} - 4S_{12} - S_{66})\sin\theta^2\cos\theta^2 + S_{66}\left(\sin\theta^4 + \cos\theta^4\right)$$

$$= 2\left(\frac{2}{E_x} + \frac{2}{E_y} + 4\frac{\mu_{12}}{E_x} - \frac{1}{G_{xy}}\right)\sin\theta^2\cos\theta^2 + \frac{1}{G_{xy}}\left(\sin\theta^4 + \cos\theta^4\right) \quad (7.47)$$

$$m_x = -S_{16}\overline{E_x}$$

$$= -E_x\left[2(2S_{11} - 2S_{12} - S_{66})\sin\theta\cos\theta^3 - (2S_{11} - 2S_{12} - S_{66})\sin\theta^3\cos\theta\right] \quad (7.48)$$

7.5 MICRO-MECHANICAL BEHAVIOR OF LAMINA

In the previous section, we have developed the stress-strain relationship for an angle lamina which was developed by using the longitudinal Young's modulus, transverse Young's modulus, Poisson's ratio, and the in-plane shear modulus. These mentioned parameters can be determined experimentally through the tensile, compressive, and shear tests. In this section, we will focus on how to develop the mathematical models to find out these material constants by introducing the concept of volume/mass fraction of fiber and matrix in composites [47–51].

Let us consider v_c, ρ_c, and w_c volume, density, and mass of the composite, respectively. Similarly, the subscripts *f* and *m* represent the fiber and matrix terms. Mathematically, the fiber volume fraction and matrix volume fraction can be defined as:

$$V_f = \frac{v_f}{v_c} \quad \& \quad V_m = \frac{v_m}{v_c} \tag{7.49a}$$

$$V_f + V_m = 1 \quad \& \quad v_f + v_m = v_c. \tag{7.50a}$$

Similarly, mass fraction for the fiber and matrix can be defined as:

$$W_f = \frac{w_f}{w_c} \quad \& \quad W_m = \frac{w_m}{w_c} \tag{7.49b}$$

$$W_f = \frac{\rho_f}{\rho_c} V_f \quad \& \quad W_m = \frac{\rho_m}{\rho_c} V_m \tag{7.49c}$$

$$W_f + W_m = 1 \quad \& \quad w_f + w_m = w_c \tag{7.50b}$$

$$\rho_c v_c = \rho_f V_f + \rho_m V_m \tag{7.51}$$

Similarly, in terms of mass fraction:

$$\frac{1}{\rho_c} = \frac{W_f}{\rho_f} + \frac{W_m}{\rho_m} \tag{7.52}$$

In experimental analysis, actual density of the composite was found to be lower than the theoretical density mainly due to the presence of voids. These voids may occur during the casting process, which can lower the mechanical strength of material. Even a 1% void content may reduce the mechanical strength of the material up to a 10% factor. Now, we have defined two densities, i.e., the experimental density (ρ_{ce}) and the theoretical density (ρ_{ct}).

$$v_c = v_f + v_m + v_v \quad \text{where} \quad v_c = \frac{w_c}{\rho_{ct}} = v_f + v_m \quad \& \quad \frac{w_c}{\rho_{ce}} = \frac{w_c}{\rho_{ct}} + v_v \tag{7.53}$$

$$v_v = \frac{w_c}{\rho_{ce}} \left(\frac{\rho_{ct} - \rho_{ce}}{\rho_{ct}} \right) \tag{7.54}$$

To determine the Young's modulus in longitudinal direction for the lamina, consider a volume element that consists of fiber surrounded by the matrix and the load is applied in the same direction as of fiber (as shown in Figure 7.4a). Herein, the

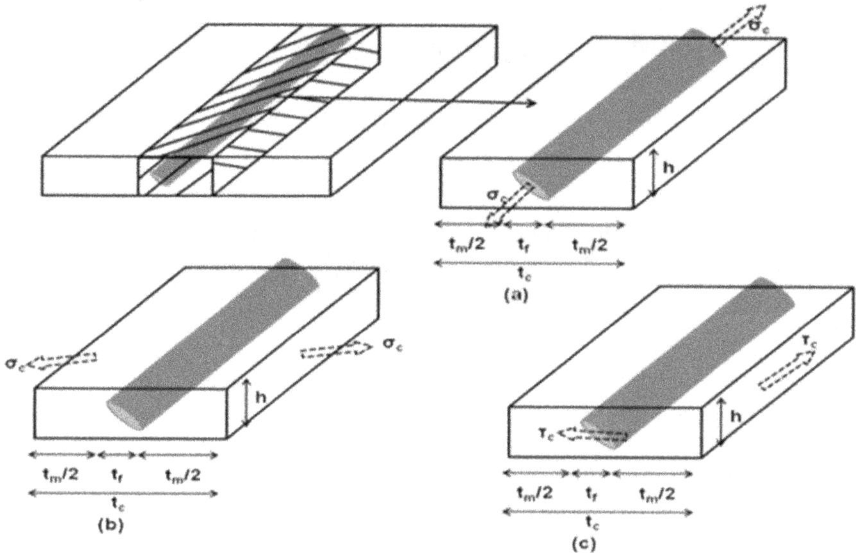

FIGURE 7.4 (a) Matrix loaded in the direction of fiber, (b) Matrix loaded in the perpendicular direction of fiber, and (c) Matrix loaded tangentially to the fiber.

material is considered to be linearly elastic, to follow Hook's law, and is free from any void. Total load on the composite (F_c) is shared by the fiber (F_f) and matrix (F_m).

$$F_c = \sigma_c A_c = F_f + F_m = \sigma_f A_f + \sigma_m A_m \tag{7.55}$$

$$E_x \epsilon_c A_c = E_f \epsilon_f A_f + E_m \epsilon_m A_m \tag{7.56}$$

$$E_x = E_f \frac{A_f}{A_c} + E_m \frac{A_m}{A_c}; \quad \text{where } \varepsilon_c = \varepsilon_f = \varepsilon_m \tag{7.57}$$

$$E_x = E_f V_f + E_m V_m = E_f V_f + E_m (1 - V_f) \tag{7.58}$$

To determine the Young's modulus in transversal direction for the lamina, the load is applied in the transverse direction to fiber, as shown in Figure 7.4b. Now, the total deflection of composite is given as:

$$U_c = t_f \varepsilon_f = U_f + U_m = t_f \varepsilon_f + t_m \varepsilon_m \tag{7.59}$$

$$t_c \frac{\sigma_c}{E_y} = t_f \frac{\sigma_f}{E_f} + t_m \frac{\sigma_m}{E_m} \tag{7.60}$$

$$\frac{1}{E_y} = \frac{t_f}{t_c}\frac{1}{E_f} + \frac{t_m}{t_c}\frac{1}{E_m} \tag{7.61}$$

$$\frac{1}{E_y} = \frac{V_f}{E_f} + \frac{V_m}{E_m} \tag{7.62}$$

Poisson's ratio for the lamina is defined as the ratio of strain of composite in transverse direction (ε_c^T) to the longitudinal direction (ε_c^L).

$$\mu_{xy} = -\frac{\varepsilon_c^T}{\varepsilon_c^L} \tag{7.63}$$

$$-t_c\mu_{xy}\varepsilon_c^L = -t_f\mu_f\varepsilon_f^L - t_m\mu_m\varepsilon_m^L \quad \text{where } \mu_f = -\frac{\varepsilon_f^T}{\varepsilon_f^L} \quad \& \quad \mu_m = -\frac{\varepsilon_m^T}{\varepsilon_m^L} \tag{7.64}$$

$$-t_c\mu_{xy}\varepsilon_c^L = -t_f\mu_f\varepsilon_f^L - t_m\mu_m\varepsilon_m^L \quad \text{where } \mu_f = -\frac{\varepsilon_f^T}{\varepsilon_f^L} \quad \& \quad \mu_m = -\frac{\varepsilon_m^T}{\varepsilon_m^L} \tag{7.65}$$

$$\mu_{xy} = \frac{t_f}{t_c}\mu_f + \frac{t_m}{t_c}\mu_m \tag{7.66}$$

$$\mu_{xy} = V_f\mu_f + V_m\mu_m \tag{7.67}$$

If the shear deformation of the lamina composite is δ_c, fiber is δ_f, and matrix is δ_m, then they are related by the following equation:

$$\delta_c = \delta_f + \delta_m \tag{7.68}$$

$$\gamma_c t_c = \gamma_f t_f + \gamma_m t_m \tag{7.69}$$

$$\frac{\tau_c}{G_{xy}}t_c = \frac{\tau_f}{G_f}t_f + \frac{\tau_m}{G_m}t_m \tag{7.70}$$

$$\frac{1}{G_{xy}} = \frac{V_f}{G_f} + \frac{V_m}{G_m} \tag{7.71}$$

If we compare the experimental results for transversal Young's modulus and the in-plane shear modulus with the Equations 7.62 and 7.71, there are some variations. To overcome these variations, Halpin and Tsai developed a new mathematical model:

$$\frac{E_y}{E_m} = \frac{1+\xi\eta V_f}{1-\eta V_f} \quad \text{where } \eta = \frac{\left(E_f/E_m\right)-1}{\left(E_f/E_m\right)+\xi} \tag{7.72}$$

where ξ is the reinforcing factor and depends on the fiber geometry, packing geometry, and loading condition. Such as ξ = 2 for circular cross-section fiber in square packing geometry, ξ = 2 (a/b) when the rectangular fiber with length a and width b (loading direction is b) is packed in the hexagonal geometry.

$$\frac{G_{xy}}{G_m} = \frac{1+\xi\eta V_f}{1-\eta V_f} \quad \text{where } \eta = \frac{\left(G_f/G_m\right)-1}{\left(E_f/G_m\right)+\xi} \tag{7.73}$$

The value of ξ = 1 for circular cross-section fiber in square packing geometry, whereas $\xi = \sqrt{3}\,\log_e\left(a/b\right)$ when the rectangular cross-section fiber with length a and width b (loading direction is b) is packed in the hexagonal geometry.

7.6 ANALYSIS OF LAMINATES

In the previous section, stress-strain relationship is developed for a single lamina, whereas in a laminate a number of such laminas with different orientations are stacked together. However, under complex loading, the unidirectional lamina will fail in the transverse direction. This problem can be overcome by stacking lamina at different orientations to improve upon the stiffness of lamina in transverse direction. Therefore, it is necessary to study the macro-mechanical analysis of laminate. In laminate, each lamina is made up of different materials and at different angles (known as ply angle). Notation of laminate is done in the square bracket in which ply angle of each lamina is separated by a slash. For example, consider [0/90/45/45/90/0] which represents a laminate composite consisting of six laminas of the same material with varying orientation with respect to the x-axis. Subscript T is also used in the notation, which represents the number of laminates at similar orientation such as $[0/90/45_2/90/0]$ (hear T is 2 for 45° ply angle lamina, i.e., two laminas at 45°). The above notation can also be denoted as $[45/90/0]_S$ which represents symmetric lamina above the mid-ply. When material of lamina is different, the notation will be done as $[0^{Gr}/\pm45^{B}]_S$ which represents six laminas: the two 0° laminas at top and bottom is made up of graphite (Gr)/epoxy whereas other laminas are made up of boron (B)/epoxy.

For the macro-mechanical analysis of material, we assume that each lamina is elastic, orthotropic, and homogeneous in nature; and after deformation, the lines that are straight and perpendicular to mid-surface remain the same, i.e., $\varepsilon_{xz} = \varepsilon_{yz} = 0$. The laminate is loaded in two dimensions with displacement less than the thickness of lamina. Also, there is no slipping between the two lamina interfaces. The midpoint of plate is at z = 0, and u_0, v_0, and w_0 are the displacements at the mid-plane whereas u, v, and w are the displacements at any point in the x, y, and z direction, respectively. As shown in Figure 7.4, by taking the cross-section in the x-y plane, displacement in the x-direction is given as:

$$u = u_0 - z\alpha = u_0 - z\frac{\partial w_0}{\partial x} \tag{7.74}$$

Similarly, taking the cross-section in the y-z plane, displacement in the x-direction is given as:

$$v = v_0 - z\alpha = u_0 - z\frac{\partial w_0}{\partial y} \tag{7.75}$$

Now, the strain in two-dimensional plane is given by:

$$\varepsilon_x = \frac{\partial u}{\partial x} = \frac{\partial u_0}{\partial x} - z\frac{\partial^2 w_0}{\partial x^2} \tag{7.76}$$

$$\varepsilon_y = \frac{\partial v}{\partial y} = \frac{\partial v_0}{\partial y} - z\frac{\partial^2 w_0}{\partial y^2} \tag{7.77}$$

$$\varepsilon_{xy} = \frac{\partial u}{\partial y} + \frac{\partial v}{\partial x} = \frac{\partial u_0}{\partial y} + \frac{\partial v_0}{\partial x} - z\frac{\partial^2 w_0}{\partial x \partial y} \tag{7.78}$$

The Equations 7.76, 7.77, and 7.78 can also be written in the matrix form:

$$\begin{Bmatrix} \varepsilon_x \\ \varepsilon_y \\ \varepsilon_{xy} \end{Bmatrix} = \begin{Bmatrix} \frac{\partial u_0}{\partial x} \\ \frac{\partial v_0}{\partial y} \\ \frac{\partial u_0}{\partial y} + \frac{\partial v_0}{\partial x} \end{Bmatrix} + z\begin{Bmatrix} \kappa_x \\ \kappa_y \\ \kappa_{xy} \end{Bmatrix} \tag{7.79}$$

Equation 7.79 can also be rewritten as:

$$\begin{Bmatrix} \varepsilon_x \\ \varepsilon_y \\ \varepsilon_{xy} \end{Bmatrix} = \underbrace{\begin{Bmatrix} \varepsilon_x^0 \\ \varepsilon_y^0 \\ \varepsilon_{xy}^0 \end{Bmatrix}}_{\text{Mid-plane strain}} - z \underbrace{\begin{Bmatrix} \frac{\partial^2 w_0}{\partial x^2} \\ \frac{\partial^2 w_0}{\partial y^2} \\ \frac{\partial^2 w_0}{\partial x \partial y} \end{Bmatrix}}_{\text{mid-plane curvature}} \tag{7.80}$$

Combining Equations 7.39 and 7.80 we get:

$$\begin{Bmatrix} \sigma_x \\ \sigma_y \\ \tau_{xy} \end{Bmatrix} = \begin{bmatrix} Q_{11} & Q_{12} & Q_{16} \\ Q_{22} & Q_{21} & Q_{26} \\ Q_{61} & Q_{62} & Q_{66} \end{bmatrix} \begin{Bmatrix} \varepsilon_x^0 \\ \varepsilon_y^0 \\ \varepsilon_{xy}^0 \end{Bmatrix} + z \begin{bmatrix} Q_{11} & Q_{12} & Q_{16} \\ Q_{22} & Q_{21} & Q_{26} \\ Q_{61} & Q_{62} & Q_{66} \end{bmatrix} \begin{Bmatrix} \kappa_x \\ \kappa_y \\ \kappa_{xy} \end{Bmatrix} \tag{7.81}$$

For laminate, the value of transformed matrix [Q] changes from each lamina to lamina as it depends upon the material and orientation of lamina. Therefore, the value of stress may vary from lamina to lamina, whereas the strain remains linear along the thickness as shown in Figure 7.4.

The laminate is made up of 'n' lamina with each lamina thickness 't.' Then, the total thickness of laminate is $h=\sum t_k$. Then, h/2 is the location of mid-plane and the location for each lamina is given as:

$$h_{k-1} = -\frac{h}{2} + \sum_{=1}^{k-1} t \left(\text{top surface}\right) \quad \left(k = 2,\ 3,\ 4,\ldots n-1\right) \tag{7.82}$$

$$h_n = \frac{h}{2} - t_n \left(\text{top surface}\right) \tag{7.83}$$

The value of resultant forces per unit length through the laminate thickness and the movements per unit length in x-y plane is obtained by integrating the global stress as follows:

$$\begin{bmatrix} F_x \\ F_y \\ F_{xy} \end{bmatrix} = \int_{-h/2}^{h/2} \begin{bmatrix} \sigma_x \\ \sigma_y \\ \sigma_{xy} \end{bmatrix} dz \tag{7.84}$$

$$\begin{bmatrix} M_x \\ M_y \\ M_{xy} \end{bmatrix} = \int_{-h/2}^{h/2} \begin{bmatrix} \sigma_x \\ \sigma_y \\ \sigma_{xy} \end{bmatrix} zdz \tag{7.85}$$

$$\begin{bmatrix} F_x \\ F_y \\ F_{xy} \end{bmatrix} = \sum_{k=1}^{n} \int_{h_{k-1}}^{h_k} \begin{bmatrix} \sigma_x \\ \sigma_y \\ \sigma_{xy} \end{bmatrix} dz \tag{7.86}$$

$$\begin{bmatrix} M_x \\ M_y \\ M_{xy} \end{bmatrix} = \sum_{k=1}^{n} \int_{h_{k-1}}^{h_k} \begin{bmatrix} \sigma_x \\ \sigma_y \\ \sigma_{xy} \end{bmatrix} zdz \tag{7.87}$$

Substituting the values of stresses from Equation 7.81 into Equation 7.87, we get

$$\begin{bmatrix} F_x \\ F_y \\ F_{xy} \end{bmatrix} = \sum_{k=1}^{n} \int_{h_{k-1}}^{h_k} \begin{bmatrix} Q_{11} & Q_{12} & Q_{16} \\ Q_{22} & Q_{21} & Q_{26} \\ Q_{61} & Q_{62} & Q_{66} \end{bmatrix} \begin{Bmatrix} \varepsilon_x^0 \\ \varepsilon_y^0 \\ \varepsilon_{xy}^0 \end{Bmatrix} dz + \sum_{k=1}^{n} \int_{h_{k-1}}^{h_k} \begin{bmatrix} Q_{11} & Q_{12} & Q_{16} \\ Q_{22} & Q_{21} & Q_{26} \\ Q_{61} & Q_{62} & Q_{66} \end{bmatrix} \begin{Bmatrix} \kappa_x \\ \kappa_y \\ \kappa_{xy} \end{Bmatrix} z dz \tag{7.88}$$

$$\begin{bmatrix} M_x \\ M_y \\ M_{xy} \end{bmatrix} = \sum_{k=1}^{n} \int_{h_{k-1}}^{h_k} \begin{bmatrix} Q_{11} & Q_{12} & Q_{16} \\ Q_{22} & Q_{21} & Q_{26} \\ Q_{61} & Q_{62} & Q_{66} \end{bmatrix} \begin{Bmatrix} \varepsilon_x^0 \\ \varepsilon_y^0 \\ \varepsilon_{xy}^0 \end{Bmatrix} z dz + \sum_{k=1}^{n} \int_{h_{k-1}}^{h_k} \begin{bmatrix} Q_{11} & Q_{12} & Q_{16} \\ Q_{22} & Q_{21} & Q_{26} \\ Q_{61} & Q_{62} & Q_{66} \end{bmatrix} \begin{Bmatrix} \kappa_x \\ \kappa_y \\ \kappa_{xy} \end{Bmatrix} z^2 dz \tag{7.89}$$

In the above equation, strains and curvature at the mid-plane are independent of z, also $[Q]$ matrix is constant for each lamina. Therefore, the above equation can be written as:

$$\begin{bmatrix} F_x \\ F_y \\ F_{xy} \end{bmatrix} = \left\{ \sum_{k=1}^{n} \begin{bmatrix} Q_{11} & Q_{12} & Q_{16} \\ Q_{22} & Q_{21} & Q_{26} \\ Q_{61} & Q_{62} & Q_{66} \end{bmatrix} \int_{h_{k-1}}^{h_k} dz \right\} \begin{Bmatrix} \varepsilon_x^0 \\ \varepsilon_y^0 \\ \varepsilon_{xy}^0 \end{Bmatrix} + \left\{ \sum_{k=1}^{n} \begin{bmatrix} Q_{11} & Q_{12} & Q_{16} \\ Q_{22} & Q_{21} & Q_{26} \\ Q_{61} & Q_{62} & Q_{66} \end{bmatrix} \int_{h_{k-1}}^{h_k} z dz \right\} \begin{Bmatrix} \kappa_x \\ \kappa_y \\ \kappa_{xy} \end{Bmatrix} \tag{7.90}$$

$$\begin{bmatrix} M_x \\ M_y \\ M_{xy} \end{bmatrix} = \left\{ \sum_{k=1}^{n} \begin{bmatrix} Q_{11} & Q_{12} & Q_{16} \\ Q_{22} & Q_{21} & Q_{26} \\ Q_{61} & Q_{62} & Q_{66} \end{bmatrix} \int_{h_{k-1}}^{h_k} z dz \right\} \begin{Bmatrix} \varepsilon_x^0 \\ \varepsilon_y^0 \\ \varepsilon_{xy}^0 \end{Bmatrix} + \left\{ \sum_{k=1}^{n} \begin{bmatrix} Q_{11} & Q_{12} & Q_{16} \\ Q_{22} & Q_{21} & Q_{26} \\ Q_{61} & Q_{62} & Q_{66} \end{bmatrix} \int_{h_{k-1}}^{h_k} z^2 dz \right\} \begin{Bmatrix} \kappa_x \\ \kappa_y \\ \kappa_{xy} \end{Bmatrix} \tag{7.91}$$

After integration, the above equation can be written in the matrix form as:

$$\begin{bmatrix} F_x \\ F_y \\ F_{xy} \end{bmatrix} = \begin{bmatrix} A_{11} & A_{12} & A_{16} \\ A_{22} & A_{21} & A_{26} \\ A_{61} & A_{62} & A_{66} \end{bmatrix} \begin{Bmatrix} \varepsilon_x^0 \\ \varepsilon_y^0 \\ \varepsilon_{xy}^0 \end{Bmatrix} + \begin{bmatrix} B_{11} & B_{12} & B_{16} \\ B_{22} & B_{21} & B_{26} \\ B_{61} & B_{62} & B_{66} \end{bmatrix} \begin{Bmatrix} \kappa_x \\ \kappa_y \\ \kappa_{xy} \end{Bmatrix} \tag{7.92}$$

$$\begin{bmatrix} M_x \\ M_y \\ M_{xy} \end{bmatrix} = \begin{bmatrix} B_{11} & B_{12} & B_{16} \\ B_{22} & B_{21} & B_{26} \\ B_{61} & B_{62} & B_{66} \end{bmatrix} \begin{Bmatrix} \varepsilon_x^0 \\ \varepsilon_y^0 \\ \varepsilon_{xy}^0 \end{Bmatrix} + \begin{bmatrix} D_{11} & D_{12} & D_{16} \\ D_{22} & D_{21} & D_{26} \\ D_{61} & D_{62} & D_{66} \end{bmatrix} \begin{Bmatrix} \kappa_x \\ \kappa_y \\ \kappa_{xy} \end{Bmatrix} \tag{7.93}$$

where

$$\left.\begin{aligned} A_{ij} &= \sum_{k=1}^{n} \left[Q_{ij}\right]_k \left(h_k - h_{k-1}\right) \quad i = 1,\ 2,\ 6\ \&\ j = 1,\ 2,\ 6 \\ B_{ij} &= \frac{1}{2}\sum_{k=1}^{n} \left[Q_{ij}\right]_k \left(h_k^2 - h_{k-1}^2\right) \quad i = 1,\ 2,\ 6\ \&\ j = 1,\ 2,\ 6 \\ D_{ij} &= \frac{1}{3}\sum_{k=1}^{n} \left[Q_{ij}\right]_k \left(h_k^3 - h_{k-1}^3\right) \quad i = 1,\ 2,\ 6\ \&\ j = 1,\ 2,\ 6 \end{aligned}\right\} \tag{7.94}$$

Matrixes A, B, and D are called the extensional, coupling, and bending stiffness matrixes, respectively. Combining the above matrix, we get

$$\begin{bmatrix} F_x \\ F_y \\ F_{xy} \\ M_x \\ M_y \\ M_{xy} \end{bmatrix} = \begin{bmatrix} A_{11} & A_{12} & A_{16} & B_{11} & B_{12} & B_{16} \\ A_{22} & A_{21} & A_{26} & B_{22} & B_{21} & B_{26} \\ A_{61} & A_{62} & A_{66} & B_{61} & B_{62} & B_{66} \\ B_{11} & B_{12} & B_{16} & D_{11} & D_{12} & D_{16} \\ B_{22} & B_{21} & B_{26} & D_{22} & D_{21} & D_{26} \\ B_{61} & B_{62} & B_{66} & D_{61} & D_{62} & D_{66} \end{bmatrix} \begin{Bmatrix} \varepsilon_x^0 \\ \varepsilon_y^0 \\ \varepsilon_{xy}^0 \\ \kappa_x \\ \kappa_y \\ \kappa_{xy} \end{Bmatrix} \tag{7.95}$$

Relation of the resultant in-plane forces to the in-plain strain is given by Matrix [A], Matrix [B] couples the force and movement terms to the mid-plane strains, and the mid-plane curvature and resultant bending movement to the plate curvature is given by Matrix [D].

7.7 VISCO-ELASTIC BEHAVIOR OF THE POLYMER MATRIX COMPOSITE

Engineering materials basically come under two categories: the first is elastic solid, which under applied load stores all the energy and during the unloading is restored

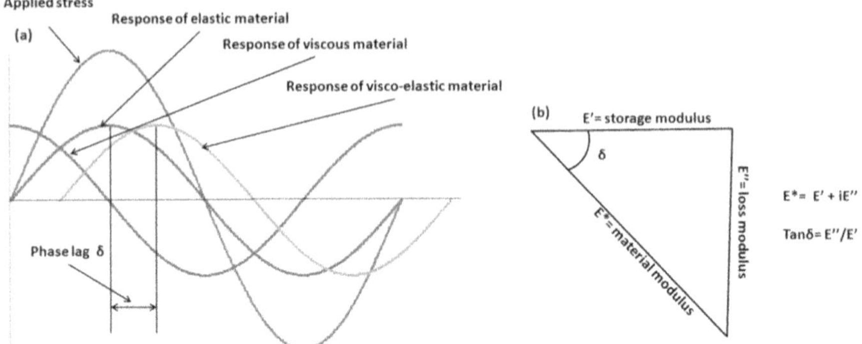

FIGURE 7.5 (a) Response of material under the application of oscillating stress, and (b) relationship between storage and loss modulus.

to its original shape. The second type is the viscous liquid, which upon the loading loses energy and does not regain its original shape. Under a low strain rate, the material behaves like a viscous material while at a high strain it behaves as elastic. Engineering applications of the polymer depend upon their stiffness. The material property of Young's modulus 'E' in the case of polymer is the function of time and temperature, unlike the case of conventional metals. The modulus of polymer composites may change by a factor of 100 whenever there is any change in the temperature values. Polymer-based composites come under the category of visco-elastic materials. They have the characteristic of 'fading memory,' i.e., they are time dependent.

Dynamic mechanical analysis (DMA) is used to determine the visco-elastic properties of composite such as the storage modulus, loss modulus, and glass transition temperature. The oscillating force/stress is applied to the specimen and response of the material is measured, as shown in Figure 7.5. Oscillating force generates the sinusoidal stress that results in a respective sinusoidal strain. Values such as the amplitude at peak and the phase lag between two sinusoidal waves are measured to determine the properties such as modulus and viscosity of the material. On the basis of behavior, materials are mainly of two types, i.e., either Hookean or elastic behavior and Newton or viscous behavior. Storage modulus, also known as the elastic modulus, is defined as the amount of energy stored in a material under the application of stress that states the elastic nature of a material. For elastic materials, response of the material is in the same phase with the applied oscillation stress as shown in Figure 7.5. Loss modulus is the amount of energy that disappears under the application of stress in the form of heat energy. For viscous material, the response of material is out of phase. In general, composite material shows visco-elastic response to the material which results in phase lag between the applied stress and strain output; this phase lag represents the viscous nature of material.

$$\text{Applied oscillating stress} \qquad \sigma(t) = \sigma_0 \sin \omega t \tag{7.96}$$

where σ_0 is maximum amplitude of stress and t and ω are time period and oscillating frequency, respectively.

$$\text{Response of the elastic material:} \qquad \sigma = E\epsilon \tag{7.97}$$

$$\epsilon(t) = \frac{\sigma}{E} = \frac{\sigma_0 \sin\omega t}{E} = \epsilon_0 \sin\omega t \tag{7.98}$$

where $\epsilon_0 = \dfrac{\sigma_0}{E}$ is the maximum stain produced in the material.

$$\text{Response of the viscous material} \qquad \sigma = \eta\gamma = \eta\left(\frac{-d\epsilon}{dt}\right) \tag{7.99}$$

$$\epsilon(t) = -\int \frac{\sigma_0 \sin\omega t \; dt}{\eta} \tag{7.100}$$

$$\epsilon(t) = \frac{\sigma_{0\cos\omega t}}{\eta} = \epsilon_0 \sin(\omega t + \pi/2) \tag{7.101}$$

where σ, E, ϵ, γ, and η are the stress, Young's modulus, strain, strain rate, and viscosity, respectively.

Another most important application of DMA is that it measures the effect of temperature on stiffness of a material which further governs its working range in the real-life conditions, i.e., at higher temperature values stiffness of composites decreases leading to serious problems [52]. In a composite, when the temperature is low, polymer molecular chains are closely packed and are in the solid state. Under the application of load, with increase in temperature free volume is increased which facilitates the stretching, bending, and rotation of molecular chain and is represented as gamma transition (T_γ). With further increase in the temperature, free volume of material increases resulting in movement of side chains known as the beta transition (T_β); this may also be associated with the movement of cross-linked branch chains [52]. At elevated temperature, the composite behaves like fluid, i.e., there is sliding/movement of main chains one over the other also known as the large-scale movement. This transition is known as glass transition (T_α). In 100% crystalline material, this transition is not seen. Further increase in the temperature results in melting of composite and this temperature is known as the melting temperature (T_m).

Creep-recovery behavior of polymer composite is another useful characterization technique to study the visco-elastic nature. In creep analysis, a constant load is applied to the sample for a certain period of time and when the load is removed, the recovery of material is also measured with it. There are mainly three stages in creep behavior that include instantaneous deformation, also known as the elastic strain (ε_e), visco-elastic strain (ε_v), and viscous strain ($\varepsilon\infty$). In the recovery phase after the removal of stress there is an instantaneous recovery of strain that represents an elastic nature of composite, then there is time-dependent recovery which is visco-elastic behavior, and there is permanent deformation Which reveals the viscous nature of composite.

$$\epsilon(t) = \epsilon_e + \epsilon_v + \epsilon_\infty \tag{7.102}$$

The time-dependent strain is quantified in term of creep compliance (J) as:

$$J(t) = \frac{\varepsilon(t)}{\sigma} \quad \text{where } \varepsilon \;\&\; \sigma \text{ are the creep strain and stress, respectively} \tag{7.103}$$

It is easy to describe the creep theoretically whereas it is difficult to do so through a mathematical approach. Elastic behavior of composites is considered analogous to a spring while the viscous behavior is analogous to a dashpot. To study the creep behavior of composites, many models that are a combination of spring and dashpot are studied as shown in Figure 7.6.

The most basic model of all is the Maxwell model which consists of one spring (having spring constant E_s) and dashpot (viscosity of dashpot η_d) in a series configuration where s and d represent the spring and dashpot, respectively. When the same amount of stress (σ_0) and strain in each element is observed, then:

$$\sigma_0 = \sigma_s = \sigma_d \tag{7.104}$$

$$\varepsilon = \varepsilon_s + \varepsilon_d \quad \text{where} \quad \varepsilon_s = \frac{\sigma_0}{E_s} \quad \& \quad \dot{\varepsilon}_d = \frac{\sigma_0}{\eta_d} \tag{7.105}$$

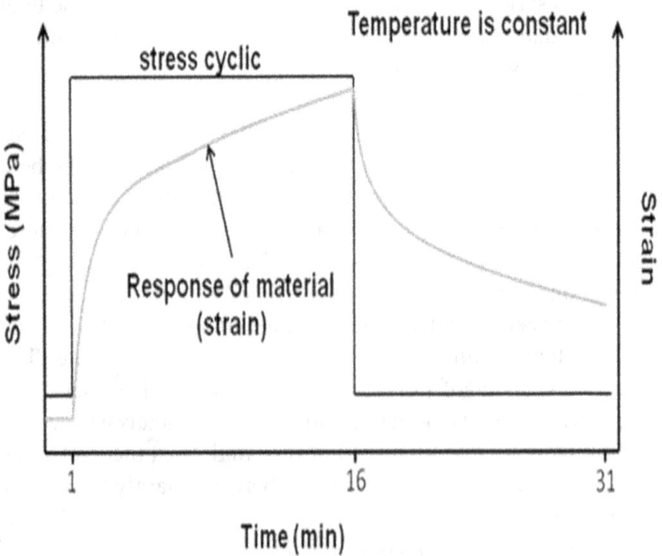

FIGURE 7.6 A typical creep and recovery analysis graph.

$$\varepsilon(t) = \frac{\sigma_0}{E_s} + \frac{\sigma_0}{\eta_d} t \tag{7.106}$$

In the Maxwell model, under stress σ_0 spring respond immediately while the dashpot takes time to react. Similarly, when stress is removed, immediate recovery is done by the spring, while the creep strain due to dashpot remains unrecovered. The creep-recovery response of the Maxwell model is presented in Figure 7.7. The creep compliance function for Maxwell model is represented as:

$$J(t) = \frac{1}{E_s} + \frac{t}{\eta_d} \tag{7.107}$$

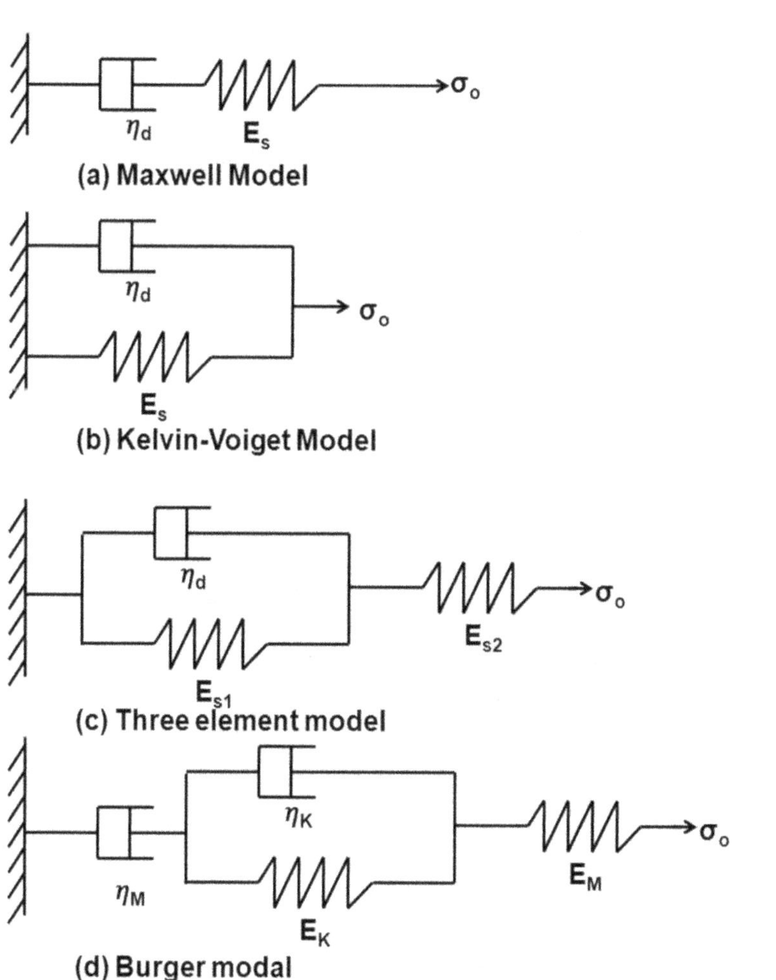

FIGURE 7.7 Different types of visco-elastic model for mathematical modeling of composite material.

Next, a two-element model is the Kelvin-voigt model, in which the spring and dashpot are connected in parallel as shown in Figure 7.7. The strain that occurs in each element is the same while the applied stress is equal to the sum of the spring and dashpot stresses individually.

$$\varepsilon = \varepsilon_s = \varepsilon_d \tag{7.108}$$

$$\sigma_0 = \sigma_s + \sigma_d \quad \text{where} \quad \varepsilon_s = \frac{\sigma_0}{E_s} \quad \& \quad \dot{\varepsilon}_d = \frac{\sigma_0}{\eta_d} \tag{7.109}$$

$$\sigma_0 = E_s\varepsilon + \eta_d\dot{\varepsilon} \tag{7.110}$$

Under the application of stress, spring will react immediately while the response of spring is restricted by the dashpot; therefore, the initial strain is transferred to the dashpot resulting in creep curve with initial slope of σ_0/η_d as shown in Figure 7.7. Then, the dashpot will take some strain and some will be transferred back to the spring. Now, slope of the creep curve becomes σ_d/η_d where σ_d is decreasing and at a limiting condition when $\sigma_d = 0$, spring will take all the stress with maximum strain of σ_0/E_s. On solving the above partial differential equations, we get:

$$\varepsilon(t) = \frac{\sigma_0}{E}\left(1 - e^{-(t/\tau)}\right) \tag{7.111}$$

$$J(t) = \frac{1}{E}\left(1 - e^{-(t/\tau)}\right) \tag{7.112}$$

where τ (η_d/E_s) is the retardation time for the Kelvin model, which is the time taken for the creep strain to accumulate. Creep strain increases rapidly for a shorter retardation time. During unloading, recovery of the spring is restricted by the dashpot. Therefore, in time dashpot comes into its original position and fully recovers. In this model, instantaneous elastic response of the composite is not represented. For this, a three-elements model in which the Kelvin model is attached with spring in series is appropriate. Then the creep strain is given as:

$$\in(t) = \frac{\sigma_0}{E_{s1}} + \frac{\sigma_0}{E_{s2}}\left(1 - e^{-t\Big/\frac{\eta_d}{E_{s2}}}\right) \tag{7.113}$$

where s1 and s2 are the spring constants for Spring 1 and 2, respectively, and η_d is the viscosity of dashpot. Moreover, this model also fails in actual production of creep strain as during the second stage of creep after visco-elastic nature, viscous behavior occurs. Therefore, in the above model additional dashpot is being applied and the model is known as a Burger modal [53, 54], as shown in Figure 7.7. It can also be seen as the combination of Maxwell and Kelvin-Voigt models in series such that the

Kelvin model comes in between the spring and dashpot of Maxwell models. The creep strain response is given as:

$$\epsilon(t) = \frac{\sigma_0}{E_M} + \frac{\sigma_0}{E_K}\left(1 - e^{-t/\frac{\eta_K}{E_K}}\right) + \frac{\sigma_0 t}{\eta_M} \tag{7.114}$$

where $\epsilon(t)$ is the creep strain, E_M and E_K represent the Maxwell and Kelvin spring elastic modulus, and η_M and η_K represent the viscosities of Maxwell and Kelvin dashpots. The time required to generate 63.2% of deformation in Kelvin unit is known as the retardation time τ.

7.8 MOLECULAR DYNAMICS TECHNIQUE TO MODEL LIGHTWEIGHT POLYMER COMPOSITES

Recently, the molecular dynamics (MD) technique has been applied to characterize the various properties of lightweight polymer composites [55]. It is a semi-empirical computer-based simulation technique in which the movement of atoms is described by Newton's second law [56–63]. Verma et al. [64] modeled the graphene-polyethylene nano-composites and evaluated the mechanical properties of it through MD. The same group also predicted the thermal properties of the nano-composite system through MD [65]. Several other studies have predicted the characterization of lightweight polymer composites through MD [66–70].

ACKNOWLEDGMENT

Monetary support from the All India Council for Technical Education (AICTE) and Department of Science and Technology (DST), India is gratefully acknowledged.

CONFLICTS OF INTEREST

There are no conflicts of interest to declare.

REFERENCES

1. Meetham, G.W., 1994. *Design Considerations for Aerospace Applications in Handbook of Polymer–Fiber Composites*, Jones F.R., Ed., Essex, UK: Longman Scientific and Technical, Chap. 5.
2. Gibson, R.F., 2016. *Principles of Composite Material Mechanics*, Boca Raton, FL: CRC Press.
3. Jones, R.M., 2014. *Mechanics of Composite Materials*, CRC Press.
4. Daniel, I.M., Ishai, O., Daniel, I.M. and Daniel, I., 1994. *Engineering Mechanics of Composite Materials* (Vol. 3, pp. 256–256). New York: Oxford university press.
5. Christensen, R.M., 2012. *Mechanics of Composite Materials*, Chelmsford, MA: Courier Corporation.
6. Herakovich, C.T., 2012. Mechanics of composites: A historical review. *Mechanics Research Communications*, *41*, pp.1–20.

7. Hill, R., 1965. A self-consistent mechanics of composite materials. *Journal of the Mechanics and Physics of Solids*, *13*(4), pp. 213–222.
8. Verma, A., Singh, C., Singh, V.K. and Jain, N., 2019. Fabrication and characterization of chitosan-coated sisal fiber–Phytagel modified soy protein-based green composite. *Journal of Composite Materials*, *53*(18), pp. 2481–2504. DOI: 10.1177%2F0021998319831748.
9. Deepmala, K., Jain, N., Singh, V.K. and Chauhan, S., 2018. Fabrication and characterization of chitosan coated human hair reinforced phytagel modified soy protein-based green composite. *Journal of Mechanical Behavior of Materials*, *27*(1–2), (pp. 1–2). DOI: 10.1515/jmbm-2018-0007.
10. Jain, N., Singh, V.K. and Chauhan, S., 2017. A review on mechanical and water absorption properties of polyvinyl alcohol based composites/films. *Journal of Mechanical Behavior of Materials*, *26*(5–6), pp. 213–222. DOI: 10.1515/jmbm-2017-0027.
11. Jain, N., Singh, V.K. and Chauhan, S., 2017. Review on effect of chemical, thermal, additive treatment on mechanical properties of basalt fiber and their composites. *Journal of Mechanical Behavior of Materials*, *26*(5–6), pp. 205–211. DOI: 10.1515/jmbm-2017-0026.
12. Verma, A., Baurai, K., Sanjay, M.R. and Siengchin, S., 2019. Mechanical, microstructural, and thermal characterization insights of pyrolyzed carbon black from waste tires reinforced epoxy nanocomposites for coating application. *Polymer Composites*, *41*(1), pp.338–349.
13. Mack, J., 1988. Advanced Polymer Composites. *Mater. Edge*, 18 January.
14. Buchanan, G.R., 1988. *Mechanics of Materials*, New York: HRW Inc.
15. Verma, A. and Singh, V.K., 2018. Mechanical, microstructural and thermal characterization of epoxy-based human hair–reinforced composites. *Journal of Testing and Evaluation*, *47*(2), pp.1193–1215.
16. Verma, A., Gaur, A. and Singh, V.K., 2017. Mechanical properties and microstructure of starch and sisal fiber biocomposite modified with epoxy resin. *Materials Performance and Characterization*, *6*(1), pp.500–520.
17. Verma, A. and Singh, V.K., 2016. Experimental investigations on thermal properties of coconut shell particles in DAP solution for use in green composite applications. *Journal of Materials Science and Engineering*, *5*(3), p.1000242.
18. Verma, A., Singh, V.K., Verma, S.K. and Sharma, A., 2016. Human hair: A biodegradable composite fiber—a review. *International Journal of Waste Resources*, *6*(206), p.2.
19. Singh, V.K., Negi, P. and Verma, A., 2018. Physical and thermal characterization of chicken feather fiber and crumb rubber reformed epoxy resin hybrid composite. *Advances in Civil Engineering Materials*, *7*(1), pp.538–557.
20. Verma, A., Negi, P. and Singh, V.K., 2018. Experimental investigation of chicken feather fiber and crumb rubber reformed epoxy resin hybrid composite: Mechanical and microstructural characterization. *Journal of the Mechanical Behavior of Materials*, *27*(3–4), pp. 3–4.
21. Verma, A., Negi, P. and Singh, V.K., 2019. Experimental analysis on carbon residuum transformed epoxy resin: Chicken feather fiber hybrid composite. *Polymer Composites*, *40*(7), pp. 2690–2699.
22. Verma, A., Joshi, K., Gaur, A. and Singh, V.K., 2018. Starch-jute fiber hybrid biocomposite modified with an epoxy resin coating: Fabrication and experimental characterization. *Journal of the Mechanical Behavior of Materials*, *27*(5–6), (pp. 5–6).
23. Verma, A., Singh, V.K. and Arif, M., 2016. Study of flame retardant and mechanical properties of coconut shell particles filled composite. *Research and Reviews: Journal of Material Sciences*, *4*(3), pp.1–5.

24. Verma, A., Budiyal, L., Sanjay, M.R. and Siengchin, S., 2019. Processing and characterization analysis of pyrolyzed oil rubber (from waste tires)-epoxy polymer blend composite for lightweight structures and coatings applications. *Polymer Engineering and Science*, *59*(10), pp. 20141–22051.
25. MIT. 2008. *On the Road in 2035: Reducing Transportation's Petroleum Consumption and GHG Emissions*, Massachusetts Institute of Technology.
26. Jain, N. and Saxena, R., 2017. Effect of self-weight on topological optimization of static loading structures. *Alexandria Engineering Journal*, *57*(2), pp. 527–535. DOI: 10.1016/j.aej.2017.01.006.
27. Jain, N., 2018. Effect of higher order element on numerical instability in topological optimization of linear static loading structure. *Journal of Theoretical and Applied Mechanics*, *48*(3), pp. 78–94. DOI: 10.2478/jtam-2018-0024.
28. Timoshenko, S.P. and Goodier, J.N., 1970. *Theory of Elasticity*, New York: McGraw–Hill.
29. Buchanan, G.R., 1988. *Mechanics of Materials*, New York: HRW, Inc.
30. Lekhnitski, S.G., 1968. *Anisotropic Plates*, New York: Gordon and Breach Science Publishers.
31. Reuter, R.C. Jr., 1971. Concise property transformation relations for an anisotropic lamina. *Journal of Composite Materials*, *6*(2), p. 270.
32. Halphin, J.C. and Pagano, N.J., 1968. Influence of end constraint in the testing of anisotropic bodies. *Journal of Composite Materials*, *2*(1), p. 18.
33. Hill, R., 1950. *The Mathematical Theory of Plasticity*, London: Oxford University Press.
34. Tsai, S.W. and Wu, E.M., 1971. A general theory of strength for anisotropic materials. *Journal of Composite Materials*, *5*(1), p. 58.
35. Pipes, R.B. and Cole, B.W., 1974. On the off-axis strength test for anisotropic materials. In: *Boron Reinforced Epoxy Systems*, Hilado C.J., Ed., Westport, CT: Technomic.
36. Geraldes, D.M. and Phillips, A.T., 2014. A comparative study of orthotropic and isotropic bone adaptation in the femur. *International Journal for Numerical Methods in Biomedical Engineering*, *30*(9), pp. 873–889. DOI: 10.1002/cnm.2633.
37. Milton, G.W., 2002. *The Theory of Composites*, Cambridge, MA: Cambridge University Press.
38. Lekhnitskii, S.G., 1963. *Theory of Elasticity of an Anisotropic Elastic Body*, San Francisco, CA: Holden-Day Inc.
39. Boresi, A.P., Schmidt, R.J. and Sidebottom, O.M., 1993. *Advanced Mechanics of Materials*, Wiley.
40. Ting, T.C.T. and Chen, T., 2005. Poisson's ratio for anisotropic elastic materials can have no bounds. *Quarterly Journal of Mechanics and Applied Mathematics*, *58*(1), pp. 73–82.
41. Ting, T.C.T., 1996. Positive definiteness of anisotropic elastic constants. *Mathematics and Mechanics of Solids*, *1*(3), pp. 301–314. DOI: 10.1177/108128659600100302.
42. Cowin, S.C. and Mehrabadi, M.M., 1995. Anisotropic symmetries in linear elasticity. *Applied Mechanics Reviews*, *48*(5), pp. 247–285.
43. Hearmon, R.F.E., 1961. *An Introduction to Applied Anisotropic Elasticity*, Oxford: Clarendon Press.
44. Hoffman, O., 1967. The brittle strength of orthotropic materials. *Journal of Composite Materials*, *1*(2), p. 296.
45. Tsai, S.W. and Hahn, H.T., 1980. *Introduction to Composite Materials*, Lancaster, PA: Technomic.
46. Chapra, S.C. and Canale, R.C., 1988. *Numerical Methods for Engineers*, 2nd ed., New York: McGraw–Hill. © 2006 by Taylor & Francis Group, LLC.

47. Hill, R., 1963. Elastic properties of reinforced solids: Some theoretical principles. *Journal of the Mechanics and Physics of Solids, 11*(5), pp. 357–372.
48. Tsai, S.W. and Hahn, H.T., 1980. *Introduction to Composite Materials*, Lancaster, PA: Technomic Publishing.
49. Hyer, M.W. and Waas, A.M. 2000. Micromechanics of linear elastic continuous fibre composites. In: *Comprehensive Composite Materials. Vol. 1: Fiber Reinforcements and General Theory of Composites*, Kelly A., Zweben C., Eds., Elsevier.
50. Herakovich, C.T., 1998. *Mechanics of Fibrous Composites*, New York: John Wiley & Sons, Inc.
51. Kaw, A.K., 2005. *Mechanics of Composite Materials*, Boca Raton, FL: CRC Press.
52. Menard, K.P. and Bilyeu, B.W., 2008. Dynamic mechanical analysis of polymers and rubbers. In: *Encyclopedia of Analytical Chemistry: Applications, Theory and Instrumentation*, Wiley.
53. Jain, N., Ali, S., Singh, V.K., Singh, K., Bisht, N. and Chauhan, S., 2019. Creep and dynamic mechanical behavior of cross-linked polyvinyl alcohol reinforced with cotton fiber laminate composites. *Journal of Polymer Engineering, 39*(4), pp.326–335.
54. Jain, N., Singh, V.K. and Chauhan, S., 2019. Dynamic and creep analysis of polyvinyl alcohol based films blended with starch and protein. *Journal of Polymer Engineering, 39*(1), pp.35–47.
55. Verma, A., Parashar, A. and Packirisamy, M., 2018. Atomistic modeling of graphene/hexagonal boron nitride polymer nanocomposites: A review. *Wiley Interdisciplinary Reviews: Computational Molecular Science, 8*(3), p.e1346.
56. Verma, A. and Parashar, A., 2017. The effect of STW defects on the mechanical properties and fracture toughness of pristine and hydrogenated graphene. *Physical Chemistry Chemical Physics: PCCP, 19*(24), pp.16023–16037.
57. Verma, A. and Parashar, A., 2018. Molecular Dynamics based simulations to study failure morphology of hydroxyl and epoxide functionalised graphene. *Computational Materials Science, 143*, pp.15–26.
58. Verma, A. and Parashar, A., 2018. Molecular Dynamics based simulations to study the fracture strength of monolayer graphene oxide. *Nanotechnology, 29*(11), p.115706.
59. Verma, A., Parashar, A. and Packirisamy, M., 2018. Tailoring the failure morphology of 2D bicrystalline graphene oxide. *Journal of Applied Physics, 124*(1), p.015102.
60. Verma, A. and Parashar, A., 2018. Reactive force field based atomistic simulations to study fracture toughness of bicrystalline graphene functionalised with oxide groups. *Diamond and Related Materials, 88*, pp.193–203.
61. Verma, A. and Parashar, A., 2018. Structural and chemical insights into thermal transport for strained functionalised graphene: A molecular dynamics study. *Materials Research Express, 5*(11), p.115605.
62. Singla, V., Verma, A. and Parashar, A., 2018. A molecular dynamics based study to estimate the point defects formation energies in graphene containing STW defects. *Materials Research Express, 6*(1), p.015606.
63. Verma, A., Parashar, A. and Packirisamy, M., 2019. Role of chemical adatoms in fracture mechanics of graphene nanolayer. *Materials Today: Proceedings, 11*, pp.920–924.
64. Verma, A., Parashar, A. and Packirisamy, M., 2019. Effect of grain boundaries on the interfacial behaviour of graphene-polyethylene nanocomposite. *Applied Surface Science, 470*, pp.1085–1092.
65. Verma, A., Kumar, R. and Parashar, A., 2019. Enhanced thermal transport across a bicrystalline graphene–polymer interface: An atomistic approach. *Physical Chemistry Chemical Physics: PCCP, 21*(11), pp.6229–6237.
66. Chaurasia, A., Verma, A., Parashar, A. and Mulik, R.S., 2019. An Experimental and Computational Study to Analyse the Effect of h-BN nanosheets on Mechanical Behaviour of h-BN/Polyethylene nanocomposite. *The Journal of Physical Chemistry C, 123*(32), pp. 20059–20070.

67. Lu, C.T., Weerasinghe, A., Maroudas, D. and Ramasubramaniam, A., 2016. A comparison of the elastic properties of graphene-and fullerene-reinforced polymer composites: The role of filler morphology and size. *Scientific Reports*, *6*, p.31735.
68. Rahman, R. and Foster, J.T., 2014. Deformation mechanism of graphene in amorphous polyethylene: A molecular dynamics based study. *Computational Materials Science*, *87*, pp. 232–240.
69. Lv, C., Xue, Q., Xia, D. and Ma, M., 2012. Effect of chemisorption structure on the interfacial bonding characteristics of graphene–polymer composites. *Applied Surface Science*, *258*(6), pp. 2077–2082.
70. Alian, A.R., Dewapriya, M.A.N. and Meguid, S.A., 2017. Molecular Dynamics study of the reinforcement effect of graphene in multilayered polymer nanocomposites. *Materials and Design*, *124*, pp.47–57.

8 Smart Lightweight Polymer Composites

Nayan Ranjan Singha, Mousumi Deb, Manas Mahapatra, Madhushree Mitra, and Pijush Kanti Chattopadhyay

CONTENTS

8.1 INTRODUCTION

The properties of smart material change significantly via nominal changes in stress, temperature, moisture, pH, ionic strength, solvency, light intensity, or the applied electric and/or magnetic field(s). For instance, colloidal microgel results in elevated performance potentials and application prospects in the biomedical fields, as these particles have inherent abilities to undergo dramatic conformational changes in response to the change(s) in surrounding environment. In smart materials, the structures and properties of the components are usually different. However, individual components in the assembled smart materials demonstrate a coupling effect, resulting in complicated responses to the physical and chemical stimuli. Notably, smart materials, such as those that are magnetostrictive and magnetorheological, electromagnetic radiation absorber, and piezoelectric, have been developed to fulfill the recent techno-commercial needs of the modern industries and civilization. Of these smart materials, the demand for lightweight flexible materials is growing day by day in making aircrafts, submarines, ships, lightweight vehicles, and other household or industrial appliances. For instance, the giant magnetostrictive materials, such as Terfenol-D, a brittle magnetostrictive alloy of terbium, dysprosium, and iron, is brittle and can withstand only small strains before failure. Therefore, these types of materials are not easy to mold into desired shapes. For avoiding such difficulty, several works have been devoted to developing flexible magnetostrictive polymer composites comprising magnetostrictive material as a basic constituent. Moreover, the response characteristics of these magnetostrictive polymer composites extend up to 100 kHz, which is much higher compared to the sub kHz noted in the monolithic magnetostrictive materials. Nowadays, it is possible to produce lightweight magnetostrictive polymer composites based on low-density polymeric foam matrices. Several researchers have reported the fabrication of lightweight smart PU-foam materials.[1–4] In fact, these magnetorheological materials are constituted of porous polymer foam-like structure filled with magnetorheological fluid. Because of the inherent porous microstructure, lightweight magnetorheological foams result in controllable modulus and excellent sound-absorbing property. The magneto-elastic properties of lightweight PU-foam matrix and magneto-sensitive microparticles have been reported, in which the structural aspects are tuned with a magnetic field during the foaming process.[1]

Recently, the electromagnetic interferences originated from the recently developed gigahertz (GHz) application devices have been damaging the entire ecosystem. Therefore, research is in progress to develop smart electromagnetic wave absorbers to protect the ecosystem from the spurious electromagnetic radiation. In this regard, several smart polymer composites acting as shielding agents of electromagnetic radiation have been developed.

The present civilization is going through a severe energy crisis. Moreover, the conventional energy resources, such as fossil fuels, are decreasing alarmingly day by day. Thus, piezoelectric materials are becoming popular because of their ability to harvest mechanical energy. Notably, for circumventing the problem relating to energy crisis, piezoelectric nanogenerator technology has been widely employed in the last decade to fabricate some energy-efficient smart piezoelectric devices. Thus, several multifunctional smart lightweight polymeric materials have been developed. Importantly, few of them have been tuned, scaled up, and commercialized to fulfill the recent demand of modern industries and societies. The rest are under trial to be adjudged for their potential for utilization in a broader spectrum. Moreover, there are some up and coming smart materials, which possess some limitations, and thus, these materials have not been commercialized for general purpose use by people. In this chapter, fabrication, properties, and applications of three major types of smart lightweight polymer composite materials, i.e., magnetostrictive/magnetorheological, electromagnetic radiation absorber, and piezoelectric materials are described, imparting special emphasis on the basic working principles involved in their functioning as stimuli-responsive materials.

8.2 MAGNETOSTRICTIVE AND MAGNETORHEOLOGICAL POLYMER COMPOSITES

The magnetostriction phenomenon is usually demonstrated by ferromagnetic materials. The magnetostriction of ferromagnetic materials remarkably alters the volume and strain in applied magnetic field. Thus, the magnetostrictive particles exert forces on the surrounding polymer matrix, leading to the overall deformation of composite, as shown in Figure 8.1. Notably, the overall deformation is guided by the resultant stress developed from the stresses accumulated in both the particle and polymer

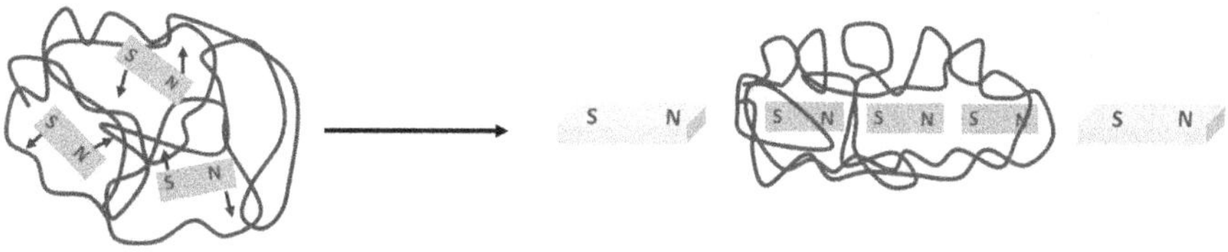

FIGURE 8.1 Ferromagnetic materials exert forces on surrounding polymer matrix in presence of applied magnetic-field, leading to change in volume and strain.

phases. Thus, magnetostrictive behavior of the polymer composite is driven by several factors including the microstructural parameters of the composite. The response of a magnetostrictive polymer composite toward the applied magnetic field is influenced significantly by the change in processing methodology and mechanical properties of the basic components. However, several works have been reported toward the development of magnetostrictive thermosetting and thermoplastic polymer composites. The basic advantage of thermosetting magnetostrictive polymer composites lies in their natural strength. The thermoplastic magnetostrictive polymer composites are processed using variable processing techniques, such as injection molding, blow molding, extrusion, and thermoforming. Moreover, the elastomers and their inherent stretching nature are exploited to maximize the changes in strain of the magnetostrictive phases within the matrix. Indeed, the performance of a magnetostrictive polymer composite is judged by the saturation magnetostriction parameter. Importantly, at the saturation magnetostriction level, the composite reaches the maximum strain, and thus, no further increase in strain follows even after the increase in the strength of the applied magnetic field. The relative performances of recently developed magnetostrictive polymer composites are summarized in Table 8.1.

TABLE 8.1
Saturation Magnetostriction Levels of Various Polymer Composites

Polymer composite(s)	Particle size (μm)	Particle content (%)	Saturation magnetostriction ($\times 10^{-6}$)	References
$Tb_{0.3}Dy_{0.7}Fe_2$/Epoxy	5–300	70 vol	720	10
$Tb_{0.3}Dy_{0.7}Fe_2$/Epoxy	0–180	97 wt.	850–1000	11
$Tb_{0.3}Dy_{0.7}Fe_2$/Epoxy	<1000	35–49 vol	1500–1600	12, 13
$Tb_{0.3}Dy_{0.7}Fe_2$/Epoxy	—	20 vol	1050	14
$Tb_{0.3}Dy_{0.7}Fe_2$/Epoxy	—	20 vol	800	14
$Tb_{0.3}Dy_{0.7}Fe_2$/Epoxy with titanate coupling agent	30–500	50 vol	~1000	16
Tb-Dy/Epoxy	<300	20 wt.	900	17
Tb-Dy/Epoxy	45000	50 vol	1270	18
Fe-Ga flakes/Epoxy	75 to 225	40 wt.	10–14	19
Fe-Ga particles/Epoxy	< 25	69.1 vol	60	20
Fe-Ga/Epoxy with 8.9% Ga	20–25	48 wt.	53.5	21
$Fe_{81}Ga_{19}$/Epoxy	<25	69.1 vol	60	22
$Fe_{80}Ga_{20}$/Epoxy	50–100	80 vol	360	22
Tb-Dy/Polyurethane	0–300	50 wt.	813	23
$Tb_{0.3}Dy_{0.7}Fe_2$/Polyurethane	212 300	50 wt.	1390	24
Carbonyl iron/Silicone rubber	5	27 vol	134	5
Silicon steel in/Silicone rubber	150–200	40–80 vol	100	25
Nickel (hollow sphere)/Vinyl ester	<25	24 vol	–24 ppm	26
Nickel (solid sphere)/Vinyl ester	<25	24 vol	–28 ppm	26
Tb-Dy/Vinyl ester	<1000	49 vol	1475	27

In the prevalent magnetic field, the long-range elasticity and low glass transition temperature of elastomer ensure easier adaptability of the elastomer composite matrix for accommodating the larger strain imposed by the magnetostrictive phases. Thus, rubber composites are frequently utilized as solid magnetorheological polymer composites.[5] Moreover, these magnetic field sensitive materials are able to change the rheological properties by a small alteration in magnetic field. In this context, the inherent facilitated reversible deformability of the elastomers is exploited for the preparation of smart materials, undergoing easy extension, rapid microstructural alterations, and easier orientation polarization of the magnetoactive fillers in the presence of external magnetic field. Thus, magnetorheological polymer composites are a kind of specialized smart materials that are usually fabricated by embedding magnetic micrometric or nanometric particles in a polymer matrix. In this regard, either viscoelastic or elastomer matrix is equipped with characteristic properties including elastic modulus and density. Previously, several magnetorheological elastomer composites composed of ethylene-propylene, acrylonitrile-butadiene, silicone, ethylene-octene, and polyoctenamer rubbers and magnetoactive fillers, such as carbonyl iron powder, γ-Fe_2O_3, and micro-/nano-Fe_3O_4 have been reported to date.[6] Moreover, the polymer-modified core-shell ferromagnetic particle in magnetorheological polymer composites produce magnetorheological fluids. Notably, the core-shell structure reduces the densities of ferromagnetic particles, and increases the static electrical repulsion between the neighboring particles, leading to better stability, dispersion, and distribution of the polymer(s)-modified ferromagnetic particles in composites. Moreover, the oxidation tendency of the dispersed hybrid becomes lesser at the surface of the particles, as surface is protected by the polymers, as shown Figure 8.2.[7] Additionally, the sedimentation of particles, usually encountered in magnetorheological fluids, is obviated using rubber-like polymer instead of the non-magnetic liquid matrix. However, compared to the magnetorheological fluids, the magnetorheological elastomers lose the magneto-controllable flexibility. Accordingly, unlike magnetorheological fluids, the particulate microstructure cannot easily be controlled by the applied magnetic field in magnetorheological elastomers, as the magnetic field-assisted motilities of the particles are impeded in viscoelastic surroundings. Moreover, the solid-like magnetorheological gels have attracted the attention of the scientific communities.

Like magnetorheological elastomers, solid-like magnetorheological gels do not suffer from problems related to the sedimentation of particles. However, there is a basic difference between solid-like magnetorheological gels and magnetorheological elastomers. For instance, the fabrication of magnetorheological solid-gels

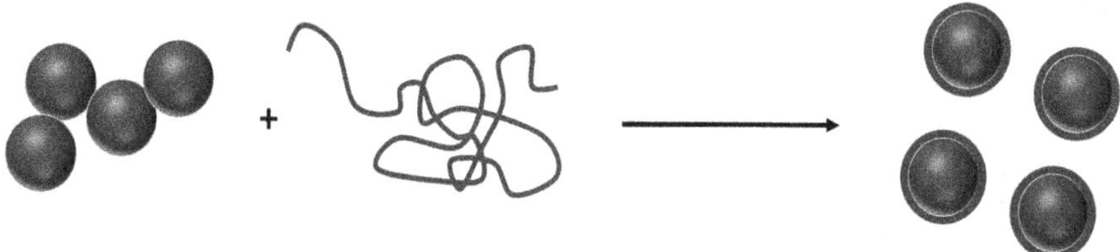

FIGURE 8.2 Morphology of polymer modified ferromagnetic particles.

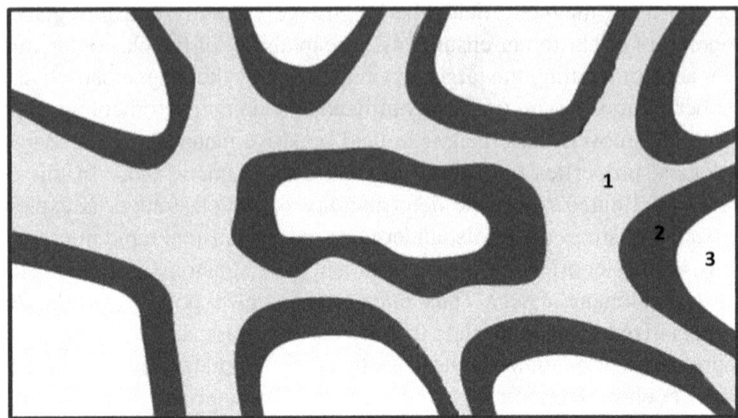

FIGURE 8.3 Structure of the magnetorheological-foam constituting of 1. microtubule-wall, 2. magnetorheological-fluid, and 3. hollow-space filled with air.

comprising micro magnetic particles dispersed in polyurethane have been reported.[8] Notably, these materials result in the inherent abilities to change into any shapes depending on the extent and direction of applied magnetic field. Thus, these materials are denoted as magnetorheological plastomers, which are as such different from magnetorheological elastomers. The magnetorheological foams are another special kind of lightweight material, which give controllable modulus and an excellent sound-absorbing property. The most likely constitution of the magnetorheological foams is represented in Figure 8.3.

8.2.1 Synthesis of Magnetostrictive and Magnetorheological Polymer Composites

For the preparation of magnetorheological foam and aligned magnetic particle-based anisotropic polymer composites, researchers usually follow the underlying procedure encompassing three basic stages: mixing of the basic ingredients under a static magnetic field to ensure the desired alignment of the magnetic particles within the system, foaming process, and curing at designated temperature.[1–4]

For the fabrication of polymeric foam + ferromagnet composites, magnetic particles, such as iron and barium ferrite, are mixed in polyalcohol by continuous stirring at 2000 rpm for 5 min to ensure the uniform dispersion of particles, followed by the addition of di-isocyanate into the dispersion, as illustrated in Figure 8.4. The mixture is then transferred into an aluminum mold attached with two sets of permanent magnets. Thereafter, a low-strength external magnetic field, i.e., >0.1 T, is applied, along with the pre-existing static magnetic field, to originate the chain-like structures manipulating the spatial distribution of the fillers. The foaming reaction is done in the presence of foaming additives. Once the foaming process is completed, the samples are cured at 40°C in an oven for 24 h.[1] Notably, a similar process is employed to produce the carbonyl-iron particles incorporated anisotropic composites.[9] Initially, the major ingredients, such as TMN-3050, a derivative of polyether

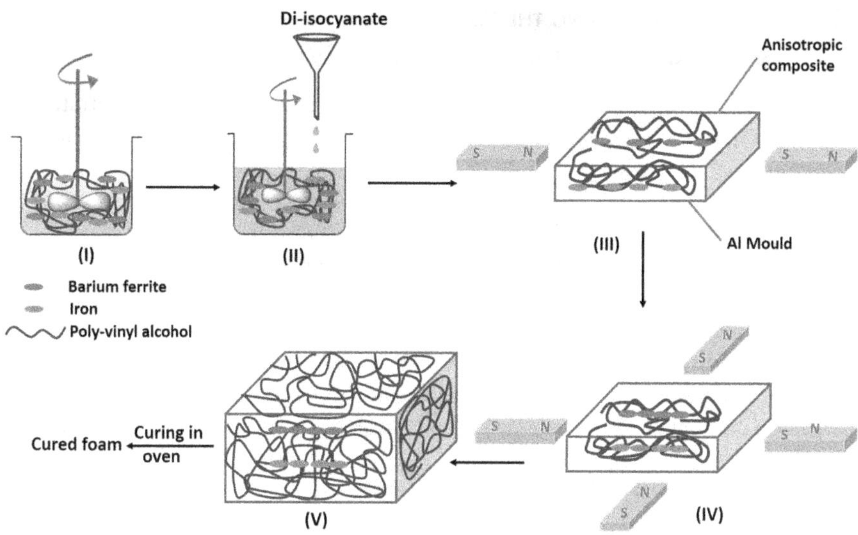

FIGURE 8.4 Preparation of cured magnetorheological foam through (I) initial dispersion and mixing, (II) addition of di-isocyanate, (III) anisotropic composite in magnetic field, (IV) chain-like structure in additional magnetic field, and (V) uncured foam.

triol, carbonyl iron, i.e., magnetic filler, glycerol, DABCO-33LV, i.e., a catalyst based on triethylenediamine, L-568, i.e., foam stabilizer, and water, are mixed and stirred at 150 rpm for 10 min through the ball milling. Thereafter, vigorous mixing is done for another 2 min during the continuous additions of dibutyltin dilaurate catalyst and isocyanate. In the next step, foaming is continued in the presence of a static magnetic field of 0.8 T magnetic flux density.

For the fabrication of magnetorheological elastomeric composites, similar steps are generally followed except the foaming step is omitted.[5, 10, 11] Herein, elastomer, filler particles of desired particle size, and the other ingredients are mixed in a magnetic stirrer at room temperature. Thereafter, the mixtures are cured with/without homogenous magnetic field to fabricate the anisotropic/isotropic composite.[5] Sometimes, size reduction of fillers is done prior to the addition of fillers in the polymeric matrix. For synthesizing the Terfenol-D + polymer composites, monolithic Terfenol-D is at first reduced into fibrils of diameter less than 1000 μm and aspect ratio >3:1,[12, 13] followed by the execution of the usual process, to ensure the required alignment of the magnetic-field sensitive fillers. Moreover, polyamines are sometimes added to react with epoxy groups of Terfenol-D + polymer composites to ensure the desired level of cross-linking during molding stage.[14, 15] For fabricating Terfenol-D + polymer composites,[16] a Terfenol-D modification step is employed just before the addition of the modified fillers into the polymer. For the modification of Terfenol-D by titanate coupling, agent Terfenol-D is initially exposed to sodium hydroxide for attaching the numerous superficial –OH functionalities. Subsequently, the –OH populated Terfenol-D is treated by titanate coupling agent to obtain the titanate modified Terfenol-D. Thereafter, the usual mixing and molding steps are followed to synthesize polymer composites filled with titanate coupling agent modified Terfenol-D.

8.2.2 Factors Governing the Performances of Magnetostrictive and Magnetorheological Polymer Composites

Magnetostriction efficiency depends on several factors, such as concentration of magnetostrictive phases, particle size, packing density, extent of alignment, and treatment by coupling agent.

8.2.2.1 Concentration of Particulate Phases

In general, the magneto-elastic response of the composites is augmented with the increasing concentrations of magnetostrictive material.[5, 17, 18] For instance, in epoxy + Tb/Dy composites, the increase in the concentrations of Tb and Dy from 20 to 50% enhanced the saturation magnetostriction from 900×10^{-6} to 1270×10^{-6}. Notably, the enhancement of magnetostrictive effect is ascribed to the continuous increase in carbonyl iron particles in carbonyl iron + silicone elastomeric composites.

8.2.2.2 Particle Size of the Discrete Phases

The magnetostrictive properties of $Tb_{0.3}Dy_{0.7}Fe_2$ + epoxy composites depend on the particle size of the discrete phase (see Table 8.1). Notably, the saturation magnetostriction is increased via continuous increase in particle size of the magnetostrictive fillers. Moreover, the magnetorheological materials consist of micro-magnetic particles suspended on a non-magnetic matrix. Importantly, a sufficiently low dimension of ferromagnetic particles, i.e., ≤ 1.5 μm iron, facilitates the uniform magnetization throughout the matrix.

8.2.2.3 Extent of Alignment

Importantly, the facilitated magnetic direction plays a decisive role in the composite of Tb + Dy alloys. For instance, the easy magnetic direction of Dy-rich alloys prefers to align along the <100> axis, whereas Tb-rich alloys favor the <111> axis. Therefore, the magnetostriction property of the composite is guided by the relative composition of Tb or Dy in Tb + Dy alloys. Thus, in Dy-rich Tb + Dy alloy-filled composites, if the magnetic field is applied along the <100> axis, the saturation magnetostriction is reached easily in spite of smaller magnetic field. In contrast, the larger magnetic field is required to obtain the saturation magnetostriction along <111> axis of applied magnetic field direction. In the case of magnetorheological polymer composites, the pre-existing randomly distributed magnetoactive particles become polarized in the presence of a sufficient magnetic field strength, and form chains within the elastomeric or viscoelastic material. Thus, the curing of these elastomeric-filled material in the presence of magnetic field produces the magnetorheological elastomeric composites, in which the magnetoactive particles are oriented along the direction of the applied magnetic field.

8.2.2.4 Treatment by Coupling Agent

The magnetostrictive properties of $Tb_{0.3}Dy_{0.7}Fe_2$ + epoxy composites are elevated modifying the magnetostrictive particulate phase, i.e., $Tb_{0.3}Dy_{0.7}Fe_2$, employing titanate coupling agent.[16] The coupling agent acts as a bonding agent between discrete and continuous phases,[28] facilitating the transfer of stress produced by the

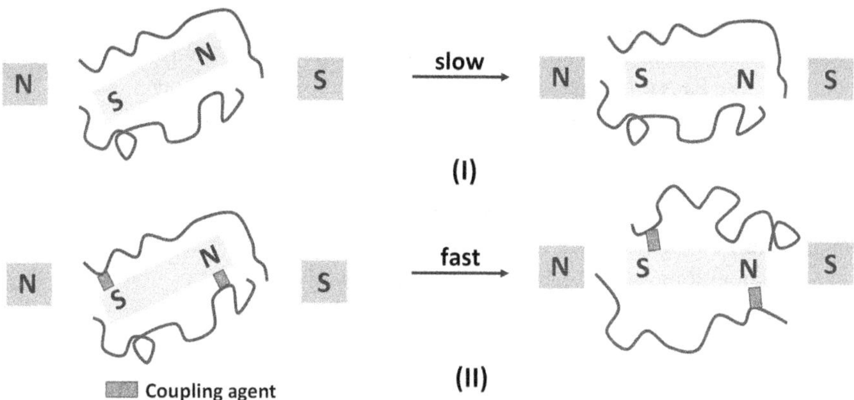

FIGURE 8.5 Remarkable shape change in the presence of coupling agent and associated stress transfer.

magnetostriction, as represented in Figure 8.5. Accordingly, if the interfacial bond strength is weak, the unwanted slipping at the interface leads to the delayed transfer of magnetostriction, restraining the smart response of composite.

8.2.2.5 Nature of the Matrix

Importantly, the flexibility or rigidity of matrix regulates the magnetostrictive properties of polymer composites. For instance, thermosetting magnetostrictive polymer composites are generally constituted of rigid microstructures, and thus, these matrices are not easily deformed in magnetic field. However, these inherently strong materials result in greater modulus compared to either thermoplastic or elastomer matrices. In contrast, the flexibility of a thermoplastic material depends strongly on the glass transition temperature.[29] Notably, the flexibility is modulated by external or internal plasticization. Thus, in magnetic field, the deformation of particulate phases is accommodated by the stretchable matrix. Moreover, thermoplastic materials result in easy processability in variable molds.

8.2.3 Applications

Magnetorheological elastomers have been widely employed in the fields of adaptive tuned vibration absorbers,[30] impact absorbers,[31] active noise abatement barrier systems,[32] vibration isolators,[33] and sensors.[34–36]

8.3 ELECTROMAGNETIC RADIATION ABSORBER MATERIAL

Recently, electromagnetic interference pollution has been becoming a serious threat to ecology, the ecosystem, and the environment.[37] Importantly, the entire ecosystem is unknowingly suffering from the increasing exposure to electromagnetic interference created by the recently developed GHz application devices, such as AC motors, digital computers, calculators, printers, modems, electromagnetic typewriters, digital circuitry, and cellular phones. Therefore, in order to reduce the levels of spurious electromagnetic radiation, continuous research is going on to develop variable electromagnetic wave absorbers, such as radar and microwave absorbing materials.[38, 39] Recently, several smart polymer composites have been developed to function as shielding agent. For instance, polymer composites containing dielectric and magnetic fillers, such as magnetite,[40] have been developed as potent lightweight low-cost flexible microwave absorbers to ensure the maximum absorption of electromagnetic energy. In fact, lightweight electromagnetic interference shielding materials are required especially in manufacturing aerospace and smart devices.[41] In this regard, although inorganic materials possess the inherent abilities to undergo large electric or magnetic loss, polymers have the sufficient tenability. Therefore, a polymer composite can be a high-performance absorber of electromagnetic radiation, if the positive aspects of inorganic materials and polymers are optimally integrated in a suitable manner.[42] For instance, ferrites are considered to be the best magnetic inorganic materials for absorbing electromagnetic wave, as these materials possess excellent magnetic and dielectric properties. However, these materials are highly expensive and dense compared to the polymers. Accordingly, ferrite materials alone cannot be used for manufacturing lightweight absorbers. Therefore, in search of finding low-cost, lightweight, flexible absorbers of electromagnetic radiation, polymer composites protruded with ferrites are fabricated,[38] wherein the ferrite particles function as absorption center of the incident electromagnetic radiation, along with the dissipating agent of absorbed energy in the form of heat. Thus, radar absorbing materials fabricated using the optimum dose of inorganic particulate fillers dispersed in suitable polymer matrix result in the minimum reflection loss in a wide frequency range. Notably, several works have been documented on magnetic materials-filled polymer composites filled with micro-fillers, such as such as Ba-ferrite,[43] iron fiber,[39] NiZn ferrite,[44] and Fe_3O_4 + YIG.[45] In this context, various polymer matrices have been utilized as the continuous phases for preparing particulate composites filled with ferrite-based fillers.[46–50] Recently, lightweight polymer or graphene nanocomposites have been established as promising electromagnetic interference shielding materials because of their enhanced electrical conductivities.[41, 51]

8.3.1 Mechanism of Electromagnetic Wave Loss

If an incident electromagnetic wave interacts with a material, the wave energy experiences reflection, absorption, and penetration, as shown in Figure 8.6.

The attenuation of electromagnetic wave through the reflection loss and absorption loss results in the shielding of electromagnetic radiation or electromagnetic interference. The shielding of electromagnetic radiation by reflection loss occurs by interaction of the electric component of electromagnetic radiation with free electrons of shielding materials through induced conduction current. In contrast, the shielding of electromagnetic radiation by the absorption loss takes place through interaction of electromagnetic radiation with electric or magnetic dipoles of shielding materials, which result in high permeability and permittivity. In contrast to the electromagnetic radiation shielding materials, electromagnetic radiation absorbers prevent reflection by the impedance or complex impedance matching between free space and material

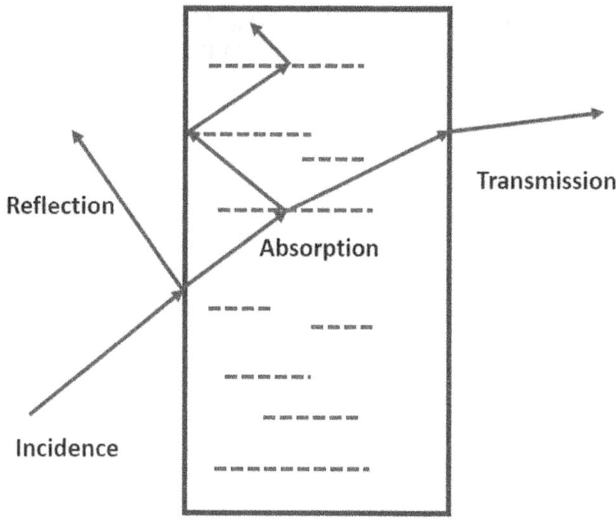

FIGURE 8.6 Interaction of electromagnetic wave with material.

surface.[42] Moreover, an important strategy behind absorption of electromagnetic radiation by these materials involves simultaneous losses of electric and magnetic components. The mechanisms for the loss of magnetic and electrical components within absorbing materials are completely different from one another. In the case of microwave absorption, the loss of absorbed microwave is contributed to by both dielectric loss (i.e., tan δ_ε) and magnetic loss (i.e., tan δ_μ), which are expressed by the following Equations 8.1 and 8.2.

$$\tan\delta_\varepsilon = \varepsilon'' / \varepsilon' \tag{8.1}$$

$$\tan\delta_\mu = \mu'' / \mu' \tag{8.2}$$

Here, ε'' and μ'' are the imaginary parts of the complex permittivity (ε^*) and complex permeability (μ^*), respectively, and these imaginary parts signify the extent of electric and magnetic component losses. In contrast, ε' and μ' are the respective real parts of ε^* and μ^*, which are related to the storage capability of absorber for electric and magnetic components. In this context, expressions for ε^* and μ^* are given by the following Equations 8.3 and 8.4.

$$\varepsilon^* = \varepsilon' - j\varepsilon'' \tag{8.3}$$

$$\mu^* = \mu' - j\mu'' \tag{8.4}$$

Therefore, the share of ε'' and μ'' within ε^* and μ^* of a highly efficient absorber of electromagnetic radiation are very, very, high, so that the absorber easily releases the absorbed energy.

8.3.1.1 Mechanism of Dielectric Loss

The electric component of the interacting electromagnetic radiation mostly undergoes three types of dielectric loss, these being conductance loss, dielectric relaxation loss, and resonance loss. Accordingly, dielectric loss is mostly composed of conductance loss tangent (i.e., tan $\delta_{\varepsilon c}$), dielectric relaxation loss tangent (i.e., tan $\delta_{\varepsilon d}$), and resonance loss tangent (i.e., tan $\delta_{\varepsilon r}$).

8.3.1.1.1 Conductance Loss

The electric component of the incident electromagnetic radiation experiences loss of energy as a result of conduction. Notably, the conductance loss depends on electrical conductivity of the material. Such conductance loss is observed in polymer composites or nanocomposites, in which the fillers produce an interconnected network structure.[52]

8.3.1.1.2 Dielectric Relaxation Loss

The propagation of electromagnetic radiation, such as microwaves, through the dielectric material generates internal electrical field, which induces translational motions of free or bound charges, such as electrons or ions, within the dielectric material, resulting in the rotation of dipoles present within the electrical field induced zone. However, dipoles resist such type of motion by means of inertial, elastic, and frictional forces, resulting in loss of energy in the form of dissipated heat. The dielectric relaxation loss is produced if the change in polarization of polarized absorbing material is slower compared to the electric field. Thus, the extent of dielectric relaxation loss depends on the rapidity of such change in polarization. In this regard, depending on the time required, polarizations are mainly classified into thermal ion polarization, dipole rotation polarization, electronic displacement polarization, and ion polarization. Of these, electronic displacement polarization and ion polarization require very short time, i.e., ~10^{-15} – 10^{-14} s. Therefore, contributions of electronic displacement polarization and ion polarization in the dielectric relaxation loss become significant if the incident electromagnetic radiation is of ultra-high frequency. Otherwise, if the frequency of the incident electromagnetic radiation is not that much higher, the dielectric relaxation loss is mostly contributed to by thermal ion polarization and dipole rotation polarization, which require appreciably higher time within 10^{-8} – 10^{-2} s. Accordingly, in the case of high frequency incident radiation, thermal ion and dipole rotation polarizations play the pivotal roles in dielectric relaxation loss.

8.3.1.1.3 Resonance Loss

The resonance loss is usually originated from the resonance effect induced by the vibration of atoms, ions, or electrons inside the material, as a result of perturbation of the incident electromagnetic radiation.

8.3.1.2 Magnetic Loss Mechanism

Unlike the dielectric loss, the magnetic component of the electromagnetic radiation within the absorbing magnetic material undergoes three types of energy losses, i.e., eddy current loss, magnetic hysteresis loss, and residual loss.

8.3.1.2.1 Eddy Current Loss

According to the Faraday's law of induction, the magnetic field induces an electrical current, i.e., eddy current, within conductors. Notably, the eddy current suffers from the losses of heat, which is proportional to $I^2 \times R$. Here, I and R are current and resistance, respectively. For an electromagnetic radiation absorber, the eddy current loss is enhanced by increasing the thickness and electrical conductivity of the absorbing material. It is anticipated that the enhanced thickness invariably increases the traveling path length of the eddy current, leading to enhanced resistance experienced by the current, which elevates the generation of heat. Similarly, the increased conductivity augments the current flow (I), and thus, the heat loss increases. Moreover, the eddy current loss is influenced by other factors, such as orientation, grain size, surface roughness, and morphology of the absorbing material.

8.3.1.2.2 Magnetic Hysteresis Loss

Magnetic hysteresis is caused by the alignment of dipoles of a ferromagnetic material in the presence of external magnetic field, followed by the change in orientation and partial nonalignment of the dipoles, if external magnetic field is switched off. Thus, as a result of magnetic hysteresis, part of the aligned dipoles remains unaltered even after the removal of external magnetic field, whereas the remaining dipoles revert back to the unaligned condition. The material experiences energy loss during the magnetic hysteresis as a result of change in rotation of magnetization and size or number of magnetic domains formed by the similarly oriented atoms in a ferromagnetic material. The extent of magnetic hysteresis loss depends mainly on magnetic properties of material, including Rayleigh constant and permeability of material. The Rayleigh constant describes Barkhausen jumps or Barkhausen effect related to the noise in the magnetic output of a ferromagnet in fluctuating magnetic field, which is caused by rapid changes in the sizes of magnetic domains.

8.3.1.2.3 Residual Loss

In addition to eddy current and magnetic hysteresis, there are other kinds of magnetic losses designated as residual loss. The characteristics of residual losses depend on the frequency of the magnetic component of electromagnetic radiation. If the incident radiation is of lower frequency, the residual loss is caused mainly by the after effect of magnetic loss, including thermal fluctuation, or the hysteresis of some electrons and ions moving to equilibrium position relative to the diffusion of applied magnetic field. Such kind of residual loss is determined by the amplitude of alternating magnetic field and relaxation time of the absorbing material. In contrast, the residual loss for high-frequency radiation is caused by different resonating effects including size resonance, ferromagnetic resonance, natural resonance, and/or domain wall resonance. In the case of residual loss, the ferromagnetic substance absorbs electromagnetic radiation if the following two basic conditions are fulfilled: (i) the incident microwaves, typically ranging within 1–35 GHz, approach and fall in a direction 90° to the static magnetic field of ferromagnetic absorbing material and (ii) the frequency of the incident microwave must be exactly the same as the effective precession frequency, i.e., Larmor frequency of the electrons of ferromagnetic material, as shown in Figure 8.7. Herein, the direction of the magnetic moment (M) of the precessing electrons lie at an angle θ with the static magnetic field (H). Once the incident microwave gets absorbed by the material, the angle between H and M becomes greater than θ. Thereafter, in the absence of microwave field, θ gets reduced as the absorbed energy dissipates in the form of heat. The natural spin resonance, a type of residual loss, depends primarily on the magnetic anisotropy, i.e., the structure and shape of magnetic particles in polymer composites. Moreover, absorption loss can occur via internal multiple reflection within the absorbing material. This internal multiple reflection is originated because of the scattering of electromagnetic waves from interfaces, defects sites, and scattering centers within the absorbing material, followed by the absorption of those waves within the material.[53] Notably, an ideal

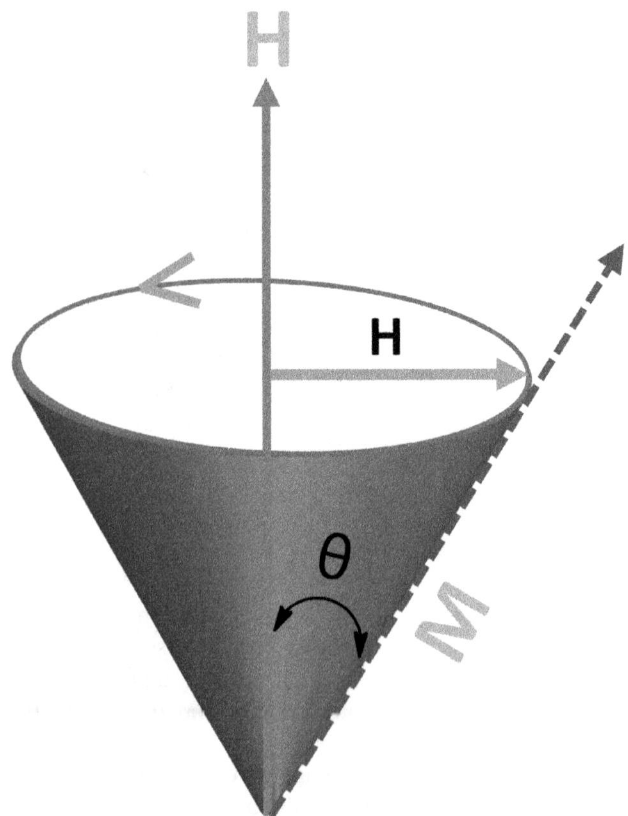

FIGURE 8.7 Precessing motion of the electrons at an angle θ with the static magnetic-field (H).

microwave absorbing material is expected to result in the following characteristics: tiny thickness, low density, wide bandwidth, and flexibility.[54]

8.3.2 Synthesis of Electromagnetic Radiation Absorber Material

Notably, several fabrication methodologies, such as mechanical mixing,[38] melt-blending,[50] in situ emulsion polymerization,[44, 55] and solution mixing[40] have been employed to fabricate the polymer composites or nanocomposites suitable as electromagnetic radiation absorbers.

8.3.2.1 Mechanical Mixing

This is the most simple and conventional way of preparing radar absorbing rubber composites.[56–58] For instance, silicone rubber composites filled with variable extents of carbonyl iron are fabricated in conventional mechanical mixing machine, followed by molding in a coaxial die.[38] Moreover, the simple mechanical mixing is utilized to prepare Z-type Ba-ferrite or polyvinylidene fluoride composites filled with $Ba_3Co_2Fe_{24}O_{41}$ ferrite powder of variable sizes.[43]

8.3.2.2 Melt-Blending

Previously, toroidal-shaped iron fiber + epoxy resin composites were fabricated as radar absorbing materials.[39] For preparing the thermoplastic natural rubber + Fe_3O_4 nanocomposites, melt-blending process is executed in a laboratory mixer.[50] Initially, major ingredients, such as polypropylene, natural rubber, Fe_3O_4-nanoparticle, and liquid natural rubber are blended in the internal mixer at 180°C at 100 rpm. Notably, the ex situ added liquid natural rubber essentially functions as the compatibilizer to ensure the homogeneous distribution of Fe_3O_4 nanoparticles throughout the blend of rubber and plastic. The as-obtained homogeneous mixture is converted into thin sheet of 3 nm thickness in mold at 180°C. The injection molding method is employed to design the toroidal-shaped samples that are ideal for performing permeability tests.[50]

8.3.2.3 In Situ Emulsion Polymerization

The in situ emulsion polymerization technique for chemical oxidation polymerization of aniline in the presence of $(NH_4)_2S_2O_8$ initiator, using the homogeneous mixture of previously prepared TiO_2 and γ-Fe_2O_3 nanofillers in dodecyl benzene sulfonic acid functioning as dopant and surfactant, has been reported by researchers.[55] In this regard, the γ-Fe_2O_3 nanofillers are synthesized previously by co-precipitation of Fe^{2+} and Fe^{3+} ions in aqueous medium by ammonium hydroxide, followed by homogenization of nano TiO_2 and γ-Fe_2O_3 fillers in 0.3 M aqueous solution of dodecyl benzene sulfonic acid. Finally, the resultant emulsion is de-emulsified by isopropyl alcohol, followed by precipitation, filtration, washing, and finally, drying at 65°C. Moreover, similar method is followed to produce polyaniline and epoxy composites containing NiZn ferrite magnetic particles, using $(NH_4)_2S_2O_8$ and dodecyl benzene sulfonic acid as initiator and dopant-cum-surfactant, respectively.[44] The only difference is that the spherical-like NiZn ferrite magnetic particles is fabricated by a sol–gel autocombustion method at room temperature. Importantly, as-obtained particle sizes, i.e., 0.2–0.5 μm, are higher compared to particles synthesized using particulate phases.

8.3.2.4 Solution Mixing

Hexahydrophtalic anhydride + boron trifluoride ethylamine complex cured composites of hybrid polyaniline + Fe_3O_4 + epoxy resin are prepared by a solution mixing process.[40] Initially, polyaniline and Fe_3O_4 are separately mixed in tetrahydrofuran solvent, followed by the mixing of individual mixtures in an ultrasonic bath, to minimize the aggregate formation during mixing. Thereafter, the epoxy resin is added with continuous stirring at moderate speed. Finally, the cross-linking is performed in a thermostatic oven vacuum in two stages, of which the first one is the baking stage at 80°C for 4 h, followed by final post-curing at 120°C for 24 h.

8.3.3 Factors Governing the Performances of Electromagnetic Radiation Absorbing Polymer Composites

Performances of polymer composites as electromagnetic radiation absorbing materials depend on various factors:

8.3.3.1 Characteristics of Filler

Traditionally, ferrites possess excellent magnetic and dielectric properties, and accordingly, these are the best magnetic inorganic materials for absorbing electromagnetic wave. Recently, these materials alone have fell out of favor because of their unfavorably higher density for manufacturing lightweight materials. For instance, the densities of MnZn ferrite and iron particle are about 5.0 and 7.8 g cm^{-3}, respectively.[42] Importantly, some recently developed polymer composites filled with substituted soft ferrites, such as Ba-ferrite,[43] iron fiber,[39] NiZn ferrite,[45] and Fe_3O_4 + YIG,[45] feature elevated electromagnetic radiation absorbing capabilities. Herein, the ferrites and polymer act as the absorption center of the incident radiation and lightweight matrix, respectively. In general, soft ferrites are of low coercivity, and thus, these materials can be demagnetized easily by changing their orientation. It is well known that natural ferromagnetic resonance absorption is the dominating magnetic loss mechanism of the ferrite absorber materials, and accordingly, two or three matching frequencies are usually observed in the absorption plots of ferrite composites, ascribed to domain wall motion and spin resonance at low and higher frequencies, respectively.[43] In comparison to polyaniline + γ-Fe_2O_3, polyaniline + TiO_2 + γ-Fe_2O_3 nanocomposite[55] demonstrates better electromagnetic radiation absorbing capability (see Table 8.2). Herein, TiO_2 dielectric filler is employed, along with the γ-Fe_2O_3 as conventionally used magnetic filler, so that both dielectric and magnetic components of electromagnetic radiation are absorbed by the ternary nanocomposite. Thus, the enhanced electromagnetic radiation absorbing capability for polyaniline + TiO_2 + γ-Fe_2O_3 nanocomposite over polyaniline + γ-Fe_2O_3 is ascribed to the large dielectric constant of TiO_2 and associated dominant dipole rotation polarization contributing to the dielectric relaxation loss. Indeed, such higher absorption driven shielding efficiency is realized from elevated saturation magnetization of polyaniline + TiO_2 + γ-Fe_2O_3 nanocomposite containing a relatively greater extent of Fe_2O_3 (see Table 8.2). Such enhanced saturation magnetization is attributed to higher polydispersity of increasingly greater proportion of γ-Fe_2O_3 in polyaniline matrix, along

TABLE 8.2
Electromagnetic Radiation Absorbing Performances of Some Polymeric Composites and Nanocomposites

Polymer composite/ Nanocomposite(s)	Polymer Matrix	Filler(s)	Filler loading/ Stoichiometric ratio	Shielding effectiveness due to reflection/ absorption (dB)	Saturation magnetization (emu g^{-1})	Absorption bandwidth (ghz)	Thickness (mm)	References
Thermoplastic natural rubber-Fe_3O_4	Thermoplastic natural rubber	Fe_3O_4 (20–30 nm)	4 wt. %	–10.79/–	—	0.6	9	50
			8 wt. %	–16.10/–	—	2.0	9	50
			12 wt. %	–25.51/–	—	2.7	9	50
Polyaniline-TiO_2-γ-Fe_2O_3	Polyaniline	TiO_2 (~70–90 nm) and γ-Fe_2O_3 (~10–15 nm)	1:1:1	–/~ 35	11.7	—	—	55
			1:1:2	–/~ 45	26.9	—	—	55
Polyaniline-γ-Fe_2O_3	Polyaniline	Γ-Fe_2O_3	—	–/~ 8.8	—	—		55
Polyaniline-TiO_2	Polyaniline	TiO_2	—	–/~ 22.4	—	—	—	55

with higher dangling bonded atoms and unsaturated coordination on the surface of polyaniline matrix.[55]

8.3.3.2 Filler Loading

The electromagnetic properties, absorption bandwidth, and resonance frequency of composites can be tuned based on the proportion of the fillers in a polymer composite or nanocomposite. Accordingly, broadening and intensification of the absorption bandwidth for a polymer composite can be possible by optimizing the ratio of polymer-to-filler. For instance, the increment of Fe_3O_4 nanofiller from 4 to 12 wt. % in thermoplastic natural rubber + Fe_3O_4 nanocomposite resulted in enlargement in the absorption bandwidth from 0.6 to 2.7 GHz. However, excess fillers including ferrites produce aggregates and agglomerates,[50] which ultimately deteriorate the mechanical property and durability of the polymer composites.[59–64] Moreover, the hysteresis loop significantly deteriorated with the increase in Fe_3O_4-nanofiller. Moreover, the coercive force decreases with increasing Fe_3O_4 nanofiller, indicating facilitated demagnetization of highly loaded nanocomposites compared to nanocomposites containing smaller amounts of magnetic nanofiller.[50] At lower concentration, alteration of orientation of magnetic nanofillers is difficult, as thermoplastic natural rubber prevents directional change of Fe_3O_4-nanofiller via strong filler-polymer interaction. However, at the higher concentrations of magnetic nanofiller, filler-polymer interaction deteriorates, resulting in easier change in alignment of the Fe_3O_4-nanofiller under external magnetic field. The extent of filler, such as γ-Fe_2O_3, has the capability to change the characteristics of electromagnetic radiation absorption of polyaniline + γ-Fe_2O_3 nanocomposites.[65] In this regard, the dielectric loss decreases with the increasing γ-Fe_2O_3 content in nanocomposite because of reduced dielectric relaxation by relatively lower proportion of polyaniline matrix. The increase in the tan δ_μ with increasing Fe_3O_4-nanofiller[50] is related to the elevated magnetic relaxation as a result of increased addition of magnetic nanofillers. However, in contrast to polyaniline + γ-Fe_2O_3-nanocomposites,[65] the dielectric loss is increased with the increasing Fe_3O_4-nanofiller in thermoplastic natural rubber.[50] In this regard, relatively higher conductivity of polyaniline matrix over thermoplastic natural rubber plays a decisive role in such an opposing trend of dielectric losses. Moreover, carbonyl iron in silicone rubber composites favorably increases tan δ_ε and tan $\delta_{\mu,}$ as documented in literature.[38] In this regard, dielectric and magnetic losses are contributed to by polarization relaxation and conductance loss, respectively. The ε'' changes with the frequency of the incident radiation. Once the frequency of 5 GHz is reached, the dramatic decrease in ε'' indicates that dielectric relaxation loss is mainly dominated by dipole rotation polarization or orientational polarization.[50] In iron fiber-filled epoxy resin composites, both the ε' and ε'' are augmented with increasing iron fiber proportions in composites.[39]

8.3.3.3 Size and Distribution of Filler

Nanofillers are always the better alternatives to fillers of bigger dimensions,[66–68] as the density of nanomaterial is relatively lower compared to the micro-fillers. Moreover, nanofillers possess significantly larger specific surface area,[69] and thus, a large number of superficially available active atoms or functional groups bring about interfacial polarization mediated huge dielectric loss at the largely available interfacial area. Investigations relating to the influence of filler size on electrical and magnetic properties of the composites are documented.[43] In this regard, composites filled with fillers of smaller dimension demonstrate lesser reflection loss. Accordingly, these materials are considered to be more potent absorbers of electromagnetic radiation. Moreover, the coercivity values of composites gradually decreases with increasing particle sizes of the soft-ferrite fillers, indicating easier demagnetization of bigger particle filled soft-ferrite composites via facilitated change in orientation of magnetic dipoles because of lesser polymer-filler interaction allowing such change in orientation. However, filler-filler interconnection or interaction plays the dominant role in determining the shielding effectiveness of the polymer composites or nanocomposites. In this regard, the enhanced interfacial polarization plays a significant role in the higher shielding efficiency of polyaniline + TiO_2 + γ-Fe_2O_3 nanocomposite, as a higher extent of interfaces are formed because of the prevalence of three different phases.[55] A group of researchers have developed remarkably enriched shielding effectiveness level > 38 dB for a magnetic, electrically conductive, flexible, and lightweight graphene + iron pentacarbonyl porous films, incorporating the optimized chitosan in the nanocomposite cellular film. Herein, the major objective behind controlled addition of chitosan is to enhance the interaction between the adjacent graphene oxide layers within the nanocomposite microstructure.[41] Such chitosan-mediated interlayer interaction of graphene oxide is obtained via intermolecular hydrogen bonding, along with the covalent bond formation through chemical reaction among amine groups of chitosan and carboxyl groups of graphene oxide layers, as illustrated in Figure 8.8. However, the addition of excess insulating chitosan chains interrupts the interconnected graphene network, leading to decreased electrical and electromagnetic interference shielding performances.

8.3.3.4 Characteristics of Polymer

To date, easily processible, flexible, lightweight, conductive polymers, such as polyaniline[54, 65, 70] and polypyrrole,[71] have been employed to fabricate the effective composite materials suitable as broad band microwave adsorbing materials. Among the conductive polymers, polyaniline is the most versatile because of its lower specific mass, better thermal and chemical stability, adjustable conductivity, and cost effectiveness. The higher polyaniline content in polyaniline-epoxy composite appreciably increases tan δ_ε,[40] suggesting enhanced dielectric loss by conduction. Moreover, the appreciable increase in both $\varepsilon''$and ε' is noted with the increasing polyaniline content. In this regard, the larger amount of polyaniline endows the higher interactions between electrons and atoms because of the greater movement of electrons through the conductive polymer, resulting in the generation of more polarons and bipolarons within the polyaniline-populated matrix. Notably, the formation of such polarons is evidenced from the peaks at 450 and 800 nm in the UV-vis spectrum of polyaniline-coated NiZn ferrite composites, ascribed to polaron-π^* and π-polaron transitions, respectively.[44] The facilitated formation of more frequent polarons and bipolarons is the major reason behind enhanced ε'' and ε'. Indeed, the values of such real and imaginary permittivities are noted to increase further with the Fe_3O_4 content. Additionally, the increasing amounts of Fe_3O_4 in the polyaniline-epoxy matrix enhanced further

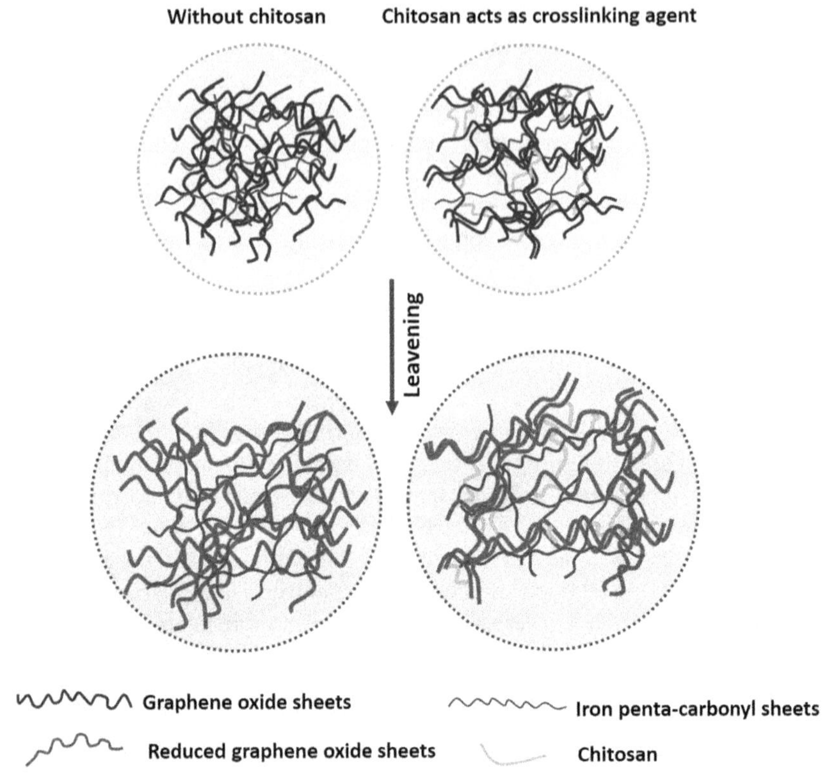

FIGURE 8.8 Chitosan acting as crosslinker within graphene/iron pentacarbonyl porous films.

both tan δ_{ε} and tan δ_{μ},[40] ultimately elevating the electromagnetic radiation absorbing capability of the Fe_3O_4-filled polyaniline + epoxy composites. In the case of Fe_3O_4-filled polyaniline + epoxy composites, the higher amounts of polyaniline produce fine network devoid of Fe_3O_4 agglomerates, facilitating the charge transfer through the orderly-arranged polyaniline-epoxy network. Moreover, once the ε' is increased by the addition of both Fe_3O_4 and polyaniline, the capacitance or the charge content of Fe_3O_4-filled polyaniline + epoxy composite is increased, and thus, the dipole density increases. Importantly, once the external electrical field of incoming electromagnetic radiation interacts with the existing dipoles, strong dipole-dipole interactions between the fillers and epoxy resin allow hopping of charges from one dipole to another throughout the polymer composite. Notably, the drop in conductivity of polyaniline significantly influences the overall dielectric relaxation process. In this regard, compared to dodecyl benzene sulfonic acid doped polyaniline, the significant drop in the conductivity of polyaniline + TiO_2 + γ-Fe_2O_3 nanocomposite is related to the creation of insulating points for preventing the conduction path within nanocomposites. Such conductivity in the polymer composite is ascribed to the formation of semi-quinone radical cations formed by hydrogen bonding between vicinal polymers.[41] Herein, the overlapping of delocalized π-orbital of aniline ring with the d-orbital of metal ion in polymer composite forms the charge transfer complex site, functioning as the localized centers acting as the source point for the hopping of charge carrier. For manufacturing the noncorrosive, multifunctional, high-strength, lightweight materials with tunable dielectric properties, polymer structure and architecture are engineered with precision controlling compositions of monomer and ingredients and reaction environment. Nowadays, low-density polymeric foams are increasingly used as matrices to minimize the density of the overall polymer composites and nanocomposites.[42] In this regard, densities of polyaniline and polypyrrole are about 1.1 and 1.2 g cm^{-3}, respectively, whereas the density of foam-like polyacetylene is only 0.04 g cm^{-3}.

8.3.3.5 Thickness of the Composite

The thickness of the composite determines the effectiveness of the polymer composite or nanocomposite as absorber of electromagnetic radiation. The conductive graphene-films prepared by compact assembly of graphene sheets are preferentially used as electromagnetic radiation absorbers.[72–74] In this context, the high-temperature annealing of graphene oxide film provides elevated shielding effectiveness of ~20 dB at 8.4 µm thickness against the electromagnetic interference.[72] Even a satisfactory level of shielding effectiveness of ~11.2 dB is obtained within 2–10 GHz for an ultrathin (i.e., ~2 µm) reduced graphene oxide film containing 56 wt. % of Fe_3O_4 nanodiscs.[75] In this regard, the shielding effectiveness level remarkably improved to 24 dB for a thicker (i.e., 0.3 mm) Fe_3O_4 + graphene paper of density = 0.78 g cm^{-3}.[74]

8.3.4 Applications

In addition to their common usage as radar absorbing materials, these materials are applied in the stealth technology of aircrafts, television image interference of high-rise buildings, and microwave dark-room. Indeed, these materials are suitable for suppressing unnecessary echo in radar signals received or sent by a ship. Similarly, such materials are employed for making anechoic chamber. Moreover, the ferrite containing polymeric materials are well suited to absorb leakage waves released from electric devices. Moreover, these materials are good candidates for manufacturing wall panels of building to reduce the television image interference and ghost images.

8.4 PIEZOELECTRIC MATERIALS

In principle, piezoelectric materials, such as crystals, ceramics, bone, DNA, and various proteins, are capable of accumulating electric charge in response to applied mechanical stress.[76] Thus, piezoelectric materials are utilized as harvesters of mechanical energy. The energy generated from a piezoelectric system is normally governed by various intrinsic functional properties and thickness of the piezoelectric material. In the last decade, piezoelectric nanogenerator technology has been widely employed for efficient harvesting of potential energy.[77] These piezoelectric nanogenerators function as energy conversion systems to overcome the day by day increasing energy crisis because of the rapidly diminishing global reserve of fossil fuels. These nanogenerators exhibit promising potential to store and utilize the mechanical energies in the form of human body movements, vehicle motion, water flow, and wind speed. In fact, piezoelectric nanogenerator technology was first reported in 2006.[78] Nowadays, the nanoscale mechanical energy is converted into electrical energy by means of piezoelectric zinc oxide nanowire arrays. In this context, piezoelectric nanogenerators can broadly be classified into inorganic, polymer, and composite-type materials. Of these, most research has been focused on developing composite-type piezoelectric nanogenerators to overcome several drawbacks of either inorganic or polymer type piezoelectric materials.[76, 77, 79] For instance, $Bi_4Ti_3O_{12}$ materials suffer from a few issues, such as leakage of current as a result of bismuth volatilization/oxygen vacancies at high temperature, along with high brittleness, leading to deteriorated energy harvesting performance.[76] In this regard, piezoelectric ceramic materials, such as barium titanate and lead zirconium titanate, possess high piezoelectric coefficients and high stiffness constants, resulting in their susceptibility to mechanical failure.[80] Therefore, brittle inorganic nanogenerators are altogether not suitable to sustain continuous and periodic mechanical deformations, such as pressing, stretching, and bending. In contrast, lightweight and highly flexible $Bi_4Ti_3O_{12}$ + polydimethylsiloxane-based composite nanogenerators are advantageous in terms of their energy harvesting performance, as these materials are fabricated at low temperature via a sol–gel technique.[76] In addition, special attention has recently been paid to make lead-free materials to set aside the lead toxicity arising from volatilization of lead at high temperature.[77] The lead zirconium titanate material has toxic effects on human health and the environment. Recently, a new sandwich structure design has been introduced in piezoelectric nanogenerators of good mechanical stability. At the same time, such highly efficient nanogenerators can be prepared with ease employing the simplified fabrication process. In fact, the basic difference between piezoelectric nanogenerators prepared by conventional Schottky contact structure and newly developed sandwich structure is that the generators bearing Schottky contact structure are sensitive to surrounding environmental conditions, such as relative humidity, temperature, and other atmospheric conditions. The key difference between newly developed sandwich structure design and the conventional nanogenerator bearing Schottky contact structure is that a group of well-organized nanoparticulate matters are mixed with dielectric materials and sandwiched between two electrodes in sandwich structure-based nanogenerators, as shown in Figure 8.9. Thus, compared to the other small-scale energy harvesting techniques, such as solar,[81, 82] thermal,[83, 84] and biomechanical energy,[85–87] the advantage of the piezoelectric nanogenerator technology is that this kind of technology-driven devices are influenced relatively less by the change in external conditions. Almost all of these flexible piezoelectric nanogenerators are constituted of different polymer matrices, such as polymethyl methacrylate and[88] polyvinylidene difluoride[89–91] as dielectric material. In addition to functioning as dielectric material, polymethyl methacrylate acts as a medium to transfer the mechanical deformation from external source to the piezoelectric materials.[88] Moreover, the composite nanogenerators contain several polymeric components, such as polyethylene terephthalate,[77] functioning as key accessories. Moreover, the electrical poling process usually aligns the piezoelectric domains in one direction to enrich the performance of the nanogenerator.[91] However,

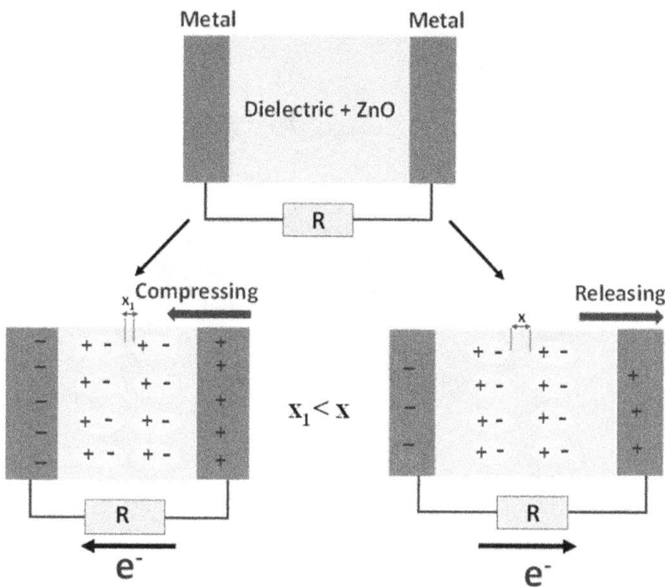

FIGURE 8.9 Conversion from mechanical to electrical energy in sandwich-type piezoelectric nanogenerators comprising of the dielectric-matrix.

the poling process prevents high electrical breakdown failure rate, current leakage, huge energy consumption, and safety issues. Nevertheless, to circumvent the electrical poling treatment, suitably designed barium titanate embedded polyvinylidene difluoride-based nanogenerator endowing the significant output performance has been fabricated, wherein the silver nanowire is incorporated as conducting supplement filler.[89] In fact, such elevated output performance of this nanogenerator is ascribed to the prevalent micro-stone-like architectures contributing to the high piezoelectric coefficient of the composite, whereas silver nanowire functions as a dispersant and conducting material in polyvinylidene difluoride. In this regard, silver nanowires of high aspect ratio reduce the internal resistance of the composite or device forming a conduction pathway. Moreover, the polyvinylidene difluoride can effectively be employed for producing the paper-based multilayer nanogenerator device, as shown in Figure 8.10, as polyvinylidene difluoride adheres with the paper,[90] along with endowing the piezoelectric property and flexibility. During the manufacturing of multilayer nanogenerator, the conductivity of silver is strategically exploited incorporating an intermediate layer comprising a paper written by silver ink pen.

In principle, the higher piezoelectric potential can be achieved, if the flow of positive charges from the positive side of nanogenerator to the negative side becomes restricted. The restrained flow can be imposed by the screening effect of free electrons introduced via doping. Thus, as a result of lack of charge flow, the negative side of the piezoelectric potential is preserved until the donor concentration or carrier density is not too high. Accordingly, the resultant piezopotential is increased because of the decrease in carrier density and increased amount of induced charges

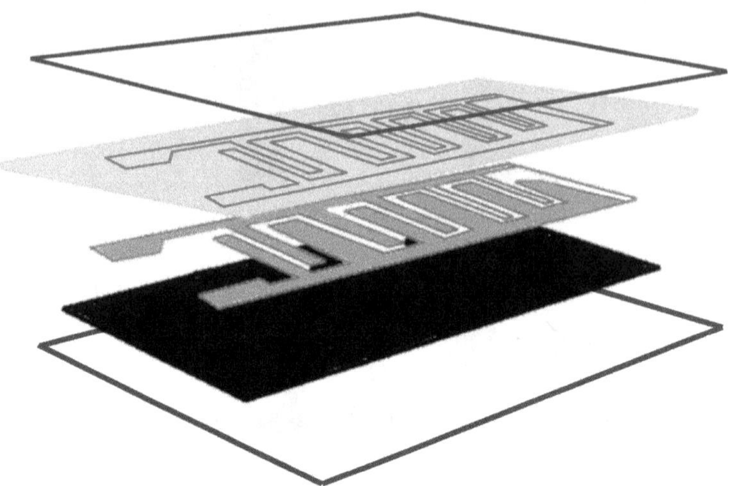

FIGURE 8.10 Structure of paper-based multilayer nanogenerator device.

in electrodes. On implementation of mechanical deformation, the strain-induced piezopotential introduces a piezopotential distribution, leading to transient flow of electrons through the connected external load from one electrode to another, as described in Figure 8.9. In the compressed structure, the electric dipole moments of ZnO-nanowires become aligned in dielectric matrix, leading to higher accumulation of induced charges at both positive and negative electrodes and associated rise in piezopotential, as shown in Figure 8.9. Conversely, when the compressive force is removed, such orderly alignment of dipoles is influenced, reducing the accumulation of induced charge at both the electrodes, and thus, the piezopotential drops appreciably.[85]

8.4.1 Synthesis of Piezoelectric Materials

8.4.1.1 Synthesis of Inorganic Particulate Matters

For fabrication of polymer nanocomposite-based high-performance piezoelectric nanogenerators, one of the major steps is to synthesize the inorganic particulate matters. In this regard, the wet chemical methods, such as sol–gel synthesis, electrochemical deposition, and hydrothermal synthesis, are relatively popular and advantageous, as these methods are least hazardous and are scalable, economic, and executed at relatively lower temperatures. Moreover, these methods are effective in controlling the size, morphology, and properties of the nanowires. Recently, a solid state method has been adopted to synthesize barium titanate-based specialized microstructure.[89]

8.4.1.1.1 Sol–Gel Synthesis

This method is a traditional wet chemical method for synthesizing $Bi_{0.5}(Na_{0.82}K_{0.18})_{0.5}TiO_3$ and $K_{0.5}Na_{0.5}Nb_{0.995}Mn_{0.005}O_3$ nanoparticles by heating the solutions.[92] For preparing $Bi_{0.5}(Na_{0.82}K_{0.18})_{0.5}TiO_3$ sol, bismuth nitrate pentahydrate, sodium nitrate, potassium nitrate, and titanium butoxide are used as the precursors, whereas potassium nitrate, sodium nitrate, niobium pentaethoxide, and manganese acetate are employed as the starting materials for preparing $K_{0.5}Na_{0.5}Nb_{0.995}Mn_{0.005}O_3$ sol. Herein, acetic acid and 2-methoxyethanol are used as solvent, and acetylacetone is added as the complexing agent to stabilize the alkoxide precursors. Both the individual solutions of $Bi_{0.5}(Na_{0.82}K_{0.18})_{0.5}TiO_3$ and $K_{0.5}Na_{0.5}Nb_{0.995}Mn_{0.005}O_3$ in acetic acid and 2-methoxyethanol are dried at 120°C for 24 h, followed by the heating at 400°C for 5 h to remove traces of water and carbon. Finally, the dried materials are calcined at a temperature within 650–800°C.

8.4.1.1.2 Electrochemical Deposition Method

The nanowires prepared by electrodeposition result in better crystallinity and piezoelectric properties compared to the nanowires grown by hydrothermal synthesis. By means of template-assisted electrodeposition, nanowires of definite orientation are grown in particular directions.[93] For instance, if ZnO-nanowires are prepared by the polycarbonate template-assisted electrodeposition method, the nanoporous polycarbonate template carrying ~200 nm pore diameter, ~12 μm pore length, and porosity within 4–20% is cleaned by deionized water in an ultrasonic bath, followed by the complete drying of the cleaned template at 100°C. Thereafter, the nanoporous polycarbonate template is sputter coated by Ag on one side, and the other side is covered by aluminum foil. Finally, the assembly is immersed into 0.1 M solution of $ZnNO_3.6H_2O$ at pH = 4–5. Thereafter, the system is heated to 75°C. Thus, the as-obtained nanogenerator comprises flexible ZnO + polycarbonate-composite with an Ag-sputtered electrode on one side and Al foil on the other.

8.4.1.1.3 Hydrothermal Synthesis Method

Previously, although it was documented that the nanowires synthesized by electrodeposition exhibit better crystallinity and piezoelectric properties compared to the nanowires grown by hydrothermal synthesis,[93] ZnO-nanowires fabricated by template-assisted hydrothermal synthesis method[80] or template-assisted electrodeposition method given in Table 8.3 envisages the identical piezoelectric property in terms of energy conversion efficiency. In fact, by means of template-assisted hydrothermal synthesis at low temperature (<100°C), it is possible to grow high-quality solid ZnO-nanowires directly within the flexible porous polymer templates, such as polycarbonates, to produce nanogenerator constituting ZnO + polycarbonate-nanocomposite structure.[80] By this method, the as-prepared nanowires become well aligned, highly ordered, and mechanically stable. Moreover, lengths and diameters of nanowires are well controlled by the pores of polycarbonate template. In contrast to the method developed by Xu et al.,[94] the template-assisted approach requires the single-growth process, wherein multiple fabrication steps are avoided. Traditionally, ZnO-crystals are grown with the major ingredients, such as hexamethylenetetramine, ammonium hydroxide, and zinc nitrate, and the growth is actuated by the following reaction:

$$Zn^{2+} + 2OH^- \leftrightarrow Zn(OH)_2 \rightarrow ZnO + H_2O$$

TABLE 8.3
Polymer Composite-Based Piezoelectric Nanogenerators

Polymer composite/Nanocomposite			Output				
Inorganic component	Polymer	Applied stress	Current(μA)/ Current density(μA cm^{-2})	Power(μW)/ Power density(μW cm^{-3})	Voltage (V)	Speciality	References
$BaTiO_3$, silver nanowire	Polyvinylidene difluoride	3 N	0.96/–	—	14	Lead free	89
		15 N	3.20/–	—	52	Lead free	89
$BaTiO_3$	Polyvinylidene difluoride	3 N	0.78/–	—	11	Lead free	89
$BaTiO_3$	Polyvinylidene difluoride, ethylene vinyl acetate	—	0.66/–	0.34/–	0.528	Paper-based multilayer nanogenerator	90
$BaTiO_3$	Polyvinylidene difluoride	—	—	—	—	Lead free	91
$Bi_4Ti_3O_{12}$ nanoparticles	Polydimethylsiloxane	—	—	—	—	Lead free	76
$BaTiO_3$	Polydimethylsiloxane	—	0.46–4.19/–	—	0.69–6.33	—	95
Acetylene black, $BaTiO_3$	Polydimethylsiloxane	—	—	–/~7.92	~7.43	Lead free	95
$K_{0.5}Na_{0.5}NbO_3$, acetylene black	Polydimethylsiloxane	—	—	—	10.55	Lead free	96
$K_{0.5}Na_{0.5}NbO_3$	Polydimethylsiloxane	—	—	—	3.22	Lead free	96
$BaTiO_3$	Polydimethylsiloxane	0.5 MPa	0.02/–	—	1.16	Lead free	98
Multi-walled carbon nanotube (MWCNT), ZnO	Polydimethylsiloxane	—	—	—	—	Lead free	108
$BaTiO_3$, $ZnSnO_3$, $Bi_{0.5}(Na_{0.82}K_{0.18})_{0.5}TiO_3$, and $K_{0.5}Na_{0.5}Nb_{0.995}Mn_{0.005}O_3$	Silk fibroin	—	–/0.12	—	2.2	Lead free	92

(Continued)

TABLE 8.3 (CONTINUED)
Polymer Composite-Based Piezoelectric Nanogenerators

Polymer composite/Nanocomposite			Output				
Inorganic component	Polymer	Applied stress	Current(μA)/ Current density(μA cm^{-2})	Power(μW)/ Power density(μW cm^{-3})	Voltage (V)	Speciality	References
ZnO-nanowires pretreated by poly (diallydimethylammonium chloride) and poly (sodium 4-styrenesulfonate)	Polymethyl methacrylate	—	6/–	–/0.20 × 10^{-6}	20	For the first time, battery part in commercial electronics	100
ZnO-nanowires pretreated by oxygen plasma	Polymethyl methacrylate	—	0.30/–	—	5	Unstable in open atmosphere	100
ZnO-nanowires pretreated by annealing in air	Polymethyl methacrylate	—	0.90/–	—	8	Stable	100
$Pb(Zr_{0.52}Ti_{0.48})O_3$	Silicone rubber	—		–/~81.25	—	Stretchable	97
Lead magnesio niobate–lead titanate and MWCNT	Silicone rubber	—	0.50/–	—	4	—	106
Lead zirconate titanate	1. Thermoplastic triblock copolymer grafted with maleic anhydride 2. Non-grafted triblock copolymer	—	1.60/–	—	65	—	107
Lead zirconate titanate	1. Thermoplastic triblock copolymer grafted with maleic anhydride	—	0.30/–	—	12	—	107

Here, hexamethylenetetramine and ammonium hydroxide provide steady supply of OH^- required for the growth of ZnO-crystals. Moreover, a combination of polyethyleneamine + ammonium hydroxide is incorporated to restrain the homogeneous nucleation in the bulk of solution, facilitating heterogeneous nucleation on the substrate. Accordingly, the formation of bulk particles of ZnO is suppressed, whereas the growth of aligned ZnO is encouraged, leading to the formation of ZnO-nanowires. However, in this process, the major drawback is the lesser diameters and lengths of ZnO-nanowires, along with the defects induced by the added polyethyleneamine and ammonium hydroxide, leading to a relatively lower piezoelectric performance. In this context, polycarbonate template pores of ~227 ± 92 nm average diameter and ~12 μm length are reported as a highly-effective support for confined growth of ZnO-nanowires avoiding the additions of polyethyleneamine and ammonium hydroxide.[80] Importantly, this method includes the minimum post-processing steps.

8.4.1.1.4 Solid State Method

A facile, cost-effective, and environmentally friendly process is developed to synthesize barium titanate microstones.[89] For fabricating these microstones, $Ba(OH)_2{\cdot}H_2O$ and TiO_2 powders are mixed in ethanol, and the homogenous mixture is put into an alumina ceramic crucible. Thereafter, the mixture is heated up to 1200°C for 4 h, followed by an intermediate heating at 700°C for 4 h, and finally, at 1000°C for 3 h, in a muffle furnace at a heating rate of 2°C min^{-1}. Once the mixture is cooled down to room temperature, the resulting powder is washed with 10% (v/v) hydrochloric acid (HCl)/water for removals of impurities. Finally, the mixture is ground finely for 10 min and used for further processing of nanogenerators. An almost similar technique is employed to prepare $BaTiO_3$-nanoparticles from TiO_2 and $BaCO_3$-nanoparticles.[95] Such a facile method is employed to synthesize orthorhombic $K_{0.5}Na_{0.5}NbO_3$ particles for manufacturing piezoelectric devices.[96] These particles are prepared by usual ball milling of K_2CO_3, Na_2CO_3, and Nb_2O_5 in absolute ethanol, followed by the calcination at higher temperature.

8.4.1.2 Fabrication of the Polymer Composites/Nanocomposites

8.4.1.2.1 Solution Mixing

For this, the polymer matrices are dissolved in a compatible solvent, followed by the addition and uniform dispersion of piezoelectric discrete phases, such as $BaTiO_3$[91] and $K_{0.5}Na_{0.5}NbO_3$[96] Moreover, other particulate phases including acetylene black are incorporated to improve the migration rate of the polarization charges.[96] Thereafter, the mixed dispersion is sometimes spin coated on a metal-coated polymeric substrate, such as Au-coated polyimide.[91] Subsequently, the coated sample is dried and undergoes a corona poling process to align the piezoelectric domains in one direction. The same technique is used to fabricate polyvinylidene difluoride + $BaTiO_3$ composite films.[91] Otherwise, piezoelectric discrete dispersion and polymer matrix is molded into sheets pouring and drying the mixture at a suitable temperature and pressure.[89] Moreover, the spin-coating technique is adopted for applying polyvinylidene difluoride + barium titanate composite-layer on a silver ink written conductive paper layer of the multilayer nanogenerator device.[90] The combination of solution-mixing and spin-coating methods produce fibroin-based piezoelectric composites.[92] Nevertheless, initially, a special treatment is required for preparing the silk fibroin solution. The solution is prepared by the initial boiling of *Bombyx mori* cocoons in a solution of sodium carbonate to remove unwanted sericin protein, followed by rinsing with distilled water and drying. Finally, a 20 wt. % aqueous solution is obtained by dissolving the fibroin in a 9.3 M lithium bromide-solution at 60°C for 4 h.

8.4.1.2.2 *Solid Phase Mixing*

For the first time, conventional solid phase mixing has been adopted to prepare the elastomer-based stretchable piezoelectric nanogenerators.[97] Initially, applying the laminar shear force transmitted via rollers of different rotational speeds, macro-agglomerates of $Pb(Zr_{0.52}Ti_{0.48})O_3$ powder are subdivided into microagglomerates. The mixing is continued until the microagglomerates break down into particles of smaller dimensions and distributes uniformly throughout the polymer matrix. Herein, utmost care must be taken to obviate the formation of unwanted bubbles during the mixing process.

8.4.1.3 Synthesis of Sandwich-Type Piezoelectric Nanogenerators

In the final step, the polymer composites/nanocomposites are sandwiched in between the conductors. The polymer composites/nanocomposites are assembled within two conductive composites composed of silver-coated glass microspheres embedded in raw rubber, followed by primary vulcanization performed at 175°C for 15 min to transform the composites into an elastic material.[97] For circumventing the residual curing agent in the composite and to stabilize the physical property of the matrix, a secondary vulcanizing process is carried out at 200°C for 2 h. Such a sandwiched nanogenerator based on polydimethylsiloxane + barium titanate is assembled by inserting polydimethylsiloxane + barium titanate layer within two parallel plate electrodes.[98] Such a sandwich-type assembly is noted in the preparation of fibroin-based piezoelectric nanogenerators, wherein the piezoelectric nanoparticle containing fibroin dispersion-layer is attached to polyethylene terephthalate-based matrix functioning as electrodes.[92]

8.4.2 Factors Governing the Performance of Piezoelectric Nanogenerators

Theoretically, the output performances of piezoelectric nanogenerators depend on several parameters, such as piezoelectric coefficient (d), Young's modulus (Y), dielectric constant (ε), area (A), thickness (t), variation in length (Δl), and original length (l) according to the following Equations 8.5 and 8.6:

$$V = \frac{d}{\varepsilon} \Upsilon t \frac{l}{l_0} \tag{8.5}$$

$$I = \frac{d \Upsilon A}{l_0} \frac{dl}{dt} \tag{8.6}$$

8.4.2.1 Crystallinity/Defects

An increase in the sizes of the crystallites within ZnO-nanowires or nanotubes[99] elevates the output voltage and current because of the lesser grain boundaries and defects in larger crystallites. The pre-treatment of ZnO-nanowires by poly (diallyldimethylammonium chloride) and poly (sodium 4-styrenesulfonate) suppress the surface defects, increasing the voltage and current output.[100] Similarly, avoiding the uses of polyethyleneamine and ammonium hydroxide, the minimization of defects in ZnO-nanowires improve the piezoelectric performance.[80]

8.4.2.2 Diameter and Length

The piezoelectric performance of a polymer-composite nanogenerator depends on the diameter and length of the inorganic nanowire, nanotube, or similar materials incorporated as discrete phases in the composite.[80] In presence of polyethyleneamine and ammonium hydroxide, diameters and lengths of ZnO-nanowires are decreased, leading to the inferior quality of nanogenerator (see Equations 8.5 and 8.6).

8.4.2.3 Thickness

Theoretically, the output voltage emanating from a piezoelectric nanogenerator depends on the thickness of the film (see Equation 8.5). For instance, by an increase in film thickness of composite piezoelectric nanogenerator from 60 to 240 μm, the output voltage of the nanogenerator is augmented to 3.61 V.

8.4.2.4 Characteristics of Piezoelectric Phase

Notably, ZnO, $ZnSnO_3$, barium titanate, $Bi_{0.5}(Na_{0.82}K_{0.18})_{0.5}TiO_3$, $K_{0.5}Na_{0.5}Nb_{0.995}Mn_{0.005}O_3$, lead magnesium niobium titanate, lead zirconate titanate, and $K_{0.5}Na_{0.5}NbO_3$ have been used to prepare lightweight polymer composites suitable for piezoelectric devices. In the case of ZnO-filled polymer composites, low piezoelectric coefficient of ZnO deteriorates the output performances. However, the advantage of ZnO over $BaTiO_3$ lies in the relative ease of synthesizing nanostructured ZnO. For lead magnesium niobium titanate and lead zirconate titanate, although piezoelectric coefficients are better compared to ZnO, the lead component imparts serious threat to the ecosystem and the environment.[101–106] In this context, both barium titanate and $K_{0.5}Na_{0.5}NbO_3$ are devoid of such shortcomings. More importantly, $K_{0.5}Na_{0.5}NbO_3$ is equipped with other advantages, such as high Curie temperature (T_c ~420°C), good piezoelectric constant (d_{33} ~160 pC/N), and excellent biocompatibility established through cell-culturing and live and/or dead cell-staining.[96]

In addition to the nature of piezoelectric phase, the proportion of piezoelectric phase in polymer composite is an important factor that directly guides the output performance of the nanogenerator. Lead magnesio niobate-lead titanate particle filled stretchable piezoelectric nanogenerators have poor performance because of the scarcity of piezoelectric particulate phase.[97, 106] Notably, the nanogenerator carrying the compact $Pb(Zr_{0.52}Ti_{0.48})O_3$ particle distribution produces more stress and higher output performance. However, the controlled addition of piezoelectric phase up to a certain extent is permissible to increase the output performance of stretchable piezoelectric nanogenerators. Otherwise, an excessive addition of piezoelectric phase restrains the stretchability of the composite device. The output voltage and current of the piezoelectric nanogenerators gradually increase with increasing densities of the piezoelectric phase from 0 to 30 wt. % in the composite films.[107] Such a phenomenon is explained by the increased polarization as a result of significant changes in the overall dielectric constant of the lead zirconate titanate filled polymer composites. However, both the piezoelectric voltage and current output decrease once the concentration of piezoelectric particulate phase exceeds 30 wt. % because of degraded electromechanical coupling effects. Usually, excess piezoelectric phase brings about unwanted inhomogeneous distribution and agglomeration of piezoelectric powders because of high viscosity of matrix and microscopic inter-particle attractive forces, such as van der Waals force, electrostatic force, and the adhesive force actuated via liquid bridge under humid conditions. In the case of $BaTiO_3$ + polydimethylsiloxane composite-based piezoelectric nanogenerator, the highest output voltage, i.e., 46 wt. %, is recorded using the optimum $BaTiO_3$ crystals.[98] In this context, the piezoelectric voltage and current output gradually increases with increasing amounts of $BaTiO_3$ in $BaTiO_3$ + polydimethylsiloxane composite.[95]

In addition, the proportions of non-piezoelectric particles, such as acetylene black, impose significant impact on the piezoelectric performance. In the case of $K_{0.5}Na_{0.5}NbO_3$ + acetylene black + polydimethylsiloxane composite-based piezoelectric nanogenerators, the maximum 10.55 V output voltage is obtained at 12 wt. % acetylene black-loading.[96] Indeed, such performance is attributed to the synergistic effect in between the conductive property of acetylene black and the piezoelectric property of $K_{0.5}Na_{0.5}NbO_3$, leading to an elevated migration rate of polarization charges with the higher proportion of acetylene black. Once the proportion of acetylene black goes above 12 wt. %, the piezoelectric property is influenced by the leakage-effect produced by an excess of acetylene black content. A similar trend is noted for $BaTiO_3$ + acetylene black + polydimethylsiloxane composite-based piezoelectric nanogenerators, wherein once the proportion of acetylene black exceeds the threshold level, i.e., 3.2 wt. %, output voltage drastically drops from 7.43 to 0.33 V.[95] Such a large drop in output voltage is ascribed to the transformation of piezoelectric composite film into a conductor via rapid escalation of conductivity in the presence of excessive acetylene black. Moreover, the Ag-nanowires are added to improve the performance of fibroin-based piezoelectric device.[92] However, similar to acetylene black-added composites, an excessive conductivity deteriorates the output voltage of the device. Therefore, in order to control and astrict the conductivity imparted by the Ag-nanowires in the device, Ag-nanowires are intentionally wrapped by polyvinyl pyrrolidone to obstruct connectivity among Ag-nanowires. Moreover, polyvinyl pyrrolidone acts as a dispersion promoter to restrain the aggregation tendencies of Ag-nanowires within the polymer matrix.

8.4.2.5 Functionalization of Piezoelectric Phase

Usually, the piezoelectric phases are functionalized to improve the output performance of the devices. Notably, few attempts are made to improve the performances of ZnO-nanowire-based composite piezoelectric nanogenerators through the pre-treatment of ZnO-nanowire arrays by oxygen plasma, annealing in air, and surface passivation with certain polymers. Of these three pre-treatment methods,[100] the maximum output voltage, current, and power density are imparted, if the surface of

ZnO-nanowire is passivated by polymers, such as positively charged poly (diallyldimethylammonium chloride) and negatively charged poly (sodium 4-styrenesulfonate) (see Table 8.3). In fact, the major strategy behind such pre-treatments is to decrease carrier density in ZnO-nanowires by means of doping. Relatively uniform dispersion of piezoelectric particulate phases in the flexible piezoelectric nanogenerators of enhanced stability is obtained, if amine-functionalized lead zirconate titanate particles are incorporated in a mixed matrix carrying the maleic anhydride-grafted thermoplastic styrene-*b*-ethylene + butylene-*b*-styrene triblock copolymer and non-grafted triblock copolymer components in optimized stoichiometric ratio.[107] In fact, the highest output voltage is obtained for the mixed matrix-based nanogenerators, when exactly equal quantity of grafted and non-grafted components are present within the mixed matrix. Indeed, such elevated uniformity and stability of dispersion results from the reaction between $-NH_2$ of amine-functionalized lead zirconate titanate particles with maleic anhydride of the grafted triblock copolymer. Notably, if the amine-functionalized lead zirconate titanate particle containing nanogenerator is subjected to bending motions, the newly-formed strong amide-bonds of the particles with the polymer matrix significantly enhance the applied stress to the nanogenerators, resulting in the induced piezoelectric potential between the two electrodes, which drives the electrons from electron-rich bottom electrode to electron-deficient top electrode to balance the developed potential difference. Subsequently, when the applied stress is released, the induced piezopotential disappears, as represented in Figure 8.9, and the accumulated electrons migrate back in the reverse direction from top to bottom electrode. In this regard, relatively inferior performance for amine-functionalized lead zirconate titanate filled maleic anhydride-grafted triblock copolymer is ascribed to very high Young's modulus, resulting in the relatively uneven stress distribution in the rigid network. Notably, an elevated deterioration results if pristine lead zirconate titanate particles are mixed with the same maleic anhydride-grafted triblock copolymer, resulting in appreciable formation of aggregates, endowing the lower piezoelectric coefficient, low output voltage and current compared to generators prepared using amine-functionalized lead zirconate titanate particles (see Table 8.3). Thus, the electric power output of a piezoelectric nanogenerator depends both on the piezoelectric coefficient and dielectric constant, along with the Young's moduli of the constituents (see Equations 8.5 and 8.6). The piezoelectric outputs are largely increased in poly(vinylidene fluoride) + $BaTiO_3$-composite containing 10 wt. % paraelectric-$BaTiO_3$ carrying the highest Young's modulus.[91]

8.4.2.6 Characteristics of the Polymer Matrix

8.4.2.6.1 Silicone Rubber

To fulfill recent demand for stretchable and more flexible nanogenerators, several researchers have developed rubber-based composites comprising variable rubbery components including silicone rubber[97, 106] filled with inorganic components, such as $Pb(Zr_{0.52}Ti_{0.48})O_3$,[97] and lead magnesio niobate-lead titanate.[106] In the case of silicone rubber-based nanogenerator filled with $Pb(Zr_{0.52}Ti_{0.48})O_3$, about 30% stretchability can still be obtained even if the proportion of $Pb(Zr_{0.52}Ti_{0.48})O_3$ is increased to 92 wt. %. Moreover, a stretchable piezoelectric nanogenerator has been fabricated by dispersing lead magnesio niobate-lead titanate particles into a translucent silicone-rubber bearing the multi-walled carbon nanotube.[106] The multi-walled carbon nanotubes function as physically dispersing, mechanically reinforcing, and electrically bridging agents in the piezoelectric composite. However, the resulting output performance is unsatisfactory because of the low percentage, i.e., 20 wt. %, piezoelectric phase or lead magnesio niobate-lead titanate particles in the composite material.

8.4.2.6.2 Polyvinylidene Difluoride

The piezoelectric properties of polyvinylidene difluoride (PVDF) are originated from the large difference in electronegativity between the fluorine and carbon atoms, resulting in the polar bonds and dipole moment directing from the fluorine atoms of the chain toward the hydrogens. However, all polyvinylidene difluorides are unable to function as piezoelectric nanogenerators. For instance, α-phase polyvinylidene difluoride is unable to result in piezoelectricity because of the net cancellation of dipoles along the chain, whereas β-phase polyvinylidene difluoride exhibits the strongest piezoelectricity, as shown in Figure 8.11.[109]

8.4.2.6.3 Polydimethylsiloxane

In addition to the inherent flexibility, polydimethylsiloxane matrix is biocompatible, and thus, can be well suited for fabricating nanogenerators attached to living systems.[108]

8.4.2.6.4 Fibroin

K. N. Kim and J. M. Baik et al. successfully employed silk fibroin as the base material for fabricating piezoelectric nanogenerators containing fibrous-protein and particulate phases, such as $ZnSnO_3$, barium titanate, $Bi_{0.5}(Na_{0.82}K_{0.18})_{0.5}TiO_3$, and $K_{0.5}Na_{0.5}Nb_{0.995}Mn_{0.005}O_3$.[92] In this regard, silk fibroin is a natural biodegradable

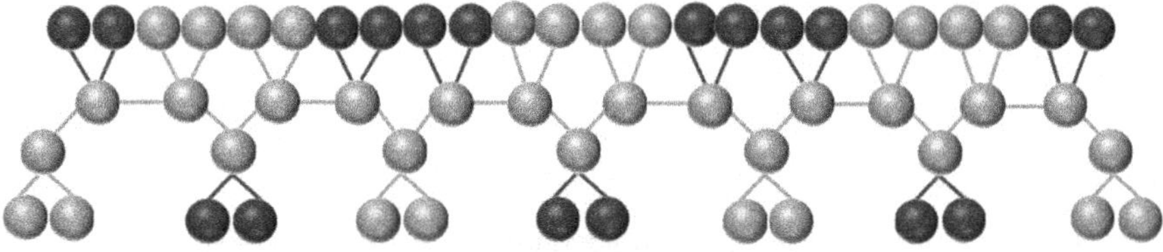

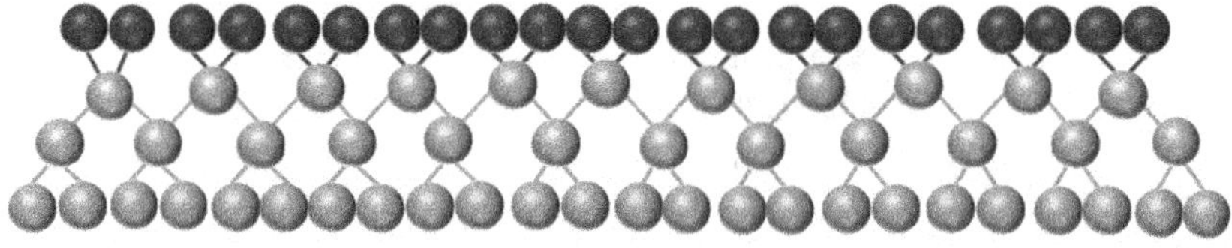

FIGURE 8.11 Molecular structures of poly-vinylidene difluorides.

polymer carrying the desired biocompatibility for diversified biomedical usages.[110–112] Moreover, fibroin has been used as platforms for transistors and various classes of photonic devices. Alongside of this, fibroin possesses excellent mechanical properties (e.g., tensile strength ~100 MPa) imparting good durability to the material undergoing various external stresses. Additionally, the silk fibroin composite is equipped with a unique time-controllable property, which is desirable for a biocompatible nanogenerator attached to human body and skin. Moreover, the water-resistance and biodegradability of fibroin are improved by adding glycerol to encourage the transformation of soluble random coil to insoluble β-sheet, accelerating the gelation of fibroin.

8.4.3 Applications

In general, piezoelectric nanogenerators are capable of functioning as sustainable independent power sources to drive electronic devices/sensors requiring low-power consumption. However, there are some dual-functional piezoelectric devices, which work as independent wearable/portable power sources to drive the commercial demand for electronic devices with self-powered sensors, i.e., battery-free sensor, to monitor various physical, chemical, biological, and optical stimuli. Because of their elevated stretchability, silicone rubber-$Pb(Zr_{0.52}Ti_{0.48})O_3$-based nanogenerator can be attached to human body for the harvesting of kinetic energy from human motion.[97] Accordingly, the harvested energy can be utilized to operate commercial electronics or may be stored in a capacitor. Moreover, the composite-type piezoelectric nanogenerators can be used as self-powered sensors/systems to monitor the fluid velocity and pH of the alkaline solution. Moreover, these piezoelectric nanogenerators can be employed as self-powered flexion sensors to classify or measure the individual finger motions and nonlinear muscular movement of human body parts. Moreover, these materials can function as acceleration sensors to measure the various accelerations of the linear motor shaft, along with driving the commercial light-emitting diodes/liquid crystal display.

8.5 CONCLUSIONS

The smart polymer composites are functioning as stimuli-responsive transducers. Accordingly, the overall performance of smart materials depends on the abilities of transducers to transform to a certain extent one form of energy into another. The efficiencies of smart composite materials depend on the size, shape, and functionalization of the individual components, and distribution of one component into other(s), along with the nature and type of interfaces influenced via cohesion and adhesion forces among individual components. Therefore, the extent of functionalizations in materials, fillers, and coupling agents alter the overall efficiency. Moreover, construction and design of binary, ternary, or multi-component, lightweight, smart polymer composites influence the overall efficiency of composite materials. In the future, polymeric foam- or aerogel-based lightweight piezoelectric nanogenerators could be a potential area of research.

REFERENCES

1. D'Auria, M.; Davino, D.; Pantani, R.; Sorrentino, L. Polymeric Foam-Ferromagnet Composites as Smart Lightweight Materials. *Smart Mater. Struct.* **2016**, *25*(5), 055014.
2. Sorrentino, L.; Aurilia, M.; Forte, G.; Iannace, S. Anisotropic Mechanical Behavior of Magnetically Oriented Iron Particle Reinforced Foams. *J. Appl. Polym. Sci.* **2011**, *119*(2), 1239–1247.
3. Sorrentino, L.; Aurilia, M.; Forte, G.; Iannace, S. Composite Polymeric Foams Produced by Using Magnetic Field. *Adv. Sci. Tech.* **2008**, *54*, 123–126.
4. Davino, D.; Mei, P.; Sorrentino, L.; Visone, C. Polymeric Composite Foams with Properties Controlled by the Magnetic Field. *IEEE T. Magn.* **2012**, *48*(11), 3043–3046.
5. Guan, X.; Dong, X.; Ou, J. Magnetostrictive Effect of Magnetorheological Elastomer. *J. Magn. Magn. Mater.* **2008**, *320*(3–4), 158–163.
6. Masłowski, M.; Zaborski, M. Smart Materials Based on Magnetorheological Composites. *Mater. Sci. Forum* **2012**, *714*, 167–173.
7. Park, B.; Fang, F.; Choi, H. Magnetorheology: Materials and Application. *Soft Matter* **2010**, *6*(21), 5246–5253.
8. Xu, Y.; Gong, X.; Xuan, S.; Zhang, W.; Fan, Y. A High-Performance Magnetorheological Material: Preparation, Characterization and Magnetic-Mechanic Coupling Properties. *Soft Matter* **2011**, *7*(11), 5246–5254.
9. Qichun, G.; Jinkui, W.; Xinglong, G.; Yanceng, F.; Hesheng, X. Smart Polyurethane Foam with Magnetic Field Controlled Modulus and Anisotropic Compression Property. *RSC Adv.* **2013**, *3*(10), 3241–3248.
10. Kaleta, J.; Lewandowski, D.; Mech, R. Magnetostriction of Field-Structural Composite with Terfenol-D Particles. *Arch. Civ. Mech. Eng.* **2015**, *15*(4), 897–902.
11. Zuo, Z.; Pan, D.; Jia, Y.; Tian, J.; Zhang, S.; Qiao, L. Enhanced Magnetoelectric Effect in Magnetostrictive/ Piezoelectric Laminates Through Adopting Magnetic Warm Compaction Terfenol D. *J. Alloy. Compd.* **2014**, *587*, 287 289.
12. Altin, G.; Ho, K. K.; Henry, C. P.; Carman, G. P. Static Properties of Crystallographically Aligned Terfenol-D/Polymer Composites. *J. Appl. Phys.* **2007**, *101*(3), 033537.
13. Ho, K. K.; Henry, C. P.; Altin, G.; Carman, G. P. Crystallographically Aligned Terfenol-D/Polymer Composites for a Hybrid Sonar Device. *Integr. Ferroelectr.* **2006**, *83*(1), 121–138.
14. Du, T.; Zhang, T.; Meng, H.; Zhou, X.; Jiang, C. A Study on Laminated Structures in Terfenol-D/Epoxy Particulate Composite with Enhanced Magnetostriction. *J. Appl. Phys.* **2014**, *115*(24), 243909.
15. Chattopadhyay, P. K.; Basuli, U.; Chattopadhyay, S. Studies on Novel Dual Filler Based Epoxidized Natural Rubber Nanocomposite. *Polym. Compos.* **2010**, *31*, 835–846.
16. Dong, X.; Qi, M.; Guan, X.; Li, J.; Ou, J. Magnetostrictive Properties of Titanate Coupling Agent Treated Terfenol-D Composites. *J. Magn. Magn. Mater.* **2012**, *324*(6), 1205–1208.
17. Duenas, T. A.; Carman, G. P. Large Magnetostrictive Response of Terfenol-D Resin Composites (Invited). *J. Appl. Phys.* **2000**, *87*(9), 4696–4701.
18. Lo, C. Y.; Or, S. W.; Chan, H. L. W. Large Magnetostriction in Epoxy-Bonded Terfenol-D Continuous-Fiber Composite with [112] Crystallographic Orientation. *IEEE T. Magn.* **2006**, *42*, 3111–3113.
19. Yoo, B.; Na, S.-M.; Pines, D. J. Influence of Particle Size and Filling Factor of Galfenol Flakes on Sensing Performance of Magnetostrictive Composite Transducers. *IEEE T. Magn.* **2015**, *51*(11).
20. Li, J.; Gao, X.; Zhu, J.; Jia, J.; Zhang, M. The Microstructure of Fe–Ga Powders and Magnetostriction of Bonded Composites. *Scr. Mater.* **2009**, *61*(6), 557–560.

21. Hong, J.; Solomon, V.; Smith, D. J.; Parker, F.; Summers, E.; Berkowitz, A. One-Step Production of Optimized Fe–Ga Particles by Spark Erosion. *Appl. Phys. Lett.* **2006**, *89*(14), 142506.
22. Walters, K.; Busbridge, S.; Walters, S. Magnetic Properties of Epoxy-Bonded Iron–Gallium Particulate Composites. *Smart Mater. Struct.* **2012**, *22*(2), 025009.
23. Rodríguez, C.; Barrio, A.; Orue, I.; Vilas, J. L.; León, L. M.; Barandiarán, J. M.; Ruiz, M. L. F.-G. High Magnetostriction Polymer-Bonded Terfenol-D Composites. *Sens. Actuat. A* **2008**, *142*(2), 538–541.
24. Rodríguez, C.; Rodríguez, M.; Orue, I.; Vilas, J. L.; Barandiarán, J. M.; Gubieda, M. L. F.; León, L. M. New Elastomer–Terfenol-D Magnetostrictive Composites. *Sens. Actuat. A* **2009**, *149*(2), 251–254.
25. Bednarek, S. The Giant Magnetostriction in Ferromagnetic Composites within an Elastomer Matrix. *Appl. Phys. Mater.* **1999**, *68*(1), 63–67.
26. Nersessian, N.; Or, S. W.; Carman, G. P.; Choe, W.; Radousky, H. B. Hollow and Solid Spherical Magnetostrictive Particulate Composites. *J. Appl. Phys.* **2004**, *96*(6), 3362–3365.
27. McKnight, G. P.; Carman, G. P. 112 Oriented Terfenol-D Composites. *Mater. Trans.* **2002**, *43*, 1008–1014.
28. Mohanty, T. R.; Bhandari, V.; Chandra, A. K.; Chattopadhyay, P. K.; Chattopadhyay, S. Role of Calcium Stearate as A Dispersion Promoter for New Generation Carbon Black-Organoclay Based Rubber Nanocomposites for Tyre Application. *Polym. Compos.* **2013**, *34*(2), 214–224.
29. Mondal, M.; Chattopadhyay, P. K.; Chattopadhyay, S.; Setua, D. K. Thermal and Morphological Analysis of Thermoplastic Polyurethane-Clay Nanocomposites: Comparison of Efficacy of Dual Modified Laponite vs. Commercial Montmorillonites. *Thermochim. Acta* **2010**, *510*(1–2), 185–194.
30. Deng, H.; Gong, X.; Wang, L. Development of an Adaptive Tuned Vibration Absorber with Magnetorheological Elastomer. *Smart Mater. Struct.* **2006**, *15*(5), N111–N116.
31. Liao, G.; Gong, X.; Xuan, S. Magnetic Field-Induced Compressive Property of Magnetorheological Elastomer Under High Strain Rate. *Ind. Eng. Chem. Res.* **2013**, *52*(25), 8445–8453.
32. Farshad, M.; Le, R. M. A New Active Noise Abatement Barrier System. *Polym. Test.* **2004**, *23*(7), 855–860.
33. Blom, P.; Kari, L. Smart Audio Frequency Energy Flow Control by Magneto-Sensitive Rubber Isolators. *Smart Mater. Struct.* **2008**, *17*(1), 015043.
34. Tian, T.; Li, W.; Deng, Y. Sensing Capabilities of Graphite Based MR Elastomers. *Smart Mater. Struct.* **2011**, *20*(2), 025022.
35. Leng, J. S.; Lan, X.; Liu, Y. J.; Du, S. Y.; Huang, W. M.; Liu, N.; Phee, S. J.; Yuan, Q. Electrical Conductivity of Thermoresponsive Shape-Memory Polymer with Embedded Micron Sized Ni Powder Chains. *Appl. Phys. Lett.* **2008**, *92*(1), 014104.
36. Bica, I. Magnetoresistor Sensor with Magnetorheological Elastomers. *J. Ind. Eng. Chem.* **2011**, *17*(1), 83–89.
37. Fadzidah, M. I.; Mansor, H.; Zulkifly, A.; Ismayadi, I.; Rodziah, N.; Idza, R. I. Recent Developments of Smart Electromagnetic Absorbers Based Polymer-Composites at Gigahertz Frequencies. *J. Magn. Magn. Mater.* **2016**, *405*, 197–208.
38. Gama, A. M.; Rezende, M. C. Complex Permeability and Permittivity Variation of Carbonyl Iron Rubber in the Frequency Range of 2 to18 GHz. *J. Aerosp. Technol. Manag.* **2010**, *2*(1), 59–62.
39. Wu, M.; He, H.; Zhao, Z.; Yao, X. Electromagnetic and Microwave Absorbing Properties of Iron Fibre-Epoxy Resin Composites. *J. Phys. D: Appl. Phys.* **2000**, *33*(19), 2398–2401.

40. Belaabed, B.; Wojkiewicz, J.; Lamouri, S.; Kamchi, N. E.; Lasri, T. Synthesis and Characterization of Hybrid Conducting Composites Based on Polyaniline/Magnetite Fillers with Improved Microwave Absorption Properties. *J. Alloy. Compd.* **2012**, *527*, 137–144.
41. Liu, J.; Zhang, H.-B.; Liu, Y.; Wang, Q.; Liu, Z.; Mai, Y.-W.; Yu, Z.-Z. Magnetic, Electrically Conductive and Lightweight Graphene/Iron Pentacarbonyl Porous Films Enhanced with Chitosan for Highly Efficient Broadband Electromagnetic Interference Shielding. *Compos. Sci. Technol.* **2017**, *151*, 71–78.
42. Huo, J.; Wang, L.; Yu, H. Polymeric Nanocomposites for Electromagnetic Wave Absorption. *J. Mater. Sci.* **2009**, *44*(15), 3917–3927.
43. Li, B. W.; Shen, Y.; Yue, Z. X.; Nan, C. W. Influence of Particle Size on Electro-Magnetic Behavior and Microwave Absorption Properties of Z-Typeba-Fer- Rite/ Polymer Composites. *J. Magn. Magn. Mater.* **2007**, *313*(2), 322–328.
44. Ting, T. H.; Yu, R. P.; Jau,Y. N. Synthesis and Microwave Absorption Characteristics of Polyaniline/Niznferrite Composites in 2–40 GHz. *Mater. Chem. Phys.* **2011**, *126*(1–2), 364–368.
45. Yusoff, A. N.; Sani, J. M.; Abdullah, M. H.; Ahmad, S. H.; Ahmad, N. Electromagnetic and Absorption Properties of Some TPNR/Fe_3O_4/YIG Microwave Absorbers and Specular Absorber Method. *Sains. Malays.* **2007**, *36*, 65–75.
46. Yusoff, A. N.; Abdullah, M. H.; Ahmad, S. H.; Jusoh, S. F.; Mansor, A. A.; Hamid, S. A. A. Electromagnetic and Absorption Properties of Some Microwave Absorbers. *J. Appl. Phys.* **2002**, *92*(2), 876–882.
47. Dosoudil, R.; Usakova, M.; Franek, J.; Slama, J.; Olah, V. RF Electromagnetic Wave Absorbing Properties of Ferrite Polymer Composite Materials. *J. Magn. Magn. Mater.* **2006**, *304*(2), e755–e757.
48. Abbas, S. M.; Dixit, A. K.; Chatterjee, R.; Goel, T. C. Complex Permittivity, Complex Permeability and Micro Wave Absorption Properties of Ferrite–Polymer Composites. *J. Magn. Magn. Mater.* **2007**, *309*(1), 20–24.
49. Zheng, H.; Yang, Y.; Zhou, M.; Li, F. Microwave Absorption and Mössbauer Studies of Fe_3O_4 Nanoparticles. *Hyperfine Interact.* **2009**, *189*(1–3), 131–136.
50. Kong, I.; Ahmad, S.; Abdullah, M.; Hui, D.; Yusoff, A. N.; Puryanti, D. Magnetic and Microwave Absorbing Properties of Magnetite-Thermoplastic Natural Rubber Nanocomposites. *J. Magn. Magn. Mater.* **2010**, *322*(21), 3401–3409.
51. Singha, N. R.; Karmakar, M.; Chattopadhyay, P. K.; Roy, S.; Deb, M.; Mondal, H.; Mahapatra, M.; Dutta, A.; Mitra, M.; Roy, J. S. D. Structures, Properties, and Performances-Relationships of Polymeric Membranes for Pervaporative Desalination. *Membranes* **2019**, *9*(5), 58.
52. Chattopadhyay, P. K.; Das, N. C.; Chattopadhyay, S. Influence of Interfacial Roughness and the Hybrid Filler Microstructures on the Properties of Ternary Elastomeric Composites. *Compos: Part A* **2011**, *42*(8), 1049–1059.
53. Kar, E.; Bose, N.; Dutta, B.; Mukherjee, N.; Mukherjee, S. Poly(Vinylidene Fluoride)/ Submicron Graphite Platelet Composite: A Smart, Lightweight Flexible Material with Significantly Enhanced β Polymorphism, Dielectric and Microwave Shielding Properties. *Eur. Polym. J.* **2017**, *90*, 442–455.
54. Hosseini, S. H.; Mohseni, S. H.; Asadnia, A.; Kerdari, H. Synthesis and Microwave Absorbing Properties of Polyaniline/$MnFe_2O_4$ Nanocomposite. *J. Alloy. Compd.* **2011**, *509*(14), 4682–4687.
55. Singh, K.; Ohlan, A.; Bakhshi, A. K.; Dhawan, S. K. Synthesis of Conducting Ferromagnetic Nanocomposite with Improved Microwave Absorption Properties. *Mater. Chem. Phys.* **2010**, *119*(1–2), 201.

56. Chattopadhyay, P. K.; Praveen, S.; Das, N. C.; Chattopadhyay, S. Contribution of Organomodified Clay on Hybrid Microstructures and Properties of Epoxidized Natural Rubber-Based Nanocomposites. *Polym. Eng. Sci.* **2013**, *53*(5), 923–930.
57. Chattopadhyay, P. K.; Chattopadhyay, S. Role of Epoxy Functionality in Microstructure–Property Relationships within Elastomeric Nanocomposites. *Plast. Rubber Compos.* **2013**, *42*(8), 340–348.
58. Chattopadhyay, P. K.; Chattopadhyay, S.; Das, N. C.; Bandyopadhyay, P. P. Impact of Carbon Black Substitution with Nanoclay on Microstructure and Tribological Properties of Ternary Elastomeric Composites. *Mater. Des.* **2011**, *32*(10), 4696–4704.
59. Praveen, S.; Chattopadhyay, P. K.; Jayendran, S.; Chakraborty, B. C.; Chattopadhyay, S. Effect of Nanoclay on the Mechanical and Damping Properties of Aramid Short Fibre-Filled Styrene Butadiene Rubber Composites. *Polym. Int.* **2010**, *59*, 187–197.
60. Praveen, S.; Chattopadhyay, P. K.; Jayendran, S.; Chakraborty, B. C.; Chattopadhyay, S. Effect of Rubber Matrix Type on the Morphology and Reinforcement Effects in Carbon Black-Nanoclay Hybrid Composites-A Comparative Assessment. *Polym. Compos.* **2010**, *31*(1), 97–104.
61. Karmakar, M.; Mahapatra, M.; Dutta, A.; Singha, N. R. Separation of Tetrahydrofuran Using RSM Optimized Accelerator-Sulfur-Filler of Rubber Membranes: Systematic Optimization and Comprehensive Mechanistic Study. *Korean J. Chem. Eng.* **2017**, *34*(5), 1416–1434.
62. Singha, N. R.; Ray, S. K. Removal of Pyridine from Water by Pervaporation Using Crosslinked and Filled Natural Rubber Membranes. *J. Appl. Polym. Sci.* **2012**, *124*(S1), E99–E107.
63. Singha, N. R.; Ray, S.; Ray, S. K. Removal of Pyridine from Water by Pervaporation Using Filled SBR Membranes. *J. Appl. Polym. Sci.* **2011**, *121*(3), 1330–1334.
64. Singha, N. R.; Parya, T. K.; Ray, S. K. Dehydration of 1, 4-Dioxane by Pervaporation Using Filled and Crosslinked Polyvinyl Alcohol Membrane. *J. Membr. Sci.* **2009**, *340*(1–2), 35–44.
65. Wang, Z.; Bi, H.; Liu, J.; Sun, T.; Wu, X. Magnetic and Microwave Absorbing Properties of Polyaniline/γ-Fe_2O_3 Nanocomposite. *J. Magn. Magn. Mater.* **2008**, *320*(16), 2132–2139.
66. Praveen, S.; Chattopadhyay, P. K.; Albert, P.; Dalvi, V. G.; Chakraborty, B. C.; Chattopadhyay, S. Synergistic Effect of Carbon Black and Nanoclay Fillers in Styrene Butadiene Rubber Matrix: Development of Dual Structure. *Compos: Part A* **2009**, *40*(3), 309–316.
67. Singha, N. R.; Das, P.; Ray, S. K. Recovery of Pyridine from Water by Pervaporation Using Filled and Crosslinked EPDM Membranes. *J. Ind. Eng. Chem.* **2013**, *19*(6), 2034–2045.
68. Samanta, H. S.; Ray, S. K.; Das, P.; Singha, N. R. Separation of Acid-Water Mixtures by Pervaporation Using Nanoparticle Filled Mixed Matrix Copolymer Membranes. *J. Chem. Technol. Biot.* **2012**, *87*(5), 608–622.
69. Mahapatra, M.; Karmakar, M.; Dutta, A.; Singha, N. R. Fabrication of Composite Membranes for Pervaporation of Tetrahydrofuran-Water: Optimization of Intrinsic Property by Response Surface Methodology and Studies on Vulcanization Mechanism by Density Functional Theory. *Korean J. Chem. Eng.* **2018**, *35*(9), 1889–1910.
70. Li, Y.; Zhang, H.; Liu, Y.; Wen, Q.; Li, J. Rod-Shaped Polyaniline–Barium Ferrite Nanocomposite: Preparation, Characterization and Properties. *Nanotechnology* **2008**, *19*(10), 105605.
71. Zhang, C.; Li, Q.; Ye, Y. Preparation and Characterization of Polypyrrole/Nano-$SrFe_{12}O_{19}$ Composites by in-situ Polymerization Method. *Synth. Met.* **2009**, *159*(11), 1008–1013.

72. Shen, B.; Zhai, W.; Zheng, W. Ultrathin Flexible Graphene Film: An Excellent Thermal Conducting Material with Efficient EMI Shielding. *Adv. Funct. Mater.* **2014**, *24*(28), 4542–4548.
73. Song, W. L.; Fan, L. Z.; Cao, M. S.; Lu, M. M.; Wang, C. Y.; Wang, J.; Chen, T. T.; Li, Y.; Hou, Z. L.; Liu, J.; Sun, Y.-P. Facile Fabrication of Ultrathin Graphene Papers for Effective Electromagnetic Shielding. *J. Mater. Chem. C* **2014**, *2*(25), 5057–5064.
74. Song, W. L.; Guan, X. T.; Fan, L. Z.; Cao, W. Q.; Wang, C. Y.; Zhao, Q. L.; Cao, M. S. Magnetic and Conductive Graphene Papers Toward Thin Layers of Effective Electromagnetic Shielding. *J. Mater. Chem. A* **2015**, *3*(5), 2097–2107.
75. Yang, Y.; Li, M.; Wu, Y.; Wang, T.; Choo, E. S. G.; Ding, J.; Zong, B.; Yang, Z.; Xue, J. Nanoscaled Self-Alignment of Fe_3O_4 Nanodiscs in Ultrathin rGO Films with Engineered Conductivity for Electromagnetic Interference Shielding. *Nanoscale* **2016**, *8*(35), 15989–15998.
76. Raj, N. P. M. J.; Alluri, N. R.; Khandelwal, G.; Kim, S.-Jae. Lead-free Piezoelectric Nanogenerator Using Lightweight Composite Films for Harnessing Biomechanical Energy. *Compos. B Eng.* **2019**, *161*, 608–616.
77. Ren, X.; Fan, H.; Zhao, Y.; Liu, Z. Flexible Lead-Free $BiFeO_3$/PDMS-Based Nanogenerator as Piezoelectric Energy Harvester. *ACS Appl. Mater. Interfaces* **2016**, *8*(39), 26190–26197.
78. Wang, Z. L.; Song, J. H. Piezoelectric Nanogenerators Based on Zinc Oxide Nanowire Arrays. *Science* **2006**, *312*(5771), 242–246.
79. Alluri, N. R.; Chandrasekhar, A.; Vivekananthan, V.; Purusothaman, Y.; Selvarajan, S.; Jeong, J. H.; Kim, S. J. Scavenging Biomechanical Energy Using High-Performance, Flexible $BaTiO_3$ Nanocube/PDMS Composite Films. *ACS Sustain. Chem. Eng.* **2017**, *5*(6), 4730–4738.
80. Ou, C.; Sanchez-Jimenez, P. E.; Datta, A.; Boughey, F. L.; Whiter, R. A.; Sahonta, S.-L.; Kar-Narayan, S. Template-Assisted Hydrothermal Growth of Aligned Zinc Oxide Nanowires for Piezoelectric Energy Harvesting Applications. *ACS Appl. Mater. Interf.* **2016**, *8*(22), 13678–13683.
81. Sargent, E. H. Colloidal Quantum Dot Solar Cells. *Nat. Photonics* **2012**, *6*, 133–135.
82. Nie, W.; Tsai, H.; Asadpour, R.; Blancon, J. C.; Neukirch, A. J.; Gupta, G.; Crochet, J. J.; Chhowalla, M.; Tretiak, S.; Alam, M. A.; Wang, H. L.; Mohite, A. D. High-Efficiency Solution-Processed Perovskite Solar Cells with Millimeter-Scale Grains. *Science* **2015**, *347*(6221), 522–525.
83. Zhang, Y.; Huang, C.; Wang, J.; Lin, G.; Chen, J. Optimum Energy Conversion Strategies of A Nano-Scaled Three-Terminal Quantum Dot Thermoelectric Device. *Energy* **2015**, *85*, 200–207.
84. Yan, Y.; Malen, J. A. Periodic Heating Amplifies the Efficiency of Thermoelectric Energy Conversion. *Energy Environ. Sci.* **2013**, *6*(4), 1267–1273.
85. Hua, Y. f.; Wang, Z. L. Recent Progress in Piezoelectric Nanogenerators as a Sustainable Power Source in Self-Powered Systems and Active Sensors. *Nano Energy* **2015**, *14*, 3–14.
86. Yuan, M.; Cheng, L.; Xu, Q.; Wu, W.; Bai, S.; Gu, L.; Wang, Z.; Lu, J.; Li, H.; Qin, Y.; Jing, T.; Wang, Z. L. Biocompatible Nanogenerators through High Piezoelectric Coefficient $0.5Ba(Zr_{0.2}Ti_{0.8})_{0.3-0.5}(Ba_{0.7}\ Ca_{0.3})TiO_3$ Nanowires for In-Vivo Applications. *Adv. Mater.* **2014**, *26*(44), 7432–7437.
87. Yang, R.; Qin, Y.; Li, C.; Zhu, G.; Wang, Z. L. Converting Biomechanical Energy into Electricity by a Muscle-Movement-Driven Nanogenerator. *Nano Lett.* **2009**, *9*(3), 1201–1205.
88. Hu, Y.; Xu, C.; Zhang, Y.; Lin, L.; Snyder, R. L.; Wang, Z. L. A Nanogenerator for Energy Harvesting from a Rotating Tire and Its Application as a Self-Powered Pressure/Speed Sensor. *Adv. Mater.* **2011**, *23*(35), 4068–4071.

89. Dudem, B.; Kim, D. H.; Bharat, L. K.; Yu, J. S. Highly-Flexible Piezoelectric Nanogenerators With Silver Nanowires and Barium Titanate Embedded Composite Films for Mechanical Energy Harvesting. *Appl. Energ.* **2018**, *230*, 865–874.
90. Chansaengsri, K.; Onlaor, K.; Tunhoo, B.; Thiwawong, T. Paper-Based Flexible Piezoelectric Nanogenerator Using Fibrous Polymer/Piezoelectric Nanoparticle Composite Material. *Electron. Lett.* **2018**, *54*(12), 772–773.
91. Kim, H. S.; Lee, D. W.; Kim, D. H.; Kong, D. S.; Choi, J.; Lee, M.; Murillo, G.; Jung, J. H. Dominant Role of Young's Modulus for Electric Power Generation in PVDF–$BaTiO_3$ Composite-Based Piezoelectric Nanogenerator. *Nanomaterials* **2018**, *8*(10), 777.
92. Kim, K. N.; Chun, J.; Chae, S.; Ahn, C. W.; Kim, I. W.; Kim, S.-W.; Wang, Z. L.; Baik, J. M. Silk Fibroin-Based Biodegradable Piezoelectric Composite Nanogenerators Using Lead-Free Ferroelectric Nanoparticles. *Nano Energy* **2015**, *14*, 87–94.
93. Boughey, F. L.; Davies, T.; Datta, A.; Whiter, R. A.; Sahonta, S.-L.; Kar-Narayan, S. Vertically Aligned Zinc Oxide Nanowires Electrodeposited Within Porous Polycarbonate Templates for Vibrational Energy Harvesting. *Nanotechnology* **2016**, *27*, 28LT02.
94. Xu, S.; Qin, Y.; Xu, C.; Wei, Y.; Yang, R.; Wang, Z. L. Self-Powered Nanowire Devices. *Nat. Nanotechnol.* **2010**, *5*(5), 366–373.
95. Luo, C.; Hu, S.; Xia, M.; Li, P.; Hu, J.; Li, G.; Jiang, H.; Zhang, W. A Flexible Lead-Free $BaTiO_3$/PDMS/C Composite Nanogenerators as Piezoelectric Energy Harvester. *Energy Technol.* **2018**, *6*(5), 922–927.
96. Xia, M.; Luo, C.; Su, X.; Li, Y.; Li, P.; Hu, J.; Li, G.; Jiang, H.; Zhang, W. KNN/PDMS/C-Based Lead-Free Piezoelectric Composite Film for Flexible Nanogenerator. *J. Mater. Sci.: Mater. El* **2019**, *30*(8), 7558–7566.
97. Niu, X.; Jia, W.; Qian, S.; Zhu, J.; Zhang, J.; Hou, X.; Mu, J.; Geng, W.; Cho, J.; He, J.; Chou, X. High-Performance PZT-Based Stretchable Piezoelectric Nanogenerator. *ACS Sustain. Chem. Eng.* **2019**, *7*(1), 979–985.
98. Zhou, X.; Xu, Q.; Bai, S.; Qin, Y.; Liu, W. Theoretical Study of the $BaTiO_3$ Powder's Volume Ratio's Influence on the Output of Composite Piezoelectric Nanogenerator. *Nanomaterials* **2017**, *7*(6), 143.
99. Stassi, S.; Cauda, V.; Ottone, C.; Chiodoni, A.; Pirri, C. F.; Canavese, G. Flexible Piezoelectric Energy Nanogenerator Based on ZnO Nanotubes Hosted in a Polycarbonate Membrane. *Nano Energy* **2015**, *13*, 474–481.
100. Hu, Y.; Lin, L.; Zhang, Y.; Wang, Z. L. Replacing a Battery by a Nanogenerator with 20 V Output. *Adv. Mater.* **2012**, *24*(1), 110–114.
101. Singha, N. R.; Mahapatra, M.; Karmakar, M.; Dutta, A.; Mondal, H.; Chattopadhyay, P. K. Synthesis of Guar Gum-*g*-(Acrylic Acid-*Co*-Acrylamide-*co*-3-Acrylamido Propanoic Acid) IPN *via In Situ* Attachment of Acrylamido Propanoic Acid for Analyzing Superadsorption Mechanism of Pb(II)/Cd(II)/Cu(II)/MB/MV. *Polym. Chem.* **2017**, *8*(44), 6750–6777.
102. Singha, N. R.; Karmakar, M.; Mahapatra, M.; Mondal, H.; Dutta, A.; Roy, C.; Chattopadhyay, P. K. Systematic Synthesis of Pectin-*g*-(Sodium Acrylate-*Co-N*-isopropylacrylamide) Interpenetrating Polymer Network for Superadsorption of Dyes/M(II): Determination of Physicochemical Changes in Loaded Hydrogels. *Polym. Chem.* **2017**, *8*(20), 3211–3237.
103. Singha, N. R.; Karmakar, M.; Mahapatra, M.; Mondal, H.; Dutta, A.; Deb, M.; Mitra, M.; Roy, C.; Chattopadhyay, P. K. An *In Situ* Approach for the Synthesis of a Gum Ghatti-*g*-Interpenetrating Terpolymer Network Hydrogel for the High-Performance Adsorption Mechanism Evaluation of Cd(II), Pb(II), Bi(III) and Sb(III). *J. Mater. Chem. A* **2018**, *6*(17), 8078–8100.

104. Karmakar, M.; Mondal, H.; Mahapatra, M.; Chattopadhyay, P. K.; Chatterjee, S.; Singha, N. R. Pectin-Grafted Terpolymer Superadsorbent via N–H Activated Strategic Protrusion of Monomer for Removals of Cd (II), Hg (II), And Pb (II). *Carbohyd. Polym.* **2019**, *206*, 778–791.
105. Mitra, M.; Mahapatra, M.; Dutta, A.; Roy, J. S. D.; Karmakar, M.; Deb, M.; Mondal, H.; Chattopadhyay, P. K.; Bandyopadhyay, A.; Singha, N. R. Carbohydrate and Collagen-Based Doubly-Grafted Interpenetrating Terpolymer Hydrogel via N–H Activated In Situ Allocation of Monomer for Superadsorption of Pb(II), Hg(II), Dyes, Vitamin-C, and P-Nitrophenol. *J. Hazard. Mater.* **2019**, *369*, 746–762.
106. Jeong, C. K.; Lee, J.; Han, S.; Ryu, J.; Hwang, G. T.; Park, D. Y.; Park, J. H.; Lee, S. S.; Byun, M.; Ko, S. H.; Lee, K. J. A Hyper-Stretchable Elastic-Composite Energy Harvester. *Adv. Mater.* **2015**, *27*(18), 2866–2875.
107. Lee, E. J.; Kim, T. Y.; Kim, S.-W.; Jeong, S.; Choi, Y.; Lee, S. Y. High-Performance Piezoelectric Nanogenerators Based on Chemically-Reinforced Composites Energy. *Environ. Sci.* **2018**, *11*, 1425–1430.
108. Jing, Q.; Kar-Narayan, S. Nanostructured Polymer-Based Piezoelectric and Triboelectric Materials and Devices for Energy Harvesting Applications. *J. Phys. D: Appl. Phys.* **2018**, *51*(30), 303001.
109. Kim, D. H.; Dudem, B.; Yu, J. S. High-Performance Flexible Piezoelectric-Assisted Triboelectric Hybrid Nanogenerator via Polydimethylsiloxane-Encapsulated Nanoflower-Like ZnO Composite Films for Scavenging Energy from Daily Human Activities. *ACS Sustain. Chem. Eng.* **2018**, *6*(7), 8525–8535.
110. Mahapatra, M.; Dutta, A.; Roy, J. S. D.; Mitra, M.; Mahalanobish, S.; Sanfui, M. D. H.; Banerjee, S.; Chattopadhyay, P. K.; Sil, P. C.; Singha, N. R. Fluorescent Terpolymers via In Situ Allocation of Aliphatic Fluorophore Monomers: Fe(III) Sensor, High-Performance Removals, and Bioimaging. *Adv. Healthc. Mater.* **2019**, *8*, 1900980.
111. Mahapatra, M.; Dutta, A.; Roy, J. S. D.; Das, U.; Banerjee, S.; Dey, S.; Chattopadhyay, P. K.; Maiti, D. K.; Singha, N. R. Multi C–C/C–N Coupled Light-Emitting Aliphatic Terpolymers: N–*H* Functionalized Fluorophore-Monomers and High-Performance Applications. *Chem: A Eur. J.* DOI: 10.1002/chem.201903935.
112. Singha, N. R.; Mahapatra, M.; Karmakar, M.; Chattopadhyay, P. K. Processing, Characterization and Application of Natural Rubber Based Environmentally Friendly Polymer Composites. *Sustain. Polym. Compos. Nanocomposites* **2019**, 855–897.

9 Carbon Fiber Reinforced Thermoplastics and Thermosetting Composites

Deepak Verma

CONTENTS

9.1 INTRODUCTION

Various attempts are being made toward the efficient utilization of fuel consumption in the world. Numerous researchers are also working toward the development of lightweight structures so that fuel can be effectively utilized. Researchers and scientists are replacing the existing structure of the vehicle whether commercial or private with the lightweight structures made up of alloys or polymer composites (thermosets, thermoplastics, or biodegradable type). Polymer composite plays an important role in the development of both interior and exterior automobile structures. Composites can be developed by utilizing the polymer matrix by the particulate or fiber reinforcements. Fiber-reinforced plastics (FRPs) showed magnificent mechanical properties. The composites reinforced with carbon fiber have great strength, weigh less, and are pervasive, compared to metals and glass fiber reinforced composites. This comparison is made not only on the basis of their superb strength-to-weight ratio, but also because of the inertness of the carbon fibers related to the environment and their corrosion-resistant property. Carbon fiber-reinforced composites feature properties of great stiffness and strength in relation to their weight and are found to be an exceptional option for a range of available engineering materials, for example, polymers, metal alloys, and foams, etc. (Alam, Mamalis, Robert, Floreani, and Brádaigh, 2019).

Carbon fiber-reinforced plastics are extensively utilized in the development of vital structural parts in the area of aerospace engineering because of their high strength-to-weight ratio (Mudhukrishnan, Hariharan, and Palanikumar, 2019).

At present it has been observed that the production of the commercial carbon fibers depends upon rayon, PAN, or pitch. In 1959, rayon-based fibers were commercially developed and found to be very good in their first applications in the defense area. Fibers developed from PAN charged the development of the carbon fiber industries in 1970, and are now extensively used in different areas of applications, for example, aircraft brakes, space structures, military and commercial planes, batteries (Li), sports goods, and structural reinforcement, etc. On the other hand, pitch-based fibers are found to be universal and have the potential to achieve ultra-high Young's modulus and thermal conductivity. Based on this, they were very well utilized in critical military and space applications. But the production of these fibers has been kept on a lower side because they were very expensive. Because of this, researchers aimed to reduce the cost of these fibers and drastically reduced the cost of the carbon fibers. Nowadays, carbon fibers are frequently utilized in automotive applications; even the entire body panels may be fabricated by utilizing them (Society, 2003).

In carbon fiber-reinforced composites, the polymer generally acts as the matrix and the carbon fiber is considered as the reinforced part. The polymer matrix may be thermoset or thermoplastic resins. In comparison with thermoset resin, thermoplastics have gained noticeable attention due to their not being required during curing stage and their low amount of dangerous chemical constituents. It has been observed that the adhesion strength of the carbon fibers and the matrix is generally poor, and ultimately affects the mechanical properties of the composites. To improve the adhesion strength, we need to perform chemical treatment of the carbon fiber. There are so many chemical treatments available like chemical oxidation, physical treatment, ozone treatment, polymer coating, and plasma treatment, which are utilized to stimulate the functional groups present on the surfaces and improve the adhesion strength between the cabon fiber and the polymer phase (Yao, Jin, Yop, Hui, and Park, 2018).

In this chapter, we mainly focus on the carbon fiber-reinforced thermoplastic or thermoset composites, their mechanical properties, and their applications.

9.2 THERMOSET AND THERMOPLASTICS POLYMERS: A BRIEF DESCRIPTION

9.2.1 Thermoset Polymers:

Thermoset polymer matrix includes polyesters, vinyl esters, epoxies, bismaleimides, cyanate esters, polyimides, and phenolics. It has been observed that the epoxies, in the present scenario, are the most utilized resins for low and moderate temperatures.

On the other hand, it has been observed that the polyesters and vinyl esters can be utilized at almost the same temperatures (like epoxies), and also are widely utilized for the commercial applications but are generally avoided for the high-performance composite applications because of their decreased properties (Composites, n.d.).

9.2.2 Thermoplastic Polymers:

Thermoplastic polymers are high-molecular-weight resins. When processed, they melt and flow and do not make cross-linking reactions. It is seen that the chains in the thermoplastic polymers are bound together by weak secondary bonds. Although, it is also observed that these resins have high molecular weight. Also, during processing, the magnitude of the viscosity of these resins is higher than that of thermosets (e.g., 10^4–10^7 poise for thermoplastics versus 10 poise for thermosets). The recyclability of these polymers is possible as thermoplastics do not cross-link during processing (Composites, n.d.).

9.3 CARBON FIBER: AN INTRODUCTION

Carbon fibers are fibers approx 5–10 micrometers in diameter and made up mostly of carbon atoms. These fibers have some good properties, such as great stiffness, great tensile strength, being lightweight, great chemical resistance, etc. Because of these properties, carbon fiber became very popular in different areas such as aerospace, civil engineering, military, motorsports, and sports.

9.3.1 Types of Carbon Fibers: General Classification

A general classification of the carbon fibers can be seen into the following categories:

a. **Based on mechanical properties**
    - Extreme-high-modulus (>450 Gpa)
    - High-modulus (between 350–450 Gpa)
    - Moderate-modulus (between 200–350 Gpa)
    - Lower modulus and high-tensile (mod <100 Gpa, tensile st. >3.0Gpa)
    - Extreme high-tensile (>4.5 Gpa)
b. **Based on precursor fiber materials**
    - PAN-based
    - Pitch-based: mesophase and isotropic
    - Rayon-based
    - Gas-phase-grown
c. **Depends on final heat treatment temperature**
    - Type-I: these are high-heat-treatment carbon fibers (HTT), with final heat treatment temperature more than 2000°C.
    - Type-II: these are intermediate-heat-treatment carbon fibers (IHT), with final heat treatment temperature above 1500°C.
    - Type-III: these are low-heat-treatment carbon fibers, with final heat treatment temperatures not above 1000°C.

9.4 PROPERTIES OF CARBON FIBER

- High strength-to-weight ratio and is very rigid
- Corrosion resistant and chemically stable
- Electrically conductive
- Good fatigue resistance
- Good tensile strength (Bhatt and Goe, 2017)

9.5 MECHANICAL PROPERTIES OF THERMOPLASTIC POLYMER-BASED COMPOSITES

Kakinuma et al. (2015) investigated the effect of feed rate on the standard of a machined hole and observed the cutting conditions at the outlet of the hole. The dissimilarity of the cutting action can be understood by looking at Figure 9.1 which shows two types of feed drilling i.e., one is extreme feed drilling at 3000 mm/min and a lower feed drilling at 50 mm/min. From Figure 9.1, it can be observed that the ultrafast feed drilling was found to be productive and all the chips were removed and almost no burrs were available at the outermost part of the hole. However, at a low feed rate of 50 mm/min, the carbon fiber-reinforced thermoplastic was slowly deformed and takes the tool shapes just before the drill penetrates, which results in development of hat-shaped chips that were not perfectly removed, resulting in the generation of the crown-shaped burrs around the edge of the hole (Kakinuma, Ishida, Koike, and Klemme, 2015).

Placet et al. (2016), used DMA for the investigation of the transverse rigidity of the laminates. Figure 9.2 exhibits the development of the storage modulus with temperature for the three different tested prevalences. It has been observed that the storage modulus, of around 2.3 GPa is fairly stable at a temperature range of 30–240°C and starts to decline at the glass transition region, which is more than 250°C. In the glass transition region, the rigidity decreases noticeably and reaches approximately 1 GPa at 270°C. This result highlights and verifies the high-temperature firmness of this polyimide polymer (Gabrion, Placet, Trivaudey, and Boubakar, 2016).

Ning et al. (2015) confirmed, as can be seen in Figure 9.3 (a) that by increasing the carbon fiber (CF) from 0 to 5 wt%, first the tensile strength value increases and then decreases by again increasing the carbon fiber value from 5 to 10 wt%. Again, it is seen that the tensile strength increased again by increasing the carbon fiber content from 10 wt% to 15 wt%. So, from Figure 9.3, it can be concluded that the highest mean value (42 MPa) was achieved at 5 wt% CF, while the minimum mean value, which is around 34 MPa, was achieved at 10wt %. Figure 9.3 (b) showed the effect of CF on Young's modulus. It has been observed from Figure 9.3 that at 7.5 wt%, there is an increase in the value of the Young's modulus, whereas when the CF content increases from 7.5 wt%, then there is an abrupt decrease in the Young's modulus value. Again, when the CF content increases from 10 to 15 wt%, an increase in Young's modulus value was observed. Also, the flexural tests were performed for the comparison of the developed specimens with and without CF. For flexural testing, ASTM D790–10 standard was adopted. It was observed that, specimens did not fail under the 5% strain limit in the flexural test. Figure 9.4 shows the stress-strain

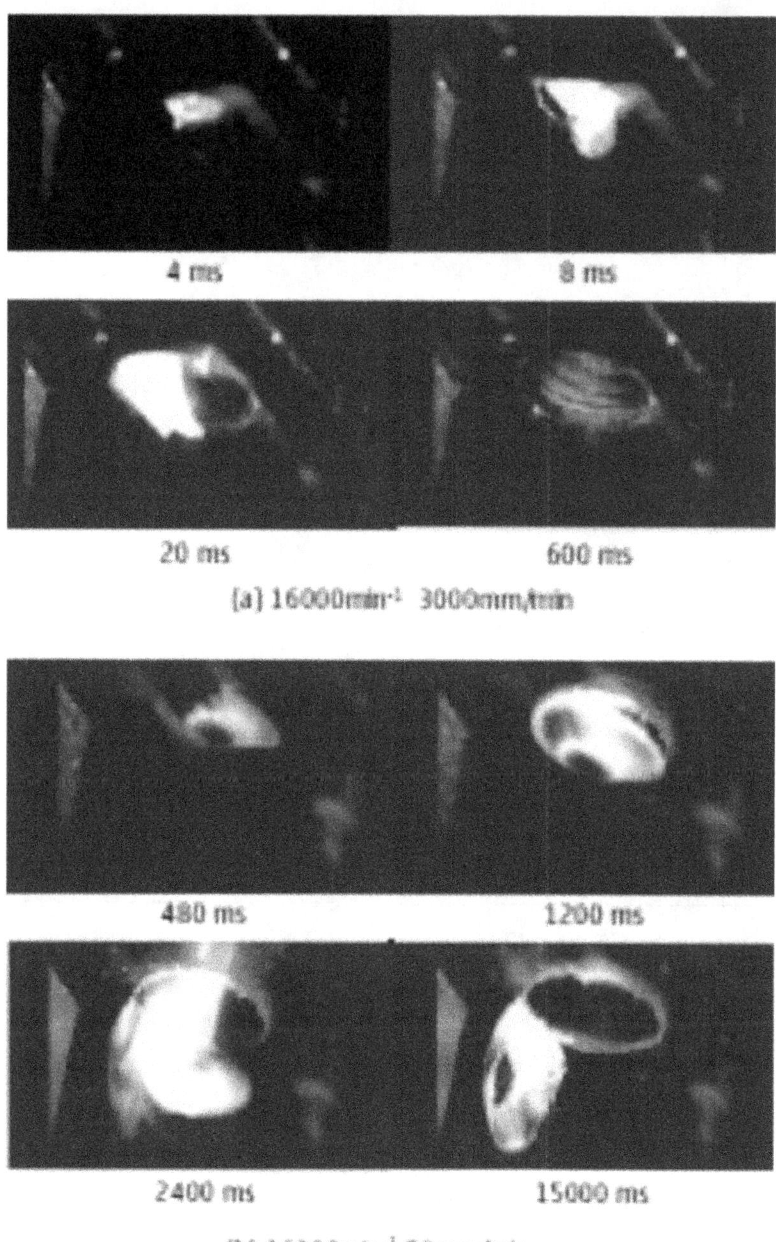

FIGURE 9.1 High-speed microscope images at the outlet of the machined holes as a result of (a) ultrafast feed drilling and (b) low feed drilling. (Reproduced with permission from Kakinuma et al., 2015.)

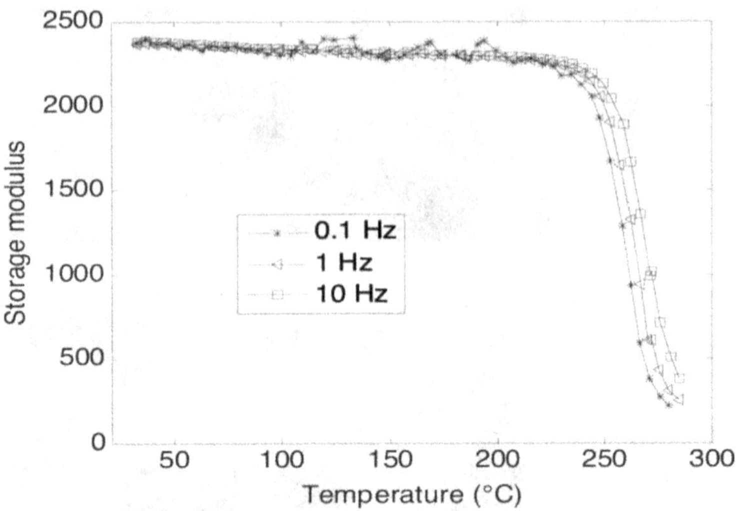

FIGURE 9.2 Storage modulus as a function of temperature, for one of the tested specimens, measured in transversal direction at three frequencies for $[0]_{10}$ laminate. (Reproduced with permission from Placet et al., 2016.)

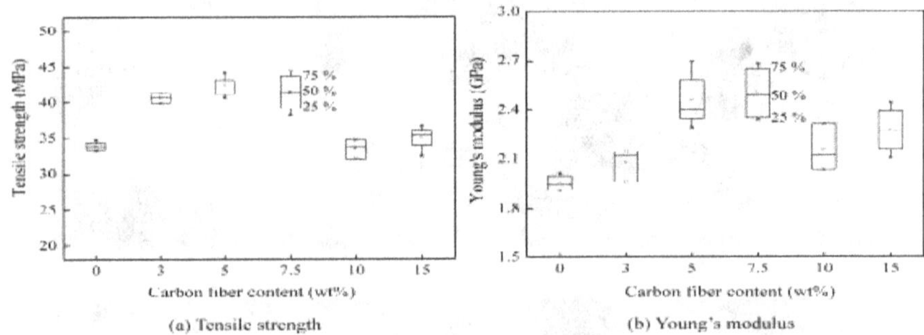

FIGURE 9.3 The effects of carbon fiber content on tensile properties (carbon fiber length is 150 μm). (Reproduced with permission from Ning et al., 2015.)

curves of the flexural strength. From Figure 9.4, it was observed that the carbon fiber composite showed great flexural stress, compared to the specimen without carbon fiber (Ning, Cong, Qiu, Wei, and Wang, 2015).

Dydek et al. (2019) showed DMA results in Figure 9.5. It is observed from Figure 9.5 (b), which shows the loss modulus results, that only one peak appeared on the reference panel curves. In the case of laminate 1T (TenCate prepreg) it appears at 126.9°C and in the case of laminate 4H (Hexcel prepreg) it appears at 177.7°C. Figure 9.5 (a) showed a change in the storage modulus values of the carbon fiber-reinforced plastics panels, which are interlayered with thermoplastic nonwovens. Because of the thermoplastic matrix, all the other laminates showed a decrease in the storage modulus values compared to the reference panel, as thermoplastic matrix has lower stiffness than epoxy resin (Dydek, Latko-Durałek, Boczkowska, Sałaciński, and Kozera, 2019).

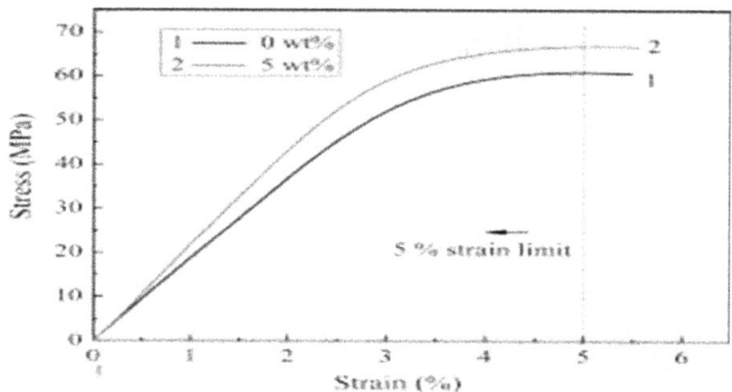

FIGURE 9.4 Typical flexural strain–stress curves for specimens with and without carbon fibers (carbon fiber length is 150 μm). (Reproduced with permission from Ning et al., 2015.)

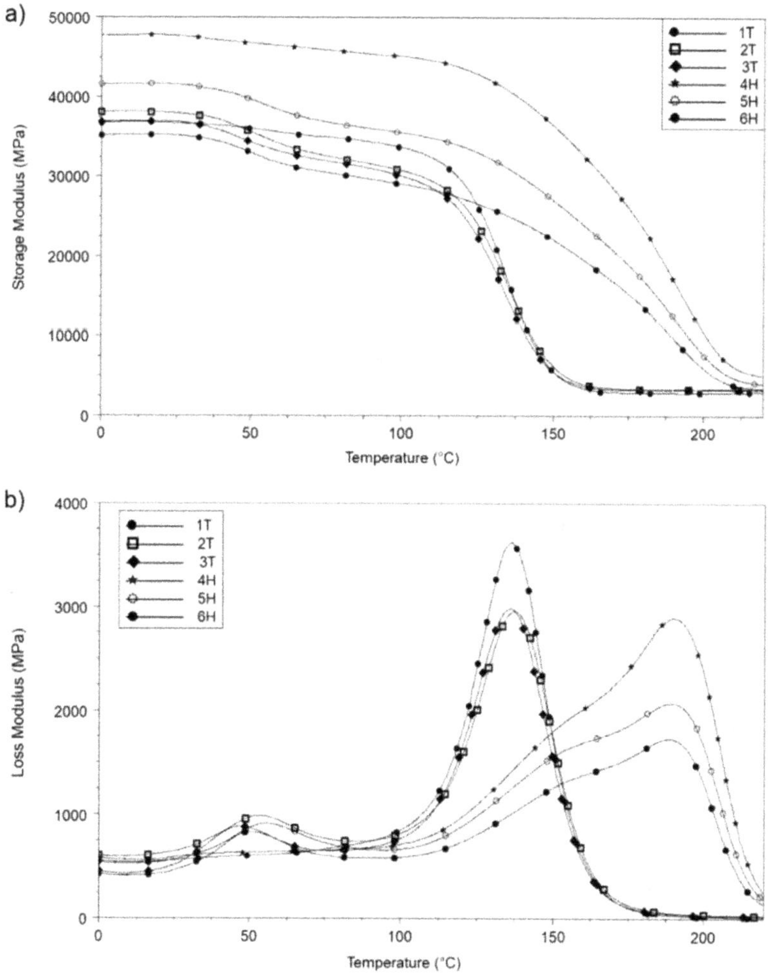

a)
Storage Modulus (MPa)
Temperature (°C)
1T
2T
3T
4H
5H
6H

b)
Loss Modulus (MPa)
Temperature (°C)
1T
2T
3T
4H
5H
6H

FIGURE 9.5 Dynamic mechanical analysis of laminates: (a) storage modulus and (b) loss modulus. (Reproduced with permission from Dydek et al., 2019.)

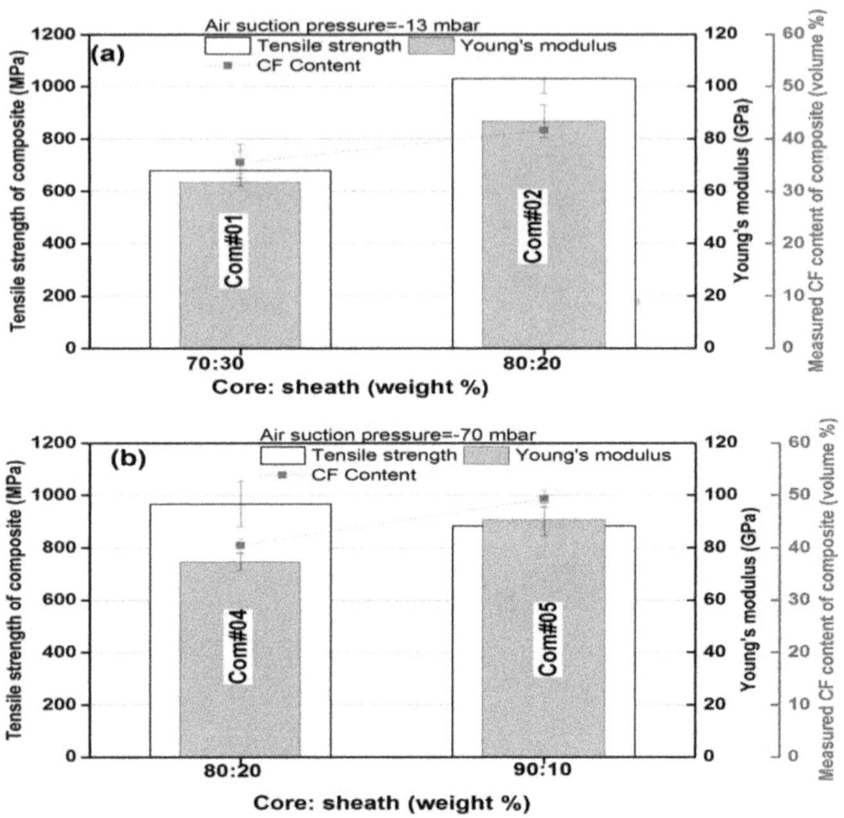

FIGURE 9.6 Tensile strength, Young's modulus, and CF content of composites manufactured with varying core-to-sheath-weight ratios and air suction pressures of (a) –13 mbar and (b) –70 mbar. (Reproduced with permission from Hasan et al. 2018.)

Hasan et al. (2018), showed the effect of the core-to-sheath ratio of hybrid yarn on the carbon fiber volume, tensile strength, and Young's modulus of unidirectional (UD) composites in Figure 9.6. The hybrid yarns shown in Figure 9.6 (a) and (b) were developed with numerous air suction pressure ranges from –13 and –70 mbar, without changing the other criterion. The maximum CF content (49.3 vol.%) was attained with the hybrid yarn with 90:10 ratio. The consistent carbon fiber contents of UD composites developed with 70:30 and 80:20 ratios were 35.5 and 41.6 vol.% at –13 mbar air suction pressure (Hasan, Nitsche, Abdkader, and Cherif, 2018).

Hun Su Lee et al. (2014) showed the tensile strengths of the parent and plasma-treated CFF reinforced composites in Figure 9.7(b). From Figure 9.7, it has been observed that there is an improvement in the tensile strength values of the carbon fiber fabric reinforced composite (CFFRC) of around 362.5% and 436.3%, in comparison to the pCBT matrix. Finite element- scanning electron microscopy (SEM) was utilized for the recognition of the fracture surface of the CFFRC tensile specimen, which is depicted in Figure 9.7c. From Figure 9.7, it is noticeable that there

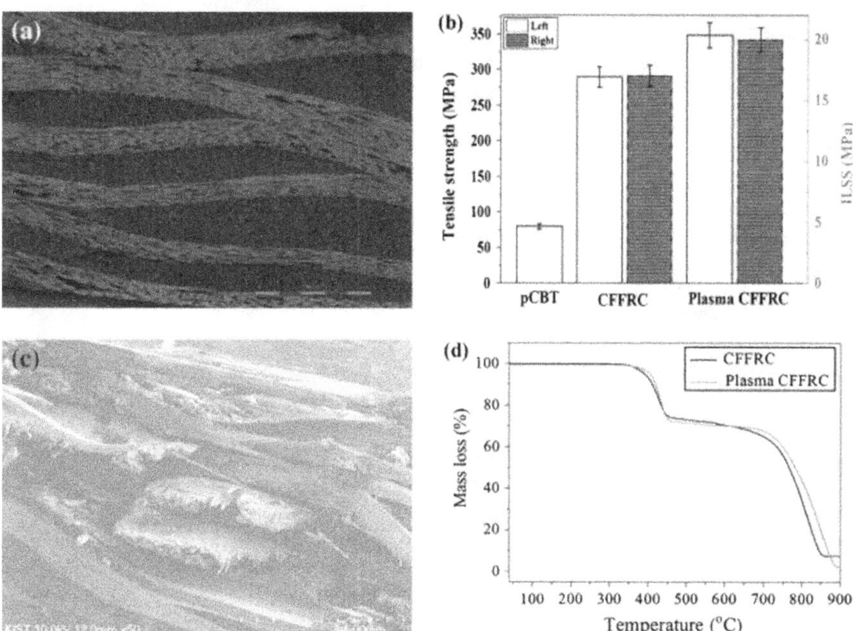

FIGURE 9.7 (a) OM image of a cross section of the compressed CFFRC tensile specimen that was molded at 250°C for 2 min. (b) Tensile strength and interlaminar shear strength (ILSS) of the CFFRC tensile specimen prepared at 250°C for 2 min. (c) FE-SEM image of the CFFRC fracture surface. (d) TGA thermogram of the CFFRC. (Reproduced with permission from Lee et al., 2014.)

are multiple fractures available in terms of the fiber breakage, delamination, and fiber debonding, etc. As per the density of each material, the carbon fiber content was measured to be close to 70 vol% evaluated by the TGA method and is shown in Figure 9.7(d) (Lee, Kim, Noh, and Kim, 2014).

Tian et al. (2016) showed the flexural strength and modulus of the specimens developed with various temperatures and displayed in Figure 9.8. From Figure 9.8, it is clear that the flexural strength and modulus are 155 MPa and 8.6 GPa, respectively at 240°C. Although, the specimen developed at the temperature of 240°C lacks the surface accuracy because of the overflow of molten polylactic acid (PLA). Based on this, the recommended temperature should be 230°C, and the values of the flexural strength and modulus at this temperature is found to be 145 MPa and 8.6 GPa, respectively. So, based on this, authors recommend the suitable process temperature to be in the range of 200–230°C (Tian, Liu, Yang, Wang, and Li, 2016).

Tian et al., 2017, recycled 3D-printed carbon fiber-reinforced thermoplastic composites and obtained an impregnated carbon fiber from it. Fiber filament properties of the basic printed carbon fiber and the recycled carbon fiber were verified by the tensile properties. From Figure 9.9, it was observed that there is a mismatch of the carbon fibers in a tow, which results in the nonsynchronous breakage, which hence decreases the tensile strength of a received carbon fiber tow.

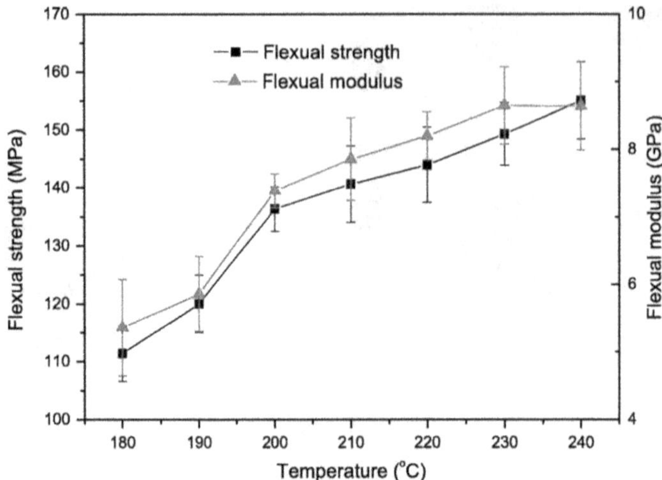

FIGURE 9.8 Influence of temperature in liquefier on the flexural strength and modulus of the 3D-printed CFR PLA composites under experimental condition of L 0.65 mm, V 100 mm/min, E 150 mm/min, H 1.2 mm. (Reproduced with permission from Tian et al., 2016.)

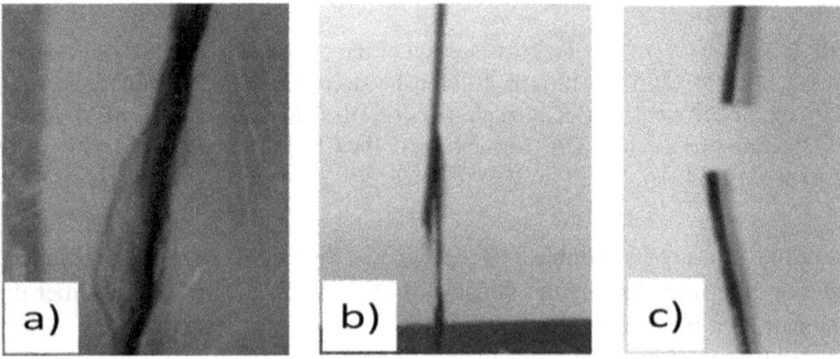

FIGURE 9.9 Fracture patterns for (a) virgin carbon fiber tows, (b) originally printed, and (c) recycled carbon fiber filament, with linear densities of 0.067 g/m, 0.98 g/m, 0.64 g/m, respectively. (Reproduced with permission from Tian et al., 2017.)

Also, authors evaluated the flexural strength of the composites shown in Figure 9.10. From Figure 9.10, it was observed that a mean flexural strength of around 263 MPa was attained for the remanufactured composite specimens. As compared to tensile strength, the remanufacturing process got an improvement in flexural strength of around 25% on the originally printed composite material (Tian et al., 2017).

Huang et al. (2018), showed the results of a tensile shear test of the FSpW joints in Figure 9.11. A gradual increase in the tensile properties was observed by increasing the rotational velocity (see Figure 9.11a). If the rotational speed is low i.e., at 800 rpm, less frictional heat is produced, leading to less efficient joining and ultimately affecting the tensile shear properties of the composites. Effective and efficient joining length and strength was achieved by increasing the rotational speed to 1000 rpm, compared to the 800 rpm. On the other hand, a sound joint with great strength

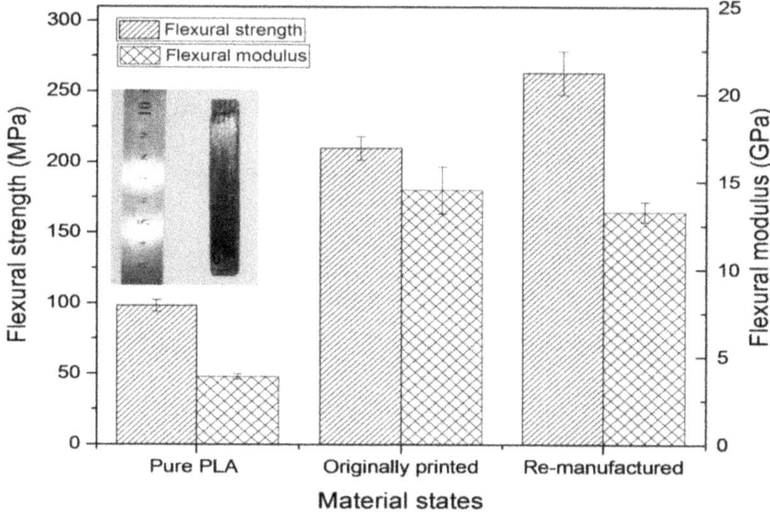

FIGURE 9.10 Flexural strength and modulus of pure PLA, originally printed and re-manufactured composites specimens. (Reproduced with permission from Tian et al., 2017.)

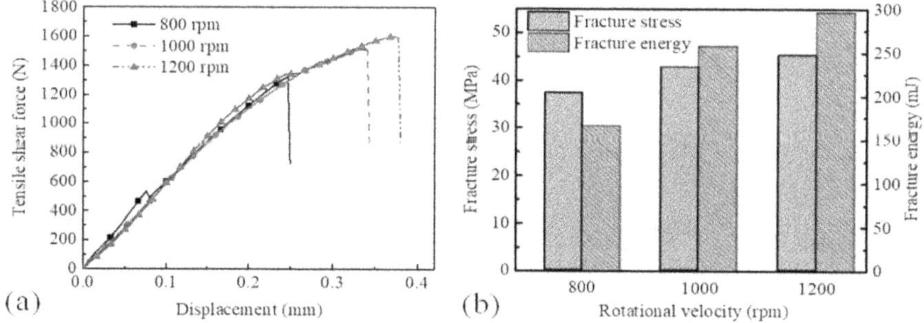

FIGURE 9.11 Tensile shear results of the FSpW joints: (a) force-displacement curves and (b) tensile shear properties. (Reproduced with permission from Huang et al., 2018.)

was obtained by further increasing the rotational speed to 1200 rpm. So, based on this, authors deduced the higher speed results in the longer joining area and the good intermixing of the shattered carbon fiber at the joining boundary. And also, a remarkable improvement in the fracture energy achieved by increasing the rotational speed which was maximum i.e., 296 mJ at 1200 rpm, provides greater toughness of the joint, as shown in Figure 9.11b (Huang et al., 2018).

Caminero et al. (2018) showed the results in Table 9.1 and Figure 9.12 and concluded that build orientation noticeably afflicted the impact resistance with an exceptional anisotropy, specifically at higher layer thickness. They also pointed out that the layer thickness is directly connected to the number of layers required to print a part and so also the printing time. It has also been observed that increased layer thickness ultimately reduces the manufacturing cost. The effect of layer thickness can also be seen on the impact strength of the composites because of the build orientation (Caminero, Chacón, García-Moreno, and Reverte, 2018).

TABLE 9.1

Average ILSS Test Results of Unreinforced Nylon Samples and Layer Thickness Ranges. Standard Deviation Is Depicted in Brackets

L_t=0.1 mm	L_t=0.125 mm	L_t=0.2 mm
10.19(0.25)	9.79(0.39)	9.33(0.38)

(Reproduced with permission from Camireno et al., 2018)

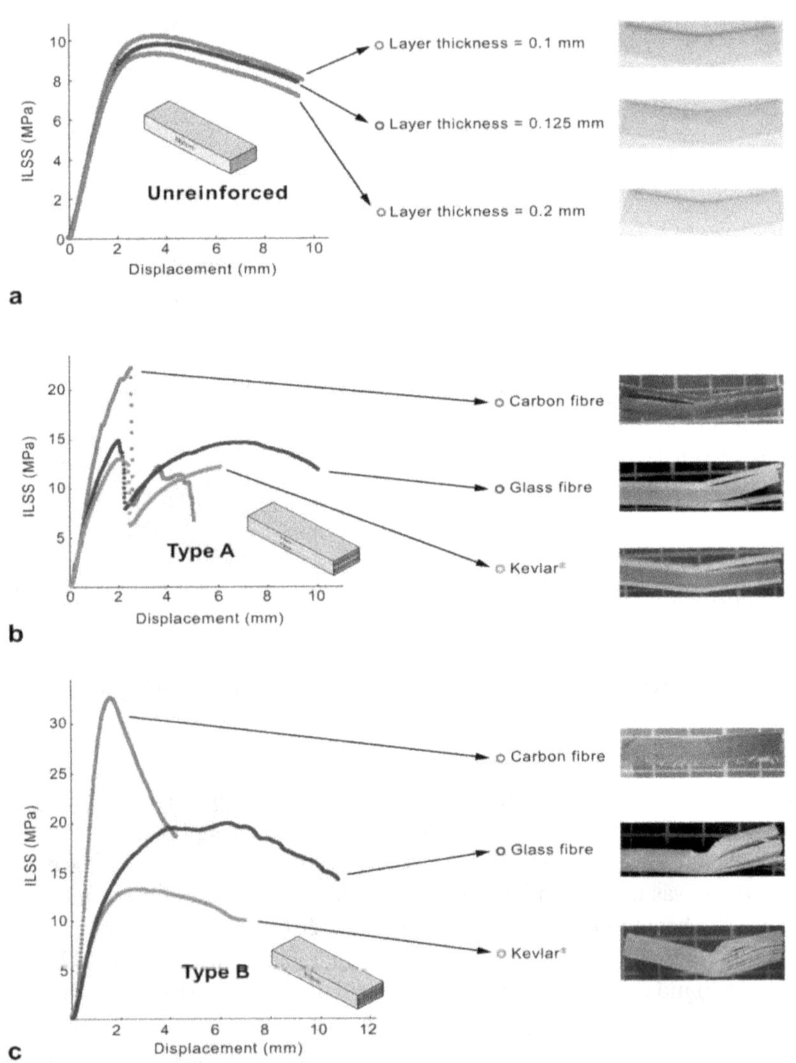

FIGURE 9.12 Average ILSS-displacement curves under different printing conditions. (Right) Cross-sectional optical micrographs showing the details of the failure modes for the different configurations. (Reproduced with permission from Camireno et al., 2018.)

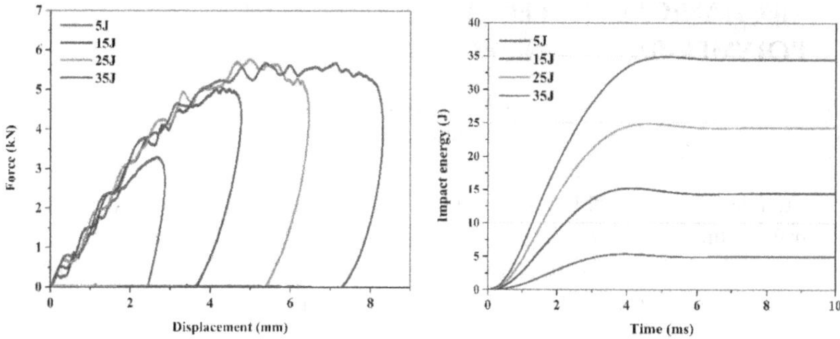

FIGURE 9.13 Force-displacement and energy-time curves for various energy levels. (Reproduced with permission from Wang et al., 2018.)

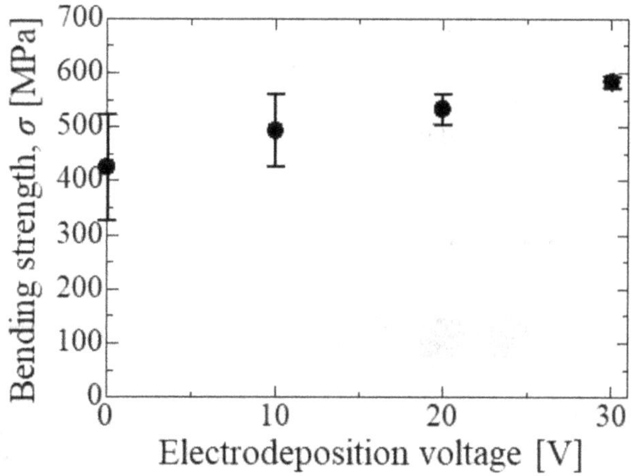

FIGURE 9.14 Effect of electrodeposition voltage on the bending strength of the Si-CFRTP. (Reproduced with permission from Yamamato et al., 2019.)

Wang et al. (2018) showed the correlation of force–displacement and energy-time. From this relationship, one can observe the samples' stiffness, maximum force, maximum displacement, impact energy, energy absorption, etc. Figure 9.13 shows a force–displacement graph and an energy-time graph for CF/polyphenylene sulfide (PPS) composites under various impact energies. From Figure 9.13, it has been observed that at 5 J, the force–displacement graph at 5 J exhibits a constant loading and unloading stage, and the stiffness indicated by the slope is approximately constant during the loading process, mentioning no evident failure to the sample (Y. Wang et al., 2018).

Yamamoto et al. (2019) evaluated the mechanical properties of the carbon fiber-reinforced thermoplastic by utilizing Si-CFs (Si-CFRTP). Authors used a tension and compression testing machine and performed three-point bending testing on it. The effect of electro deposition voltage on the flexural strength of the Si-CFRTP is shown in Figure 9.14. From Figure 9.14, it has been observed that by increasing voltage, the flexural strength ultimately increases (Yamamoto, Yabushita, Irisawa, and Tanabe, 2019).

9.6 MECHANICAL PROPERTIES OF THERMOSET POLYMER-BASED COMPOSITES

From various research, it has been now observed that the mechanical properties of thermoset resins used for the development of the carbon fiber composites mainly depend on their development conditions. In resin, the utilization of the fibers generally influences the thermal balance, makes chemical inhomogeneities, and arranged structural evolutions in the polymer in the vicinity of surface of the fiber.

Kieffer et al. (2014) used the Brillouin spectra for the analysis and observed that both the epoxies, in the middle of the layers and the bulk epoxy, have almost the similar average longitudinal modulus i.e., 8.81 GPa and 8.77 GPa, respectively. On the other hand, from Figure 9.15, a less longitudinal modulus was observed when measured at 90^0 and 45^0 to the fiber axis. However, a tangential modulus, measures to the fiber circumference and at 90^0 to the fiber axis was found to be around 8.63 GPa. This value showed a 2.04% loss in modulus in comparison to the epoxy in the middle of the layers (Aldridge, Waas, and Kieffer, 2014).

Wang et al. (2015) carbonized a carbon fiber-composite specimen in nitrogen surroundings at higher temperature. They protected the carbon fibers from the oxidation while the resin was carbonized in the nitrogen environment. The resin carbonization

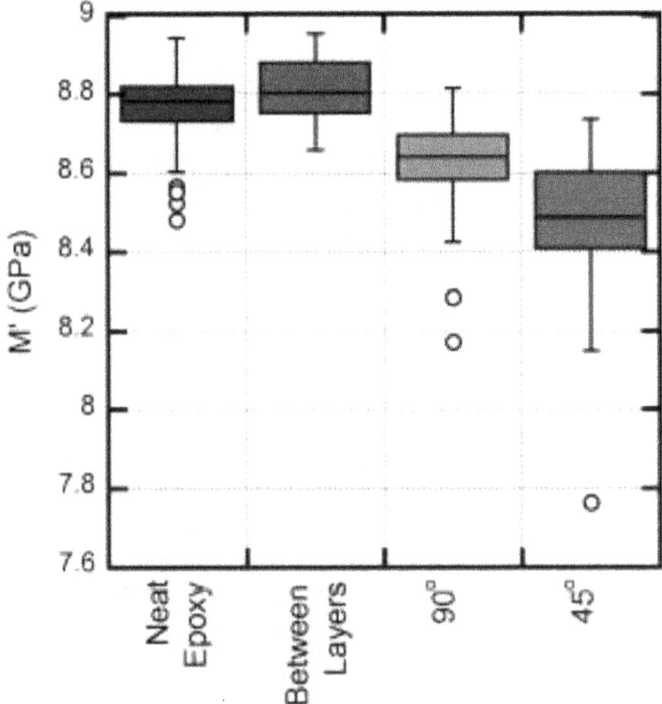

FIGURE 9.15 Box plot of the elastic moduli for three regions within the composite as compared to bulk epoxy. (Reproduced with permission from Kieffer et al., 2014.)

TABLE 9.2
The Fiber Content Results for One Batch of Carbon Epoxy Samples

Sample	Total mass (g)	Carbon fiber mass (g)	Residue mass (g)	Nominal fiber content (%)	Fiber content from carbonization in N_2 (%)	Deviation (%)
C/Epoxy-1	4.505	1.783	1.961	39.6	39.9	0.3
C/Epoxy-2	4.095	1.730	1.887	42.2	42.6	0.4
C/Epoxy-3	3.977	1.897	2.000	47.7	47.1	−0.6
C/Epoxy-4	3.998	1.806	1.956	45.2	45.6	0.5
C/Epoxy-5	4.348	1.811	1.954	41.6	41.4	−0.3
Neat epoxy	5.667	—	0.343	—	—	—

(Reproduced with permission from Wang et al., 2015)

TABLE 9.3
Comparison of Mechanical Properties and Flexural Strength before and after Lightning Damage between the CF/PANI and CF/epoxy Composites

Strength	Applied current	CF/PANI	CF/epoxy
Flexural strength (Mpa)		267	610
Flexural stiffness (Gpa)		54.8	52.5
ILSS (Mpa)		20.0	66.9
Flexural strength after lightning (Mpa) (residual ratio)	−40 kA	240 (0.90)	147 (0.24)
	−100 kA	239 (0.90)	N/A

(Reproduced with permission from Hirano et al., 2016)

rate was calibrated by using the neat resin sample for the calculation of the amount of the resin matrix in the composite sample.

Table 9.2 shows the measured fiber constituents of a combination of carbon fiber composites samples. From Table 9.2, a deviation was observed between the as-received nominal fiber volume and the fiber volume from the carbonization-in-nitrogen process and found to be lesser, as reported in Table 9.2 (Q. Wang, Ning, Vaidya, Pillay, and Nolen, 2015).

Hirano et al. (2014) used a four-point flexural testing method for the examination of residual strength after lightning damage. Both entire and deteriorated samples were made. Two deteriorated samples were excerpted from a single lightning test sample so that every beam sample consists of deteriorated area at its center. Table 9.3 shows the results of the four-point bending tests. It is observed that the residual strength of the CF/epoxy sample (−40 kA) is reduced to 24% of that of the entire sample. On the other hand, the CF/polyaniline (PANI) sample kept 90% of the earlier strength on the application of the lightning current at either −40 or −100 kA. It is seen that the deteriorated region of the CF/PANI sample tested at simulated

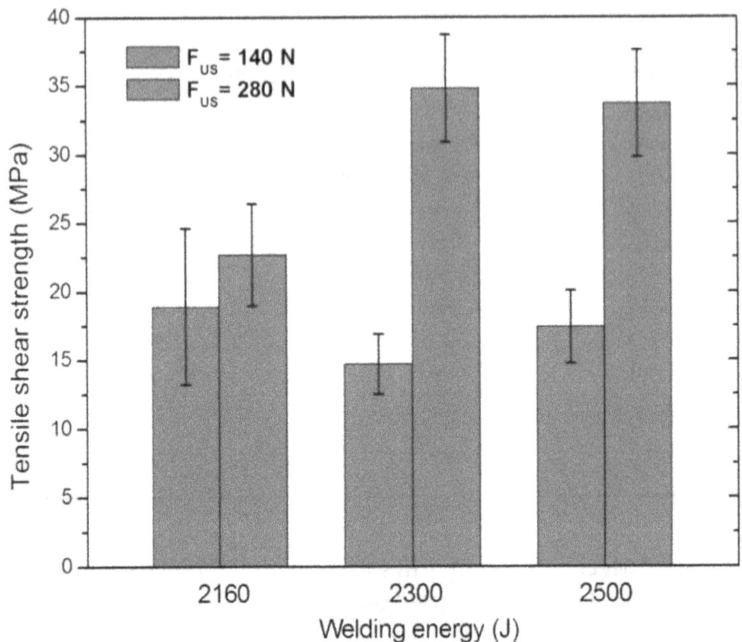

FIGURE 9.16 Tensile shear strength on CF/epoxy-PA6-AA5754 joints after ultrasonic metal welding at different ultrasonic welding forces and energies (ultrasonic amplitude 40 μm). (Reproduced with permission from Lionetto, 2017.)

lightning current of −100 kA, was 6.6 times higher, compared to −40 kA. Hence, the novel CF/PANI composite showed exceptional lightning deterioration resistance, compared to the current CF/epoxy composite (Hirano et al., 2016).

Lionetto et al. (2017) used ultrasonic metal welding for joining aluminum AA5754 sheets to a carbon fiber-reinforced epoxy resin (CF/epoxy) composite. Before curing, a thermoplastic film of polyamide 6 was utilized as a surface layer of the CF/epoxy stack. Figure 9.16 shows the influence of the welding variables on the tensile strength, where the described values are the average of the 12 samples. Also, the CF/epoxy-PA6-AA5754 joints welded at 140 N showed tensile strengths in the range of 14 to 19 MPa, which is solely dependent on the welding energy used. The optimized welding parameters such as welding force (140 N) and welding energy (2160 J), result in tensile strength of 19 MPa on CF/epoxy-PA6-AA5754 joints, compared to a value of 31.5 MPa without them (Lionetto, Balle, and Maffezzoli, 2017).

Yourdkhani et al. (2018) used a resin film infusion (RFI) method for the development of composite materials and investigated the deterioration of CNT dispersion amid high-temperature development of a thermoset fiber composite. Impact strength and the compression after impact tests were performed for the evaluation of the mechanical properties of the composites. Figure 9.17a shows the load-time graphs for impact tests. Comparable impact strengths were attained for all specimens, with inconsequential difference in the maximum force and absorbed energy between specimens. Damage caused by the impacts was evaluated by C-scan

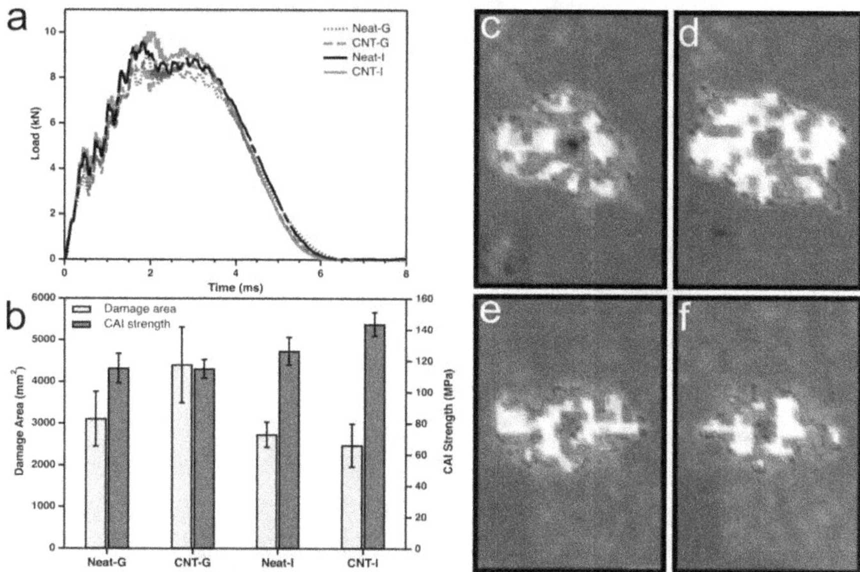

FIGURE 9.17 Results of impact and compression-after-impact (CAI) experiments. (a) Representative impact load-time curves. (b) Impact damage area and residual compressive strength of the four samples. (c–f) Representative C-scan images from impacted specimens: (c) Neat-G, (d) CNT-G, (e) Neat-I, (f) CNT-I. Specimens are 10 × 15 cm^2. (Reproduced with permission from Yourdkhani et al., 2018.)

measurements. Representative C-scan images of each specimen are depicted in Figure 9.17 c–f. Higher destructive regions were formed in composites developed by the grouped layup process than those developed by the interleaved layup process (see Figure 9.17b). The inclusion of CNTs to the resin resulted in a higher destructive region for the grouped layup process (Yourdkhani, Liu, Baril-Gosselin, Robitaille, and Hubert, 2018).

Mamalis et al. (2018) showed transverse flexural results of all developed composites. From Figure 9.18a, it is observed that the evaluated flexural strengths for the 50C (~114 MPa) and 60E (~105 MPa) composites were noticeably higher, compared to the F0E (~76 MPa) composites. The fiber matrix interfacial strength of a UD composite can be obtained by the transverse flexural test. So, from results, one can notice that the 50C and 60E composites showed durable interface bonding and had greater flexural properties. Figure 9.18b represents the flexural modulus of the three samples (Mamalis, Flanagan, and Ó Brádaigh, 2018).

Liu et al. (2019) reinforced shorts and continuous fibers into neat shape memory epoxy for the development of the shape memory epoxy composites (SMEPCs) with exceptional mechanical properties. Figure 9.19 showed stress-strain (σ-ε) curves of the discontinuous filler-based SMEPCs at both RT (25°C) and Tsw (170°C). The comparative data of modulus E, breaking stress (σb), and breaking strain (εb) are shown in Figure 9.19b–d and f. However, the specimen with 25 wt% SCF_3 and 35 wt% SCF_3/Al_2O_3 showed good mechanical performances. The availability of higher

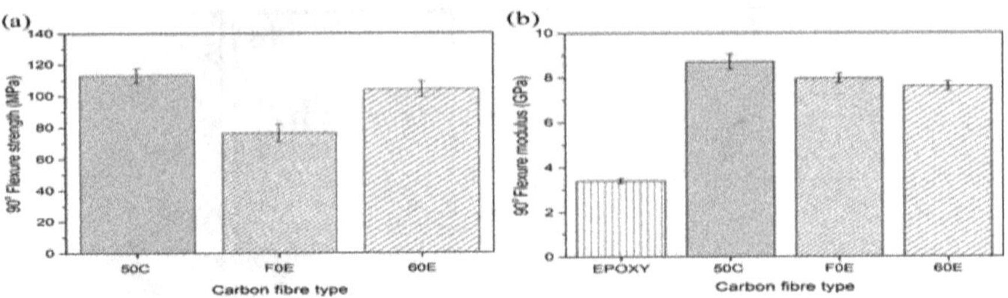

FIGURE 9.18 Transverse flexural results (a) Strength and (b) Modulus, obtained for T700S family composites, under tensioning conditions, as well as for the neat (powder) epoxy blocks. The flexural values are averages from 15 tests for each case. (Reproduced with permission from Mamalis et al., 2018.)

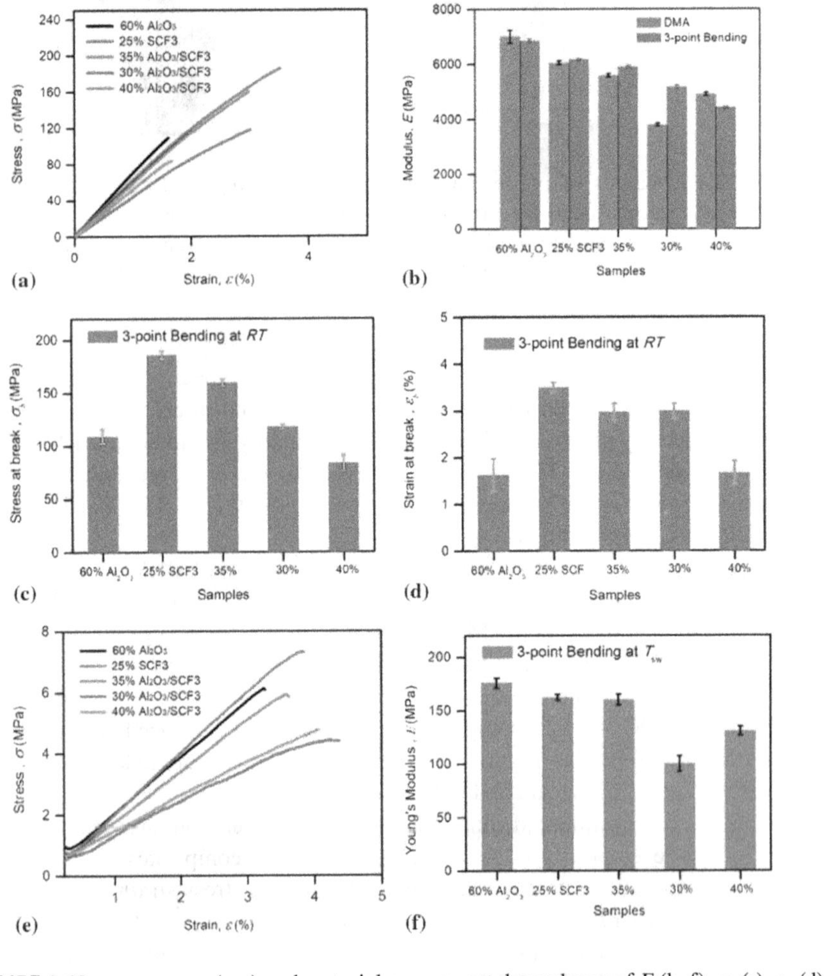

FIGURE 9.19 σ-ε curves (a, e) and material component dependence of E (b, f), σ_b (c), ε_b (d) for SMEPCs with non-continuous fillers. Among them, testing temperatures of a–d and e–f are 25°C and 170°C, respectively. (Reproduced with permission from Liu et al., 2019.)

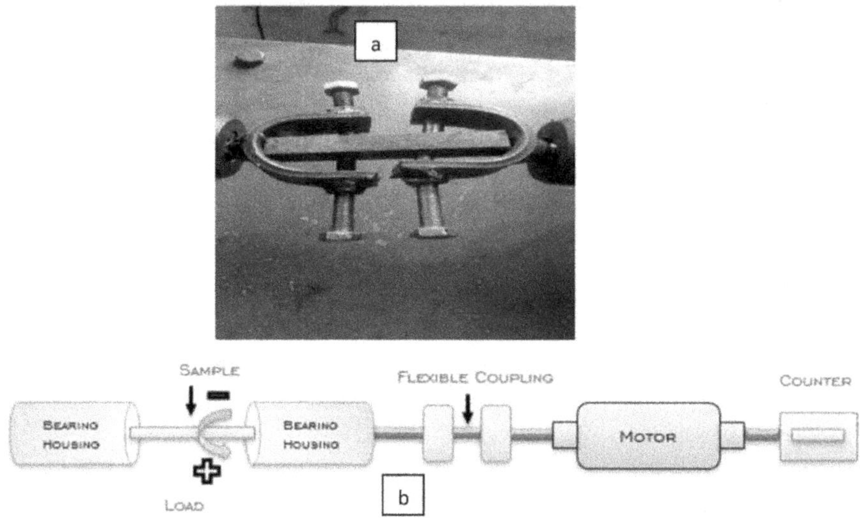

FIGURE 9.20 (a) CFRPC sample fixed in a cyclic fatigue strength (CFS) testing machine (b) Schematics of experimental set-up. (Reproduced with permission from Vedrtnam, 2019.)

Al_2O_3 particles ultimately enhances the stress along with a decreasing strain (Liu et al., 2019).

Vedrtnam et al. (2019) reported a new method which ultimately improves the fatigue life of CFRPC composites by 96.56% using interfacial strengthening. A standard bending cyclic fatigue test was performed to evaluate the fatigue life of the composites. Figure 9.20a shows the cast iron fixtures, which were used for carrying the CFRPC specimens rigidly. To ensure the proper gripping, the rubber pads were utilized for the protection of CFRPC samples from damage because of the contact with the fixture (Vedrtnam, 2019).

Knappich et al. (2019) studied a potential solvent-based recycling process for three carbon fiber-reinforced plastics (CFRP) samples consisting of polyamide 6, polyurethane resin, and epoxy resin. To soften polymeric matrix and detach the inert carbon fibers, various proprietary CreaSolv Formulations were used at a laboratory under thermodynamically subcritical conditions. Figure 9.21 shows a graph of the tensile strength of the recycled fibers. It is observed that the values of the tensile strength of CF-polyurethane rigid foam (PUR) fibers improve persistently through the dissolution processes. The numbers attained were from 104% of the as-received fiber strength (first dissolution @200°C) to 118% (after fourth dissolution) and for further dissolution at 190°C from 98% (first dissolution) to 123% (after fourth solvent extraction) (Knappich, Klotz, Schlummer, Wölling, and Mäurer, 2019).

9.7 CONCLUSION

Carbon fiber-reinforced plastics (CFRP) are some of the strongest materials which have replaced metal matrix composites (MMCs) because of their weight constraint. Carbon fiber-reinforced plastics are light in weight and provide exceptional

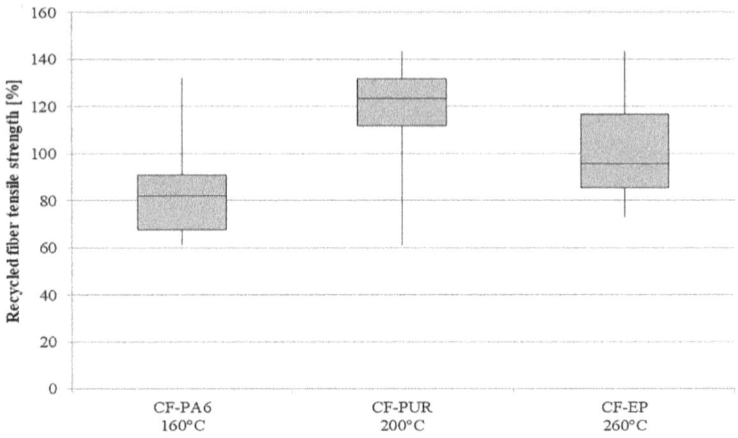

FIGURE 9.21 Box plot of the tensile strength of recovered fiber samples from one selected process series comprising four solvent extraction steps for each CFRP (in total 3 h for CFRP-PA6, 2.5 h for CFRP-PUR, and 3.5 h for CFRP-EP). 100% corresponds to the virgin fiber strength. (Reproduced with permission from Knappich et al., 2019.)

mechanical strength and also consume less fuel, compared to the available MMCs. CFRPs are mostly utilized not only in the automotive field for the development of the various components like bonnet, bumpers, and the whole automotive body panels, but also in the aviation industries for the development of the turbines, turbine blades, wings of the aircrafts, etc. If we are concerned about the mechanical strengths, then CFRPs showed very good mechanical properties, such as high compression strength, tensile strength, toughness, and flexural strengths, compared to the available metal matrix components. In this chapter, we briefly summarized the recent research done on the utilization of carbon fiber in the polymer matrix i.e., both in thermosets and thermoplastics. We mainly focused on the mechanical properties of the carbon fiber in both of the polymer matrix types and listed some results from the past research.

REFERENCES

Alam, P., Mamalis, D., Robert, C., Floreani, C., & Brádaigh, C. M. Ó. (2019). The fatigue of carbon fi bre reinforced plastics—A review. *Composites Part B*, *166*(December 2018), 555–579. doi:10.1016/j.compositesb.2019.02.016

Aldridge, M., Waas, A., & Kieffer, J. (2014). Spatially resolved, in situ elastic modulus of thermoset polymer amidst carbon fibers in a polymer matrix composite. *Composites Science and Technology*, *98*, 22–27. doi:10.1016/j.compscitech.2014.03.002

American Chemical Society (2003). *High Performance Carbon Fibers*. Washington, DC: American Chemical Society.

Bhatt, P., & Goe, A. (2017). Carbon fibres: Production, properties and potential use. *Material Science Research India*, *14*(1), 52–57. doi:10.13005/msri/140109

Caminero, M. A., Chacón, J. M., García-Moreno, I., & Reverte, J. M. (2018). Interlaminar bonding performance of 3D printed continuous fibre reinforced thermoplastic composites using fused deposition modelling. *Polymer Testing*, *68*(April), 415–423. doi:10.1016/j.polymertesting.2018.04.038

Campbell, F.C. (2004), Thermoset resins: The glue that holds the strings together, Eds. F.C. Campbell, *Manufacturing Processes for Advanced Composites*, Elsevier Science, 63–101. https://doi.org/10.1016/B978-185617415-2/50004-6

Dydek, K., Latko-Durałek, P., Boczkowska, A., Sałaciński, M., & Kozera, R. (2019). Carbon Fiber Reinforced Polymers modified with thermoplastic nonwovens containing multi-walled carbon nanotubes. *Composites Science and Technology*, *173*(November 2018), 110–117. doi:10.1016/j.compscitech.2019.02.007

Gabrion, X., Placet, V., Trivaudey, F., & Boubakar, L. (2016). About the thermomechanical behaviour of a carbon fibre reinforced high-temperature thermoplastic composite. *Composites Part B: Engineering*, *95*, 386–394. doi:10.1016/j.compositesb.2016.03.068

Hasan, M. M. B., Nitsche, S., Abdkader, A., & Cherif, C. (2018). Carbon fibre reinforced thermoplastic composites developed from innovative hybrid yarn structures consisting of staple carbon fibres and polyamide 6 fibres. *Composites Science and Technology*, *167*(February), 379–387. doi:10.1016/j.compscitech.2018.08.030

Hirano, Y., Yokozeki, T., Ishida, Y., Goto, T., Takahashi, T., Qian, D., Ito, S., Ogasawara, T., & Ishibashi, M. (2016). Lightning damage suppression in a carbon fiber-reinforced polymer with a polyaniline-based conductive thermoset matrix. *Composites Science and Technology*, *127*, 1–7. doi:10.1016/j.compscitech.2016.02.022

Huang, Y., Meng, X., Xie, Y., Lv, Z., Wan, L., Cao, J., & Feng, J. (2018). Friction spot welding of carbon fiber-reinforced polyetherimide laminate. *Composite Structures*, *189*(November 2017), 627–634. doi:10.1016/j.compstruct.2018.02.004

Kakinuma, Y., Ishida, T., Koike, R., & Klemme, H. (2015). Ultrafast feed drilling of carbon fiber-reinforced thermoplastics. *Procedia CIRP*, *35*, 91–95. doi:10.1016/j.procir.2015.08.074

Knappich, F., Klotz, M., Schlummer, M., Wölling, J., & Mäurer, A. (2019). Recycling process for carbon fiber reinforced plastics with polyamide 6, polyurethane and epoxy matrix by gentle solvent treatment. *Waste Management*, *85*, 73–81. doi:10.1016/j.wasman.2018.12.016

Lee, H. S., Kim, S. Y., Noh, Y. J., & Kim, S. Y. (2014). Design of microwave plasma and enhanced mechanical properties of thermoplastic composites reinforced with microwave plasma-treated carbon fiber fabric. *Composites Part B: Engineering*, *60*, 621–626. doi:10.1016/j.compositesb.2013.12.064

Lionetto, F., Balle, F., & Maffezzoli, A. (2017). Hybrid ultrasonic spot welding of aluminum to carbon fiber reinforced epoxy composites. *Journal of Materials Processing Technology*, *247*(December 2016), 289–295. doi:10.1016/j.jmatprotec.2017.05.002

Liu, Y., Guo, Y., Zhao, J., Chen, X., Zhang, H., Hu, G., Yu, X., & Zhang, Z. (2019). Carbon fiber reinforced shape memory epoxy composites with superior mechanical performances. *Composites Science and Technology*, *177*(October 2018), 49–56. doi:10.1016/j.compscitech.2019.04.014

Mamalis, D., Flanagan, T., & Ó Brádaigh, C. M. (2018). Effect of fibre straightness and sizing in carbon fibre reinforced powder epoxy composites. *Composites—Part A: Applied Science and Manufacturing*, *110* (April), 93–105. doi:10.1016/j.compositesa.2018.04.013

Mudhukrishnan, M., Hariharan, P., & Palanikumar, K. (2019). ScienceDirect delamination analysis in drilling of carbon fiber reinforced polypropylene (CFR-PP) composite materials. *Materials Today: Proceedings*, *16*, 792–799. doi:10.1016/j.matpr.2019.05.160

Ning, F., Cong, W., Qiu, J., Wei, J., & Wang, S. (2015). Additive manufacturing of carbon fiber reinforced thermoplastic composites using fused deposition modeling. *Composites Part B: Engineering*, *80*, 369–378. doi:10.1016/j.compositesb.2015.06.013

Tian, X., Liu, T., Wang, Q., Dilmurat, A., Li, D., & Ziegmann, G. (2017). Recycling and remanufacturing of 3D printed continuous carbon fiber reinforced PLA composites. *Journal of Cleaner Production*, *142*, 1609–1618. doi:10.1016/j.jclepro.2016.11.139

Tian, X., Liu, T., Yang, C., Wang, Q., & Li, D. (2016). Interface and performance of 3D printed continuous carbon fiber reinforced PLA composites. *Composites—Part A: Applied Science and Manufacturing*, *88*, 198–205. doi:10.1016/j.compositesa.2016.05.032

Vedrtnam, A. (2019). Novel method for improving fatigue behavior of carbon fiber reinforced epoxy composite. *Composites Part B: Engineering*, *157*(July 2018), 305–321. doi:10.1016/j.compositesb.2018.08.062

Wang, Q., Ning, H., Vaidya, U., Pillay, S., & Nolen, L. A. (2015). Development of a carbonization-in-nitrogen method for measuring the fiber content of carbon fiber reinforced thermoset composites. *Composites—Part A: Applied Science and Manufacturing*, *73*, 80–84. doi:10.1016/j.compositesa.2015.02.025

Wang, Y., Zhang, J., Fang, G., Zhang, J., Zhou, Z., & Wang, S. (2018). Influence of temperature on the impact behavior of woven-ply carbon fiber reinforced thermoplastic composites. *Composite Structures*, *185*(November 2017), 435–445. doi:10.1016/j.compstruct.2017.11.056

Yamamoto, T., Yabushita, S., Irisawa, T., & Tanabe, Y. (2019). Enhancement of bending strength, thermal stability and recyclability of carbon-fiber-reinforced thermoplastics by using silica colloids. *Composites Science and Technology*, *181*(April), 107665. doi:10.1016/j.compscitech.2019.05.022

Yao, S., Jin, F., Yop, K., Hui, D., & Park, S. (2018). Recent advances in carbon- fi ber-reinforced thermoplastic composites: A review. *Composites Part B: Engineering, 142*(November 2017), 241–250. doi:10.1016/j.compositesb.2017.12.007

Yourdkhani, M., Liu, W., Baril-Gosselin, S., Robitaille, F., & Hubert, P. (2018). Carbon nanotube-reinforced carbon fibre-epoxy composites manufactured by resin film infusion. *Composites Science and Technology*, *166*, 169–175. doi:10.1016/j.compscitech.2018.01.006

10 Glass Fiber Thermoset and Thermoplastic Composites

T P Sathishkumar and Gobinath Velu Kaliyannan

CONTENTS

10.1 INTRODUCTION

The glass wool is also called glass fiber and it was invented between 1932 and 1933 by a US engineer Russell Games Slayter, who worked the global company of Owens-Corning. The glass fiber-reinforced polymer composites have been used extensively in various applications for reducing total weight of a system and reducing excessive fuel consumption. The composites are made of different forms, namely random-oriented glass fiber polymer composites, laminated composites using random and

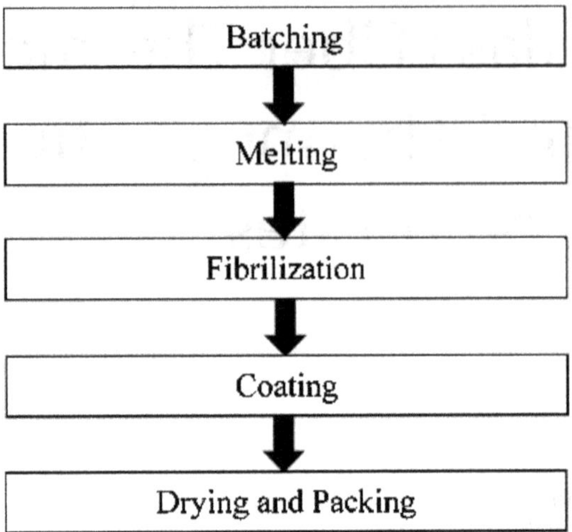

FIGURE 10.1 Making process of the glass fiber.

woven mats; also, these forms are reinforced with natural fiber and other synthetic fiber such as carbon and Kevlar for preparing the hybrid fibers polymer composites for various structural applications [2, 3]. Normally, two types of glass are prepared such as glass and glass fiber. The simple glass is brittle in nature and during preparation of glass, the silicon and oxygen atom are not backed linearly. During making of glass fiber, the application of slow cooling process is used to get the micron size glass fiber filament. Then the silicon and oxygen atoms are slowly backed in line into crystalline arrangements. Figure 10.1 shows the making process of the glass fiber. Different processes are used to make the glass fiber such as a) Batching, b) Melting, c) Fibrilization, d) Coating, e) Drying and Packing.

10.1.1 Batching Process

The commercial glass fiber is made up of silicon material alone, which gives it a brittle nature. By adding different ingredients to the pure glass fiber, various glass fibers can be produced. Adding SiO_2, Al_2O_3, CaO, and MgO in the glass fiber processing, produces E-glass fiber which is used for electrical appliances. Adding a small quantity of SiO_2 in the glass fiber makes S-glass fiber, which is used for applications requiring higher tensile strength. During the initial stage of glass fiber manufacturing, the ingredient materials should be carefully weighed and mixed toughly by mechanical starrier; this is called batching [1].

10.1.2 Melting Process

After preparing the batch, the pneumatic or electric conveyor is used to send the batch mixture to a high temperature furnace. The furnace melts the batch by increasing the temperature up to 1400°C with the help of natural gas. The furnace section is divided into three sections that receive the batch, molten metal flows, and forehearth.

In the first section, the batch is received, melted, and the bubbles are removed. In the second stage, a refiner receives the molten glass metal and reduces the temperature down to 1370°C. In the third section, the forehearth process, this has four to seven series bushing, then the extruder process is used to extrude the molten glass into the micron size of glass fibers.

10.1.3 Fibrilization

The fibrilization is called glass fiber formation. During the extrusion process, the bushing process is applied to the molten glass and the glass fiber is made, which will cause erosion in the bushing. For reducing this erosion, the electronical heater is used to control the glass fiber temperature and maintain the glass viscosity. The process of mechanical drawing of molten glass forming the fibrous filaments, is called fibrilization.

10.1.4 Coating

The final stage of glass fiber manufacturing is coating or sizing. The sizing is the adding of 0.5 to 2 lubricants, binders, or coupling agents in the glass fiber filaments. Avoiding abrading and breaking of filament, the adequate percentage of lubricant is added to the glass fiber. The coupling agent helps to increase the wettability with industrial resin and increases the adhesive bonding strength.

10.1.5 Drying and Packing

Finally, the sized or coated filaments are collected as a small bundle which is called a strand. Each strand has 51 to 1624 filaments of glass fibers. This stand is wound in a steel drum and backed. The forming package is still wet in water for cooling and sizing. After being dried, the package is transported to cut the chopped fiber, from woven fiber mat with the help of a textile process and from random fiber mat with the help of a needle punching process.

The prepared chopped fibers and mats forms are used to prepare the various thermoset and thermoplastic polymer composites for various applications. These glass fibers are reinforced with synthetic fibers such as carbon and Kevlar, and natural fiber such as sisal, jute, snake grass, coir, banana, etc., to prepare the hybrid fibers polymer composites.

Figure 10.2 shows the types of glass fiber materials used for various application [2]. The glass fiber is in different forms such as chopped random mat, cloth, woven roving, knits, and roving. Figures 10.3, 10.4, and 10.5 show the types of glass fibers used for different applications.

10.2 PREPARATION OF THERMOSET GLASS FIBER COMPOSITES

10.2.1 Resin Transfer Mold (RTM)

The resin transfer molding process was used to make the glass/epoxy-laminated composite with cavity size of 648 × 483 mm. The random mat of M113/150, U720/450,

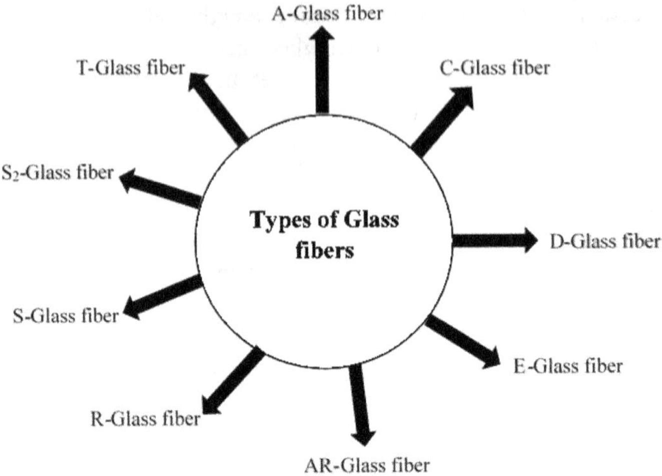

FIGURE 10.2 Types of glass fiber.

FIGURE 10.3 A-glass fiber material applications.

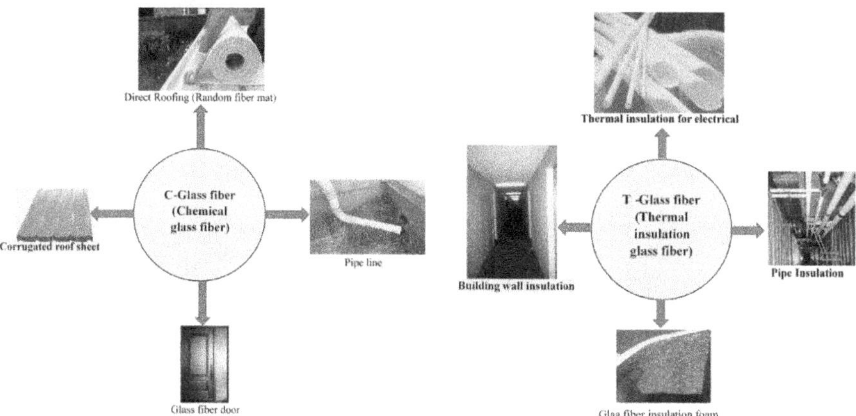

FIGURE 10.4 C- and T-glass fiber material applications.

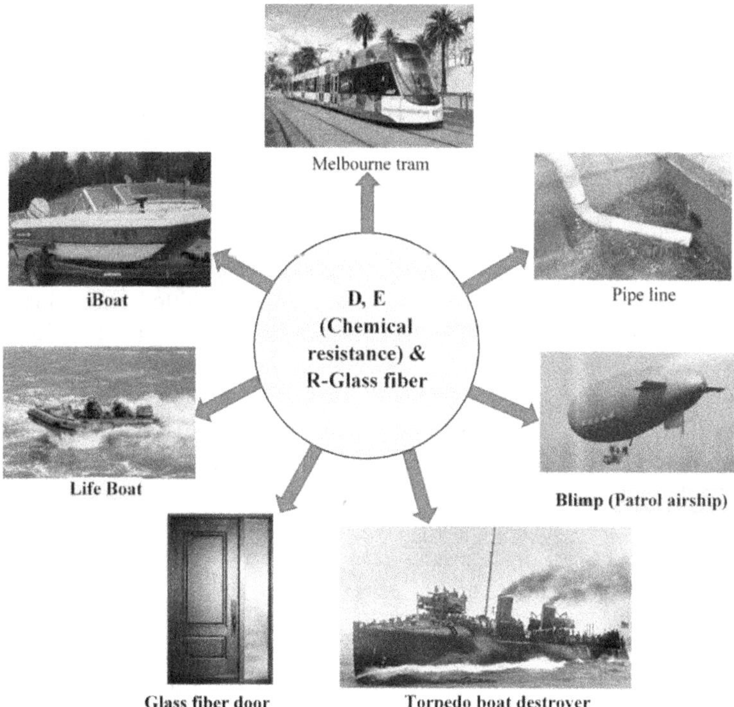

FIGURE 10.5 D- E- and R-glass fiber material applications.

and U720/300, and woven mat of R12/450 are used to prepare the composites. Initially, the glass fiber layers were placed on the female cavity of the mold and male die was used to close the mold. Then, an injection unit was used for injecting the epoxy resin to fill the cavity. The first type of composite was prepared by following laminate sequence (M113/150)/ (U720/300)/(U720/450)/(U720/450)/(M113/150)

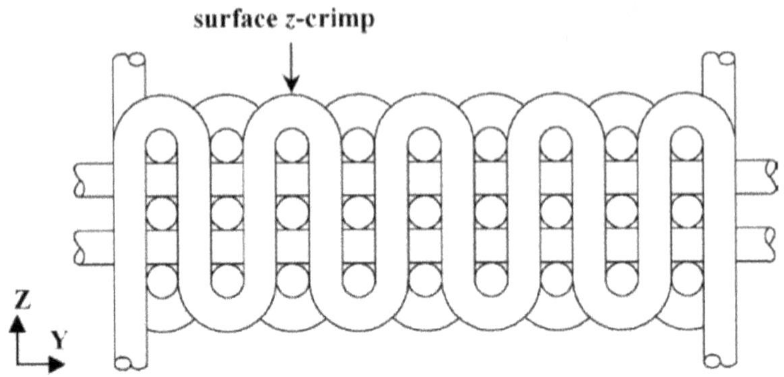

FIGURE 10.6 3D orthogonal fabric preform (Hameed et al., 2007).

and the second type of composite was prepared by following laminate sequence (M113/150)/ (U720/300)/(U720/450)/ (R12/450)/(M113/150) [4]. Three E-glass fibers were used to prepare the three-dimensional (3D) orthogonal (3D monolithic) weave as shown in Figure 10.6. Three types of glass fiber mat such as 3D monolithic, 2D plane weave (2D-E-24x4), and 4-ply biaxially reinforced warp-knit were used to prepare the vinyl ester composites by RTM method for 4 to 8 hours post curing at 85°C [9].

10.2.2 Hany Lay-up Method

The three types of textile-knitted glass fiber fabrics were used to prepare the epoxy-laminated composites. They are plain, Milano and rib knitted glass fiber fabrics. The CY225 grade epoxy and HY225 hardener was used and eight piles of all fabric's composites were prepared by hand lay-up method and hot laminated press. The pressure and temperature of hot press was maintained at 8 MPa and 110°C for 100 min. After that, the composite plate was cooled at room temperature for obtaining the complete solidification. The thickness of composites plate was maintained as 3 mm [6]. The unidirectional E-glass fiber epoxy composites were prepared by using two different types of tasking sequence such as $[0/90/0/90]_s$ (cross ply) and $[0/90/+45/-45]_s$ (angle ply). The hot press was used to prepare the composites under 15 MPa pressure at 120°C for 2 hours and of approximately 3 mm thickness. The E-glass fiber weight fraction in the composite was 65%. The density of the E-glass fiber was measured as 509 g/m^3. The glass fiber fabrics of 300×300 mm^2 were dipped in epoxy resin and laminated between two Teflon sheets which removed the excessive resin by rolling a steel roller on top sheet. The composite was allowed to cure for three days at room temperature and after that the cure composites plates were post cured at 60°C in a hot air oven for 3 hours. The final composite had a thickness of 1.5 mm and a 47% weight fraction of fiber [5]. The fiber glass waste was collected, separated, and cut for composites preparation. The 1% of methyl ethyl ketone (MEK) catalyzer was mixed with unsaturated polyester resin for curing. Initially segregated, the fiberglass was spread over female die by hand and the polyester resin was poured on the glass fiber. Finally, the male die was used to close the mold and was kept under a compression load of 88.3 kN [8].

10.3 RANDOMLY ORIENTED GLASS FIBER THERMOSET COMPOSITES

10.3.1 Polyester Composites

Putic et al. [2] have prepared the random/woven glass fiber mat reinforced polyester composite using hand lay-up method and suited the interlaminar shear strength with three-layer and eight layers patterns. The three resins, namely bisphenolic, water-resistant, and acid-resistant resins were used to prepare the composites with different densities of glass fabric. The internal layers of the composites were made of woven mat fiber and external layers were made of short glass fabric. The thickness of external layers was around 1 mm and of intermediate layer was around 0.5–0.8 mm. The pattern model P6 [glass mat density (240 g/m^2)—woven mat (0/90) (800 g/m^2)—glass mat (240 g/m^2)] had greater interlaminar shear strength relative to other pattern models.

Agarwal et al. [3] have reported the mechanical behavior of randomly oriented E-glass fiber (GF)/polyester composite with varying environmental conditions like brine, ganga water, acid solution, kerosene oil, and freezing conditions. The experiment was carried out at various time intervals such as 64 h, 128 h, and 256 h. After every interval of time, the percentage reduction in tensile strength was gradually reduced. The highest percentage reduction was discovered on NaOH solution, and the lowest percentage reduction was discovered in freezing conditions.

Hossain et al. [4] have reported the flexural characteristics of woven glass fiber/polyester composite with adding of different wt% of carbon nano filler (CNF). The experiment was conducted on CNF-filled composites with 0.1, 02, 03, and 0.4 wt% of CNF. From the flexural stress versus strain curve, the highest flexural properties were achieved by 0.2wt% content of CNF-filled composite. This was owing to good and uniform dispersion of nano filler in the entire composite. Also, a good improvement was found in compressive strength and modulus due to better and stronger fiber-matrix interfacial interaction.

10.3.2 Epoxy Composites

Atas et al. [5] have reported the impact response of woven glass fiber mat/epoxy composites with non-orthogonal and orthogonal fabric, with a weaving angle of 20°, 30°, 45°, 60°, 75°, and 90°, respectively from a vertical position (warp direction). With decreasing the weaving angle between interweaving yarns, the energy absorption was gradually improved. Woven composites with a lower weaving angle of 20° and 30° between interlacing yarns have longer contact period, lesser peak force, high energy absorbed, and greater deflection than the bigger weaving angles of 60°, 75°, and 90°. The [0/20°] woven composite showed more energy absorption compared to [0/90°] woven composite.

Iba et al. [6] investigated the mechanical characteristics of GF/epoxy composite prepared with three different fiber diameters, of 18, 37, and 50 μm, corresponding the fiber volume fraction from 0.25 to 0.45. The stress-strain graph revealed that the tensile strength and longitudinal Young's modulus of the composite improved by increasing the fiber volume fraction. The maximum mechanical modulus and

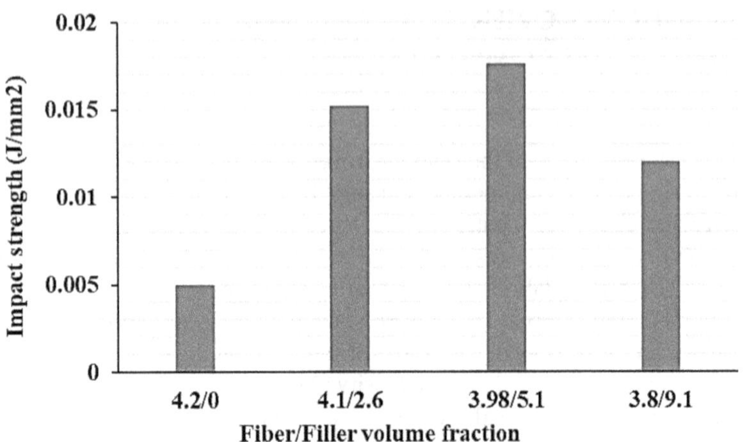

FIGURE 10.7 Impact strength of the E-GF/epoxy composites with and without fillers.

strength were observed for composites conditioning the 18 μm fiber diameter at 0.45 fiber volume fraction.

Gupta et al. [7] have reported the impact and compressive behavior of discontinuous E-GF/epoxy composites with inclusion of fillers like fly ash and calcium carbonate. Figure 10.7 shows the impact strength of the GF/epoxy with and without fillers. The fly ash filler content is 2.6 and 5.1 volume fraction, and the $CaCO_3$ is 9.1 volume fraction. Compared to the calcium carbonate filler in the composites, fly ash particles resulted in decreased impact strength and compressive strength of the composites. The aspect ratio of the fiber decreased the impact strength and improved the compressive strength. The presence of fly ash and calcium carbonate particles in the composite was improved bonding between fiber and matrix which showed higher impact strength. The compatibility between fly ash and matrix was found to be better than epoxy and glass fiber interaction bonding. The plastic deformation epoxy and fiber pull-out on the surface of impact test sample was clearly observed and it reflected the properties of the composites.

Yang et al. [8] have reported the compression, bending, and shear behavior of woven mat glass fiber mat/epoxy composites which were prepared by resin transfer molding technique. The composites were prepared with various fabrics such as biaxial non-crimp, uniaxial stitched plain weave, and unstitched plain weave fabrics. From the results, they found that the Z-directional stitching fibers enhanced the resistance to delamination, decreased impact failure, and reduced the composite bending. The non-crimp laminate's compressive strength was15 % greater than that of woven fabric composite.

Hameed et al. [9] have reported the mechanical and thermal behavior of E-glass fiber/modified epoxy (styrene-co-acrylonitrile) composite with various fiber volume fractions from 10 to 60%. The viscoelastic characteristics were evaluated at the frequency of 1 Hz under three-point bending mode and samples were heated to a heating rate of up to 250°C per minute. The storage modulus was reduced by increasing the surrounding temperature. The storage modulus versus the temperature plot showed the maximum storage module (15,606 GPa) for the 50 vol. % fiber content

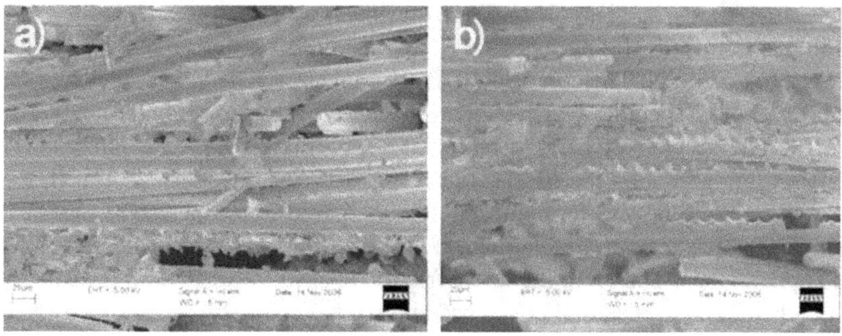

FIGURE 10.8 a) Failed surface of glass fiber composites with neat epoxy resin and b) Epoxy resin modified with SAN (Hameed et al., 2007).

of composites. Loss modulus versus temperature plot depicted that the loss modulus improved at fiber content of 50 vol %. The loss modulus reduced further increasing the fiber content. Figure 10.8 a) and b) depicts the failed surface of GF/modified epoxy composites. This showed that the failure of fiber had occurred uniformly for epoxy resin modified with SAN compared to neat epoxy resin glass fiber composites.

Patnaik et al. [10] have prepared the randomly oriented E-GF/epoxy composite with particulate-filler such as aluminum oxide (Al_2O_3), silicon carbide (SIC), and pine bark dust. They studied mechanical properties such as flexural, tensile, interlaminar shear strength, and impact strength. The various compositions of composite such as 50 wt% of GF + 50 wt% of epoxy, 50 wt% of GF + 40 wt% of epoxy + 10 wt% of Al_2O_3, 50 wt% of GF + 40 wt% of epoxy + 10 wt% of pine bark dust, 50 wt% of GF + 40 wt% of epoxy + 10 wt% of SiC were used to prepare the composites. The results exhibited that the GF + epoxy composite had the highest flexural strength (368 MPa) and tensile strength (249.6 MPa), GF + epoxy + pine bark dust composition had the greater interlaminar shear strength (23.46 MPa), and GF + epoxy + SiC composite had the maximum impact strength (1.840 J) and maximum hardness (42 Hv).

10.3.3 Vinyl Ester Composites

LeBlanc et al. [11] have reported the compressive properties of woven mat S-2 GF/epoxy/vinyl ester composites with varying preform areal weights such as 93, 98, 100, and 190 oz/yd^2. The maximum compressive strength was found for composites containing 93 oz preform at weft and warp direction, but the composites containing 93 and 190 oz preform showed a gradually decreasing trend, and 100 oz preform did not any improvement. The maximum compressive strength at warp and weft direction of 93 oz preform composites was 330 MPa and 259 MPa, respectively.

Chauhan et al. [12] have reported the friction and sliding wear behavior of GF/vinyl ester composite. The GF/vinyl ester composites were prepared with bi-directional woven S-glass fibers (65 wt%) reinforced by vinyl ester resin with various commoners like butyl acrylate and methyl acrylate. Three different types of composition were fabricated, including composite A with GF + (vinyl ester + methyl crylate), composite B with GF + (vinyl ester + styrene), and composite C with

GF + (vinyl ester + butyl acrylate). The experiment was conducted under dry sliding conditions at various sliding speeds of 1, 2, 3, and 4 m/s and different loads such as 10, 20, 30, and 40 N. The composite GF + (vinyl ester + butyl acrylate) was found to have a greater specific wear rate and reduced coefficient friction at lower sliding velocity. The specific wear rate was increased by increasing the applied load and decreased by increasing the sliding velocities. Among all results, the specific wear rate was found to be less for composite C, compared to composites A and B. The coefficient of friction was gradually decreased with increasing the sliding velocities and applied loads. Therefore the lower friction coefficient was found for composite C at higher sliding speed of 4 m/sec and applied load of 40 N. According to surface morphology, the composites C at 20N applied load and 2 m/s sliding velocity showed that on the wear track less quantity of glass fiber was showing at the center of the specimen.

10.4 LAMINATED GLASS FIBER THERMOSET COMPOSITES

10.4.1 Polyester Composites

Aramide et al. [13] investigated the mechanical properties of glass fiber (woven mat) reinforced polyester composite. The composites were prepared by hand lay-up technique with varying percentage of GF volume fractions such as 5, 10, 15, 20, 25, and 30. The tensile strength and Young's modulus of the GF/polyester composite increased with increasing the GF volume fraction. The strain of the GF/polyester composite was found at 25% of GF and then found to be reduced. The maximum tensile strength (2.030 N/mm^2) and modulus of elasticity (110.32 N/mm^2) were found to be 30 % GF volume fraction.

Erden et al. [14] investigated the mechanical and vibrational properties of woven roving glass fabric/polyester composite with addition of matrix agent oligomeric siloxane at 1, 2, and 3% of weight content. The total volume fraction was maintained as 37%. The mixing of oligomeric siloxane into polyester resin enhanced mechanical characteristics like interlaminar shear, tensile, and flexural strength, elasticity modulus, and 1st, 2nd, and 3rd natural frequency values. The strongest mechanical properties were found for woven roving glass fabric composites with 3 wt% of oligomeric siloxane composite compared to 1 and 2 wt% composites. This may be because of better interfacial adhesion between the glass fiber surface with polyester. The tensile strength of GF/polyester composites improved from 41.5 MPa to 395.8 MPa while the modulus of elasticity was slightly altered.

Gul Hameed Awan et al. [15] investigated the tensile characteristics of glass fiber woven fabric-reinforced unsaturated polyester composites with different forms like 175 g/sq.m, 200 g/sq.m, 400 g/sq.m, and 600 g/sq.m and varying the number of plies such as 1-ply, 2-ply, and 3-ply, respectively. The higher density glass fiber fabric 600 g/sq.m with 3-ply laminated composite showed higher tensile strength. The tensile strength of 600 g/sq.m with 1-ply was 119.6 MPa, the tensile strength of 2-ply was 143.9 MPa, and the tensile strength of 3-ply increased to 189.4 MPa.

Al-alkawi et al. [16] have reported the fatigue behavior of woven glass fiber/polyester composites at different temperature conditions, that of 40°C, 50°C, and 60°C, respectively. In all fatigue experiments, the stress ratio was maintained constant at R=–1 and 33% of constant fiber volume fraction. The S-N curve showed that the tensile strength and fatigue strength decreased at 33% of fiber volume fraction with increasing temperature up to 60 °C. The fatigue strength factor percentage decreasing factor was greater than the tensile strength factor percentage decreasing factor at all temperature levels.

Yuanjian et al. [17] investigated the low-velocity impact and tension–tension fatigue behavior of woven glass fabric/polyester composites with two different geometries of fabric, namely $[\pm 45°]_4$ and $[0/90°]_{2s}$ which are stitch-bonded glass fiber fabric. The results show that the residual tensile strength and stiffness of $[\pm 45°]_4$ laminated composites were gradually deceased with increasing the impact energy. In $[0/90°]_{2s}$ laminated composites, the tensile strength and stiffness was not affected up to a 10 J test, and then decreased. For the two geometries, the effect damage was comparable. Low-impact energy of composites reinforced by GF may damage the matrix. At multiple impact damage energies of 1.4 J, 5 J, and 10 J, the tension–tension fatigue failure test was conducted and as seen in the S-N graph, the fatigue lifetimes were gradually reduced and at 1.4 J, the maximum fatigue stress at $[0/90°]_{2s}$ and $[\pm 45°]_4$ was found and then the fatigue lifetime was abruptly decreased.

Faizal johari et al. [18] have reported the tensile properties of plain-weave woven E-glass fiber/polyester composites with various curing pressures of 35.8 kg/m^2, 70.1 kg/m^2, 104 kg/m^2, and 138.2 kg/m^2. The composites were prepared using different lay-ups, that is, with symmetrical and non-symmetrical structure. The stress-strain graph showed that the tensile modulus declined for both symmetrical and non-symmetrical lay-up with increased curing pressure. The stiffness in the symmetrical lay-up was found to be less, compared to the stiffness of other composites. With greater curing pressure for non-symmetric arrangement, the ductility improved and symmetric arrangement was misaligned.

The fracture toughness and critical energy release of glass fiber-reinforced unsaturated polyester composite were reported by Leonard et al. [19]. This composite was prepared with various fiber volume fractions such as 12%, 24%, 36%, 48%, and 60%, respectively. The stress-strain graph revealed that 60% fiber volume fraction of glass fiber composite had the highest tensile strength of 325 MPa, Young's modulus of 13.9 GPa, 20-fold fracture toughness, and 1200-fold critical power release rate.

Visco et al. [20] investigated the mechanical characteristics of glass fiber/polyester composites which were pre and post dipped in seawater. Two types of polyester resins like orthophthalic and isophthalic, and two distinct kinds of laminates were utilized for the fabrication of GF/polyester composite. The experimental consequence indicated that flexural strength, modulus, and shear modulus reduced with rise in seawater dipping time. Isophthalic resin was stronger than orthophthalic resin because of binding with GFs being higher, which caused the resistance to the seawater absorption.

10.4.2 Epoxy Composites

The impact behavior of unidirectional E-glass fiber/epoxy composite with two distinct stacking patterns of [0/90/0/90] and [0/90/+45/–45] were investigated by

Aktas et al. [21]. Damage method and damage formation on the laminates under various impact force from 5 J to 80 J have been explored. The indenter penetration on composite [0/90/+45/–45] was found to be lower and showed lower circumferential damage, compared to [0/90/0/90] and the [0/90/0/90] laminate.

Karakuzu et al. [22] reported the impact behavior of unidirectional E-GF/epoxy composite with [0°/30°/60°/90°] stacking sequence. Four varying impact energies 10 J, 20 J, 30 J, and 40 J were used for testing with four impact weights of 5 kg, 10 kg, 15 kg, and 20 kg. Increasing the impact energy increased the contact force, deflection, and contact time, depending on equal mass and velocity. The delamination area was found to be higher for impact weight of 15 kg due to more fiber fracture and it was found to be lower for 5kg and 20 kg.

Aktas et al. [23] analyzed the post impact behavior of E-GF/epoxy-laminated composite with various knitted fabrics, that of Rib, Milano, and Plain. Various impact energy levels ranged from 5 J to 25 J. The test results indicated that the Rib knitted system had the highest contact force, found at 22.5J impact loading. The penetration deflection was found to be higher for Milano knitted structure at 10J impact load.

10.4.3 Vinyl Ester Composites

Baucom et al. [24] have reported that the low-velocity impact of woven E-GF/vinyl ester composites with different laminates such as two dimensional (2D) plain-woven laminate,3D orthogonally woven monolith, and a biaxial reinforced warp-knit. The consequence of the drop-weight system showed that the 3D composites were more resistant to penetration. The maximum energy distribution of 140 J was found for 3D composites.

Karakuzu et al. [25] studied the pin loaded tensile behaviors of woven mat GF/vinyl ester composite and analyzed net-tension, shear-out, and bearing mode failures. The composite fiber volume fraction was maintained as 63%, the distance between the free edge of plate and hole dimeter ratio was 1, 2, 3, 4, and 5, respectively. The maximum tension load was found for composites with a width-to-diameter ratio of 5 and edge-to-diameter ratio of 3.

10.5 GLASS FIBER THERMOPLASTIC COMPOSITES

Ke Wang et al. [26] investigated the transcrystallization role on mechanical and thermal behavior of polypropylene (PP)/GF composite using a twin screw extruder machine with dynamic packing system. The shear stress was discovered to play a major part in the injection molding transcrystallization of PP/GF composites. Figure 10.9 shows the polarized optical microscopy images of single glass fiber polypropylene composites with and without shear. Without shear, the transcrystallization did not occur, and with small shear, the transcrystallization phase was observed around the fiber. The glass fiber composites has transcrystallization phase. So that the shearing process will be applied to solidify the glass fiber and PP matrix.

FIGURE 10.9 Polarized optical microscopy images experiment on single glass fiber/PP composite (a) Without shear and (b) With shear (Wang et al., 2007).

10.6 HYBRID FIBER-REINFORCED POLYMER COMPOSITES

10.6.1 Thermoset Composites

Vineet Kumar Bhagat et al. [27] investigated the physical and mechanical characteristics of short coir with E-glass epoxy hybrid composite at different fiber lengths and weights. The hand lay-up technique was used to prepare the composites, including C1 (epoxy (75 wt%) + glass fiber (20 wt%) + coir fiber (fiber length 5 mm) (5 wt%)), C2 (epoxy (75 wt%) + glass fiber (20 wt%) + coir fiber (fiber length 10 mm) (5 wt%)), C3 (epoxy (75 wt%) + glass fiber (20 wt%) + coir fiber (fiber length 15 mm) (5 wt%)), C4 (epoxy (75 wt%) + glass fiber (20 wt%) + coir fiber (fiber length 20 mm) (5 wt%)), C5 (epoxy (75 wt%) + glass fiber (20 wt%) + coir fiber (fiber length 5 mm) (10 wt%)), C6 (epoxy (75 wt%) + glass fiber (20 wt%) + coir fiber (fiber length 10 mm) (10 wt%)), C7 (epoxy (75 wt%) + glass fiber (20 wt%) + coir fiber (fiber length 15 mm) (10 wt%)), and C8 (epoxy (75 wt%) + glass fiber (20 wt%) + coir fiber (fiber length 20 mm) (10 wt%)). The experimental results showed that the theoretical density of all composites was found to be higher than the experimental density. The least density was found for C8 composites. The surface micro-hardness was shown to be higher when the fiber length and weight content were increased. The maximum tensile strength was found for composites containing 15 mm fiber length with 10 and 20% weight content of coir and glass fiber. This may be because of better distribution of fibers and adhesion between fibers and matrix. The flexural strength of 15-mm fiber length composites was found to be of 63 MPa. The tensile fractured specimen images are clearly explaining the coir and GF in epoxy hybrid composite. The fiber matrix adhesion was found to be better for 15-mm fiber composites than those with the 20-mm fiber length. Because, the more fiber pull-out showed in 20 mm fiber length composites.

Daiane Romanzini et al. [28] investigated the mechanical and dynamic mechanical properties of GF/ramie/polyester hybrid composite with different fiber contents of 10, 21, and 32 vol.% and the relative volume fraction of GF/ramie fibers were taken as 0:100, 25:75, 50:50, and 75:25, respectively. The maximum fibers content and greater ramie fiber fraction in the GF/ramie/polyester hybrid composite had lower weight composites and greater water absorption. Figures 10.10 and 10.11 show the

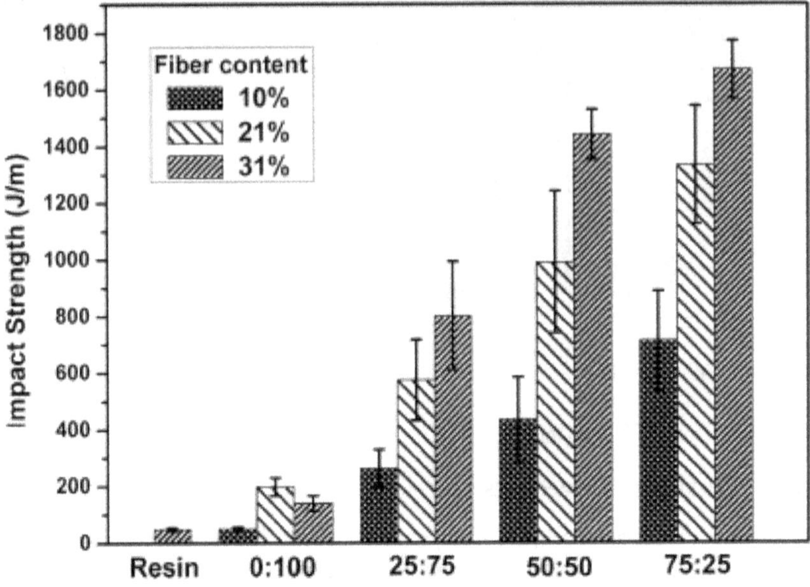

FIGURE 10.10 Impact strength versus relative volume fraction of fibers (Daiane Romanzini et al. 2013).

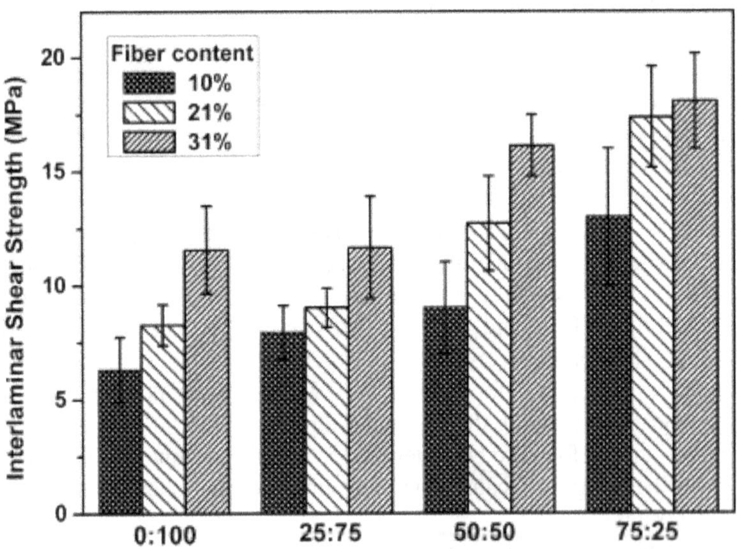

FIGURE 10.11 Interlaminar shear strength versus relative volume fraction of fibers (Daiane Romanzini et al., 2013).

impact and interlaminar shear strength of the hybrid composites. The impact and shear strength were gradually increased by increasing the glass fiber content, compared to pure ramie composites. The maximum properties were found at 31% of total fibers volume fraction with 75:25 relative volume fraction of glass and ramie fibers. In dynamic mechanical testing, the maximum storage modulus at initial stage and after reaching viscoelastic stage was found to be higher at 31% of total fibers volume fraction with 75:25 relative volume fraction of glass and ramie fibers, and also the damping factor was found to be higher.

Ramesh et al. [29] investigated the mechanical properties, such as tensile strength, flexural strength, and impact strength, of sisal–jute–glass fiber-reinforced polyester hybrid composite. The composites were fabricated using hand lay-up techniques. The hybrid composite was prepared with different compositions such as GF + sisal fiber + polyester resin, GF + jute fiber + polyester resin, and GF + sisal fiber + jute fiber + polyester resin composites. The experimental results showed that the GF + jute fiber + polyester resin composite showed a maximum tensile strength of 229.54 MPa, GF + sisal fiber + jute fiber + polyester resin composites showed a maximum flexural strength with 14.2 mm flexural displacement, and GF + jute fiber + polyester resin composition showed a maximum impact strength of 18.67 joules.

Mishra et al. [30] have prepared biofibers like pineapple leaf and sisal fiber with glass fiber polyester composites and studied the mechanical properties. Before the preparation of composites, the sisal and pineapple leaf fiber were washed with detergent power with water solution at 70°C temperature for 1 hour. Also, the sisal fiber was chemically treated with alkali, cyanoethylated, and acetylated. Among the various chemically treated sisal fibers, 5 wt% of alkali-treated fiber composites showed optimum tensile and impact properties due to increasing the bonding area and adhesion strength.

10.6.2 Thermoplastic Composites

Fu et al. [31] reported that the carbon short fiber (SCF) and glass short fiber (SGF) PP composites granules were prepared by twin screw extruder at six heat zones of 230, 230, 220, 220, 220, and 220°C. Thereafter, the granules were injected into the mold to prepare the dumbbell-shaped specimen for experimentation. Figure 10.12 shows the tensile stress versus tensile strain of SCF/PP and SGF/PP thermoplastic composites. The curves G1, G2, and G3 have 25, 16, and 8 vol% glass fibers SGF/pp composites and C1, C2, and C3 have 25, 16, and 8 vol% carbon fibers SCF/pp composites. It is clear that the tensile stress was linearly increased with tensile strain. The SGF/PP composites showed higher strain for failure compared to SCF/PP composites. Then, the maximum tensile stress was found for SCF/PP composites and both fiber PP composites exhibited brittle failure. The critical length of the glass fiber and carbon fiber were found to be 887.92 and 813.87 mm.

Husic et al. [32] reported the mechanical characteristics of untreated E-glass /polyurethane (PU) composite. The composites were prepared with two polyurethanes; soypolyol and petrochemical polyol Jeffol resin. The results about the mechanical properties showed that lower flexural, tensile, and interlaminar shear strength were found for soypolyol-based composite compared to the Jeffol-based petrochemical

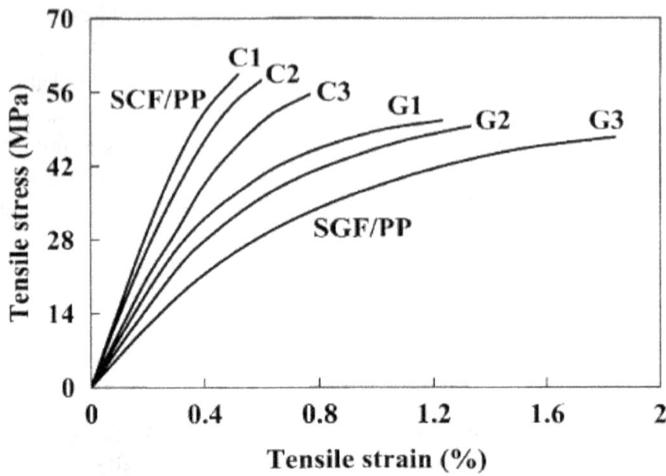

FIGURE 10.12 Tensile stress versus tensile strain of carbon and glass fiber pp composites (Fu et al., 2000).

polyol composite. This was due to reduced cross-link density of the polymer and the existence in the matrix of side-dangling chains. For both resin composites, the interlaminar shear strength was comparable.

Kasama Jarukumjorn et al. [33] investigated the effect of glass fiber hybridization on tensile, flexural, and impact properties of sisal/polypropylene composite. The mechanical strength was improved by addition of the compatibilizer owing to the increased interfacial bond between the sisal fiber and PP matrix. Inclusion of glass fiber into the sisal/PP composite increased tensile, flexural, and impact strength. But the flexural and tensile moduli did not change by addition of glass fiber. Moreover, adding of GF increased the thermal stability and the water resistance.

Samal et al. [34] prepared the hybrid composite PP reinforced with bamboo fiber and glass fiber by using a twin screw extruder and injection-molding method. By increasing the binding between fibers with PP, the maleic anhydride grafted PP was used for preparing the hybrid composites. Tensile, flexural, and impact properties were studied according to American Society for Testing and Materials (ASTM) standard. The maximum mechanical properties for tensile, flexural, and impact strength were found in composites containing 15:15 ratio of bamboo/glass fiber in hybrid composites with 2wt% of maleic anhydride grafted PP. Compared to those of virgin PP, the tensile strength and impact strength of hybrid composites were increased to nearly 69, 86, and 83%.

Haihong Jiang et al. [35] prepared the L- and S-glass fiber-reinforced poly(vinyl chloride) (PVC)/red Oak wood flour thermoplastic composites. The unnotched and notched impact strength of PVC/wood flour/glass fiber hybrid composites were gradually improved, but without reducing flexural properties by incorporating L-glass fiber and more than 40% of PVC. The improvement was not yet seen while using S-glass fiber. The impact strength of PVC/wood flour/glass fiber hybrid composites was improved at 50% of PVC content by increasing L-glass fiber content.

Bikiaris et al. [36] reported that the three functional copolymers and two functional silanes (organic), namely ethylene/vinyl alcohol, ethylene/acrylic acid, ethylene-g-maleic anhydride, g-methacryloxypropyltrimethoxy, and cationic styryl, respectively were utilized as coupling agent for preparing the glass fiber-reinforced polyethylene composites. The low-density polyethylene (LDPE) glass fiber thermoplastic composites were prepared and the tensile and impact properties were studied. The adhesive agent of ethylene-g-maleic anhydride (copolymer) attained better mechanical results compared to other copolymers and organic silanes. The tensile strength was increased and impact strength was decreased by increasing the glass fiber content in LDPE composites.

Hufenbach et al. [37] reported the tribological properties of polypropylene after addition of glass fiber. The short, unidirectional, and woven fabric glass fiber was used to prepare the polypropylene composites. The coefficient of fraction was found to be higher for 90° unidirectional fiber composites and the specific wear rate was 0° in unidirectional fiber composites. With the incorporation of glass fiber in PP composites, the coefficient of friction was gradually decreased by increasing the sliding distance. Among two different PP matrix types, homopolymer and copolymer, the PP homopolymer showed good wear resistance at 30% fiber loading compared to that found in pure PP results.

Hufenbach et al. [38] prepared the multi-layered flat bed weft-knitted glass fabrics (MKF), non-crimp glass fabrics (NCF), and woven glass fiber fabric for fabricating the PP composites. The textile fiber on warf and weft directions of fiber density varied by number of yarn counts. The tensile strength was measured at 23 and 80°C. The maximum tensile strength was found in PP composites with MKF fabric at 1 m/s testing speed. Among two environmental conditions, the maximum tensile strength was found in PP composite at 90° fiber mat orientation, in 23°C environment, and at 1 m/s testing speed. The pin load tensile strength was found higher for woven glass fiber/PP composites with 90° fiber orientation.

Thomason et al. [39] investigated the effect of fiber surface coating on interface bonding between glass fiber and polypropylene. The strength of interfacial bonding between fiber and PP was measured by glass fiber pull-out at fracture surface of the specimen. The combination of silane and other components with matrix highly influenced the bonding and increased the interfacial strength. The flexural strength of glass fiber composites was associated with the shear strength, which showed during fiber pull-out tests. The shear strength was measured by fiber pull-out tests and it depended on fiber coating.

10.7 CONCLUSION

The thermoset and thermoplastic glass fiber-reinforced composites have been prepared by various manufacturing processes. The handy lay-up method, resin transfer mold process, vacuum-assisted mold method and compression mold methods were used to prepare the thermoset composites, and the screw extruder and compression mold machine were used to prepare the thermoplastic composites. The epoxy, polyester, vinyl ester, PP, PU, PVC, LDPE, and other polymers have been used as matrix. The different forms of glass fiber such as random mat, woven mat, unidirectional mat, short and long fibers are reinforced with carbon and natural fiber to prepare the composites. By increasing the volume or weight fraction of glass fiber, the tensile, flexural, impact, interlaminar shear strength, dynamic mechanical properties, and wear resistance were improved and incorporation of coupling agents in polymer increased the above properties considerably. Therefore, glass fiber is one of the best synthetic fibers for developing various components and parts for various applications.

REFERENCES

1. Composites World. The making of glass. Available at: www.compositesworld.com/articles/the-making-of-glass-fiber.
2. Putić, S., et al., The interlaminar strength of the glass fiber polyester composite. *Chemical Industry and Chemical Engineering Quarterly/CICEQ*, 2009. **15**(1): p. 45–48.
3. Agarwal, A., et al., Tensile behavior of glass fiber reinforced plastics subjected to different environmental conditions, 2010. **17**(6): p. 471–476.
4. Hossain, M.K., et al., Flexural and compression response of woven *E*-glass/polyester–CNF nanophased composites. *Composites—Part A: Applied Science and Manufacturing*, 2011. **42**(11): p. 1774–1782.
5. Atas, C. and D. Liu, Impact response of woven composites with small weaving angles. *International Journal of Impact Engineering*, 2008. **35**(2): p. 80–97.
6. Iba, H., T. Chang, and Y. Kagawa, Optically transparent continuous glass fibre-reinforced epoxy matrix composite: Fabrication, optical and mechanical properties. *Composites Science and Technology*, 2002. **62**(15): p. 2043–2052.
7. Gupta, N., B.S. Brar, and E. Woldesenbet, Effect of filler addition on the compressive and impact properties of glass fibre reinforced epoxy. *Bulletin of Materials Science*, 2001. **24**(2): p. 219–223.
8. Yang, B., et al., Bending, compression, and shear behavior of woven glass fiber–epoxy composites. *Composites Part B: Engineering*, 2000. **31**(8): p. 715–721.
9. Hameed, N., et al., Morphology, dynamic mechanical and thermal studies on poly (styrene-co-acrylonitrile) modified epoxy resin/glass fibre composites. *Composites—Part A: Applied Science and Manufacturing*, 2007. **38**(12): p. 2422–2432.
10. Patnaik, A., A. Satapathy, and S. Biswas, Investigations on three-body abrasive wear and mechanical properties of particulate filled glass epoxy composites. *Malaysian Polymer Journal*, 2010. **5**(2): p. 37–48.
11. LeBlanc, J., et al., Shock loading of three-dimensional woven composite materials. *Composite Structures*, 2007. **79**(3): p. 344–355.
12. Chauhan, S., A. Kumar, and I. Singh, Study on friction and sliding wear behavior of woven *S*-glass fiber reinforced vinylester composites manufactured with different comonomers. *Journal of Materials Science*, 2009. **44**(23): p. 6338–6347.
13. Aramide, F., P. Atanda, and O. Olorunniwo, Mechanical properties of a polyester fibre glass composite. *International Journal of Composite Materials*, 2012. **2**(6): p. 147–151.
14. Erden, S., et al., Enhancement of the mechanical properties of glass/polyester composites *via* matrix modification glass/polyester composite siloxane matrix modification. *Fibers and Polymers*, 2010. **11**(5): p. 732–737.
15. Ramzan, E. and E. Ehsan, Effect of various forms of glass fiber reinforcements on tensile properties of polyester matrix composite. Facilities Engineering Technology, 2009. **16**: p. 33–39.
16. Al-alkawi, J.H., S.D. Al-Fattal, and H.A.-J. Ali, Fatigue behavior of woven glass fiber reinforced polyester under variable temperature, *Elixir. Mechanics in Engineering*, 2012. **53**.

17. Yuanjian, T. and D. Isaac, Combined impact and fatigue of glass fiber reinforced composites. *Composites Part B: Engineering*, 2008. **39**(3): p. 505–512.
18. Faizal, M.A., Y.K. Beng, and M.N. Dalimin, Tensile property of hand lay-up plain-weave woven e glass/polyester composite: Curing pressure and ply arrangement effect. *Borneo Science*, 2006. **19**: p. 27–34.
19. Leonard, L., et al., Fracture behaviour of glass fibre-reinforced polyester composite. *Proceedings of the Institution of Mechanical Engineers, Part L: Journal of Materials: Design and Applications*, 2009. **223**(2): p. 83–89.
20. Visco, A.M., L. Calabrese, and P. Cianciafara, Modification of polyester resin based composites induced by seawater absorption. *Composites—Part A: Applied Science and Manufacturing*, 2008. **39**(5): p. 805–814.
21. Aktaş, M., et al., An experimental investigation of the impact response of composite laminates. *Composite Structures*, 2009. **87**(4): p. 307–313.
22. Karakuzu, R., E. Erbil, and M. Aktas, Impact characterization of glass/epoxy composite plates: An experimental and numerical study. *Composites Part B: Engineering*, 2010. **41**(5): p. 388–395.
23. Aktaş, A., et al., Investigation of knitting architecture on the impact behavior of glass/epoxy composites. *Composites Part B: Engineering*, 2013. **46**: p. 81–90.
24. Baucom, J.A. and M. Zikry, Low-velocity impact damage progression in woven *E*-glass composite systems. *Composites—Part A: Applied Science and Manufacturing*, 2005. **36**(5): p. 658–664.
25. Karakuzu, R., T. Gülem, and B.M. İçten, Failure analysis of woven laminated glass–vinylester composites with pin-loaded hole. *Composite Structures*, 2006. **72**(1): p. 27–32.
26. Wang, K., et al., Facilitating transcrystallization of polypropylene/glass fiber composites by imposed shear during injection molding. *Polymer*, 2006. **47**(25): p. 8374–8379.
27. Bhagat, V.K., S. Biswas, and J. Dehury, Physical, mechanical, and water absorption behavior of coir/glass fiber reinforced epoxy based hybrid composites. *Polymer Composites*, 2014. **35**(5): p. 925–930.
28. Romanzini, D., et al., Influence of fiber content on the mechanical and dynamic mechanical properties of glass/ramie polymer composites. *Materials and Design*, 2013. **47**: p. 9–15.
29. Ramesh, M., K. Palanikumar, and K.H. Reddy, Mechanical property evaluation of sisal–jute–glass fiber reinforced polyester composites. *Composites Part B: Engineering*, 2013. **48**: p. 1–9.
30. Mishra, S., et al., Studies on mechanical performance of biofibre/glass reinforced polyester hybrid composites. *Composites Science and Technology*, 2003. **63**(10): p. 1377–1385.
31. Fu, S.-Y., et al., Tensile properties of short-glass-fiber-and short-carbon-fiber-reinforced polypropylene composites. *Composites—Part A: Applied Science and Manufacturing*, 2000. **31**(10): p. 1117–1125.
32. Husić, S., I. Javni, and Z.S. Petrović, Thermal and mechanical properties of glass reinforced soy-based polyurethane composites. *Composites Science and Technology*, 2005. **65**(1): p. 19–25.
33. Jarukumjorn, K. and N. Suppakarn, Effect of glass fiber hybridization on properties of sisal fiber–polypropylene composites. *Composites Part B: Engineering*, 2009. **40**(7): p. 623–627.
34. Samal, S.K., S. Mohanty, and S.K. Nayak, Polypropylene—Bamboo/glass fiber hybrid composites: Fabrication and analysis of mechanical, morphological, thermal, and dynamic mechanical behavior. *Journal of Reinforced Plastics and Composites*, 2009. **28**(22): p. 2729–2747.

35. Jiang, H., et al., Mechanical properties of poly (vinyl chloride)/wood flour/glass fiber hybrid composites. *Journal of Vinyl and Additive Technology*, 2003. **9**(3): p. 138–145.
36. Bikiaris, D., et al., Use of silanes and copolymers as adhesion promoters in glass fiber/polyethylene composites. *Journal of Applied Polymer Science*, 2001. **80**(14): p. 2877–2888.
37. Hufenbach, W.A., et al., Tribo-mechanical properties of glass fibre reinforced polypropylene composites. *Tribology International*, 2012. **49**: p. 8–16.
38. Hufenbach, W., et al., Polypropylene/glass fibre 3*D*-textile reinforced composites for automotive applications. *Materials and Design*, 2011. **32**(3): p. 1468–1476.
39. Thomason, J.L. and G. Schoolenberg, An investigation of glass fibre/polypropylene interface strength and its effect on composite properties. *Composites*, 1994. **25**(3): p. 197–203.

11 Inorganic Nanofillers-Based Thermoplastic and Thermosetting Composites

M. Chandrasekar, T. Senthil Muthu Kumar, K. Senthilkumar, N. Mohd Nurazzi, M.R. Sanjay, N. Rajini, and Suchart Siengchin

CONTENTS

11.1 INTRODUCTION

Polymers are usually reinforced with fillers to enhance their functional properties as well as to reduce costs [1]. Fillers are materials typically in the form of particulates used as reinforcements in polymer matrix systems to enhance certain properties of the resultant composites for a particular application. These fillers are embedded with the polymer matrix to improve the functionalities of the composites in terms of mechanical, thermal, optical, dielectric, magnetic, and antibacterial behavior, etc. [2]. The conventional inorganic fillers are in the range of several microns and not in nano scale. Inorganic nanofillers are gaining more attention in recent years in the formulation of nanocomposites [3, 4].

Some of the inorganic nanofillers used in formulation of nanocomposites include metal and metal alloys, minerals, oxides, and carbon-based materials [5]. The properties of the ensuing composites infused with inorganic nanofillers depend on several factors such as the filler size and shape, fraction of filler, and the loading ratio [6, 7]. Apart from the size and shape, their interaction with the polymer phase could have significant effect on the mechanisms and kinetics of crystallization, nucleation of crystallization, and other physicochemical phenomena of the resultant nanocomposite which is closely related to the mechanical properties [8]. It is very difficult to characterize polymer-filler interaction, which comprises the interfacial forces between the filler and the polymer, the orientation of the polymer in the instant vicinity of the filler surface [9]. Functionalization of polymers with nanofillers infused can be successful only if the particle agglomerations are avoided or prevented. For this, they must be combined in a way that results in the isolation of the nanoparticles in the polymer matrix. Furthermore, the aim is also to modify the materials which can potentially increase the aspect ratio of the filler particles and also to improve their compatibility and interfacial bonding with the chemically different polymer and the filler [10].

Synthesis of inorganic fillers can be classified into two techniques; viz. bottom-up and top-down methods [4]. Generally, the bottom-up technique uses liquid-phase chemical reactions to obtain the nanofillers where the chemical compositions and reaction conditions are precisely controlled, and this makes the bottom-up technique a versatile nanofiller production technology. It includes sol–gel reaction, hydrothermal reaction, and intercalation chemistry [11–13]. Synthesis of filler particles was found to be difficult with traditional methods such as flame pyrolysis and precipitation. The main disadvantage in these techniques is that the aggregates formed in the synthesis cannot be broken further. Furthermore, these fillers are mostly hydrophilic and hence, their surface must be functionalized to prevent agglomeration when integrated with a hydrophobic matrix [4]. In order to overcome this problem, some researchers have successfully attempted the in situ generation and phase transfer techniques for the synthesis of inorganic nanoparticles in the polymer matrix systems [14–16]. Even though the infusion of nanofillers into the polymer matrix enhances certain properties, there are challenges in achieving uniform dispersion of nanoparticles within the matrix as the agglomeration can lead to reduction in properties [17].

The most common end-use markets of inorganic filler-infused composites are construction, automotive, consumer products, furniture, industrial machinery, electronics, and packaging. The nanofillers featuring antibacterial properties, such as nanoscale titanium- and silver-based particles, could be applicable in filtration, hygiene, and hospital disposables applications or in consumer products [18]. They also find applications in packaging of microelectronics, bone tissue engineering, dental filling, and other biomedical applications [19–21]. With these inputs, this chapter focuses on the influence of various inorganic nanofillers on the properties of composites based on thermoplastics and thermosetting plastics.

11.2 TYPES OF INORGANIC NANOFILLERS

Fillers are not only used to reduce the cost of the composites but also to enhance certain functional properties which may not be achieved by using the polymer alone

TABLE 11.1
Commonly Used Inorganic Fillers in the Polymeric Composites

Chemical family	Examples
Oxides	Glass, magnesium oxide, silicon dioxide, antimony trioxide Aluminum dioxide
Hydroxides	Aluminum hydroxide, magnesium hydroxide
Salts	Calcium carbonate, barium sulfate, calcium sulfate phosphates
Silicates	Talc, mica, montmorillonite, nanoclay, kaolin
Metals	Boron, steel

or any other reinforcements. Fillers may be classified into two types based on their origin, i.e., organic fillers and inorganic fillers. Recently, the use of inorganic fillers particularly in nano-scaled size is getting more attention due to their superior properties. Numerous inorganic nanofillers have been investigated to date [22] and some of the most commonly used inorganic fillers are presented in Table 11.1.

In general, these inorganic nanofillers exhibit enhanced mechanical and thermal properties when compared with the polymer matrix. Hence, most of the inorganic nanofillers have a potential to enhance the mechanical and thermal properties of the resultant nanocomposites. In addition to this, based on the type of filler used, various other functionalities such as optical, dielectric, electrical, and magnetic properties can be achieved [13].

11.2.1 Gold Nanoparticles (GNPs)

Gold nanoparticles are one of the most studied inorganic nanoparticles due to their stability even at nano scale. They can also be easily functionalized with organic materials. The significant properties GNPs are that they are basically inert and non-toxic, sizes range from 1 nm to 150 nm, and they can be easily synthesized [23]. GNPs with varying core sizes are prepared by the reduction of gold salts in the presence of suitable stabilizing agents that prevent particle aggregation. GNPs have been extensively used in bio-nanotechnology based on their inimitable characters and numerous surface functionalities [24]. The major applications of GNPs include electronics (used as conductors in electronic chips), sensors (to identify suitability of food for consumption), probes (biological imaging applications), diagnostics (detect biomarkers), and catalysis [25].

11.2.2 Quantum Dots (QDs)

Quantum dots are zero-dimensional materials with three-dimensional confined carriers. Crystals in nano scale size are often termed as quantum dots. Size of QD typically ranges from 2 nm to 20 nm. However, their diameter should be strictly below 10 nm [26]. The QDs can be synthesized by traditional methods such as chemical vapor deposition (CVD), sol–gel, molecular beam epitaxy (MBE), and liquid phase epitaxy (LPE), etc. The main advantage in using quantum dots is that because of their controlled size, it is possible to have very precise control over the conductive properties of the material [27]. The superior transport and optical properties of the QDs can be attributed to the sharper density of state.

11.2.3 Silver Nanoparticles (AgNPs)

Silver nanoparticles are among the most explored inorganic nanofillers, due to their large surface area, thermal stability, high conductivity, better catalytic performance, and a broad range of antimicrobial activity [14, 28]. Apart from these properties, silver nanoparticles are known to be inhibitory against different fungi and viruses [14, 29]. They are also found in the form of oxides or other ionic silver compounds with similar antimicrobial activity. Silver nanoparticles are also evolving as potential agents for cancer therapy. AgNPs are synthesized by different methods like in situ generation and hydrothermal method. They are used in different applications such as in biosensor materials, wound healing, and as packaging materials for enhanced food shelf life [15, 30]. Infusion of AgNPs not only improves the mechanical properties and barrier properties but also the enzyme mobilization of the material.

11.2.4 Iron Oxide Nanoparticles (INPs)

The use of iron in a polymer matrix as oxygen scavengers is recognized in the packaging industry [31]. The growth of nano-sized scavenger particles based on iron is therefore of increasing interest since it can be integrated in very thin layers such as multilayered systems and in much lower amounts than in standard applications like sachets [32].

11.2.5 Carbon Nano Tubes (CNTs)

Carbon nanotubes (CNTs) are allotropes of carbon and are derived from rolled graphene planes with a length-to-diameter ratio greater than 1,000,000 [33]. Carbon nanotubes are tiny tubes with diameters of a few nanometers and lengths of several microns made of carbon atoms [34]. The sp2 of carbon-carbon bond in the basal plane of graphite is among the strongest of all chemical bonds, making graphite very useful as a structural material. CNTs have attracted significant consideration due to their inherent mechanical, electrical, and biocidal properties. As the size of CNT reaches the nanometer level (1 to 100 nm), the interactions at the interfaces become considerably large with respect to the size of the inclusion, and thus show significant change in the final properties [35]. In CNT, the nanotubes composed of a single layer of graphene area are called single-wall carbon nanotubes (SNWCNT), and feature a diameter in the order of 1 nm and length in the order of 100 nm to several μm. Multi-wall carbon nanotubes (MWCNT) have larger diameters of 10 to hundreds of nm. Intermediate between these two are the so-called few-walled carbon nanotubes (FWCNT) that have only a few layers, such as double-wall carbon nanotubes (DWCNT).

The most outstanding properties of CNT are their low density, small size, high aspect ratio (1000–10000) [36], high rigidity with Young modulus as high as 1 TPa

[37], high tensile strength from 10 to 200 GPa [38], excellent electrical conductivity (>105 S/m for MWCNT) [39], and thermal conductivity (>3000 W/mk for MWCNT) [40], which has resulted in CNT being extensively used in the development of nanocomposites. The combination of the highlighted performance makes the CNT perfect candidates as ideal reinforcing fillers in high-strength, lightweight polymer nanocomposites that are high performance and multifunctional [38, 41, 42]. Furthermore, due to the unique and extraordinary performance of CNT, many structural and smart applications of CNT have been proposed, including quantum wires, tiny electronic devices, hetero-junction devices, and electron emitters as well as lighter, smaller, and higher performance structures for aerospace and other structural applications [38, 43]. CNTs can also be conjugated with various biological molecules including drugs, proteins, and nucleic acid to afford bio-functionalities [44]. They possess numerous exciting properties, such as high aspect ratio, ultra-light weight, high strength, high thermal conductivity, and outstanding electronic properties ranging from metallic to semiconducting [45].

However, the use of CNTs in nanocomposites to date has been limited by challenges in processing, dispersion, and health considerations, besides their high cost [46]. The amazing properties of CNT can only be realized in composite materials so long as the CNTs are uniformly dispersed in the polymer matrix [47]. CNTs usually tend to get entangled or aggregate toward each other because of the strong van der Walls force, estimated at 500 eV/μm and chemical inertness caused by their unique sp2 bonding in the graphene layers [48, 49]. Besides, due to the large surface area of the nanoparticles and mostly high aspect ratio, they tend to agglomerate greatly, which reduces the capacity for achieving the expected properties [50]. Besides the challenge in obtaining homogenous dispersion of CNTs in the polymer matrix, the interfacial interaction between the polymer matrix and CNT structure also affects the efficient load transfer in the matrix. Without chemical bonding, load transfer between the CNTs and the matrix mainly comes from electrostatic and van der Walls interactions [51].

11.2.5.1 Methods in Optimizing CNT as Reinforcement

There are several methods conducted previously by the researchers to fully exploit the potential of CNT and to avoid the agglomeration, such as chemical functionalization [52], ultrasonication [53], melt-mixing [54], solution casting [55], and most recently, by hybridization [47, 56]. Chemical functionalization of the CNT is a potential method to improve dispersion of the CNT but covalent functionalization of CNT can distort the CNT structure [57]. According research [58], ultrasonication is one of the most used processing methods for dispersing carbon nanotubes in polymer matrix; however, it is only effective in low viscosity matrix, e.g., matrix for in situ polymerization or diluted with solvent [59, 60], where the low viscosity of matrix aids CNT dispersion during ultrasonication. Using the solution casting method, high-energy sonication of the CNT suspensions over prolonged periods of time or the use of ultrasonic head is usually necessary to produce uniformly dispersed CNT suspensions before they are subsequently mixed with polymer solutions. Furthermore, researchers [61] have demonstrated that the presence of non-ionic surfactants during processing (containing 1 wt.% of MWCNT) improves the dispersion and strengthens the interactions between the nanotubes and the matrix. Besides that, researchers [62] reported that in order to prevent agglomeration during the fabrication of CNT/Polyvinyl alcohol (PVA) by solution mixing, the CNTs were chemically treated to produce an electrostatically stabilized dispersion which was then mixed with aqueous PVA solutions. In their study, by Kumari et al. and Ahmad et al. [63, 64] reported that the CNT used along with the inorganic fillers significantly improved the mechanical properties compared to the single filler system. There are a few conventional methods for the hybridization of CNT such as milling and hot press sintering. However, both techniques may cause damage to the CNT structure due to the long milling duration and involvement of high compression; also, the CNT does not attach physically to the hybrid-filler surface, thus resulting in a non-uniform dispersion and it leads to the agglomeration of CNT during the fabrication of composites [65].

11.2.6 Silica Nanoparticles

Silica nanoparticles are considered to be promising agents for biological applications due to their exceptional biocompatibility, low toxicity, good thermal stability, simple synthetic route, and large-scale synthetic availability. They are classified as mesoporous and nonporous, both of which bear amorphous silica structure [66]. The particle size, crystallinity, porosity, and shape can be exactly manipulated, aiding the capacity for use of silica nanoparticles for various applications. Silica nanoparticles can be synthesized by different approaches such as reverse micro emulsion, sol–gel method, spray-drying, thermal vaporization, hydrolysis, hard templating, and soft templating, etc. [67]. Moreover, several surface modifications of silica nanoparticles allow for control of surface chemistry to achieve drug loading and good dispensability. These properties make silica nanoparticles have potential in biomedical imaging, detecting, therapeutic delivery, monitoring, and ablative therapies. They also have wide-spread application in other areas such as energy source, electronic, sensor, and catalysis purposes [68]. Multifunctional silica nanoparticles (NPs) with enhanced optical, chemical, and magnetic properties, all combined in one single nanostructure can also be developed [69]. Silica nanoparticles also have potential to improve the thermal, mechanical, and/or barrier properties of various polymer matrices [70].

11.3 INFLUENCE OF METALLIC NANOPARTICLES ON THE THERMOSET AND THERMOPLASTIC-BASED COMPOSITES

In this section, thermal, mechanical, physical, and chemical properties of the composites filled with the inorganic fillers (such as gold nanoparticles, silver nanoparticles, aluminum (Al), etc.) mixed with the thermoset and thermoplastic-based composites are discussed. These composites with inorganic fillers have the potential to be used in fields such as those of (i) biomedical, (ii) drug resistance, (iii) biological, and (iv) chemical engineering. Table 11.2 summarizes the changes found in thermal, mechanical, physical, chemical, etc. properties due to the addition of inorganic fillers in the thermoset and thermoplastic matrix from the literature.

TABLE 11.2
Effect of Inorganic Fillers on Various Properties of the Thermoset and Thermoplastic-Based Composites

Filler	Details of characterization	Details of composite	Findings	Reference
Gold nanoparticle	TGA and DSC (thermal analysis)	a) Gold nanoparticle/epoxy; The content of gold nanoparticles in epoxy varied in the range of 0.025wt%, 0.05wt%, and 0.1wt%.	i. Thermal stability decreased in the order: 1. pure epoxy > 0.025wt% > 0.05wt% > 0.1wt%. ii. In DSC analysis, the glass transition temperature (Tg) increased in the order: pure epoxy < 0.025wt% < 0.05wt% < 0.1wt%.	[71]
Gold nanoparticle	Dielectric analysis	Gold nanoparticle/epoxy;	i. Permittivity and loss factor increased with increasing amount of the nanoparticle. ii. When increasing the content of nanoparticle for twice the times in epoxy, the permittivity improved by 140%. Likewise, the ionic conductivity also increased with increasing the amount of nanoparticle.	[72]
Silver nanoparticle	Tensile, TGA, DSC, and contact angle	a) Cellulose/banana peel powder (BPP)/silver nanoparticle; b) Cellulose solution was prepared from the cotton linter pulp. The BPP was varied from 5wt–25wt% (order of 5wt%).	i. Tensile strength and tensile modulus of the hybrid composite increased with increasing the BPP in cellulose/silver nanoparticle. ii. Thermal stability increased in the order: cellulose < hybrid composite < BPP. iii. Tg of hybrid composites decreased with increasing the amount of BPP. iii. The contact angle was increased until the addition of 20wt% of BPP in cellulose/silver nanoparticle hybrid composites. The highest contact angle of 90° was observed at cellulose/BPP (20wt%)/silver nanoparticle-based hybrid composites.	[14]

(Continued)

TABLE 11.2 (CONTINUED)
Effect of Inorganic Fillers on Various Properties of the Thermoset and Thermoplastic-Based Composites

Filler	Details of characterization	Details of composite	Findings	Reference
	TGA and tensile property	a) Poly(propylene)carbonate (PPC)/tamarind seed polysaccharide (TSP)/silver nanoparticles hybrid composites; silver nanoparticle was varied from 1 mM to 5 mM (order of 1 mM).	i. Thermal stability was increased with increasing the amount of TSP in PPC/silver nanoparticle composites. ii. Tensile strength and tensile modulus were improved with the addition of TSP and silver nanoparticle in PPC matrix.	[15]
Aluminum, copper, and iron	Tensile property	a) Aluminum/high-density polyethylene (HDPE) b) Copper/HDPE c) Iron/HDPE d) The filler was ranged between 0–55 vol% in HDPE.	Maximum tensile strength was observed at optimum filler concentration as shown below: i. Aluminum/HDPE—20 vol% ii. Copper/HDPE—10 vol% iii. Iron/HDPE—10 vol%	[73]
micro-sized rubber particles and silica nanoparticle (20 nm diameter)	Thermal study	a) Glass fiber/epoxy b) Glass fiber/rubber/epoxy c) Glass fiber/silica nanoparticle/epoxy d) Glass fiber/rubber/silica nanoparticle/epoxy Curing temperature: 120°C and 180°C, respectively.	i. Tg of 180°C-cured composites exhibited higher values than the 120°C-cured composites. ii. The highest Tg (155°C) was observed for glass fiber/silica nanoparticle/epoxy and glass fiber/rubber/silica nanoparticle/epoxy composites, cured at 180°C.	[74]
Aluminum hydroxide (Al (OH_3))	Mechanical property	a) Al (OH_3)/polypropylene (PP)/polypropylene grafted with acrylic acid (FPP). The content of Al (OH_3) was varied between 0–60 wt%.	i. Mechanical properties were reduced after the addition of Al (OH_3) in the PP matrix. However, these properties were improved by the addition of FPP in Al (OH_3)/PP composites.	[75]

(Continued)

TABLE 11.2 (CONTINUED)
Effect of Inorganic Fillers on Various Properties of the Thermoset and Thermoplastic-Based Composites

Filler	Details of characterization	Details of composite	Findings	Reference
Carbon nanotube (CNT), a synthetic diamond (SND), boron nitride (BN), and copper (Cu)	Mechanical and thermal property	a) PP b) CNT/PP c) SND/PP d) BN/PP e) Cu/PP Fillers were varied from 0–4 vol%.	i. Tensile strengths were improved in the order: 1. CNT/PP < Cu/PP < BN/PP < SND/PP. ii. The % of weight loss decreased in the order: SND/PP > Cu/PP > BN/PP > CN/PP > PP.	[76]
Graphite	Mechanical and tribological properties	a) Graphite/nylon-6 Graphite was varied from 5–15wt%.	i. Tensile strength, elongation at break, impact strength, and specific wear rate improved with decreasing graphite content in nylon-6 composites.	[77]
Calcium carbonate ($CaCO_3$)	Mechanical property	a) $CaCO_3$/PP b) $CaCO_3$ was ranged between 0–70wt%. c) $CaCO_3$ was subjected to chemical treatment by titanate. d) The content of titanate was varied in the order of 0%, 0.2%, 0.3% and 0.4%.	i. The mechanical properties of untreated $CaCO_3$/PP composites decreased by varying the filler content. Conversely, the properties were found to be enhanced with the addition of treated fillers in the PP matrix.	[78]

Baris et al. [71] performed thermogravimetric analysis (TGA) and differential scanning calorimetry (DSC) for epoxy matrix with various percentages of gold nanoparticles. The existence of filler in the matrix during the thermal decomposition led to reduction in thermal stability of composites compared to the neat epoxy. However, glass transition temperature (Tg) of gold nanoparticle/epoxy composites were improved by increasing the concentration of nanoparticles. The improvement was credited to (i) reduction in chain flexibility and (ii) the free volume of epoxy matrix upon addition of nanoparticles.

In another study, Indira devi et al. [72] studied the thermal properties using TGA and tensile properties of polypropylene carbonate (PPC)/tamarind seed polysaccharide (TSP)/silver nanoparticle hybrid composites. The fabricated composites had lower thermal stability than the neat PPC at lower temperatures. Further increasing the temperature beyond 350°C, the thermal stability of PPC/TSP/silver nanoparticle hybrid composites was found to be improved. It was attributed to the (i) existence of polyphenols in TSP and (ii) the incorporation of silver nanoparticles. Furthermore, the tensile properties were also improved (i) by a proper dispersion of fillers in situ generated stiff silver nanoparticle in hybrid composites and (ii) the effective stress transfer.

Nurazreena et al. [73] used three types of fillers such as aluminum, copper, and iron in high-density polyethylene (HDPE) matrix. The fillers had different shapes such as flaky, irregular, dendritic, and spherical. The results depicted that the tensile properties were influenced by (i) the shape of the filler (ii) the degree of crystallinity, and (iii) the effect of bonding with the inorganic fillers/HDPE. Further, Young's modulus of these composites enhanced with increasing the content of aluminum, copper, and iron in the HDPE matrix.

Micro and nano-sized fillers were mixed with the epoxy/glass fiber composites [74]. According to the results obtained, the silica nanoparticle was not influenced by the Tg of composites since the movement of the polymer was not affected. Moreover, the hybrid composites such as rubber/glass fiber and rubber/silica nanoparticle/glass fiber epoxy composites did not show any significant improvement or changes in the Tg values.

Nurul and Mariatti [76] analyzed the mechanical and thermal properties of four different types of fillers (CNT, SND, BN, Cu) in the PP matrix. They reported superior mechanical and thermal properties for the CNT/PP-based composites. The maximum tensile strength was observed until 2vol% of filler while Young's modulus of these composites was found to be increased with filler loading. Likewise, the thermal stability was also improved by increasing the filler loading.

Huseyin et al. [77] reported that the addition of graphite to nylon-6 between 5–10 wt% produced acceptable improvements in mechanical and tribological properties due to the finer crystal structure. However, addition of graphite reduced the bonding force between the filler and matrix resulting in a decrease in the mechanical properties. At 5wt% graphite, tribological properties of the composites improved by nearly 100%. It was attributed to the existence of transfer film which had a significant role in influencing the wear resistance properties.

Doufnoune et al. [78] compared the mechanical properties of treated (titanate) and untreated $CaCO_3$/PP composites. As expected, the untreated filler-based composites showed lesser improvements in mechanical properties than the treated filler-based composites. This was attributed to (i) the weak bonding between the $CaCO_3$ and PP and (ii) the poor distribution of $CaCO_3$ in the PP matrix. Nevertheless, the

TABLE 11.3
Flammability, Viscosity, and Other Characteristics of the Various Inorganic Fillers in Thermoset- and Thermoplastic-Based Matrices

Filler/reinforcement/polymer	Details of characterization	Reference
Magnesium hydroxide (MH)/ Acrylonitrile-butadiene styrene	Flammability test conducted by two methods such as (i) UL 94 and (ii) cone calorimeter	[79]
MH/PP	Flame retardancy	[80]
Carbon nanotube/PP	Flame retardant	[81]
Graphene/epoxy	Flame retardant	[82]
$CaCO_3$/PP	Viscosity	[83]
Lead powder/polyurethane	Viscosity	[84]
(i) Glass fiber/polytetrafluoroethylene (PTFE)/polyoxymethylene (ii) Nylon 66/glass fiber (iii) Glass fiber/polycarbonate	Tensile strength retaining after exposure to γ-radiation at 3.5 MRad.	[85]
Talc, zinc borate, aluminum hydroxide/ polyurethane	Physical property	[86]

treated $CaCO_3$ was found to reduce the wetting angle of the matrix, resulting in a noticeable improvement in mechanical properties.

As discussed above, various fillers have been incorporated with the thermoset- and thermoplastic-based polymers. Their concentration was found to govern their mechanical, thermal, chemical, and electrical properties. Besides these characterizations, the filler-infused composites were also subjected to other characterization techniques such as flammability (fire resistance test), rheology (i.e., viscosity determination for materials), exposure to different environmental and physical properties. Table 11.3 displays some of the research works on the above-mentioned aspects with the thermoset and/or thermoplastic matrix.

Apart from using the thermoset- and thermoplastic-based polymers, researchers have also utilized biodegradable polymers to avoid dependency on the synthetic polymers. For example, Senthil et al. [4] studied the mechanical and thermal properties by varying the content of banana peel powder (BPP) in cellulose solution, whereas silver nanoparticles were also infused within the composite. They reported that the silver nanoparticle and BPP acted effectively as reinforcements in the hybrid composite films. The hybrid composite exhibited higher thermal stability than the BPP/cellulose composites. This was attributed to (i) the presence of silver nanoparticles and (ii) the existence of polyphenols in BPP.

11.4 INFLUENCE OF CNT ON THE THERMOSET- AND THERMOPLASTIC-BASED COMPOSITES

Study done by Zakaria et al. [47] on the CNT–Al_2O_3 hybrid epoxy composites showed higher tensile and thermal properties using chemical vapor deposition than the CNT–Al_2O_3 physically mixed epoxy composites. The increment was associated with the homogenous dispersion of CNT–Al_2O_3 particle filler as observed under a field emission scanning electron microscope (SEM). It was demonstrated that the CNT–Al_2O_3 hybrid epoxy composites are capable of increasing the tensile strength by up to 30%, giving a tensile modulus of 39%, thermal conductivity of 20%, and a glass transition temperature value of 25%, when compared to a neat epoxy composite. MWCNT/polymethylmethacrylate (PMMA) composites have been fabricated by melt blending. The dynamic mechanical behavior of the composites shows that the storage modulus of the composites is significantly increased by the incorporation of nanotubes. particularly at high temperatures and the tan δ shows some broadening and moves to a slightly higher temperature [87].

A finite element analysis of carbon nanotube-reinforced, epoxy-based composite subjected to thermo-mechanical loading was studied [88]. A coupled field analysis of a hexagonal representative volume element (RVE) with long and short CNT embedded in an epoxy polymer matrix was carried out to analyze the effect of CNT reinforcement on the thermo-mechanical properties of the epoxy polymer. The analysis showed that long CNTs proved to be better reinforcement than the short CNTs for the nanocomposites which are subjected to thermo-mechanical loading due to a greater volume fraction of 6.2%, compared to 2.6% in the case of short CNT in the matrix.

MWCNTs were modified using acid functionalization (H_2-SO_4:HNO_3 = 1:3 by volume) and then the mechanical properties of reinforced epoxy polysulfide resin by the both pure and treated MWNTs have been characterized by Hadavand et al. [89]. Different weight percentages of pure and treated MWCNT (0.1–0.3 wt%) were dispersed in the epoxy polysulfide resin separately and then mixed with curing agent. Results indicated a significant difference between the acid-treated and untreated MWCNTs in the mechanical properties of epoxy polysulfide nanocomposites. Nanocomposite with 0.1–0.3% acid-treated MWCNTs exhibited increase in Young's modulus from 458 to 723 MPa, tensile strength from 5.29 to 8.83 Mpa, and fracture strain from 0.16 to 0.25%. The field emission electron microscopy (FESEM) results showed better dispersion of modified carbon nanotube than the unmodified filler in the polymeric matrix.

Lee et al. [90] studied the effect of CNT modification on the tensile and thermal properties of CNT/basalt/epoxy composites. CNT/basalt/epoxy composites were fabricated by impregnating woven basalt fibers into epoxy resin mixed with CNTs. Three groups of composites were fabricated, consisting of unmodified, acid-modified (oxidized), and silane-modified (silanized) CNTs, each at 1 wt.%. The silanized CNT/basalt/epoxy composites showed better tensile properties when compared to the unmodified and oxidized CNT/basalt/epoxy composites. In particular, the tensile strength and Young's modulus of silanized composites were 34 and 60% greater, respectively, than that of unmodified composites. In addition, the silanized composites showed better thermal stability with a higher storage modulus and Tg than the unmodified and oxidized composites. Synthetically, the enhancement of mechanical and thermal properties of silanized CNT/basalt/epoxy composites is attributed to the obstruction of molecular mobility and the formation of a strong cross-linking network, both made possible by good dispersion and strong interfacial interaction between the silane functionalized CNTs and the epoxy in the basalt fabric/epoxy composites.

The effect of CNT additives on the elastic behavior of the composite was performed [38]. The materials consist of three phases, namely carbon fibers fabric, epoxy matrix, and carbon nanotubes. Different volume fractions of CNTs were used (0% as reference, 0.5%, 1%, 2%, and 4%). A set of mechanical tests such as open-hole tension, shear beam test, and flatwise tension tests were performed. The experimental results show an increase in the mechanical performance of the composite with up to 2% of CNT additives. However, beyond this value, the material strength shows a significant decline. The critical volume fraction threshold was estimated to be between 0.5 and 2% of CNTs-reinforced textile composites. The decrease and degradation of the mechanical behavior at 4% can be explained by the effect of CNTs distributions and the existence of an upper limit which starts from 2%. At high CNT concentration, the viscosity of nanocomposites increases drastically and during the composite fabrication process, air bubble tends to get trapped leading to the formation of porosity which has a major effect on mechanical properties reduction. In addition, according to Zhuang et al. [91], CNT concentration from 0.5 wt% produced agglomerations, and according to the authors, the CNT content should be smaller to ensure the stable dispersion of the tubes. 0.1 wt% of CNTs was found to be the critical tube content giving homogenous distribution.

Single-wall nanotubes (SWNTs) which were functionalized by sonicating with nitric and sulfuric acids were used to fabricate a SWNT/epoxy composite [92]. Composites with functionalized (f-SWCNTs) showed a significant improvement in the Young's modulus and tensile strength when compared to the neat epoxy and non-functionalized SWCNT composites. Functionalization enhances the dispersion and interfacial bonding between CNTs and polymer matrix. It is suggested that functionalization of CNT can play an important role in the further development of CNT composites.

Researchers [93] characterized the comparative effect of tube and sheet-like nanocarbon on the structure–property relationships of fiber-reinforced composites. Graphene nanosheets and MWCNTs were dispersed into commercial glass fiber fabric (Gf) to obtain composite through the vacuum-assisted resin infusion process. The experimental results indicated that oxidized MWCNTs or graphene could ensure excellent dispersion on the fiber surface and ultimately enhance the mechanical properties and thermal stability of the resultant composites. Graphene oxide (GO) with a wrinkled and roughened texture was shown to be superior to MWCNTs in terms of toughening the fiber/matrix interface and delaying the deformation or failure in the epoxy matrix. Under the same dosage of MWCNTs, the inter-laminar shear strength of GO-Gf-reinforced composites was raised by approximately 12%, and the relevant onset thermal-decomposition temperature was increased by 11°C compared to the carboxyl MWCNT-Gf-reinforced composites. Moreover, the GO-GfCs exhibited superior static and dynamic mechanical properties compared to those of Mc-GfCs.

Electrical, mechanical, thermal, and thermo-mechanical properties of epoxy/CNT and epoxy/CNT/calcium carbonate ($CaCO_3$) nanocomposites produced via in situ polymerization assisted by ultrasonication without solvent was characterized [58]. The epoxy/CNT composites showed very low percolation threshold at 0.05 wt % and nanocomposites with higher contents of CNT presented further increase in electrical conductivity. The addition of $CaCO_3$ in epoxy/CNT nanocomposites increased the electrical conductivity, due to volume exclusion phenomena. Due to the low content of the CNT and $CaCO_3$, no changes in Tg were observed. DMA results showed no significant changes in thermo-mechanical properties once the contents of CNT and $CaCO_3$ are below stiffness threshold. However, an increase of flexural modulus by adding CNT and $CaCO_3$ was observed.

The effects of reinforcement of MWCNTs with different volume fractions on thermal properties of epoxy resin-based composites were studied [94]. The thermal conductivity increased linearly with nanotube concentration to a maximum increase of 40% at 0.4 vol.% carbon nanotubes. The effect of filler loading on processing and thermo-physical properties of poly(thiourea-azo-ether)/SWCNT was investigated by the researcher [95]. A filler content from 1 to 5 wt.% increased the electrical conductivity from 2.3 to 4.8 Scm^{-1}. Values of tensile strength increased up to 5 wt.% MWCNT and excess of filler above this weight percent caused aggregation of the nanotube phase and therefore reduced compatibility between the acid-treated MWCNT and poly(thiourea-azo-ether). Ultimate tensile strength of functional hybrids of 37.39 to 41.23 MPa was improved relative to non-functional multi-walled CNT in matrix. Furthermore, the tensile modulus considerably increased from 9.9 to 13.3 GPa. Finally, poly (thiourea-azo-ether)/MWCNT hybrids showed improved thermal, electrical, and mechanical properties because the interfacial forces are the stronger physical bonds.

Investigation on MWCNT-reinforced epoxy composites by adding various concentrations (0%, 0.1%, 0.3%, and 0.5%) of CNTs showed that both the flow stress and fracture strain increased. Furthermore, presence of the MWCNT was found to nucleate crystallization in the epoxy. This crystal growth is thought to enhance the strength of composite. The fracture observations show that the flow stress increases relative to the inhibition of the slip band by MWNT [96].

The tensile properties of SWCNTs-reinforced polypropylene (PP) were studied [97]. Solvent processing was used to disperse SWNTs in a PP. The percentage of SMCNT used is 0.5 and 1 wt%. For 1 wt % loading of SWCNT, the fiber tensile strength increased 40% (from 9.0 to 13.1 g/denier). At the same time, the modulus increased by 55% (from 60 to 93 g/denier). In addition, at 1 wt% loading level, the fibers had tensile strengths that were intermediate between high-strength industrial PP and Kevlar fibers.

The influence of MWCNT on thermal stability, thermal conductivity, and cross-linking density on reinforced-epoxy matrix was determined [98]. Three kinds of MWCNT were used: non-modified with 1 and 1.5 μm length, and 1 μm length modified with amino groups. The obtained results proved that the addition of carbon nanotubes with amino groups leads to the increase of the Tg as well as cross-linking density as an effect of the reaction between amino and hydroxyl reactive groups. Although the weight fraction of amino groups on the MWCNT surface seems to be negligible in comparison to the weight fraction of curing agent, their influence is noticeable due to the high specific surface area of nanotubes. Addition of MWCNT led to the decrease in thermal stability of epoxy matrix as an effect of increase in the thermal conductivity. Thermal diffusivity of nanocomposites increased with the increase of carbon nanotubes' weight fraction and their length. This lowers the energy needed for the decomposition of nanocomposites.

Epoxy composite specimens reinforced with MWCNT filler were fabricated using shear mixer and ultra-sonication processor [99]. The results show that the specimens with 0.6 wt% nanotube content had better dispersion and higher strength than the other specimens. Young's modulus of the specimens increased as the contents of the nanotube filler in the matrix were increased from 2.75 GPa to 3.11 GPa. The specimen with 0.6 wt% nanotube filler content showed higher thermal conductivity than the other specimens. Specimens with 0.4 and 0.6 wt% filler contents showed a lower value of thermal expansion than those of the other specimens.

11.5 CONCLUSION

In this chapter, the reinforcing effect of various inorganic fillers on the thermal, physical, mechanical, and other properties of the thermoset- and thermoplastic-based composites were reviewed. The following are the observations made from the studies in the literature:

- Since the use of thermoset- and thermoplastic-based applications increased, the possibilities of using the fillers have also expanded.
- Addition of fillers in the composite material could improve the thermal properties, strength, and modulus of composites, reduce the possibility of mold shrinkage, and it can control the viscosity of the resin. However, the improvement depends on the filler concentration in the composite and maximum enhancement occurs at the optimum filler concentration.
- At higher filler concentrations, the filler-infused composites displayed inferior performance. The agglomeration of fillers at higher concentration was found to affect the properties.
- If the filler size is larger with a low level of interaction with the surrounding materials, the fillers may be called degrading fillers. Hence not all fillers may be useful to improving the intrinsic properties of materials.
- The filler properties can be modified and re-synthesized in the presence of other materials. This allows room for development of custom-made fillers according to the requirements of an application.

REFERENCES

1. L. Cabedo and J. Gamez-Pérez, "Inorganic-based nanostructures and their use in food packaging," In: *Nanomaterials for Food Packaging*, Elsevier, 2018, pp. 13–45.
2. S. Kobayashi and K. Müllen, *Encyclopedia of Polymeric Nanomaterials—With 2021 Figures and 146 Tables*, Springer, 2015.
3. M. Sanjay, S. Siengchin, C. I. Pruncu, M. Jawaid, T. S. M. Kumar, and N. Rajini, "Biomedical applications of polymer/layered double hydroxide bionanocomposites," In: *Nanostructured Polymer Composites for Biomedical Applications*, Elsevier, 2019, pp. 315–322.
4. H. Althues, J. Henle, and S. Kaskel, "Functional inorganic nanofillers for transparent polymers," *Chemical Society Reviews*, 36(9), pp. 1454–1465, 2007.
5. S. Li, M. Meng Lin, M. S. Toprak, D. K. Kim, and M. Muhammed, "Nanocomposites of polymer and inorganic nanoparticles for optical and magnetic applications," *Nanotechnology Reviews*, 1, p. 5214, 2010.

6. R. Hadal, A. Dasari, J. Rohrmann, and R. Misra, "Susceptibility to scratch surface damage of wollastonite-and talc-containing polypropylene micrometric composites," *Materials Science and Engineering: Part A*, 380(1–2), pp. 326–339, 2004.
7. H. Sepet, N. Tarakcioglu, and R. Misra, "Effect of inorganic nanofillers on the impact behavior and fracture probability of industrial high-density polyethylene nanocomposite," *Journal of Composite Materials*, 52(18), pp. 2431–2442, 2018.
8. N. Korivi, "Preparation, characterization, and applications of poly (ethylene terephthalate) nanocomposites," In: *Manufacturing of Nanocomposites with Engineering Plastics*, Elsevier, 2015, pp. 167–198.
9. K. Ziegel and A. Romanov, "Modulus reinforcement in elastomer composites. I. Inorganic fillers," *Journal of Applied Polymer Science*, 17(4), pp. 1119–1131, 1973.
10. N. G. McCrum, C. Buckley, C. B. Bucknall, and C. Bucknall, *Principles of Polymer Engineering*, Oxford University Press, Oxford, UK, 1997.
11. K. Byrappa and T. Adschiri, "Hydrothermal technology for nanotechnology," *Progress in Crystal Growth and Characterization of Materials*, 53(2), pp. 117–166, 2007.
12. C. J. Brinker and G. W. Scherer, *Sol-Gel Science: The Physics and Chemistry of Sol-Gel Processing*, Academic Press, 2013.
13. T. J. Pinnavaia and G. W. Beall, *Polymer-Clay Nanocomposites*, John Wiley, 2000.
14. S. M. K. Thiagamani, N. Rajini, S. Siengchin, A. V. Rajulu, N. Hariram, and N. Ayrilmis, "Influence of silver nanoparticles on the mechanical, thermal and antimicrobial properties of cellulose-based hybrid nanocomposites," *Composites Part B: Engineering*, 165, pp. 516–525, 2019.
15. D. MP, N. Nallamuthu, N. Rajini, S. Siengchin, V. Rajulu, and N. Hariram, "Antimicrobial properties of poly (propylene) carbonate/Ag nanoparticle-modified tamarind seed polysaccharide with composite films,"" *Ionics*, 25(7), pp. 3461–3471, 2019.
16. M. I. Devi, N. Nallamuthu, N. Rajini, T. S. M. Kumar, S. Siengchin, A. V. Rajulu, N. Ayrilmis, "Biodegradable poly (propylene) carbonate using in-situ generated CuNPs coated Tamarindus indica filler for biomedical applications," *Materials Today Communications*, 19, pp. 106–113, 2019.
17. R. M. Shahroze, M. R. Ishak, M. S. Salit, Z. Leman, M. Asim, and M. Chandrasekar "Effect of organo-modified nanoclay on the mechanical properties of sugar palm fiber-reinforced polyester composites," *BioResources*, 13(4), pp. 7430–7444, 2018.
18. M. Zanetti, G. Camino, P. Reichert, and R. Mülhaupt, "Thermal behaviour of poly (propylene) layered silicate nanocomposites," *Macromolecular Rapid Communications*, 22(3), pp. 176–180, 2001.
19. V. M. Rusu, C.-H. Ng, M. Wilke, B. Tiersch, P. Fratzl, and M. G. Peter, "Size-controlled hydroxyapatite nanoparticles as self-organized organic–inorganic composite materials," *Biomaterials*, 26(26), pp. 5414–5426, 2005.
20. G.-W. Lee, M. Park, J. Kim, J. I. Lee, and H. G. Yoon, "Enhanced thermal conductivity of polymer composites filled with hybrid filler," *Composites—Part A: Applied Science and Manufacturing*, 37(5), pp. 727–734, 2006.
21. S. Klapdohr and N. Moszner, "New inorganic components for dental filling composites," *Monatshefte für Chemie/Chemical Monthly*, 136(1), pp. 21–45, 2005.
22. S. Kango, S. Kalia, A. Celli, J. Njuguna, Y. Habibi, and R. Kumar, "Surface modification of inorganic nanoparticles for development of organic–inorganic nanocomposites—A review," *Progress in Polymer Science*, 38(8), pp. 1232–1261, 2013.
23. S. Manju and K. Sreenivasan, "Functionalised nanoparticles for targeted drug delivery," In: *Biointegration of Medical Implant Materials*, Elsevier, 2010, pp. 267–297.
24. Y.-C. Yeh, B. Creran, and V. M. Rotello, "Gold nanoparticles: Preparation, properties, and applications in bionanotechnology," *Nanoscale*, 4(6), pp. 1871–1880, 2012.

25. M. Ali, U. Hashim, S. Mustafa, Y. Man, and K. N. Islam, "Gold nanoparticle sensor for the visual detection of pork adulteration in meatball formulation," *Journal of Nanomaterials*, 2012, p. 1, 2012.
26. A. Ferancová and J. Labuda, "DNA biosensors based on nanostructured materials," In *Nanostructured Materials in Electrochemistry*, pp. 409–434, 2008.
27. L. Tong, F. Qiu, T. Zeng, J. Long, J. Yang, R. Wang, *et al.*, "Recent progress in the preparation and application of quantum dots/graphene composite materials," *RSC Advances*, 7(76), pp. 47999–48018, 2017.
28. M. A. Busolo, P. Fernandez, M. J. Ocio, and J. M. Lagaron, "Novel silver-based nanoclay as an antimicrobial in polylactic acid food packaging coatings," *Food Additives and Contaminants*, 27(11), pp. 1617–1626, 2010.
29. T. V. Duncan, "Applications of nanotechnology in food packaging and food safety: Barrier materials, antimicrobials and sensors," *Journal of Colloid and Interface Science*, 363(1), pp. 1–24, 2011.
30. D. Qu, W. Sun, Y. Chen, J. Zhou, and C. Liu, "Synthesis and in vitro antineoplastic evaluation of silver nanoparticles mediated by Agrimoniae herba extract," *International Journal of Nanomedicine*, 9, p. 1871, 2014.
31. L. Vermeiren, L. Heirlings, F. Devlieghere, and J. Debevere, "Oxygen, ethylene and other scavengers," In *Novel Food Packaging Techniques*, pp. 22–49, 2003.
32. Z. Foltynowicz, A. Bardenshtein, S. Sängerlaub, H. Antvorskov, and W. Kozak, "Nanoscale, zero valent iron particles for application as oxygen scavenger in food packaging," *Food Packaging and Shelf Life*, 11, pp. 74–83, 2017.
33. W. Krätschmer, L. D. Lamb, K. Fostiropoulos, and D. R. Huffman, "Solid C60: A new form of carbon," *Nature*, 347(6291), p. 354, 1990.
34. M. M. Shokrieh, A. Saeedi, and M. Chitsazzadeh, "Mechanical properties of multi-walled carbon nanotube/polyester nanocomposites," *Journal of Nanostructure in Chemistry*, 3(1), p. 20, 2013.
35. B. Kadhim, "Thermal Properties of Carbon Nano Tubes Reinforced Epoxy Resin Nano CompositesBahjat B. Kadhim1," *Al-Mustansiriyah Journal of Science*, 22, pp. 276–290, 2011.
36. C. Laurent Peigney, E. Flahaut, A. Rousset, A. Rousset, "Carbon nanotubes in novel ceramic matrix nanocomposites," *Ceramics International*, 26(6), pp. 677–683, 2000.
37. Y. Wang Demczyk, J. Cumings, M. Hetman, W. Han, A. Zettl, A. Zettl, R. O. Ritchie, "Direct mechanical measurement of the tensile strength and elastic modulus of multi-walled carbon nanotubes," *Materials Science and Engineering: Part A*, 334(1–2), pp. 173–178, 2002.
38. M. Tarfaoui, K. Lafdi, and A. El Moumen, "Mechanical properties of carbon nanotubes based polymer composites," *Composites Part B: Engineering*, 103, pp. 113–121, 2016.
39. Y. Ando, X. Zhao, H. Shimoyama, G. Sakai, and K. Kaneto, "Physical properties of multiwalled carbon nanotubes," *International Journal of Inorganic Materials*, 1(1), pp. 77–82, 1999.
40. M. Biercuk, M. Llaguno, M. Radosavljevic, J. Hyun, A. Johnson, and J. Fischer, "Carbon nanotube composites for thermal management," *Applied Physics Letters*, 80(15), pp. 2767–2769, 2002.
41. J. Roberts, T. Imholt, Z. Ye, C. Dyke, D. Price Jr, and J. Tour, "Electromagnetic wave properties of polymer blends of single wall carbon nanotubes using a resonant microwave cavity as a probe," *Journal of Applied Physics*, 95(8), pp. 4352–4356, 2004.
42. M. Irshidat, M. Al-Saleh, and M. Al-Shoubaki, "Using carbon nanotubes to improve strengthening efficiency of carbon fiber/epoxy composites confined RC columns," *Composite Structures*, 134, pp. 523–532, 2015.

43. T. Liu, I. Phang, L. Shen, S. Chow, and W. Zhang, "Morphology and mechanical properties of multiwalled carbon nanotubes reinforced nylon-6 composites," *Macromolecules*, 37(19), pp. 7214–7222, 2004.
44. Z. Liu, S. M. Tabakman, Z. Chen, and H. Dai, "Preparation of carbon nanotube bioconjugates for biomedical applications," *Nature Protocols*, 4(9), p. 1372, 2009.
45. J. E. N. Dolatabadi, A. A. Jamali, M. Hasanzadeh, and Y. Omidi, "Quercetin delivery into cancer cells with single walled carbon nanotubes," *International Journal of Bioscience, Biochemistry and Bioinformatics*, 1, p. 21, 2011.
46. H. Velichkova, I. Petrova, S. Kotsilkov, E. Ivanov, N. K. Vitanov, and R. Kotsilkova, "Influence of polymer swelling and dissolution into food simulants on the release of graphene nanoplates and carbon nanotubes from poly (lactic) acid and polypropylene composite films," *Journal of Applied Polymer Science*, 134(44), p. 45469, 2017.
47. M. Zakaria, H. Akil, M. Kudus, and S. Saleh, "Enhancement of tensile and thermal properties of epoxy nanocomposites through chemical hybridization of carbon nanotubes and alumina," *Composites—Part A: Applied Science and Manufacturing*, 66, pp. 109–116, 2014.
48. S. Bal and S. Samal, "Carbon nanotube reinforced polymer composites—A state of the art," *Bulletin of Materials Science*, 30(4), p. 379, 2007.
49. J. Du, J. Bai, and H. Cheng, "The present status and key problems of carbon nanotube based polymer composites," *Express Polymer Letters*, 1(5), pp. 253–273, 2007.
50. C. Laurent Peigney, E. Flahaut, R. Bacsa, A. Rousset, A. Rousset, "Specific surface area of carbon nanotubes and bundles of carbon nanotubes," *Carbon*, 39(4), pp. 507–514, 2001.
51. K. Liao and S. Li, "Interfacial characteristics of a carbon nanotube–polystyrene composite system," *Applied Physics Letters*, 79(25), pp. 4225–4227, 2001.
52. S. Banerjee, T. Hemraj-Benny, and S. Wong, "Covalent surface chemistry of single-walled carbon nanotubes," *Advanced Materials*, 17(1), pp. 17–29, 2005.
53. Y. Tan and D. Resasco, "Dispersion of single-walled carbon nanotubes of narrow diameter distribution," *The Journal of Physical Chemistry: Part B*, 109(30), pp. 14454–14460, 2005.
54. P. Pötschke, T. Fornes, and D. Paul, "Rheological behavior of multiwalled carbon nanotube/polycarbonate composites," *Polymer*, 43(11), pp. 3247–3255, 2002.
55. B. Safadi, R. Andrews, and E. Grulke, "Multiwalled carbon nanotube polymer composites: Synthesis and characterization of thin films," *Journal of Applied Polymer Science*, 84(14), pp. 2660–2669, 2002.
56. T. Makris, L. Giorgi, R. Giorgi, N. Lisi, and E. Salernitano, "CNT growth on alumina supported nickel catalyst by thermal CVD," *Diamond and Related Materials*, 14(3–7), pp. 815–819, 2005.
57. B. Coto, I. Antia, M. Blanco, I. Martinez-de-Arenaza, E. Meaurio, J. Barriga, J. Sarasua, "Molecular Dynamics study of the influence of functionalization on the elastic properties of single and multiwall carbon nanotubes," *Computational Materials Science*, 50(12), pp. 3417–3424, 2011.
58. E. Backes, T. Sene, F. Passador, and L. Pessan, "Electrical, thermal and mechanical properties of epoxy/CNT/calcium carbonate nanocomposites," *Materials Research*, 21(1), 2018.
59. P. Ma, N. Siddiqui, G. Marom, and J. Kim, "Dispersion and functionalization of carbon nanotubes for polymer-based nanocomposites: A review," *Composites—Part A: Applied Science and Manufacturing*, 41(10), pp. 1345–1367, 2010.
60. S. Sabet, H. Mahfuz, J. Hashemi, M. Nezakat, and J. Szpunar, "Effects of sonication energy on the dispersion of carbon nanotubes in a vinyl ester matrix and associated thermo-mechanical properties," *Journal of Materials Science*, 50(13), pp. 4729–4740, 2015.

61. X. Gong, J. Liu, S. Baskaran, R. Voise, and J. Young, "Surfactant-assisted processing of carbon nanotube/polymer composites," *Chemistry of Materials*, 12(4), pp. 1049–1052, 2000.
62. M. Shaffer and A. Windle, "Fabrication and characterization of carbon nanotube/poly (vinyl alcohol) composites," *Advanced Materials*, 11(11), pp. 937–941, 1999.
63. L. Kumari, T. Zhang, G. H. Du, W. Z. Li, Q. W. Wang, A. Datye, *et al.*, "Synthesis, microstructure and electrical conductivity of carbon nanotube–alumina nanocomposites," *Ceramics International,* 35, pp. 1775–1781, 2009.
64. I. Ahmad, M. Unwin, H. Cao, H. Chen, H. Zhao, A. Kennedy, *et al.*, "Multi-walled carbon nanotubes reinforced Al2O3 nanocomposites: Mechanical properties and interfacial investigations," *Composites Science and Technology,* 70, pp. 1199–1206, 2010.
65. M. Kudus, H. Akil, H. Mohamad, and L. Loon, "Effect of catalyst calcination temperature on the synthesis of MWCNT–alumina hybrid compound using methane decomposition method," *Journal of Alloys and Compounds*, 509(6), pp. 2784–2788, 2011.
66. I. I. Slowing, J. L. Vivero-Escoto, C.-W. Wu, and V. S.-Y. Lin, "Mesoporous silica nanoparticles as controlled release drug delivery and gene transfection carriers," *Advanced Drug Delivery Reviews*, 60(11), pp. 1278–1288, 2008.
67. L. Tang and J. Cheng, "Nonporous silica nanoparticles for nanomedicine application," *Nano Today*, 8(3), pp. 290–312, 2013.
68. J. Wu, Z. Ye, G. Wang, and J. Yuan, "Multifunctional nanoparticles possessing magnetic, long-lived fluorescence and bio-affinity properties for time-resolved fluorescence cell imaging," *Talanta*, 72(5), pp. 1693–1697, 2007.
69. C. Barbe, J. Bartlett, L. Kong, K. Finnie, H. Q. Lin, M. Larkin, *et al.*, "Silica particles: A novel drug-delivery system," *Advanced Materials*, 16(21), pp. 1959–1966, 2004.
70. H. Lv, S. Song, S. Sun, L. Ren, and H. Zhang, "Enhanced properties of poly (lactic acid) with silica nanoparticles," *Polymers for Advanced Technologies*, 27(9), pp. 1156–1163, 2016.
71. B. Demir, K.-Y. Chan, D. Yang, A. Mouritz, H. Lin, B. Jia, *et al.*, "Epoxy-gold nanoparticle nanocomposites with enhanced thermo-mechanical properties: An integrated modelling and experimental study," *Composites Science and Technology*, 174, pp. 106–116, 2019.
72. M. P. Indira Devi, N. Nallamuthu, N. Rajini, Suchart Siengchin, Varada Rajulu, and N. Hariram. "Antimicrobial properties of poly (propylene) carbonate/Ag nanoparticle-modified tamarind seed polysaccharide with composite films," *Ionics* 25(7), pp. 3461–3471, 2019.
73. L. B. Hussain Nurazreena, H. Ismail, M. Mariatti, M. Mariatti, "Metal filled high density polyethylene composites–electrical and tensile properties," *Journal of Thermoplastic Composite Materials*, 19(4), pp. 413–425, 2006.
74. S. A. Ngah and A. C. Taylor, "Fracture behaviour of rubber-and silica nanoparticle-toughened glass fibre composites under static and fatigue loading," *Composites—Part A: Applied Science and Manufacturing*, 109, pp. 239–256, 2018.
75. K. Mai, Z. Li, Y. Qiu, and H. Zeng, "Mechanical properties and fracture morphology of Al (OH) 3/polypropylene composites modified by PP grafting with acrylic acid," *Journal of Applied Polymer Science*, 80(13), pp. 2617–2623, 2001.
76. M. Nurul and M. Mariatti, "Effect of thermal conductive fillers on the properties of polypropylene composites," *Journal of Thermoplastic Composite Materials*, 26(5), pp. 627–639, 2013.
77. H. Unal, K. Esmer, and A. Mimaroglu, "Mechanical, electrical and tribological properties of graphite filled polyamide-6 composite materials," *Journal of Polymer Engineering*, 33(4), pp. 351–355, 2013.

78. R. Doufnoune, F. Chebira, and N. Haddaoui, "Effect of titanate coupling agent on the mechanical properties of calcium carbonate filled polypropylene," *International Journal of Polymeric Materials*, 52(11–12), pp. 967–984, 2003.
79. N. F. Attia, E. S. Goda, M. Nour, M. Sabaa, and M. Hassan, "Novel synthesis of magnesium hydroxide nanoparticles modified with organic phosphate and their effect on the flammability of acrylonitrile-butadiene styrene nanocomposites," *Materials Chemistry and Physics*, 168, pp. 147–158, 2015.
80. M. Wang, X.-F. Zeng, J.-Y. Chen, J.-X. Wang, L.-L. Zhang, and J.-F. Chen, "Magnesium hydroxide nanodispersion for polypropylene nanocomposites with high transparency and excellent fire-retardant properties," *Polymer Degradation and Stability*, 146, pp. 327–333, 2017.
81. Q. He, T. Yuan, X. Yan, D. Ding, Q. Wang, Z. Luo, *et al.*, "Flame-retardant polypropylene/multiwall carbon nanotube nanocomposites: Effects of surface functionalization and surfactant molecular weight," *Macromolecular Chemistry and Physics*, 215(4), pp. 327–340, 2014.
82. B. Yu, Y. Shi, B. Yuan, S. Qiu, W. Xing, W. Hu, *et al.*, "Enhanced thermal and flame retardant properties of flame-retardant-wrapped graphene/epoxy resin nanocomposites," *Journal of Materials Chemistry A*, 3(15), pp. 8034–8044, 2015.
83. R. Gendron, L. E. Daigneault, J. Tatibouëut, and M. M. Dumoulin, "Residence time distribution in extruders determined by in-line ultrasonic measurements," *Advances in Polymer Technology*, 15(2), pp. 111–125, 1996.
84. J. L. Caillaud, S. Deguillaume, M. Vincent, J. C. Giannotta, and J. M. Widmaier, "Influence of a metallic filler on polyurethane formation," *Polymer International*, 40(1), pp. 1–7, 1996.
85. G. Wypych, *Handbook of Fillers*, vol. 92, ChemTec Pub, Toronto, 2010.
86. G. Sung and J. H. Kim, "Influence of filler surface characteristics on morphological, physical, acoustic properties of polyurethane composite foams filled with inorganic fillers," *Composites Science and Technology*, 146, pp. 147–154, 2017.
87. Z. Jin, K. Pramoda, G. Xu, and S. Goh, "Dynamic mechanical behavior of melt-processed multi-walled carbon nanotube/poly (methyl methacrylate) composites," *Chemical Physics Letters*, 337(1–3), pp. 43–47, 2001.
88. S. Harsha Singh, A. Parashar, A. Parashar, "Finite Element Analysis of CNT reinforced epoxy composite due to Thermo-mechanical loading," *Procedia Technology*, 23, pp. 138–143, 2016.
89. B. S. Hadavand, K. M. Javid, and M. Gharagozlou, "Mechanical properties of multiwalled carbon nanotube/epoxy polysulfide nanocomposite," *Materials and Design*, 50, pp. 62–67, 2013.
90. J. Lee, K. Rhee, and S. Park, "The tensile and thermal properties of modified CNT-reinforced basalt/epoxy composites," *Materials Science and Engineering: Part A*, 527(26), pp. 6838–6843, 2010.
91. G. Zhuang, G. Sui, Z. Sun, and R. Yang, "Pseudoreinforcement effect of multiwalled carbon nanotubes in epoxy matrix composites," *Journal of Applied Polymer Science*, 102(4), pp. 3664–3672, 2006.
92. K. Wong, S. Shi, and A. Lau, "Mechanical and thermal behavior of a polymer composite reinforced with functionalized carbon nanotubes," In: *Key Engineering Materials*, pp. 705–708, 2007.
93. S. Zeng, M. Shen, S. Chen, L. Yang, F. Lu, and Y. Xue. "Mechanical and thermal properties of carbon nanotube-and graphene-glass fiber fabric-reinforced epoxy composites: A comparative study," *Textile Research Journal*, 89(12), pp. 2353–2363, 2019.
94. B. B. Kadhim, "Thermal Properties of Carbon Nano Tubes Reinforced Epoxy Resin Nano CompositesBahjat B. Kadhim1," *Al-Mustansiriyah Journal of Science*, 22, pp. 276–290, 2011.

95. A. Kausar and S. Hussain, "Effect of multi-walled carbon nanotube reinforcement on the physical properties of poly (thiourea-azo-ether)-based nanocomposites," *Journal of Plastic Film and Sheeting*, 29(4), pp. 365–383, 2013.
96. Y. Yang, T. Hsieh Lee, T. Chen, T. Cheng, T. Cheng, "Mechanical property of multiwall carbon nanotube reinforced polymer composites," *Polymers and Polymer Composites*, 26(1), pp. 99–104, 2018.
97. J. Kearns and R. Shambaugh, "Polypropylene fibers reinforced with carbon nanotubes," *Journal of Applied Polymer Science*, 86(8), pp. 2079–2084, 2002.
98. E. Ciecierska, A. Boczkowska, K. Kurzydlowski, I. Rosca, and S. Van Hoa, "The effect of carbon nanotubes on epoxy matrix nanocomposites," *Journal of Thermal Analysis and Calorimetry*, 111(2), pp. 1019–1024, 2013.
99. M. Koo and G. Lee, "Thermal property of multi-walled-carbon-nanotube reinforced epoxy composites," *World Academy of Science, Engineering and Technology, International Journal of Chemical, Molecular, Nuclear, Materials and Metallurgical Engineering*, 9, pp. 25–29.

12 Applications of Thermoplastic and Thermosetting Polymer Composites

Sanjay Remanan, Sabyasachi Ghosh, Tushar Kanti Das, and Narayan Chandra Das

CONTENTS

12.1 INTRODUCTION

Polymers are 21st-century materials, which have brought an enormous number of applications in various fields, ranging from the toroid-shaped tire for the automotive industry to the silicone heart valves in the medical industry (Coulter et al., 2019). From the applications perspective, polymers can be classified under various groups such as biomedical, agriculture, engineering, and commodity types. Plastics such as polyethylene (PE), polypropylene (PP), and polystyrene (PS) are classified as commodity polymers while polyether ether ketone (PEEK), poly(vinylidene fluoride) (PVDF), and acrylonitrile butadiene styrene (ABS) are known as engineering polymers. Polymers, in general, are lightweight, flexible, electrically insulating, optically transparent, and low-density materials which can be easily processed into any desired shapes. Intricate design and structures can be produced with minimal effort compared to the processing of metal or glass materials. Depending upon the main-chain composition and the presence of functional groups, properties of the polymer change significantly and this is reflected in the final selection of the material for specific applications. For example, styrene-butadiene rubber (SBR) has good abrasive property attributes in the presence of rigid aromatic groups, whereas polychloroprene rubber has good thermal stability due to the presence of halogen group in the polymer chain and so on.

Hence, material properties largely depend upon the chemical composition of the polymer chain. Highly rigid structures such as aromatic rings in the main chain decrease the flexibility but increase the high-temperature stability. Presence of -C–S- linkage increases the material flexibility but at the same time decreases the thermal stability. -Si– O- bond in silicone polymer not only increases the high-temperature stability but also enhances the high material flexibility. Presence of polar groups can influence the dielectric and mechanical properties. Inter/intra-molecular crosslinking through the polar-polar interactions between the functional groups or hydrogen bonding can increase the material's mechanical properties. The presence of pendant groups on both sides of the main chain increases the gas barrier properties and decreases the material crystallization. Increase in the pendant-group chain length can cause a more significant reduction in the degree of crystallization. This decrease in crystallization changes the polymer's optical and gas barrier properties. Hence, tailoring of the chemical composition of the main chain and presence of pendant groups can significantly influence the final material properties.

Now, here comes the discussion about the material behavior in response to the applied heat. Depending upon the response to the heat, polymers can be classified into two major groups called thermosetting and thermoplastic types. Thermoplastic polymers are the materials which soften and melt under the application of heat and turn to a rigid structure when they cool down. Reshaping and remolding of a finished product under the application of heat can be done several times. Thermosetting polymer softens during the application of heat and permanently deforms to a shape which cannot be remolded and reused in the presence of heat. A thermosetting polymer forms the permanent structure with chemical cross-links and the change is irreversible. Two significant advantages of thermoplastic polymers are their high impact resistance and their remolding ability makes them favorable to market. (Commonly used thermoplastic and thermosetting polymers and their thermo-mechanical properties are listed in Table 12.1). On the other hand, for the fabrication of high-strength polymer composites, incorporation of reinforcing filler in the polymer is necessary. From this point of view, the thermosetting polymer is far better as the incorporation process is easier than that of thermoplastic polymers and also less expensive. Thermoplastic polymers are mostly plastics, while thermosetting types are mostly rubbers.

For targeted applications, incorporation of filler particles within the polymer matrix is a simple strategy to ameliorate the properties (Parameswaranpillai et al., 2016). Some examples are, addition of reinforcing carbon black fillers in elastomer matrix to achieve the high abrasion tendency, clay-based nanocomposite to improve fire retardancy, UV-resistant additives for weather protection, conducting nanofillers to increase electrical properties, incorporation of antifouling and antibacterial nanofillers to achieve a strong antibacterial property for membrane separation, incorporation of various nanofillers to achieve good physico-mechanical properties, and there

TABLE 12.1
Thermo-Mechanical Properties of the Commonly Used Thermoplastic and Thermosetting Polymers

Polymer	Density (g/cm^3)	T_m (°C)[a]	T_g (°C)[b]	Tensile strength (MPa)	Tensile modulus (MPa)
HDPE	0.94–0.96	140	−110	15–35	814–950
LDPE	0.91–0.93	110	−110	9–14	297–540
PP	0.90	160	−10	19–70	1200–6500
PVC	1.3–1.4	160	85	13.5–23.6	2620
PS	1.05	85[c]	100	36–45	3400
PVDF	1.72	167	−32	47–55	2000
ABS	1.05–1.25	108–124	95	37–45	1950–2600
PC	1.24	164	146.35	62–65	1800–2500
PET	1.3–1.6	238	79	140	~10000
Nylon 6	1.07–1.56	215	75	80	2900
NR	0.93	—	−72	27	—
SBR	0.93	—	−73	21–26	15.6
NBR	0.988–1.04	—	−35 to −25	26.5–31.8	—
EVA	0.93–0.95	73–101	−33	9–17	72
Ethylene methyl acrylate copolymer (EMA)	0.92–0.95	85	−33	9–12	55[d]
Poly(ether sulfones (PES)	1.37	220[c]	230	84	2.4
PPS	1.30	279–295	86–92	66	3.3
PEEK	1.31	340	150–170	70	3.8
Polyethylenimine (PEI)	1.27	242[c]	217–225	105	3.0
Polymide (PI)	1.37	260[c]	256	138	3.4

[a] T_m = melting temperature, [b]T_g = glass transition temperature, [c]vicat softening temperature, [d]secant modulus.

are many more (Geethamma et al., 2005; Remanan et al., 2017). Different types of nanoparticles such as nanorods, nanospheres, nanosheets, and carbon dots are incorporated to achieve the enhanced properties for various applications (Tiwari et al., 2015; Pulikkalparambil et al., 2017; Parameswaranpillai et al., 2019; Das P. et al., 2019). Carbonaceous materials like carbon nanotubes, graphene, and graphene-derived materials are the primary research interest in the last two decades. The filler is not only the principal component for an application but also selection of polymer is also crucial for getting the desired end property. Sometimes, two polymers are mixed (physically or chemically) to attain the property required for a particular application; then the polymer mixture prepared is called a blend. Polymer blending can be either a mixture of two thermoplastic polymers, or two thermosetting polymers, or a mixture of thermoplastic and thermosetting polymer. This physical mixing indicates that for targeted applications, the suitable choice of filler as well as of polymer matrix are crucial. The broad spectrum of applications such as in electronic, biomedical, separation, automotive, and in aviation industries are benefited from the polymer nanocomposites and myriads of studies report on the application of these nanocomposites (Sengupta et al., 2011).

Depending upon their practical usages, thermoplastic polymeric materials are classified into the following categories:

- Commodity thermoplastic (such as PE, PP, polyvinyl chloride (PVC), PS, ethylene-vinyl acetate (EVA), etc.)
- Quasi-commodity thermoplastic (such as ABS, polymethyl methacrylate (PMMA), polyethylene terephthalate (PET), styrene acrylonitrile resin (SAN), etc.)
- Engineering thermoplastic (such as polyamide (PA), polycarbonate (PC), polyphenylene sulfide (PPS), cellulose acetate, etc.)
- Specialty thermoplastic (such as polytetrafluoroethylene (PTFE), phenyl phosphine oxide (PPO), PEEK, polysulfone (PSU), etc.)
- Emerging thermoplastic polymer (such as functionalized thermoplastic polymer and various thermoplastic nanocomposite)

Commodity thermoplastic consumes 50% of the market volume due to their properties and cost-effectiveness. Although the properties of quasi-commodity thermoplastic are better than commodity thermoplastic, the higher cost of the former reduces their usage in final applications. Engineering thermoplastic materials have good physico-mechanical and thermal properties, which makes them a potential candidate for various applications and for replacing traditionally heavier and fragile materials such as glass. The consumption of specialty thermoplastic is much lower as compared to other thermoplastics due to their high cost. The last type, emerging thermoplastic polymer is fabricated depending upon the applications. An overview of a few commercially available polymer materials and their manufacturer details are listed in Table 12.2.

Herein this chapter focuses on various applications of the thermosetting and thermoplastic materials including the engineering, biomedical, and separation applications, and discuss the material properties required for these applications. A

TABLE 12.2
Commercial Names of a Few Thermoplastic and Thermosetting Polymers with Manufacturer Details

Sl. No	Polymer name	Commercial name	Manufacturer
Thermoplastic polymers	Polyethylene (HDPE)	Relene (45GP004)	Reliance Industries Limited
	Polyethylene (LDPE)	ACCUGUARD	LyondellBasell Industries
	Polyethylene (LLDPE, adhesive resin)	ADPOLY EM-400	Lotte Chemical Corporation
	Polypropylene (PP)	Repol (AMI20N)	Reliance Industries Limited
	Polystyrene (HIPS)	ALPHALAC 51SF	LG Chem Ltd.
	Polycarbonate (PC)	ALCOLOR® PC	ALBIS PLASTIC GmbH
	Polymethylmethacrylate (PMMA)	Acrigel AFP	Unigel Plásticos
	Polyvinylidene fluoride (PVDF)	Kynar	Arkema
	Polyvinyl chloride (PVC)	Reon	Reliance Industries Limited
	Polyether etherketone (PEEK)	KetaSpire KT-820	Solvay Specialty Polymers
	Ethylene methyl acrylate copolymer (EMA)	Elvaloy	DuPont Packaging & Industrial Polymers
Thermosetting polymers	Polybutadiene rubber/ butadiene rubber (PBR)	Relflex CISAMER PBR	Reliance Industries Limited
	Styrene-butadiene rubber (SBR)	Relflex Stylamer	Reliance Industries Limited
	Acrylonitrile butadiene rubber (NBR)	Chemisat® LCH-7272	Zeon Chemicals L.P.
	Ethylene propylene diene terpolymer (EPDM)	Dutral® TER 4047	Versalis S.p.A.
Thermoplastic urethane (TPU)	Thermoplastic polyurethane elastomer	Elastollan	BASF Corporation
Thermoplastic elastomer (TPE)	Styrene–ethylene–butylene–styrene (SEBS)	Kraton A1535 H	Kraton

description of a wide spectrum of applications ranging from aerospace, tissue engineering, hydrogel, and polymer membranes are included in this chapter.

12.2 APPLICATION OF THERMOPLASTIC AND THERMOSETTING POLYMERS

12.2.1 Automotive Applications

Polymers serve a broad spectrum of applications in the automotive industry; right from the exterior parts (tire, dashboard, and seat cover) to the interior parts (engine and fuel compartments) and their usage has significantly increasing in recent years

(Patil, Patel, and Purohit, 2017). The major reasons for employing the polymers for automotive applications is because they are lightweight, enable greater freedom in designing the intricate parts to create comfort and safety, they can be integrated into parts easily, and they incur lower manufacturing costs than other conventional materials. Major automotive parts which are made from the polymers can be included in dashboards, seat cover, steering wheel, door handle, window trim seal, convertible top cover, mirror case, wiper and shock absorber, seals, and braking system. Depending upon the environment, different polymer materials and their blends can be employed for automotive applications. Other important advantages of using the polymers are reduced fuel consumption, aesthetic improvements, and economy in the manufacture.

Polymers such as polyvinyl chloride, acrylonitrile butadiene styrene, polycarbonate, polymethyl methacrylate, and many other polymers are used for automotive applications.

PVC is mainly used for the internal lining and coating material for the electric cables inside the vehicles. The material is selected due to its low flammability, the fact that it is easy to weld and a good absorbent of impact and vibrations. The low flammability is induced due to the presence of a halogen group and active radical formation at the vinyl position of the ethylene chain. PVC-based sheathing is also used for the manufacture of the seat covers. ABS is another important polymer, with excellent impact resistance, high toughness, and rigidity. After molding, these materials have an excellent surface finish and the well-polished surface avoids the need for post-surface finishing operations. ABS polymer is typically used for the manufacture of the dashboard, steering wheel, and housings. It is also used for the manufacture of grills, wheel cover, and rearview mirror cases. Another interesting polymer material is polycarbonate, which is usually used for the manufacture of front and rear lights and lenses owing to their transparent and high-strength nature. PC has high modulus, dimensional stability, and good electrical insulation properties. Due to the presence of aromatic groups, the material has excellent high-temperature stability up to 140 °C and is resistant to temperature up to around 220 °C. Its high optical transparence property is advantageous for the manufacture of parts like lights and lenses.

Other polymers such as polyurethane (PU) and thermoplastic polyurethane (TPU), ethylene propylene diene rubber (EPDM), polyamide (PA), polyesters, and styrene-ethylene-butylene-styrene (SEBS) are also widely used in the automobile industry (Anagha and Naskar, 2019; Gopalan and Naskar, 2019). PU-based seat cushioning, door-side fittings, steering elements, convertible top covers, bonnets, and floor lining are the essential applications. PU has good recovery after deformation, a good impact and vibration absorbance property, and lower crack propagation tendency. PU also shows excellent resistance to oils, grease, and other chemicals which favors it as a suitable choice for a wide variety of automotive applications. Polyesters find application in bumpers, radiator grills, rearview mirror housings, and door handles. Polyesters are polymers that are stable at high temperatures (~200 °C), that have high strength, hardness, and excellent electrical insulation properties. They are opaque and have a high modulus and low absorption of humidity. Other polymers such as EPDM can be used as the window liner and floor liner due to its excellent weather resistance property. In EPDM, the unsaturation is present in the side group for sulfur curing reaction and it is not a part of polymer backbone. As a result, EPDM possesses excellent ozone resistance which increases the outdoor stability in applications such as weatherstripping for windows and doors. Elastomers such as acrylonitrile butadiene rubber (NBR) finds application in machine elements, seals, and fuel, oil, and air ducts, flanges, and dust protector applications. NBR has a very high oil-resistance property which arises from the presence of pendant acrylonitrile group in the main chain. Increase in the acrylonitrile content changes the rubbery nature of the material, and >33% is not preferred for elastomer-based applications. Chloroprene rubber is used for the manufacture of the wiper, driving belts, air ducts, and flanges. Chloroprene elastomers have excellent weather (sunlight and ozone) resistance, oil and flame resistance and are suitable for outdoor applications (Štrumberger, Gospočić, and Bartulić, 2005). A wide operating temperature can be achieved when chloroprene rubber is suitably compounded. Neoprene™ and HYPALON® (chlorosulfonated polyethylene, CSM) are two classes of chloroprene rubber. Neoprene rubber find applications such as ignition wire jackets, specialty seals, radiator hoses etc. in automotive industry. Oil resistance of the CSM increases with the increase in chlorine percentage in rubber. CSM rubber shows excellent oxygen or ozone resistance, good electrical resistance, solvent resistance, and excellent outdoor color stability when it is suitably pigmented. Incorporation of different nanoparticles or blending with a second polymer is also a strategy to get the desired final properties.

The use of polymer nanocomposite containing natural fiber is one of the emerging approaches in the fabrication of automotive components. Thermoplastic polymers such as PP, PE, and nylons and thermosets such as vinyl esters, epoxy, and polyesters are used as matrix polymers. Natural fibers such as coir, sisal, hemp, flax, and kenaf are the typical fillers used for the reinforcement. Accessories such as door panel, seat covering, trunk panel, insulation, and exterior floor panels can be made using the natural fiber polymer composites (Holbery and Houston, 2006; Koronis, Silva, and Fontul, 2013). The fibers used for the preparation of green composite have several advantages, including high strength and stiffness, low CO_2 emission, biodegradability, and being low cost. Comparison between polymer composite containing glass fiber (E-glass) and polymer composite containing natural fiber shows comparable flexural modulus (the ratio of stress range to corresponding strain range for a test specimen loaded in flexure), flexural strength (the maximum stress at the outer surface of a flexure test specimen corresponding to the peak applied force prior to flexural failure), and elongation at break. Hence, wood-plastic composite (WPC) or the green polymer composite is a sustainable solution for the preparation of automotive parts (Ashori, 2008; T.K. Das, Prosenjit, and N.C. Das, 2019).

Polymer materials have good processability and transform into required shapes with good surface finish compared to other conventional materials like metals. This advantage eliminates or reduces the need for surface finishing operations. Weather, electrical, and chemical resistant properties of the material also help to eliminate the need for other surface treatment operations. Hence, these lightweight polymeric materials help the automotive industries to achieve greater fuel economy, greater design flexibility, good aesthetics, and longer service life.

12.2.2 Aerospace Applications

Polymers are used for a variety of applications in space environments: seals, coatings, adhesives, dampers, and thermal insulations (Cheng, 2019). Commonly used spacecraft polymers are polyimide, silicone, and fluorinated polymers (Kapton® and Teflon®) attributed to their low density, flexibility, high-temperature stability, electrical and optical properties (De Groh et al., 2008). The extreme environment and commercialization of the space industry demand stringent requirements from the polymer material properties. The material properties that last for longer durations are favorable due to the high cost of the aerospace programs. The property requirements can slightly vary as the vehicle changes either to polar or geostationary satellites (Krishnamurthy, 1995). The typical degradation behavior in the space environment is due to the following reasons,

- High dose electron or proton cloud exposure
- Exposure to ultraviolet radiation
- Change in the atmospheric pressure
- Thermal cycling
- Presence of micrometeoroid
- Exposure to atomic oxygen

The atomic oxygen presence can cause more violent reactions on the surfaces of the polymer and changes the optical, thermal, and mechanical properties. The atomic oxygen is arising from the photo-dissociation of the molecular oxygen. However, fluorinated and silicone polymers show higher atomic oxygen resistivity. During degradation, the C-F bond in fluorinated polymer requires very high bond dissociation energy whereas silicone rubbers forms a glassy protective layer on the surface, which prevents the further degradation of the layer beneath. A recent study by NASA on 41 different polymeric samples reveals the effect of atomic oxygen on mass loss of samples exposed to the harsh space environment for nearly four years. The experiment was conducted in the lower earth orbit (LEO) under identical conditions (De Groh et al., 2008). Presence of micrometeoroid can cause the fracture and deteriorates the polymer mechanical properties. UV and electron or proton radiations change the molecular structure of the macromolecular chains. It can either cause the chain scission or can cross-link the polymeric network depending upon the degree of radiation. Increase in altitude decreasing the atmospheric pressure can cause the diffusion of additives onto the polymer surface, volatilization, and can even poison the surrounding area. Volatilization of additives is highly undesirable in the case of a human-crewed mission.

Latest developments in the shape memory materials pave the way for new inventions in space applications (Santo et al., 2020). Shape memory polymers (SMP) are the materials that regain original shape by remembering the structure after deformation to a specific external stimulus (Jose et al., 2020). SMP and their composite (SMPC) finds aerospace applications in different components that meet the need of space-deployable structures (Liu et al., 2014; Mu et al., 2018). This SMP application in different parts of the satellite varies from the solar arrays, antennas, and deployable panels. Incorporation of various nanoparticles in the SMPs imparts property enhancement which improves the applicability, depending upon the nature of the nanoparticles. The main attraction of the SMP is the variation of their properties in the vicinity of the transition temperature or melting temperature. Near to T_g, SMP modulus changes two order of magnitude, rapidly changing the viscoelastic properties. SMP can be classified into two groups: thermoplastic and thermosetting SMPs. Thermoplastic SMPs exhibit inferior properties after several repeated cycles of experiment and eventually lose their shape memory effect whereas cross-linked thermoset SMPs show high material stiffness, weatherability, and wide range of T_g makes this an important shape memory material. Several materials have been reported recently, such as epoxy, polyaspartimide–urea, cyanate–ester–epoxy–poly(tetramethylene oxide), ethylene octane copolymer (EOC), polyurethane, and EOC-EPDM blends for shape memory applications (Xie and Rousseau, 2009; Biju, Gouri, and Nair, 2012; Shumaker, McClung, and Baur, 2012; Behera, Mondal, and Singha, 2018; Chatterjee and Naskar, 2019; Chatterjee et al., 2019). Other approaches include the use of phenyl phosphine oxide containing perfluorocyclobutyl (PFCB) polymers are being developed and have been found to have high temperature stability (~450 °C) or high T_g (~150 °C) for space applications (Jin et al., 2003). The major disadvantage of using SMPs is the limited material availability for the composites' fabrication to work effectively in the space environment.

12.2.3 Polymer Membranes

Membranes are the porous structures which facilitate the separation and mass transfer of a component from the mixture of solutions. The membrane can be used for various applications including separation of biological proteins, gases and water filtration, energy storage, and the clarification of juice and whey proteins. Polymer-based membranes are found to be an effective solution for the separation application, majorly for water filtration, fuel cell, and gas separation (Mulder, 2012). Most of the membrane materials are thermoplastic. Various polymeric materials such as cellulose acetate, PVDF, polysulfone, polyethersulfone, and polyethylene (Remanan et al., 2018) are widely used polymer materials for the membrane fabrication.

The essential criteria behind the selection of the membrane material lie in the field of application. The application demands the tailoring of the properties to meet the required separation efficiency of the membrane. For example, the membrane used for distillation applications mainly requires the presence of a highly hydrophobic property so that water molecules can repel by the surface and allow only the water vapors through the membrane. Generally, membranes used for the pressure-driven applications favors hydrophilic surfaces to avoid the concentration polarization by pollutants such as microbes and other proteins present in the water bodies and allow the permeation of water molecules across the membrane.

PVDF is one of the widely studied thermoplastic membrane materials for pressure-driven and membrane distillation (MD) processes. Membranes are prepared by phase inversion process owing to the simplicity in the preparation procedure. Figure 12.1 shows the typical immersion precipitated membrane morphology prepared by the phase inversion mechanism. The primary requirements for the polymer

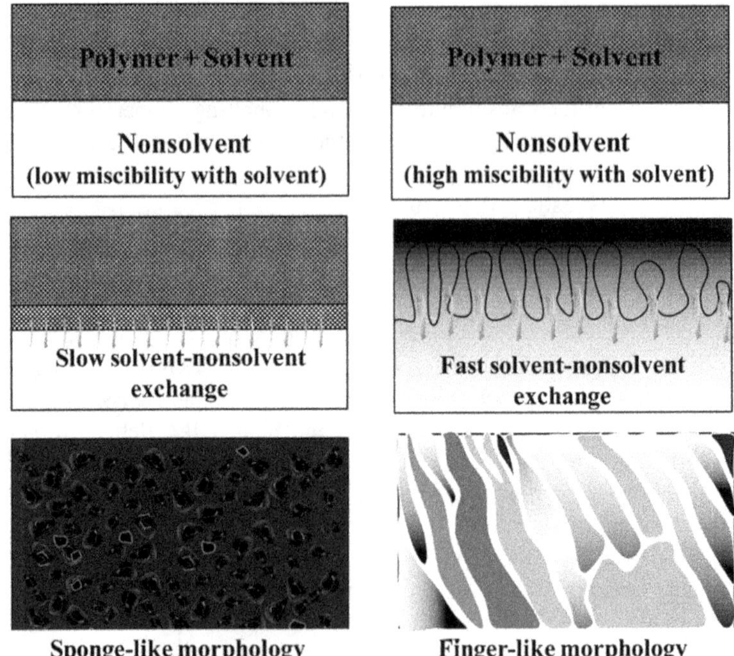

FIGURE 12.1 Different membrane morphologies caused by different types of demixing. (Adapted from Guillen et al., 2011, with permission from American Chemical Society.)

fuel cell membranes are the high proton conductivity, low methanol-water permeability, and moderate price (Vijayalekshmi and Khastgir, 2018a). Recent studies have reported on the preparation of the fuel cell membrane from various polymeric materials, which includes chitosan, PEEK, polybenzimidazole, polyimide, and perfluorinated-based membranes and sulfonated modifications for the better power density (Kim, Lee, and Nam. 2017; Vijayalekshmi and Khastgir, 2018b).

12.2.4 Electromagnetic Interference Shielding

The growth of the electronics industry has led to a new world of electronic devices that operate at the high-frequency range used for industrial, consumer, and military utilities. The high-frequency range operated devices create a new form of pollution called radio frequency interference (RFI) or electromagnetic interference (EMI) (N.C. Das et al., 2000; Geetha et al., 2009; Ghosh, et al., 2018). Our atmosphere is filled with EM radiations and this is a new threat to human lives. High-frequency signals used in the microprocessor-controlled devices may be transmitted out and cause the malfunctioning of the nearby electronic devices. The EMI consist of unwanted signals that perturb the operation of nearby electronic devices and other communication systems. This interference is the reason behind the 'turn-off' request of electronic and communication devices of the passengers just before the take-off of airplanes to avoid the frequency disturbances in communication between the

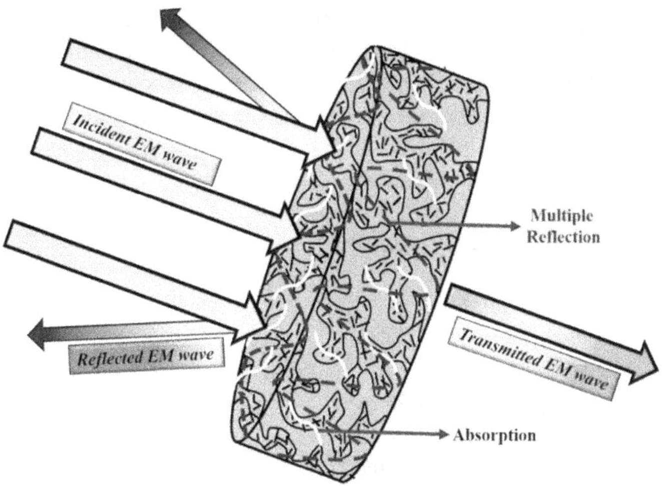

FIGURE 12.2 Mechanism of EMI shielding by a polymer nanocomposite. (Adapted from Ravindren et al., 2019, with permission from Elsevier.)

pilot and air traffic controller. The EMI is also the reason behind the flash in television display and spark noises in radio. EMI has also been found to have a significant effect on human health, which causes insomnia, headache, and nervousness. Hence, effective capturing or attenuation of the unwanted electromagnetic waves has become a primary research topic in recent times (Ganguly, Bhawal, et al., 2018; Ghosh, Ganguly, Remanan, et al., 2019).

Electromagnetic waves have two components, namely magnetic field and electric field, and their direction of propagation is at right angles to each other. When the electromagnetic waves reach on a surface, it can reflect from the surface or be absorbed on to it and can transmit a portion of the waves (see Figure 12.2). The EMI shielding effectiveness is the ratio of electric and magnetic field components before and after attenuation. Different materials such as metals, polymers, and woods are reported to be used for shielding. Metals suffer the disadvantages of corrosiveness and processing problems. Recently, polymer materials were proven to be excellent shielding materials that have the advantage of being lightweight, with improved processability and flexibility (Pawar et al., 2015). Polymers are in general insulators, and have very high resistivity. Hence, to make the polymer conductive and to allow the sufficient absorption of EM waves can be achieved via incorporation of conducting fillers. Metallic, carbonaceous, and magnetic particles were added in different polymeric matrices and the shielding effectiveness of the system was studied (Biswas et al., 2017). Different polymer systems were used as the base material for shielding. Especially, thermoplastic materials such as PP, PVDF, ABS, PC, EMA, EPDM, EVA, PU can be used as the matrix polymers (N.C. Das, Chaki, and Khastgir, 2002; Biswas, Kar, and Bose, 2015; Kar, Biswas, and Bose, 2016; Ghosh, Remanan, et al., 2018; Ghosh, Ganguly, et al., 2018; Ravindren et al., 2019). Thermosetting polymers such as natural rubber (NR), ethylene octane copolymer (EOC), and chlorinated polyethylene were also reported as matrices for EMI shielding. Various polymer blend systems were also reported as potential low-cost shielding material. Carbonaceous fillers such as multi-walled carbon nanotubes (MWCNT) and graphene derivatives are also recently gaining increased significance as conducting fillers. Developments in polymer blending improved the design of the shielding material in which distribution of fillers can be controlled. The added fillers can either be effectively distributed in the particular phase (double percolation) or at the interface of the two immiscible polymers to achieve the optimum filler loading and effective shielding.

12.2.5 Cable Insulation

The higher volume resistivity and lower dielectric constant (permittivity) are the prime desired properties required for a cable-insulating material. The dielectric constant is related to the polarity of the material. Volume resistivity can be calculated using the standard ASTM D257-66 and from the following formula,

$$\text{Volume resistivity (ohm cm)} = \frac{AR}{t} \tag{12.1}$$

where A is the area of the upper electrode, and R is the resistance between the upper and lower electrode, and t is the sample thickness. The dielectric constant of the material is defined as the ratio of the capacitance of a capacitor in which given material acts as a dielectric medium to the capacitance of the same capacitor using vacuum as dielectric medium. Usually, polymers have low dielectric constant, albeit some natural polymers or polar polymers showing high value. Generally, polyethylene type of materials such as high-density polyethylene (HDPE) and cross-linked polyethylene (XLPE) can be used for insulation and covering for electrical wire and cables (Dao et al., 2010). For low-voltage electrical applications, HDPE, and medium-density polyethylene (MDPE) are used depending upon the applications and XLPE used as insulating cable where high thermal resistance is required. Dielectric constant increases with an increase in polarity that leads to increased heat dissipation and sheathing may burn out due to the very high heat energy. This increased heat accumulation is an unfavorable event in the cable sheath insulation and may damage the sheathing material. Hence, lower polarity materials showing low heat generation with excellent flame resistance can be used for designing the sheathing compound. This material property is the reason why very low-density polyethylene cannot be taken as a sheathing material (for high-voltage applications) while cross-linked polyethylene can be used. Without cross-linking, PE gets softened on the transmission line during the heat generation whereas cross-linked PE withstands this condition. PVC can also be used as a sheathing material owing to its excellent flame resistance property. Following are the properties required for a cable insulating material (Naskar, Mohanty, and Nando, 2007),

- Low permittivity
- High dielectric strength

- High volume resistance
- Low dielectric loss
- Low thermal conductivity
- Preferably non-hygroscopic
- Chemical stability over a wide range of temperature
- Higher durability and longer service in exposed environments

Recently, new generation polymeric materials have been developed for cable-insulation applications. Halogen-free cable sheath compound developed from the polyolefine-based blends (Naskar, Mohanty, and Nando, 2007) such as blends of Engage® (ethylene octane copolymer, EOC), and ethylene-vinyl acetate have been used for cable sheath compound preparation. Dutta, Ramachandran, and Naskar, 2016, recently reported on the preparation of cable sheath materials from thermoplastic vulcanizates (TPV). The TPV based on ethylene-vinyl acetate/thermoplastic polyurethane (EVA/TPU) was developed and shows excellent mechanical, thermal, and remarkable oil resistance, and a volume resistivity of 1013 Ωcm, which is suitable for cable sheathing. Padmanabhan et al. have reported on the preparation and development of cable-insulating material from the ethylene octane copolymer (EOC)/polydimethylsiloxane (PDMS) blends. The developed blend shows higher volume resistivity and lower permittivity (Ramachandran, Naskar, and Nando, 2017; Padmanabhan, Naskar, and Nando, 2017).

12.2.6 Biomedical Applications

Tissue engineering is a rapidly growing branch that aims to repair and regenerate cells, damaged tissues, and organs in the human body. In tissue engineering, stable support or scaffold (see Figure 12.3) is required to provide mechanical support, chemical stimuli, and biological signals to cells.

Polymer-based materials are recently being explored for tissue engineering, drug delivery, regenerative medicine, and biosensor applications (Dhandayuthapani et al., 2011). Chitosan, fibroin, and collagen are being used for various biomedical applications, and synthetic polymers such as polycaprolactam also find application as scaffolds material. Different nanoparticles such as graphene and graphene-derived materials, carbon nanotubes, fullerenes, and hydroxyapatite-based nanocomposites are expected to increase the targeted property (Alegret, Dominguez-Alfaro, and Mecerreyes, 2019). Also, how graphene-derived materials can influence the interactions of the material surface with cells, biomolecules, and tissues (S. Kumar and Chatterjee, 2016) is widely studied. The conductive "volatile organic compound" sensor is developed from the nanocomposite composed of poly(lactic acid) as the matrix and MWCNT as the conducting medium, and detection of saccharide using poly(aniline boronic acid)/SWCNT composite have been widely investigated in literature (; B. Kumar, Castro, and Feller, 2012; Badhulika, Tlili, and Mulchandani, 2014). Polycaprolactam/GO nanocomposite-based two-dimensional (2D) and three-dimensional (3D) scaffolds were developed for the pre-osteoblast proliferation and found better proliferation on GO composites on 2D scaffolds than in neat PCL.

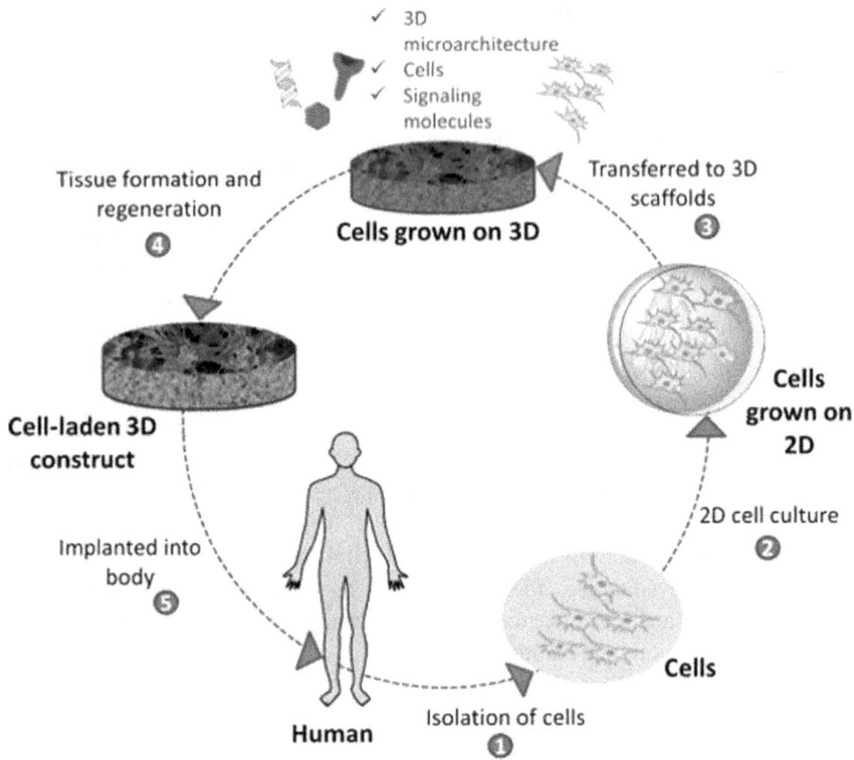

FIGURE 12.3 Schematic of tissue engineering approach. (Adapted from Dasgupta et al., 2019, with permission from Taylor and Francis.)

Incorporation of GO changes the scaffold morphology of the composite by generating the rough surface with sharp protruding edges of the nanofillers, which influences the pre-osteoblast proliferation (S. Kumar et al., 2016). Other approaches such as bone tissue regeneration on graphene-incorporated polymer composite have been widely reported in literature (Meka, Kumar, and Chatterjee, 2016; Jaidev, Kumar, and Chatterjee, 2017; Padmavathy et al., 2017).

12.2.7 Hydrogel

Hydrogels are the three-dimensional network structure that can retain water many times the weight of its polymeric weight. Typical three-dimensional morphology of the hydrogel is shown in Figure 12.4. The high water absorption capacity is attributed to the polar groups present in the polymers, which makes hydrogels hydrophilic material that attracts water molecules to the voids leading to the occupancy of a large amount of water molecules. The cross-links present in the three-dimensional structure help to avoid the disintegration and give strength to the network structure. The presence of cross-links makes the hydrogel structure an insoluble material in many solvents.

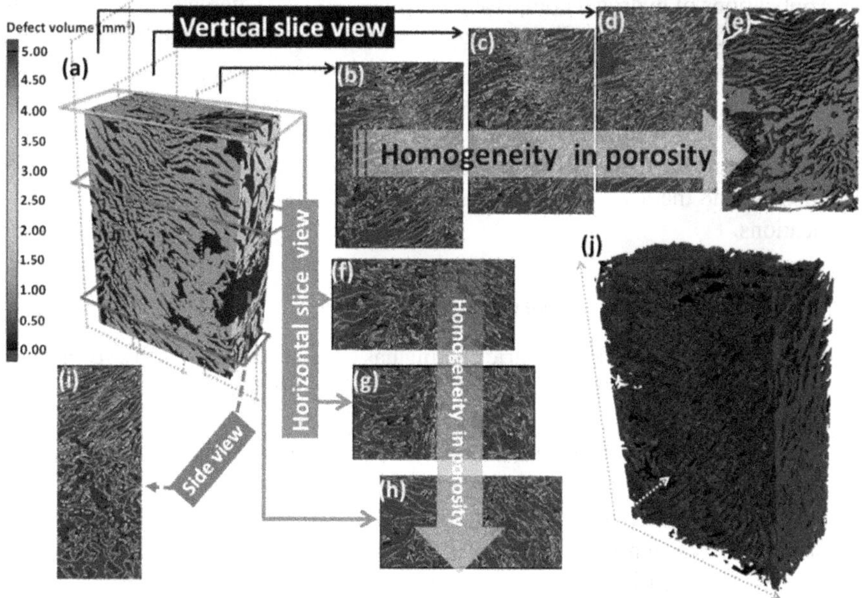

FIGURE 12.4 (a) 3D view of reduced graphene oxide (RGO)-hydrogel monolith. (b–e) Vertical slice view after image processing by software. (f–h) Horizontal slice view. (i) Side view. (j) 3D porous structure with connected and nonconnected pores. (Adapted from Ganguly et al., 2018, with permission from American Chemical Society.)

Hence, non-soluble and high solvent swelling hydrogels can be made by suitably tailoring the polymer materials with desired hydrophilic groups and cross-linking ability. Hydrogels can be made from the homopolymers and are polymerized from the single monomer and forms a cross-linked network structure. Another strategy is to make the structure from the network polymer comprising the two or more monomers form the randomly copolymerized network or ordered network structure through controlled polymerization. Interpenetrating polymeric network (IPN) structure from the two polymeric materials that form mutual networks is also a preparation method for hydrogel. In the full-IPN structure, both the polymers form the independent cross-linked structure whereas in semi-IPN structure one polymer is cross-linked over the other polymer and forms the network structure. Hydrogels can be classified based on the presence of charge; nonionic or neutral, ionic, ampholytic, and zwitterionic types.

Stimuli response hydrogels are developed, and under the influence of external stimuli, hydrogels can undergo swelling or deswelling. Temperature, electric and magnetic field, light, pressure, and sound are physical types of stimuli. pH, ionic strength, and solvent compositions are classified as chemical stimuli. Various polymerization techniques such as bulk, solution, and radiation types are used for the hydrogel preparation. Mainly polar monomer is used for the hydrogel preparation and acrylic-based hydrogels structures are widely studied and reported in the literature (Ganguly, Maity, et al., 2018; Ganguly, Mondal, et al., 2018).

Applications of hydrogel comprise the drug delivery, hygienic products, agriculture, food additive, tissue engineering, biosensor, regenerative medicines, wound dressing, and separation of biomolecules, and biomedical applications (Ganguly and Das, 2017; Ganguly and Das, 2018; Ganguly, Maity, et al., 2018; Ganguly, Das, et al., 2018). Their advantages such as the highest water absorption capacity even under various external stimuli, structural integrity, biodegradability, rewetting ability, and the lower cost makes them an in-demand structure for the various applications.

12.2.8 Antibacterial Polymeric Surfaces

Recently, the antibacterial polymeric coating has been developed to achieve bacterial resistance for various applications such as in water filtration and biomedical fields (Samantaray et al., 2018). Antibacterial property can be imparted in two ways, such as by the incorporation of suitable nanoparticle in the polymer matrix or through the development of antibacterial polymers.

Silver nanoparticle is the most commonly exploited antibacterial agent, and its property is known since ancient times. Varying the silver nanoparticle size influences the antimicrobial activity and was used for developing the different polymer nanocomposite systems in applications such as smart textiles, membranes, and coatings (Ghosh, Ganguly, Das, et al., 2019). Also, incorporation of two-dimensional nanoparticles such as graphene or graphene oxide in the polymer matrix was found to have a considerable impact on the antibacterial property. The sharp edges of the nanosheets can break the cell membrane and cause the cytoplasmic leakage; various proposed mechanisms for the antibacterial activity of the nanosheets were reported in literature (Tu et al., 2013). A plausible mechanism for the antimicrobial activity of the nanocomposite system was shown in Figure 12.5.

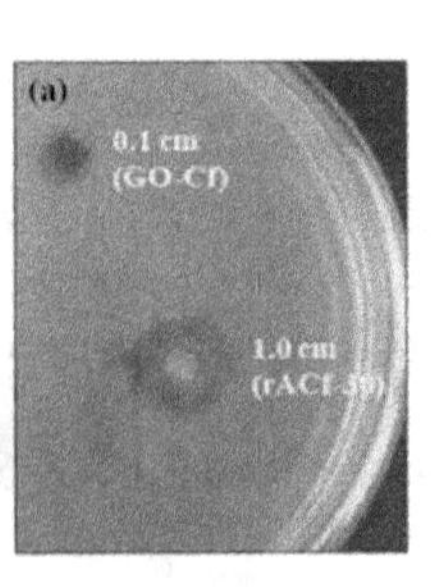

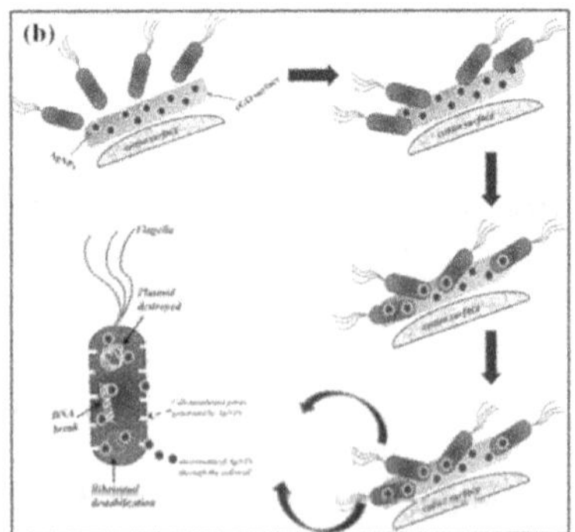

FIGURE 12.5 (a) Digital photographic image of the agar plate showing the inhibition zone of GO-Cf and rACf-30 against E. coli and (b) antibacterial mechanism performed by rGO/Ag decorated cotton fabric against E. coli. (Adapted from Ghosh et al., 2019, with permission from Springer Nature.)

Low-molecular-weight antimicrobial agents have many disadvantages, including pollution to the environment by leaching and reduced activity after a while. The antimicrobial polymer can be a promising answer to reduce the environmental problems and providing prolonged lifetime for various applications. There are different methods to prepare the antimicrobial polymers that include,

1. Addition of organic or inorganic biocides during the processing of polymers (Bugatti et al., 2011)
2. Modification of polymer surface with suitable antibacterial agents (Kenawy et al., 1998; Eknoian, Worley, and Harris, 1998; Abd El-Rehim et al., 2004; Chen and Han, 2011)
3. Preparation of antimicrobial polymer from modified biocide active monomers (Bankova et al., 1996; Bankova et al., 1997; Kienberger et al., 2012)

Kenawy Worley, and Broughton, 2007, described the basic requirements for a polymer to act as antibacterial as follows,

1. Should not be soluble in water
2. Should be reusable after a prolonged service
3. Can be applicable for a broad spectrum of biocides
4. Should have a good service life
5. Stable at ambient temperature and high-temperature application
6. Should not decompose to produce toxic effects in the medium or during handling

Several antimicrobial active polymers and their structures are listed in Table 12.3.

12.3 SUMMARY

This chapter is meticulously focused on the different applications of the thermoplastic- and thermosetting-based composites. Various thermoplastic- and thermosetting-based composites in engineering, biomedical, and aerospace applications were discussed. New polymeric materials with tailored properties can be synthesized by altering the chemical compositions or blending two different polymers. The incorporation of nanofillers can enhance the properties for EMI shielding, aerospace, and cable insulation applications. The optimum amount of nanofillers is an important aspect because they are required to achieve (a network of interconnected fillers in the polymer matrix) the better thermal, electrical, and physicomechanical properties. Other recent approaches, such as polymers containing CNT and graphene-derived nanoparticles for biomedical and tissue engineering applications are also gaining significant attention. Hence, the fabrication of polymer material with tunable properties can be achieved with the incorporation of nanofillers or by modifying the polymer chains, which is a simple strategy for lightweight and flexible device applications.

TABLE 12.3
Antimicrobial Polymers and Their Structures

Sl. no.	Polymer	Structure	Bacteria used	Reference
1	Poly(hexamethylene biguanide chloride) (PHMB)	$-[(CH_2)_6-NH-C(=NH)-NH-C(=NH.HCl)-NH]_n-$	*Escherichia coli* (*E. coli*)	Broxton, Woodcock, and Gilbert, 1983
2	Polymethacrylate containing pendant biguanide	$-[C(CH_3)(C(=O)-O-CH_2-CH_2-C_6H_4-NH-C(=NH)-NH-C(=NH)-NH-C_6H_4-Cl)-CH_2]_n-$	*Staphylococcus aureus* (*S. aureus*) and *E. coli*	Ikeda, Yamaguchi, and Tazuke, 1984

(*Continued*)

TABLE 12.3 (CONTINUED)
Antimicrobial Polymers and Their Structures

Sl. no.	Polymer	Structure	Bacteria used	Reference
3	Poly(vinylbenzyl ammonium chloride)	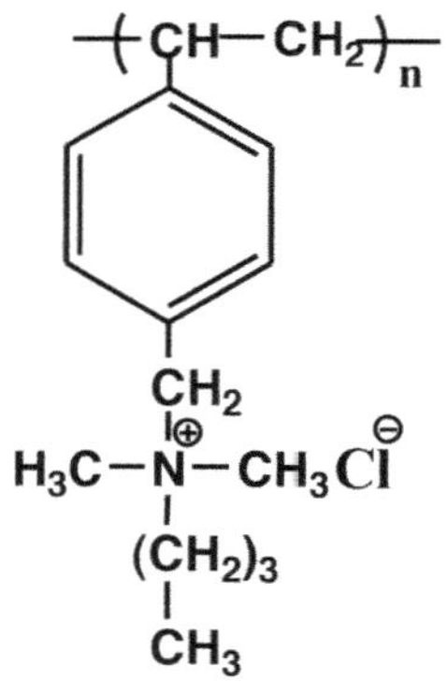	*S. aureus*	Ikeda, Tazuke, and Suzuki, 1984
4	Vinyl derivatives	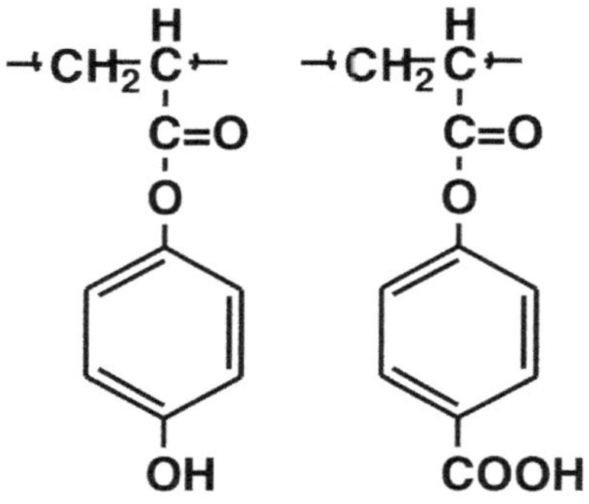	*S. aureus* *P. aeruginosa* *Penicillium-pinophilum* *Aspergillus-fumigatus*	Park et al., 2001

(*Continued*)

TABLE 12.3 (CONTINUED)
Antimicrobial Polymers and Their Structures

Sl. no.	Polymer	Structure	Bacteria used	Reference
5	N-hexylated+methylated high molecular weight polyethylenimine (PEI)	–	Airborne *S. aureus* *Candida albicans* *E. coli*	Lin et al., 2003
6	Primary ammonium ethyl methacrylate homopolymers (AEMPs)	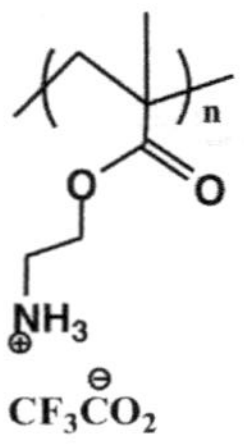 n, O, O, $\overset{\oplus}{N}H_3$, $CF_3\overset{\ominus}{C}O_2$	*Pseudomonas-aeruginosa* *S. aureus*	Thoma, Boles, and Kuroda, 2014

REFERENCES

Abd El-Rehim, H. A., A. El-Hag Ali, T. B. Mostafa, and Hala A. Farrag. 2004. "Anti-microbial activity of anhydride copolymers and their derivatives prepared by ionizing radiation." *European Polymer Journal* 40(9): 2203–2212. doi: 10.1016/j.eurpolymj.2004.01.044.

Alegret, Nuria, Antonio Dominguez-Alfaro, and David Mecerreyes. 2019. "3D scaffolds based on conductive polymers for biomedical applications." *Biomacromolecules* 20(1): 73–89. doi: 10.1021/acs.biomac.8b01382.

Anagha, M. G., and Kinsuk Naskar. 2019. "Augmentation of performance properties of maleated SEBS/TPU blends through reactive blending." *Journal of Applied Polymer Science* 48727. doi: 10.1002/app.48727.

Ashori, Alireza. 2008. "Wood–plastic composites as promising green-composites for automotive industries!" *Bioresource Technology* 99(11): 4661–4667. doi: 10.1016/j.biortech.2007.09.043.

Badhulika, Sushmee, Chaker Tlili, and Ashok Mulchandani. 2014. "Poly(3-aminophenylboronic acid)-functionalized carbon nanotubes-based chemiresistive sensors for detection of sugars." *Analyst* 139(12): 3077–3082. doi: 10.1039/C4AN00004H.

Bankova, M., N. Manolova, N. Markova, T. Radoucheva, K. Dilova, and I. Rashkov. 1997. "Hydrolysis and antibacterial activity of polymers containing 8-quinolinyl acrylate." *Journal of Bioactive and Compatible Polymers* 12(4): 294–307. doi: 10.1177/088391159701200403.

Bankova, M., Ts. Petrova, N. Manolova, and I. Rashkov. 1996. "Homopolymers of 5-chloro-8-quinolinyl acrylate and 5-chloro-8-quinolinyl methacrylate and their copolymers with acrylic and methacrylic acid." *European Polymer Journal* 32(5): 569–578. doi: 10.1016/0014-3057(95)00167-0.

Behera, Prasanta Kumar, Prantik Mondal, and Nikhil K. Singha. 2018. "Polyurethane with an ionic liquid crosslinker: A new class of super shape memory-like polymers." *Polymer Chemistry* 9(31): 4205–4217. doi: 10.1039/C8PY00549D.

Biju, R., C. Gouri, and C. P. Reghunadhan Nair. 2012. "Shape memory polymers based on cyanate ester-epoxy-poly (tetramethyleneoxide) co-reacted system." *European Polymer Journal* 48(3): 499–511. doi: 10.1016/j.eurpolymj.2011.11.019.

Biswas, Sourav, Injamamul Arief, Sujit Sankar Panja, and Suryasarathi Bose. 2017. "Absorption-dominated electromagnetic wave suppressor derived from ferrite-doped cross-linked graphene framework and conducting carbon." *ACS Applied Materials and Interfaces* 9(3): 3030–3039. doi: 10.1021/acsami.6b14853.

Biswas, Sourav, Goutam Prasanna Kar, and Suryasarathi Bose. 2015. "Attenuating microwave radiation by absorption through controlled nanoparticle localization in PC/PVDF blends." *Physical Chemistry Chemical Physics: PCCP* 17(41): 27698–27712. doi: 10.1039/C5CP05189D.

Broxton, P., P. M. Woodcock, and P. Gilbert. 1983. "A study of the antibacterial activity of some polyhexamethylene biguanides towards Escherichia coli ATCC 8739." *Journal of Applied Bacteriology* 54(3): 345–353. doi: 10.1111/j.1365-2672.1983.tb02627.x.

Bugatti, Valeria, Giuliana Gorrasi, Francesca Montanari, Morena Nocchetti, Loredana Tammaro, and Vittoria Vittoria. 2011. "Modified layered double hydroxides in polycaprolactone as a tunable delivery system: In vitro release of antimicrobial benzoate derivatives." *Applied Clay Science* 52(1): 34–40. doi: 10.1016/j.clay.2011.01.025.

Chatterjee, Tuhin, and Kinsuk Naskar. 2019. "Thermo-sensitive shape memory polymer nanocomposite based on polyhedral oligomeric silsesquioxane (POSS) filled polyolefins." *Polymer-Plastics Technology and Materials* 58(6): 630–640. doi: 10.1080/03602559.2018.1493127.

Chatterjee, Tuhin, Sven Wiessner, Y. K. Bhardwaj, and Kinsuk Naskar. 2019. "Exploring heat induced shape memory behaviour of alpha olefinic blends having dual network structure." *Materials Science and Engineering B: Solid-State Materials for Advanced Technology* 240: 75–84. doi: 10.1016/j.mseb.2019.01.015.

Chen, Yong, and Qiuxia Han. 2011. "Designing N-halamine based antibacterial surface on polymers: Fabrication, characterization, and biocidal functions." *Applied Surface Science* 257(14): 6034–6039. doi: 10.1016/j.apsusc.2011.01.115.

Cheng, Tianze. 2019. "Review of novel energetic polymers and binders–high energy propellant ingredients for the new space race." *Designed Monomers and Polymers* 22(1): 54–65. doi: 10.1080/15685551.2019.1575652.

Coulter, Fergal B., Manuel Schaffner, Jakob A. Faber, Ahmad Rafsanjani, Robin Smith, Harish Appa, Peter Zilla, Deon Bezuidenhout, and André R. Studart. 2019. "Bioinspired heart valve prosthesis made by Silicone Additive Manufacturing." *Matter* 1(1): 266–279. doi: 10.1016/j.matt.2019.05.013.

Dao, N. L., P. L. Lewin, I. L. Hosier, and S. G. Swingler. 2010. "A comparison between LDPE and HDPE cable insulation properties following lightning impulse ageing." *2010 10th IEEE International Conference on Solid Dielectrics*, 4–9 July 2010. doi: 10.1109/ICSD.2010.5567944.

Das, N. C., T. K. Chaki, and D. Khastgir. 2002. "Effect of processing parameters, applied pressure and temperature on the electrical resistivity of rubber-based conductive composites." *Carbon* 40(6): 807–816. doi: 10.1016/S0008-6223(01)00229-9.

Das, N. C., D. Khastgir, T. K. Chaki, and A. Chakraborty. 2000. "Electromagnetic interference shielding effectiveness of carbon black and carbon fibre filled EVA and NR based composites." *Composites—Part A: Applied Science and Manufacturing* 31(10): 1069–1081. doi: 10.1016/S1359-835X(00)00064-6.

Das, P., S. Ganguly, P. P. Maity, H. K. Srivastava, M. Bose, Dhara, B. Sharba, A. K. Das, B. Susanta, and N. C. Das. 2019. "Converting waste Allium sativum peel to nitrogen and sulphur co-doped photoluminescence carbon dots for solar conversion, cell labeling, and photobleaching diligences: A path from discarded waste to value-added products." *Journal of Photochemistry and Photobiology, Part B: Biology* 197: 111545. doi: 10.1016/j.jphotobiol.2019.111545.

Das, T. K., G. Prosenjit, and N. C. Das. 2019. "Preparation, development, outcomes, and application versatility of carbon fiber-based polymer composites: A review." *Advanced Composites and Hybrid Materials*: 1–20. doi: 10.1007/s42114-018-0072-z.

Dasgupta, Queeny, Giridhar Madras, and Kaushik Chatterjee. "Biodegradable polyol-based polymers for biomedical applications." *International Materials Reviews* 64, no. 5 (2019): 288–309. doi: 10.1080/09506608.2018.1505066.

De Groh, Kim K., Bruce A. Banks, Catherine E. Mccarthy, Rochelle N. Rucker, Lily M. Roberts, and Lauren A. Berger. 2008. "MISSE 2 PEACE polymers atomic oxygen erosion experiment on the International Space Station." *High Performance Polymers* 20(4–5): 388–409. doi: 10.1177/0954008308089705.

Dhandayuthapani, Brahatheeswaran, Yasuhiko Yoshida, Toru Maekawa, and D. Sakthi Kumar. 2011. "Polymeric scaffolds in tissue engineering application: A review." *International Journal of Polymer Science* 2011. doi: 10.1155/2011/290602.

Dutta, Joyeeta, Padmanabhan Ramachandran, and Kinsuk Naskar. 2016. "Scrutinizing the influence of peroxide crosslinking of dynamically vulcanized EVA/TPU blends with special reference to cable sheathing applications." *Journal of Applied Polymer Science* 133(29). doi 10.1002/app.43706.

Eknoian, M. W., S. D. Worley, and J. M. Harris. 1998. "New biocidal N-Halamine-PEG polymers." *Journal of Bioactive and Compatible Polymers* 13(2): 136–145. doi: 10.1177/088391159801300205.

Ganguly, Sayan, Poushali Bhawal, Revathy Ravindren, and Narayan Chandra Das. 2018. "Polymer nanocomposites for electromagnetic interference shielding: A review." *Journal of Nanoscience and Nanotechnology* 18(11): 7641–7669. doi: 10.1166/jnn.2018.15828.

Ganguly, Sayan, and Narayan Ch Das. 2017. "Water uptake kinetics and control release of agrochemical fertilizers from nanoclay-assisted semi-interpenetrating sodium acrylate-based hydrogel." *Polymer-Plastics Technology and Engineering* 56(7): 744–761. doi: 10.1080/03602559.2016.1233268.

Ganguly, Sayan, and Narayan Ch Das. 2018. "Synthesis of mussel inspired polydopamine coated halloysite nanotubes based semi-IPN: An approach to fine tuning in drug release and mechanical toughening." *Macromolecular Symposia* 382(1): 1800076. doi: 10.1002/masy.201800076.

Ganguly, Sayan, Poushali Das, Priti Prasanna Maity, Subhadip Mondal, Sabyasachi Ghosh, Santanu Dhara, and Narayan Ch Das. 2018. "Green reduced graphene oxide toughened semi-IPN monolith hydrogel as dual responsive drug release system: Rheological, physicomechanical, and electrical evaluations." *The Journal of Physical Chemistry. Part B* 122(29): 7201–7218. doi: 10.1021/acs.jpcb.8b02919.

Ganguly, Sayan, Priti Prasanna Maity, Subhadip Mondal, Poushali Das, Poushali Bhawal, Santanu Dhara, and Narayan Ch Das. 2018. "Polysaccharide and poly(methacrylic acid) based biodegradable elastomeric biocompatible semi-IPN hydrogel for controlled drug delivery." *Materials Science and Engineering C: Biomimetic Materials Sensors and Systems* 92: 34–51. doi: 10.1016/j.msec.2018.06.034.

Ganguly, Sayan, Subhadip Mondal, Poushali Das, Poushali Bhawal, Priti Prasanna Maity, Sabyasachi Ghosh, Santanu Dhara, and Narayan Ch Das. 2018. "Design of psyllium-g-poly(acrylic acid-co-sodium acrylate)/cloisite 10A semi-IPN nanocomposite hydrogel and its mechanical, rheological and controlled drug release behaviour." *International Journal of Biological Macromolecules* 111: 983–998. doi: 10.1016/j.ijbiomac.2018.01.100.

Geetha, S., K. K. Satheesh Kumar, Chepuri R. K. Rao, M. Vijayan, and D. C. Trivedi. 2009. "EMI shielding: Methods and materials—A review." *Journal of Applied Polymer Science* 112(4): 2073–2086. doi: 10.1002/app.29812.

Geethamma, V. G., G. Kalaprasad, Gabriël Groeninckx, and Sabu Thomas. 2005. "Dynamic mechanical behavior of short coir fiber reinforced natural rubber composites." *Composites—Part A: Applied Science and Manufacturing* 36(11): 1499–1506. doi: 10.1016/j.compositesa.2005.03.004.

Ghosh, Sabyasachi, Sayan Ganguly, Poushali Das, Tushar Kanti Das, Madhuparna Bose, Nikhil K. Singha, Amit Kumar Das, and Narayan Ch Das. 2019. "Fabrication of reduced graphene oxide/silver nanoparticles decorated conductive cotton fabric for high performing electromagnetic interference shielding and antibacterial application." *Fibers and Polymers* 20(6): 1161–1171. doi: 10.1007/s12221-019-1001-7.

Ghosh, Sabyasachi, Sayan Ganguly, Sanjay Remanan, and Narayan Ch Das. 2019. "Fabrication and investigation of 3D tuned PEG/PEDOT: PSS treated conductive and durable cotton fabric for superior electrical conductivity and flexible electromagnetic interference shielding." *Composites Science and Technology* 181. doi: 10.1016/j.compscitech.2019.107682.

Ghosh, Sabyasachi, Sayan Ganguly, Sanjay Remanan, Subhadip Mondal, Subhodeep Jana, Pradip K. Maji, Nikhil Singha, and Narayan Ch Das. 2018. "Ultra-light weight, water durable and flexible highly electrical conductive polyurethane foam for superior electromagnetic interference shielding materials." *Journal of Materials Science: Materials in Electronics* 29(12): 10177–10189. doi: 10.1007/s10854-018-9068-2.

Ghosh, Sabyasachi, Sanjay Remanan, Subhadip Mondal, Sayan Ganguly, Poushali Das, Nikhil Singha, and Narayan Ch Das. 2018. "An approach to prepare mechanically robust full IPN strengthened conductive cotton fabric for high strain tolerant electromagnetic interference shielding." *Chemical Engineering Journal* 344: 138–154. doi: 10.1016/j.cej.2018.03.039.

Gopalan, Anagha M., and Kinsuk Naskar. 2019. "Ultra-high molecular weight styrenic block copolymer/TPU blends for automotive applications: Influence of various compatibilizers." *Polymers for Advanced Technologies* 30(3): 608–619. doi: 10.1002/pat.4497.

Guillen, Gregory R., Yinjin Pan, Minghua Li, and Eric M.V. Hoek. "Preparation and characterization of membranes formed by nonsolvent induced phase separation: A review." *Industrial & Engineering Chemistry Research* 50, no. 7 (2011): 3798–3817. doi: 10.1021/ie101928r.

Holbery, James, and Dan Houston. 2006. "Natural-fiber-reinforced polymer composites in automotive applications." *JOM* 58(11): 80–86. doi: 10.1007/s11837-006-0234-2.

Ikeda, Tomiki, Shigeo Tazuke, and Yasuzo Suzuki. 1984. "Biologically active polycations, 4. Synthesis and antimicrobial activity of poly(trialkylvinylbenzylammonium chloride)s." *Die Makromolekulare Chemie* 185(5): 869–876. doi: 10.1002/macp.1984.021850503.

Ikeda, Tomiki, Hideki Yamaguchi, and Shigeo Tazuke. 1984. "New polymeric biocides: Synthesis and antibacterial activities of polycations with pendant biguanide groups." *Antimicrobial Agents and Chemotherapy* 26(2): 139–144. doi: 10.1128/AAC.26.2.139.

Jaidev, L. R., Sachin Kumar, and Kaushik Chatterjee. 2017. "Multi-biofunctional polymer graphene composite for bone tissue regeneration that elutes copper ions to impart angiogenic, osteogenic and bactericidal properties." *Colloids and Surfaces, Part B: Biointerfaces* 159: 293–302. doi: 10.1016/j.colsurfb.2017.07.083.

Jin, Jianyong, Dennis W. Smith, Chris M. Topping, S. Suresh, Shengrong Chen, Stephen H. Foulger, Norman Rice, Jon Nebo, and Bob H. Mojazza. 2003. "Synthesis and characterization of phenylphosphine oxide containing Perfluorocyclobutyl aromatic ether polymers for potential space applications." *Macromolecules* 36(24): 9000–9004. doi: 10.1021/ma035241g.

Jose, Seno, Jinu Jacob George, Suchart Siengchin, and Jyotishkumar Parameswaranpillai. 2020. "Introduction to shape-memory polymers, polymer blends and composites: State of the art, opportunities, new challenges and future outlook." In: *Shape Memory Polymers, Blends and Composites*, 1–19. Springer. doi: 10.1007/978-981-13-8574-2_1.

Kar, Goutam Prasanna, Sourav Biswas, and Suryasarathi Bose. 2016. "Tuning the microwave absorption through engineered nanostructures in co-continuous polymer blends." *Materials Research Express* 3(6): 064002. doi: 10.1088/2053-1591/3/6/064002.

Kenawy, El-Refaie, Fouad I. Abdel-Hay, Abd El-Raheem R. El-Shanshoury, and Mohamed H. El-Newehy. 1998. "Biologically active polymers: Synthesis and antimicrobial activity of modified glycidyl methacrylate polymers having a quaternary ammonium and phosphonium groups." *Journal of Controlled Release* 50(1–3): 145–152. doi: 10.1016/S0168-3659(97)00126-0.

Kenawy, El-Refaie, S. D. Worley, and Roy Broughton. 2007. "The chemistry and applications of antimicrobial polymers: A state-of-the-art review." *Biomacromolecules* 8(5): 1359–1384. doi: 10.1021/bm061150q.

Kienberger, Julia, Nadja Noormofidi, Inge Mühlbacher, Ingo Klarholz, Carsten Harms, and Christian Slugovc. 2012. "Antimicrobial equipment of poly (isoprene) applying thiol-ene chemistry." *Journal of Polymer Science Part A: Polymer Chemistry* 50(11): 2236–2243. doi: 10.1002/pola.26001.

Kim, Deuk Ju, Byeol-Nim Lee, and Sang Yong Nam. 2017. Characterization of highly sulfonated PEEK based membrane for the fuel cell application." *International Journal of Hydrogen Energy* 42(37): 23768–23775. doi: 10.1016/j.ijhydene.2017.04.082.

Koronis, Georgios, Arlindo Silva, and Mihail Fontul. 2013. "Green composites: A review of adequate materials for automotive applications." *Composites Part B Engineering* 44(1): 120–127. doi: 10.1016/j.compositesb.2012.07.004.

Krishnamurthy, V. N. 1995. "Polymers in space environments." In: *Polymers and Other Advanced Materials*, 221–226. Springer. doi: 10.1007/978-1-4899-0502-4_23.

Kumar, Bijandra, Mickaël Castro, and Jean-François Feller. 2012. "Poly (lactic acid)–multi-wall carbon nanotube conductive biopolymer nanocomposite vapour sensors." *Sensors and Actuators. Part B: Chemical* 161(1): 621–628. doi: 10.1016/j.snb.2011.10.077.

Kumar, Sachin, Dilkash Azam, Shammy Raj, Elayaraja Kolanthai, K. S. Vasu, A. K. Sood, and Kaushik Chatterjee. 2016. "3D scaffold alters cellular response to graphene in a polymer composite for orthopedic applications." *Journal of Biomedical Materials Research, Part B: Applied Biomaterials* 104(4): 732–749. doi: 10.1002/jbm.b.33549.

Kumar, Sachin, and Kaushik Chatterjee. 2016. "Comprehensive review on the use of graphene-based substrates for regenerative medicine and biomedical devices." *ACS Applied Materials and Interfaces* 8(40): 26431–26457. doi: 10.1021/acsami.6b09801.

Lin, Jian, Shuyi Qiu, Kim Lewis, and Alexander M. Klibanov. 2003. "Mechanism of bactericidal and fungicidal activities of textiles covalently modified with alkylated polyethylenimine." *Biotechnology and Bioengineering* 83(2): 168–172. doi: 10.1002/bit.10651.

Liu, Yanju, Haiyang Du, Liwu Liu, and Jinsong Leng. 2014. "Shape memory polymers and their composites in aerospace applications: A review." *Smart Materials and Structures* 23(2): 023001. doi: 10.1088/0964-1726/23/2/023001.

Meka, Sai Rama Krishna, Sachin Kumar, and Kaushik Chatterjee. 2016. "Nanofibrous scaffolds for the regeneration of bone tissue." In: *Biomaterials and Nanotechnology for Tissue Engineering*, 53–92. CRC Press. doi: 10.1201/9781315368955.

Mu, Tong, Liwu Liu, Xin Lan, Yanju Liu, and Jinsong Leng. 2018. "Shape memory polymers for composites." *Composites Science and Technology* 160: 169–198. doi: 10.1016/j.compscitech.2018.03.018.

Mulder, J. 2012. *Basic Principles of Membrane Technology*. Springer Science & Business Media. doi: 10.1007/978-94-009-1766-8.

Naskar, K., S. Mohanty, and G. B. Nando. 2007. "Development of thin-walled halogen-free cable insulation and halogen-free fire-resistant low-smoke cable-sheathing compounds based on polyolefin elastomer and ethylene vinyl acetate blends." *Journal of Applied Polymer Science* 104(5): 2839–2848. doi: 10.1002/app.25870.

Padmanabhan, R., Kinsuk Naskar, and Golok B. Nando. 2017. "Influence of octene level in EOC–PDMS thermoplastic vulcanizates for cable insulation applications." *Polymer-Plastics Technology and Engineering* 56(3): 276–295. doi: 10.1080/03602559.2016.1227841.

Padmavathy, Nagarajan, L. R. Jaidev, Suryasarathi Bose, and Kaushik Chatterjee. 2017. "Oligomer-grafted graphene in a soft nanocomposite augments mechanical properties and biological activity." *Materials and Design* 126: 238–249. doi: 10.1016/j.matdes.2017.03.087.

Parameswaranpillai, Jyotishkumar, Nishar Hameed, Thomas Kurian, and Yu Yingfeng. 2016. *Nanocomposite materials: Synthesis, properties and applications*. CRC Press. doi: 10.1201/9781315372310.

Parameswaranpillai, Jyotishkumar, Harikrishnan Pulikkalparambil, M. R. Sanjay, and Suchart Siengchin. 2019. "Polypropylene/high-density polyethylene based blends and nanocomposites with improved toughness." *Materials Research Express* 6(7): 075334. doi: 10.1088/2053-1591/ab18cd.

Park, Eun-Soo, Woong-Sig Moon Min-Jin Song, Mal-Nam Kim, Kyoo-Hyun Chung, and Jin-San Yoon. 2001. "Antimicrobial activity of phenol and benzoic acid derivatives." *International Biodeterioration and Biodegradation* 47(4): 209–214. doi: 10.1016/S0964-8305(01)00058-0.

Patil, Akshat, Arun Patel, and Rajesh Purohit. 2017. "An overview of polymeric materials for automotive applications." *Materials Today: Proceedings* 4(2): 3807–3815. doi: 10.1016/j.matpr.2017.02.278.

Pawar, Shital Patangrao, Dhruva A. Marathe, K. Pattabhi, and Suryasarathi Bose. 2015. "Electromagnetic interference shielding through MWNT grafted Fe3O4 nanoparticles in PC/SAN blends." *Journal of Materials Chemistry A* 3(2): 656–669. doi: 10.1039/C4TA04559A.

Pulikkalparambil, Harikrishnan, J. Parameswaranpillai, J. J. George, K. Yorseng, and S. Siengchin. 2017. Physical and thermo-mechanical properties of bionano reinforced poly (butylene adipate-co-terephthalate), hemp/CNF/Ag-NPs composites. *AIMS Materials Science* 4(3): 814–831. doi: 10.3934/matersci.2017.3.814.

Ramachandran, Padmanabhan, Kinsuk Naskar, and Golok B. Nando. 2017. "Exploring the effect of radiation crosslinking on the physico-mechanical, dynamic mechanical and dielectric properties of EOC–PDMS blends for cable insulation applications." *Polymers for Advanced Technologies* 28(1): 80–93. doi: 10.1002/pat.3862.

Ravindren, Revathy, Subhadip Mondal, Krishnendu Nath, and Narayan Ch Das. 2019. "Prediction of electrical conductivity, double percolation limit and electromagnetic interference shielding effectiveness of copper nanowire filled flexible polymer blend nanocomposites." *Composites Part B Engineering* 164: 559–569. doi: 10.1016/j.compositesb.2019.01.066.

Remanan, Sanjay, Maya Sharma, Suryasarathi Bose, and Narayan Ch Das. 2018. "Recent advances in preparation of porous polymeric membranes by unique techniques and mitigation of fouling through surface modification." *ChemistrySelect* 3(2): 609–633. doi: 10.1002/slct.201702503.

Remanan, Sanjay, Maya Sharma, Priyadarshini Jayashree, Jyotishkumar Parameswaranpillai, Thomas Fabian, Julie Shih, Prasad Shankarappa, Bharath Nuggehalli, and Suryasarathi Bose. 2017. "Unique synergism in flame retardancy in ABS based composites through blending PVDF and halloysite nanotubes." *Materials Research Express* 4(6). doi: 10.1088/2053-1591/aa7617.

Samantaray, Paresh Kumar, Sonika Baloda, Giridhar Madras, and Suryasarathi Bose. 2018. "A designer membrane tool-box with a mixed metal organic framework and RAFT-synthesized antibacterial polymer perform in tandem towards desalination, antifouling and heavy metal exclusion." *Journal of Materials Chemistry A* 6(34): 16664–16679. doi: 10.1039/C8TA05052J.

Santo, L., F. Quadrini, D. Bellisario, and L. Iorio. 2020. "Applications of shape-memory polymers, and their blends and composites." In: *Shape Memory Polymers, Blends and Composites*, 311–329. Springer. doi: 10.1007/978-981-13-8574-2_13.

Sengupta, Rajatendu, Mithun Bhattacharya, S. Bandyopadhyay, and Anil K. Bhowmick. 2011. "A review on the mechanical and electrical properties of graphite and modified graphite reinforced polymer composites." *Progress in Polymer Science* 36(5): 638–670. doi: 10.1016/j.progpolymsci.2010.11.003.

Shumaker, J. A., A. J. W. McClung, and J. W. Baur. 2012. "Synthesis of high temperature polyaspartimide-urea based shape memory polymers." *Polymer* 53(21): 4637–4642. doi: 10.1016/j.polymer.2012.08.021.

Štrumberger, Nada, Alen Gospočić, and Čedomir Bartulić. 2005. "Polymeric materials in automobiles." *Promet-Traffic and Transportation* 17(3): 149–160. doi: http://traffic.fpz.hr/index.php/PROMTT/article/view/630.

Thoma, Laura M., Blaise R. Boles, and Kenichi Kuroda. 2014. "Cationic methacrylate polymers as topical antimicrobial agents against Staphylococcus aureus Nasal colonization." *Biomacromolecules* 15(8): 2933–2943. doi: 10.1021/bm500557d.

Tiwari, Shivam, Asish Ghosh, Angshuman Santra, Tuhin Das, and Sabyasachi Ghosh. 2015. "Emerging tunable fluorescence in nitrogen doped carbon quantum dot." *The Journal of Materials Science and Mechanical Engineering* 2: 58–60.

Tu, Yusong, Min Lv, Peng Xiu, Tien Huynh, Meng Zhang, Matteo Castelli, Zengrong Liu, Qing Huang, Chunhai Fan, Haiping Fang, and Ruhong Zhou. 2013. "Destructive extraction of phospholipids from Escherichia coli membranes by graphene nanosheets." *Nature Nanotechnology* 8(8): 594. doi: 10.1038/nnano.2013.125.

Vijayalekshmi, V., and Dipak Khastgir. 2018a. "Chitosan/partially sulfonated poly(vinylidene fluoride) blends as polymer electrolyte membranes for direct methanol fuel Cell Applications." *Cellulose* 25(1): 661–681. doi: 10.1007/s10570-017-1565-6.

Vijayalekshmi, V. and Dipak Khastgir. 2018b. "Fabrication and comprehensive investigation of physicochemical and electrochemical properties of chitosan-silica supported silicotungstic acid nanocomposite membranes for fuel cell applications." *Energy* 142: 313–330. doi: 10.1016/j.energy.2017.10.019.

Xie, Tao, and Ingrid A. Rousseau. 2009. "Facile tailoring of thermal transition temperatures of epoxy shape memory polymers." *Polymer* 50(8): 1852–1856. doi: 10.1016/j.polymer.2009.02.035.

13 Life Cycle Assessment of Thermoplastic and Thermosetting Composites

Alexander Kaluza, Johanna Sophie Hagen, Antal Dér, Felipe Cerdas, and Christoph Herrmann

CONTENTS

13.1 INTRODUCTION

The living standards of modern societies and the associated high demand for industrial and consumer goods are strongly linked to environmental concerns (Allwood et al., 2011). For instance, industrial production and transport contributed to global CO_2 emissions related to energy and processes with shares of 25% and 23% in 2014, while increasing absolute numbers were observed (IPCC, 2018). An exponential population growth and increasing affluence lead to further increases of man-made environmental impacts. Therefore, with the goal to counteract that development, industrial products and processes have to be engineered with knowledge of related environmental effects (Hauschild, Herrmann, and Kara, 2017).

Fiber-reinforced plastics (FRP) are high-performance engineering materials that are applied to manufacture industrial goods in different sectors. In general, FRP consist of fiber materials that are embedded in polymer matrices (Fleischer et al., 2018). Glass, carbon, and natural fibers are the prevalent fiber reinforcements, while materials are distinguished between thermoplastic and thermoset materials. In general, fiber and matrix materials can origin from fossil as well as renewable resources. Fossil resources cover crude oil, that serves as a basic product of industrial polymer manufacturing (Plastics Europe, 2019). Renewable resources include natural feedstock, such as sugarcane, corn, or ligno-cellulosics (Spierling et al., 2018). A wide range of industrial processes is available for manufacturing FRP. Thereby, thermoset process chains rely on chemical curing for solidification, whereas thermoplastic materials solidify when cooling down after melting during manufacturing (Fleischer et al., 2018).

Advantages of FRP over other engineering materials, e.g., plastics or metals, include a high specific strength and stiffness, low thermal expansion, corrosion resistance, and high energy absorption capacity (Fleischer et al., 2018). FRP could therefore enable the realization of high-performance designs or the reduction of a product's weight while retaining or improving mechanical performance. In the first case, industrial products might demand for a specific material profile. For example, metrology devices or inspection machines require maximum dynamic stiffness and damping capacity (Möhring, 2017). In the second case, FRP's strength and stiffness properties could be leveraged for reducing inertia and mass of moved elements, e.g., machine tools (Möhring, 2017) or reducing use stage energy demands in transportation, e.g., in automotive or aviation industries (Timmis et al., 2015; Koffler and Rohde-Brandenburger, 2010).

Material substitution in general and the use of FRP in specific requires a life cycle perspective if the overall goal of reduced environmental impacts should be achieved and problem shifting should be avoided. The life cycle assessment (LCA) methodology provides foundational steps toward fulfilling that task. LCA aims at quantifying environmental impacts of products and processes. Based on an inventory of all material and energy flows of a product system, resulting impacts are calculated for different impact categories such as climate change measured in equivalents of carbon dioxide emissions (CO_2-eq).

The present chapter discusses basic methods and specific challenges when performing LCA for FRP composites. The following subsection will introduce an LCA system description for performing FRP LCA. The subsequent sections further detail methods and challenges to be addressed within single life cycle stages.

13.2 LCA METHODOLOGY FOR THERMOPLASTIC AND THERMOSETTING COMPOSITES

The present section will introduce a system understanding for performing LCA for FRP composites. This includes a description of major life cycle stages and an explanation of terms related to LCA-based system understanding. According to ISO

14040, the LCA methodology comprises four major steps: i) goal and scope definition, ii) life cycle inventory (LCI), iii) life cycle impact assessment (LCIA), and iv) interpretation. However, a description of the system is discussed first.

13.2.1 System Description for LCA of Thermoplastic and Thermosetting Composites

Figure 13.1 presents a high-level system description for performing LCA of thermoplastic and thermosetting composites. The figure is organized with composite life cycle stages being in the center and depicts energy and material flows along the composite life cycle. Further, the composite life cycle is embedded into the LCA methodology, where LCI is derived from the accumulation of energy and material flows, LCIA covers the analysis of associated environmental impacts to those flows, and interpretation covers the application of gained insights in engineering or decision-making situations. The depicted stages include all processes that are directly linked or allocated to a composite's life cycle. Following the description by Fleischer et al., four main stages are distinguished for FRP life cycles (Fleischer et al., 2018):

- **Materials production**: This initial stage includes the manufacturing of fibers and thermoplastic or thermoset matrix materials that are applied as an input to the manufacturing of composite semi-products and parts.

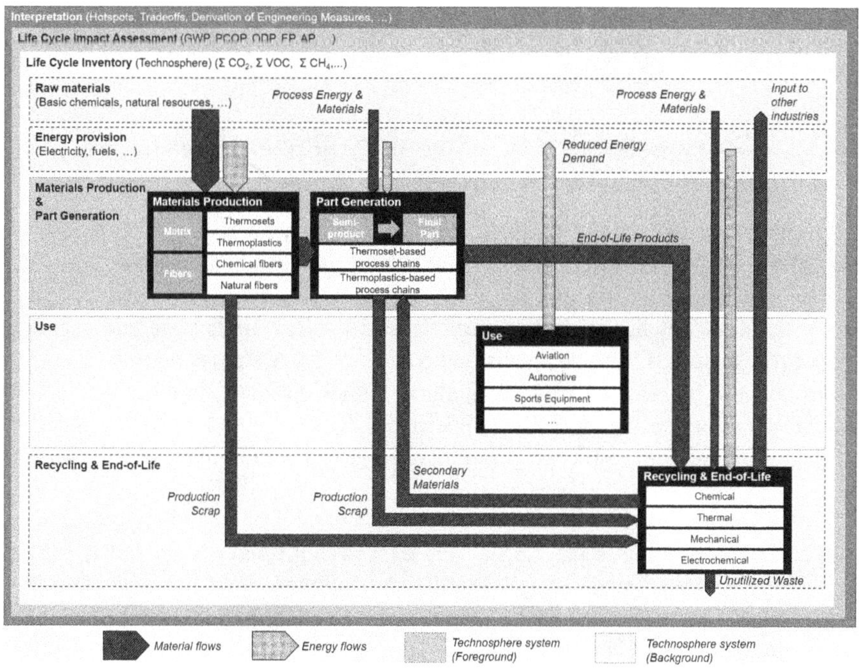

FIGURE 13.1 LCA system description for assessing thermoplastic and thermosetting composites.

- **Part generation**: Steps starting from matrix and fiber materials toward obtaining a functional composite product are subsumed. Typically, thermoset- and thermoplastics-based process chains are distinguished due to different processing properties. Yield losses in manufacturing and the handling of production wastes might have a significant influence on composite environmental impacts.
- **Use**: FRP composites can be applied for a wide range of different products, typically as components or structural parts of more complex products. Thus, the use stage needs to be evaluated in context of that larger product system. A well-discussed example is the quantification of energy savings in automotive applications through reducing vehicle mass by introducing a composite structure.
- **Recycling and end-of-life**: In comparison to conventional engineering materials, composites pose specific challenges to recycling and end-of-life treatment. Most prevalent tasks in end-of-life handling occur due to potential quality losses of fibers in recycling processes (chemical, mechanical, thermal, thermochemical) as well as the reintroduction of recyclates to primary production.

LCA studies can comprise the full life cycle of FRP or only selected stages. Studies that include all life cycle stages are referred to as "cradle-to-grave." Other approaches take a manufacturer's perspective and only include the generation of an FRP part as well as all associated upstream processes, so-called "cradle-to-gate" studies. The definition of system boundaries needs to follow the goal and scope of a specific LCA study (see Section 13.2.2).

13.2.2 Goal and Scope of LCA Studies for Thermoplastic and Thermosetting Composites

An overview of the state of research in FRP LCA is analyzed in order to identify typical goals and system boundaries. Table 13.1 lists and characterizes nine LCA studies that focus on LCA of FRP. Categorizations are made for materials, manufacturing processes for composite parts as well as their application and evaluated end-of-life scenarios. Further, goals and scopes of LCA studies as well as studied environmental impacts are listed. Evaluated studies discuss composite use in automotive, aviation, marine, or civil engineering.

Regarding the goal and scope, different observations can be made from Table 13.1:

- All studies target a comparative assessment of environmental impacts. This either encompasses the comparison of a design option to another design applying other engineering materials, the comparison of manufacturing or end-of-life process chains, or the variation of key parameters within the life cycle.
- A large number of studies perform hotspot and sensitivity analyses. Hotspot analyses enable the contribution of life cycle stages or sub-components to overall environmental impacts to be determined. Sensitivity analyses aim

TABLE 13.1
Overview of the State of Research in LCA of FRP

Study	(Hohmann et al., 2018)	(Delogu et al., 2018)	(Kelly et al., 2015)	(Timmis et al., 2015)	(Witik et al., 2013)	(Duflou et al., 2012)	(Witik et al., 2011)	(Das, 2011)	(Kara and Manmek, 2009)
Matrix materials	TS, TP	TS, TP	TS	TP	TS, TP	TS, TP	TS, TP	TS, TP	TS
Fiber materials	GF, CF	GF, CF, NF	CF	CF	CF, GF	GF, CF, NF	GF, CF	CF (PAN, lignin)	CF, GF
Manufacturing	9 process chains	Multiple process chains	—	Fiber placement, autoclave	SMC, RTM	SMC, RTM, pultrusion, autoclave, injection molding	Stamping, press molding, reaction injection molding	SMC, Programmable powdered pre- forming process	Pultrusion, SMC, molding
Application	Automotive (ICE)	Automotive (ICE)	Automotive (ICE)	Aviation	Automotive	Automotive, aviation, civil engineering	Automotive (ICE)	Automotive (ICE)	Aviation, marine, civil engineering
End-of-Life	Recycled content	Shredding & ASR incineration/ landfill	Recycled content	Landfill	Landfill, recycling, incineration with energy recovery	Landfill, recycling, incineration with energy recovery	Shredding & shredder residue incineration, energy credit	Thermal recycling, material credit	Landfill

(*Continued*)

TABLE 13.1 (CONTINUED)
Overview of the State of Research in LCA of FRP

Study	(Hohmann et al., 2018)	(Delogu et al., 2018)	(Kelly et al., 2015)	(Timmis et al., 2015)	(Witik et al., 2013)	(Duflou et al., 2012)	(Witik et al., 2011)	(Das, 2011)	(Kara and Manmek, 2009)
LCA goal Comparative assessment	Manufacturing process chains	Lightweight designs to reference designs	Material substitution strategies	Material substitution strategies	End-of-life routes for CFRP wastes	Review for different sectors	Material substitution strategies	Alternative CF precursor materials and production technologies	FRP designs to reference designs
Hotspots	x	x	x	x	x		x	x	x
Sensitivities	x	x	x	x	x			x	
LCA scope	Cradle-to-gate + use scenarios	Cradle-to-grave	Cradle-to-gate + use scenarios	Cradle-to-grave + use scenarios	Cradle-to-gate, gate-to grave	Cradle-to-grave	Cradle-to-grave	Cradle-to-grave	Cradle-to-grave
LC impact assessment (LCIA)	CED, GWP, ADP	GWP, ADP, HTP, MAETP, FAETP	GWP	GWP, Eco-indicator	GWP, resource depletion, human health, ecosystem quality	CED, GWP	GWP, resource depletion, human health, ecosystem quality	CED, GWP, human health (air pollutants), Smog	CED, GWP, Eco-indicator

Abbreviations

TS	Thermoset	PAN	Polyacrylonitrile	GWP	Global warming potential
TP	Thermoplastics	SMC	Sheet molding compound	ADP	Abiotic depletion potential
GF	Glass fiber	RTM	Resin transfer molding	HTP	Human toxicity potential
CF	Carbon fiber	ICE	Internal combustion engine	MAETP	Marine aquatic eco toxicity potential
NF	Natural fiber	CED	Cumulative energy demand	FAETP	Freshwater aquatic eco toxicity potential

to determine the influence of selected parameters to overall impacts as well as to identify trade-offs between parameters.

- Common system boundaries include both cradle-to-gate and cradle-to-grave approaches as well as one study on gate-to-grave, that focuses on the evaluation of end-of-life technologies.
- A variety of environmental impacts is evaluated within the studies. Analyses are limited to midpoint indicators (see Section 13.2.3). Global warming potential (GWP) is measured in CO_2-equivalents. Almost all studies examine cumulated energy demands for the observed composite lifecycles. Only few studies extend the scope to other impacts related to resource depletion, human health, or ecosystem quality.
- End-of-life is considered in most studies and systems include shredding, incineration including energy recovery, landfill as well as processes to recover materials by recycling.

13.2.3 Life Cycle Inventory and Life Cycle Impact Assessment for Thermoplastic and Thermosetting Composites

The products and processes associated with the life cycle stages of a FRP constitute the technosphere of an LCA system (see Figure 13.1). The life cycle inventory includes all energy and material flows within the technosphere, the depletion of resources from the biosphere, i.e., earth's natural resources, as well as emissions into the biosphere, i.e., emissions into air, soil, or water.

Further, LCI modeling distinguishes foreground and background systems. The foreground system covers all processes under direct influence of the decision-maker that performs an LCA study or is the principal addressee of a study (UN Environent Life Cycle Iniative, 2019). Figure 13.1 places materials production and part generation to be in the foreground system and thus in the focus of engineers and decision-makers related to product design and manufacturing. Typically, energy and material flows in foreground systems require collecting primary data, e.g., by measurement campaigns, expert interviews, or other means. In contrast, the so-called background system is not in the direct focus of addressed decision-makers. This allows to use also secondary data. In LCA, this could originate from industrial associations or commercially as well as non-commercially available inventory databases. Figure 13.1 places use stage and end-of-life in the background system, as these processes are likely to be outside the direct focus of domains related to product design and manufacturing. This is as well reflected by the review of the state of research (see Table 13.1). Further, process energy and material provision are part of the background system. Associated environmental impacts can show a strong variability in terms of technological, spatial, and temporal influences of fore- and background systems. One prominent example is the share of renewable electricity sources in energy provision that varies widely over countries and increases over time for most of them.

Based on the LCI, the LCA methodology aims to characterize all flows between the studied technosphere system and the biosphere toward obtaining quantified environmental impacts. This stage is called life cycle impact assessment and is also depicted in Figure 13.1. Within LCIA, single energy and resource flows are assigned to impact categories. In general, impact categories can be distinguished between midpoint and endpoint. Whereas end-point categories aim to measure influences on the biosphere at the end of a cause-effect chain between inventory and impacts, e.g., human life expectancy, midpoint indicators look at intermediate steps of this chain. For example, an increased concentration of substances in air, soil, or water can be expressed as midpoint impacts. Due to their good comparability to other technical systems, midpoint indicators are predominantly applied in LCA. Table 13.1 lists typical midpoint indicators in the focus of composite LCA.

13.2.4 LCA Interpretation for Thermoplastic and Thermosetting Composites

LCIA results need to be interpreted to answer the questions that were asked in the goal definition. The interpretation considers the results of LCI and LCIA. To validate the conclusions, the robustness of the conclusions, and to identify the focus points for improving the technosphere system, sensitivity and uncertainty analyses have to be performed (Hauschild, Rosenbaum, and Olsen, 2017).

LCA interpretation requires coping with complex interdependencies between product models, the LCI within the different life cycle stages of the foreground system, and impacts of the background system (Cerdas et al., 2017). The evaluation of FRP over the entire life cycle thus requires analysis of a larger number of scenarios, including potential advancements in technology. Figure 13.2 exemplarily shows results of an LCA study of thermoset-based carbon fiber-reinforced plastics (CFRP) parts production based on the works of Hohmann et al. (Hohmann, Albrecht, Lindner, Wehner, et al., 2018). The base scenario (V1) follows a cradle-to-gate approach covering the production of carbon fibers, matrix materials, textile products, as well as manufacturing processes for part production and losses due to cut-offs. It can be observed that carbon fiber production contributes most to overall greenhouse gas emissions. On that basis, scenarios for introducing renewable electricity sources as well as technology optimizations in manufacturing are analyzed. The aforementioned measures can lead to variations in greenhouse gas emissions between 13 and 40 kg CO_2-eq of a specific composite part taking a cradle-to-gate perspective.

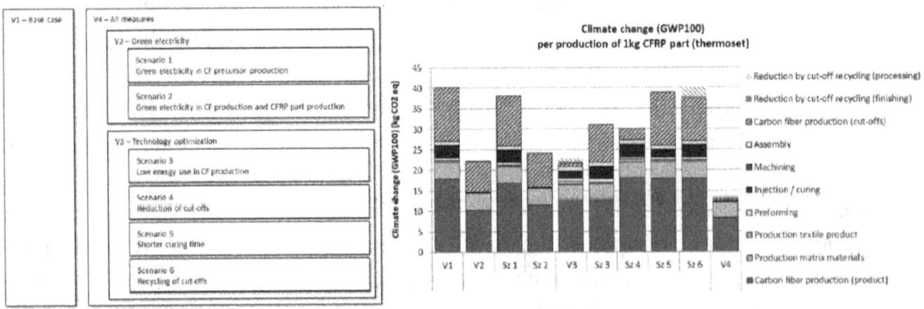

FIGURE 13.2 GWP comparison of scenarios and variants for thermoset CFRP parts for the production of 1 kg of thermoset-based CFRP (Hohmann, Albrecht, Lindner, Wehner, et al., 2018).

13.3 MATERIALS PRODUCTION

In general, different production routes for fibers and matrix materials can be distinguished. Whereas conventional routes rely on chemical processes, fibers and matrix materials from renewable resources increasingly enter the market. To provide a spotlight over methodical challenges, the following sections give a deeper insight into LCI of carbon fiber manufacturing, thermoset and thermoplastics materials as well as natural fibers and bioplastics as replacement for matrix materials.

13.3.1 Carbon Fiber Manufacturing

Carbon fiber-reinforced plastics are important engineering materials due to their excellent mechanical properties and strength-to-weight ratios. The overall market demand for CFRPs is steadily increasing and reached nearly 130 thousand tons in 2018 (Witten et al., 2018). The most important sectors are aviation, military, automotive, wind energy, and sports leisure. Carbon fibers contain at least 92 weight % carbon and can be classified according to their mechanical properties, most commonly by modulus and tensile strength. Material properties can be tailored within a vast range to comply with industrial sector requirements (Das et al., 2016; Huang, 2009; Newcomb, 2016). Those properties are realized by variations of process conditions and processing times, which result in diverse energy intensities per kg carbon fiber (Huang, 2009). In general, two major process steps are distinguished for industrial carbon fiber manufacturing: precursor production in the white line and conversion of the precursor to carbon fiber in the black line. Precursor production is described for manufacturing polyacrylonitrile (PAN) as the most important precursor in industry. However, research is being carried out for introducing alternative precursors in industrial carbon fiber manufacturing, e.g., hardwood lignin (Mainka et al., 2015).

Before performing an LCI of carbon fiber production, a systems analysis of all relevant energy and material flows of the manufacturing process is required. Therefore, Figures 13.3 and 13.4 show relevant energy and resource flows of an

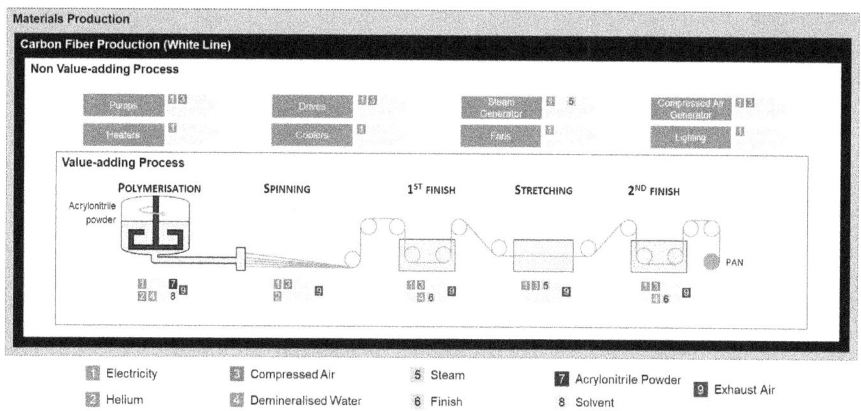

FIGURE 13.3 Energy and resource flows in carbon fiber manufacturing (white line).

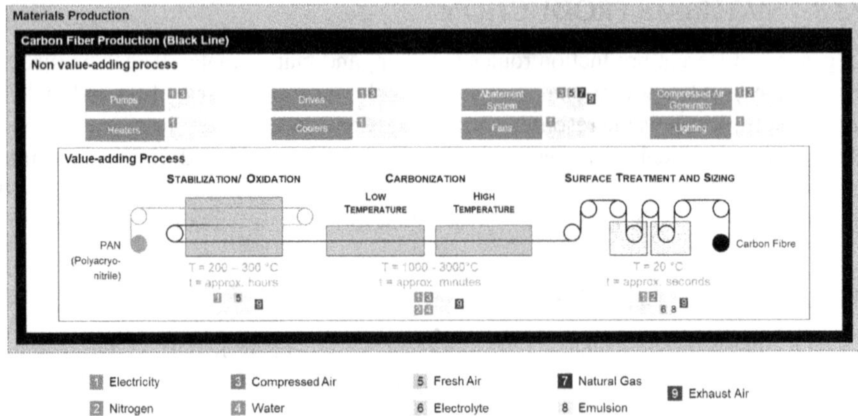

FIGURE 13.4 Energy and resource flows in carbon fiber manufacturing (black line).

industrial carbon fiber manufacturing process. The precursor production in white line undergoes polymerization, spinning, finishing, and stretching processes. The black line process chain is divided into stabilization and oxidation, carbonization and surface treatment, and sizing. Different energy and material flows are required to perform the value-adding steps in both parts of the process chain (see Figures 13.3 and 13.4). This includes energy carriers, e.g., electricity, natural gas, or compressed air, as well as processing gases and materials, e.g., nitrogen, helium, or solvents. Comparing both processes, it is apparent that black line processes require a larger amount of thermal energy compared to white line processes. Stabilization and oxidation are performed over a longer processing time of one to more hours, whereas carbonization is performed in a shorter timeframe, but at temperature levels above 1000 degrees Celsius. In the current example, electric heaters provide the required process heat.

Further, Figures 13.3 and 13.4 emphasize the importance of analyzing manufacturing equipment that is not directly associated with a value-adding process of interest. Carbon fiber manufacturing relies on a closely interlinked interplay between the value-adding processes as well as associated technical building services (TBS). For example, black line stabilization and oxidation is an exothermal process that requires abatement and ventilation systems to transport waste airflows in accordance to the current state of processing. A major challenge while modeling LCIs lies in the allocation of process and TBS-related energy and resource demands to the product of interest. An LCI requires including information on the plant operation itself, e.g., shift times or process utilization rates, as well as information on the quality of manufactured products (Dér et al., 2018). As described previously, access to the manufacturing plant would allow for process data acquisition based on primary data, whereas energy supply as well as processing gases and materials are typically modeled based on secondary inventory data, as a thorough analysis of those production steps is typically outside of the decision scope of product and process engineers in carbon fiber manufacturing.

As mentioned before, a typical goal of a composite LCA study could lie in a comparison of a composite product with a product made from conventional engineering materials, e.g., steel. In terms of carbon fiber-reinforced composites, the mechanical performance of those products is dominated by the properties of the fiber reinforcement. Thus, carbon fiber LCI data should provide information on fiber properties in relation to energy and resource flows associated with manufacturing those fibers. In this context, Figures 13.5, 13.6, and Table 13.2 provide an overview of previously executed LCI studies on carbon fiber manufacturing within the white and black line of manufacturing. Thereby, white line processes have been analyzed regarding their

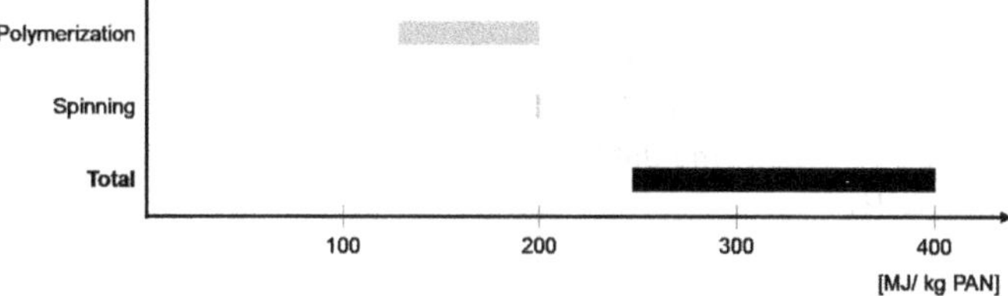

FIGURE 13.5 Reported energy intensities in the white line for PAN-based precursor manufacturing, based on Das, 2011 (cradle-to-gate), Liddell et al., 2016 (gate-to-gate), Liddell et al., 2017 (gate-to-gate).

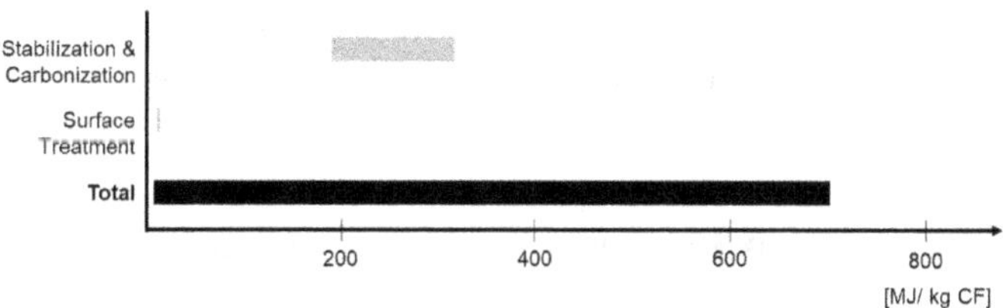

FIGURE 13.6 Reported energy intensities in the black line for carbon fiber manufacturing, based on De Vegt and Haije, 1997; Suzuki and Takahashi, 2005; Griffing and Overcash, 2009; Das, 2011; Liddell et al., 2016; Liddell et al., 2017; Arnold et al., 2018.

TABLE 13.2
Reported Fiber Properties and System Boundaries in Studies Evaluating the Energy Intensities in the Black Line for Carbon Fiber Manufacturing

	(De Vegt and Haije, 1997)	(Suzuki and Takahashi, 2005)	(Griffing and Overcash, 2009)	(Das, 2011)	(Liddell et al., 2016)	(Liddell et al., 2017)	(Arnold et al., 2018)
Fiber properties	N/A	N/A	High strength	N/A	N/A	N/A	N/A
System boundary	Gate-to-gate	N/A	Gate-to-gate	Primary energy	Gate-to-gate	Gate-to-gate	Gate-to-gate

required energy demands per kg of PAN, whereas the black line is analyzed regarding the energy demand per kg of carbon fiber. Several major observations can be drawn from this example:

- Typically, studies focus on gate-to-gate energy demands of manufacturing sites, omitting other energy and resource flows. This partial LCI information also omits information from background systems, e.g., on the sources of electricity, and thus is not sufficient for performing an LCIA with multiple impact categories.
- Carbon fiber properties are only provided by one of the studies on black line processes.
- Observed energy demands vary widely for white and black line processes. Whereas a range between 245 and 395 MJ per kg PAN is reported for white line processes, black line data varies between 4.52 and 704 MJ per kg carbon fiber (CF).

Especially the latter observation illustrates the importance of carefully selecting and discussing information from secondary data sources for performing a composite LCA. Addressing the identified gap, commercial LCI databases provide insights on environmental impacts of fiber manufacturing. However, while being certified to comply with ISO 14040 standards, those databases tend to apply aggregated LCI models that limit users to distinguish between different value-adding steps, associated technical building services, and other drivers of impacts.

13.3.2 Thermoset and Thermoplastics Matrix Manufacturing

Matrix materials are divided between thermoset resins, such as epoxy or polyester, and thermoplastics such as polypropylene or polyamide. While thermosets are still the dominant matrix materials, thermoplastics have an increased relevance due to shorter cycle times in manufacturing and improved recyclability (Duflou et al., 2012). However, manufacturing equipment for thermoplastics tends to be more complex due to the higher viscosity of thermoplastic materials and interfacial fiber/matrix bonds tending to be weaker (Swolfs, 2015; Herrmann et al., 2018). The environmental impacts of manufacturing thermoplastics and thermosets are caused by their energy demands in the synthesis processes. The energy demand comprises the extraction of mineral oil, separation and refining, characterization, and polymerization. Figure 13.7 exemplarily presents processes for the production of polyamide 6.6 and liquid epoxy resin. A review on greenhouse gas impacts of different matrix materials is provided by Duflou et al. (Duflou et al., 2012).

A special challenge occurs in the LCA methodology for matrix materials. Syntheses processes in chemical industry are typically multi-functional processes that produce multiple valuable chemical products and co- and by-products in parallel (Plastics Europe, 2019). In terms of performing LCA, ISO 14040 provides two options regarding the system definition:

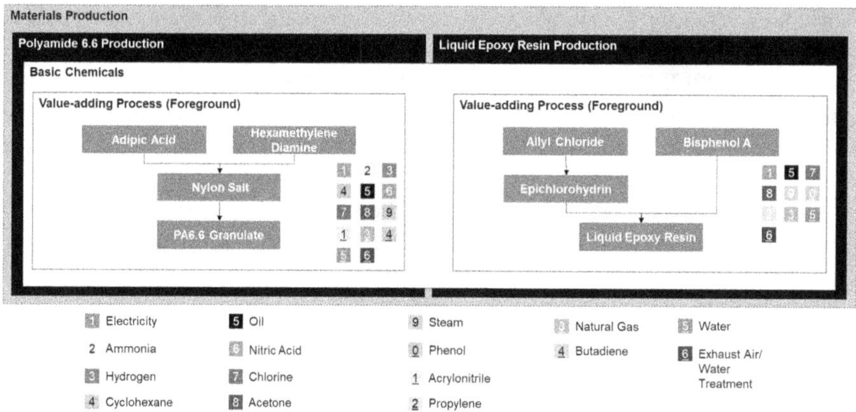

FIGURE 13.7 Energy and material flows for PA6.6 and liquid epoxy resin production. (Based on Plastics Europe, 2019.)

- **Allocation**: Environmental impacts of processes are allocated to different products and co-products. This could be, for example, based on an economic approach, where the economic value of products and co-products determines the share of environmental impacts that is associated with the product of interest, i.e., a thermoset or thermoplastic material.
- **System expansion**: In system expansion, co-products are considered alternatives to other products on the global market. This means that, impacts for the product of interest are credited by avoided environmental impacts due to a valuable application of co- and by-products that are not manufactured by other primary processes. To allow the described crediting approach, the initial system boundary needs to be extended by an LCA study of the avoided primary production.

To enable an industry-wide comparison of LCA information and underlying methods, data is provided by industry associations like Plastics Europe (Plastics Europe, 2019). Thereby, associations are enabled to define a superordinate LCI system of a specific region that includes all relevant industry products as well as co- and by-products of the industry sector. Allocation and system expansion rules can thus be applied on actual material, energy, and product flows between plastics industry and other industry sectors.

13.3.3 Fibers and Matrix Materials from Renewable Resources

Aside from fiber and matrix materials that originate from chemical processes, materials from renewable resources are increasingly being researched and are entering the market. This encompasses matrices and fiber reinforcements.

Biobased matrix materials for engineering applications include biodegradable materials like polylactic acid (PLA), modified starch or polyhydroxyalkanoates (PHAs) as well as durable materials like bio-polyethylene, bio-polyamides or bio polyurethanes.

Depending on the application of the composite, biodegradability might be a beneficial engineering property (Mohanty, Misra, and Hinrichsen, 2000; Spierling et al., 2018; Duflou et al., 2012; Koronis, Silva, and Fontul, 2013). Environmental challenges of biobased matrix materials include higher production cost and potential competition with the food supply chain in sourcing feedstock. This includes sugarcane, corn, or ligno-cellulosics (Spierling et al., 2018). However, there is ongoing development of biobased polymers that do not compete with the feedstock, such as wheat gluten (Vo Hong et al., 2015). Spierling et al. did a comprehensive review of biobased plastics regarding their environmental, social, and economic impact assessment. The review showed that the LCA of bioplastics is subject to a wide range of methodological choices. This encompasses impact assessment methods, allocation methods, credits for by-products, the inclusion of biogenic carbon as well as the background of data (e.g., R&D, industry). To get a more robust comparison, guidelines or category rules for biobased materials would be required (Spierling et al., 2018).

In the case of fiber reinforcements, relevant materials in technical applications encompass flax (64% of the market), jute (11%), hemp (10%), and sisal (7%), while market shares correlate with mechanical properties of the fibers (Deng, 2014). While representing a renewable material, also technical drivers and low cost promote the use of natural fibers (Deng, 2014; Herrmann et al., 2018). Natural fibers show a competitive density in comparison to glass fibers and can compete with respect to their stiffness-to-weight ratio as well as superior damping properties (Duflou et al., 2014; Prabhakaran et al., 2014; Herrmann et al., 2018), while drawbacks include their natural variability of mechanical properties, processing temperature limitations, interfacial fiber/matrix bonds, and moisture absorption (Dicker et al., 2014; Stamboulis et al., 2000). Alkali-based treatments and the steam-based duralin process can be used to improve their performance, yet both come with a non-negligible increase in environmental impact (Dicker et al., 2014).

An issue for further research is the evaluation of lifetimes of natural fiber composites in comparison to glass fibers. As pointed out by Duflou et al., a shortened service life and increased replacement ratio could lead to different scenarios of environmental impacts, e.g., if a premature end-of-life requires the exchange of a whole assembly (Duflou et al., 2014). Further environmental concerns for natural fiber reinforcements include emissions of nitrogen and phosphorus to waterways over large areas of arable land. While studying the changes in environmental impacts when replacing glass fibers with kenaf or natural cellulose in automotive applications, Boland et al. point out the lower weight of natural fibers compared to glass alternatives and similar processing efforts between all fibers. Moreover, the function of natural fibers as biogenic carbon storage is highlighted (Boland et al., 2015). This means that plants that are used for fiber or matrix production from renewable resources store CO_2 during their growth period. This CO_2 is only released in the end-of-life phase, depending on the treatment processes.

13.4 PART GENERATION

Manufacturing routes for composites can be distinguished between processes being suitable for thermoset or thermoplastics matrix systems or both, as listed in Table 13.3,

TABLE 13.3
Manufacturing Routes Overview Based on Fleischer et al., 2018 and Extended by Factory Influencing Life Cycle Inventory

Manufacturing process		Process characterization (Fleischer et al., 2018)			Influence on LCI			
		Matrix materials	Market maturity	Production quantity	Material utilization	Specific Energy demand	Process materials	Process emissions
		(Fleischer et al. 2018)				(Hohmann, Albrecht, Lindner, Wehner, et al., 2018; Neitzel, Mitschang, and Breuer, 2014) and expert interviews		
Compression molding	SMC	TS	⬀	⬀	⬀	⇒	⬀	⬀
	BMC	TS	⬀	⬀	⬀	⇨	⬀	⬀
Liquid composite molding	RTM	TP/TS	⬀	⬀	⬂	⬀	⬀	⬀
	Wet pressing	TS	⇨	⬀	⬂	⬀	⬀	⬀
	Rotational molding	TP/TS	⇨	⇨	⬀	⬀	⬀	⬀
Fiber deposition	Tape laying	TP/TS	⬀	⇨	⬀	⬂	⬂	⬂
	Fiber placement	TP/TS	⬀	⇨	⬀	⬂	⬂	⬂
Thermoforming		TP	⇨	⬀	⬂	⬀	⬀	⬀
Pultrusion		TP/TS	⬀	⬀	⬀	⇨	⬂	⬀
Filament winding	Rotating mandrel	TS	⬀	⇨	⬀	⇨	⬂	⬂
	Flexible robot	TS	⬂	⇨	⬀	⇨	⬂	⬂
Injection molding		TP/TS	⬀	⬀	⬀	⬀	⬀	⬀
Scale			⬀ High ⇨ Middle ⬂ Low	⬀ >10.000 ⇨ <10.000 ⬂ <100	⬀ High ⇨ Middle ⬂ Low	⬀ Heating & high pressure ⇨ heating or high pressure ⬂ None	⬀ Additional processing materials required (chemicals, …) ⬂ None	⬀ Dedicated abatement and/ or protective equipment required ⇨ ⬂ None

that is based on a review by Fleischer et al. (Fleischer et al., 2018). The manufacturing processes can be further distinguished regarding their technological maturity, realizable production quantities or production quantities. As identified within the system description (see Section 13.2.1), the environmental evaluation is based on an LCI including all relevant material and energy flows as well as associated emissions. Therefore, Table 13.3 extends the work by Fleischer et al. with a qualitative overview on specific energy demands, required process materials, and process emissions to air. Also, material utilization is listed as a relevant factor as it directly influences generated production wastes that need to be further handled.

The specific energy demands in manufacturing processes are qualitatively evaluated regarding high pressures and/ or high temperatures required for the specific manufacturing processes. For example, the MAI carbon cluster analyzed the LCI of ten thermoset and thermoplastics process chains that apply CFRP materials. Tool heating, injection/curing, and press processes were identified as a major driver of process-related energy demands for the majority of the processes. In addition, low material utilizations and thus high material yield losses show a major impact on the production-related environmental impacts (Hohmann, Albrecht, Lindner, Wehner, et al., 2018). As exemplarily shown for two resin transfer molding (RTM) process chains, part-related (e.g., part size, part thickness), and process-related (e.g., press utilization, curing time) parameters influence process energy demands significantly. Considering a total of 20 variable parameters, a reduction of 77% or an increase of 615% could be observed for dry-fiber-placement-RTM (Hohmann, Albrecht, Lindner, Voringer, et al., 2018).

In general, LCI studies for composite manufacturing chains are based on different scales of technology implementation depending on the market maturity level of process chains. Therefore, measured environmental impacts should be analyzed with respect to potential technological improvements when increasing market volumes of a production technology. In relation to composite manufacturing chains, this could encompass an increase in material utilization, shorter cycle times, or more efficient and effective recycling processes (Hohmann, Albrecht, Lindner, Wehner, et al., 2018). To enable the derivation of LCI scenarios using scale-up scenarios for manufacturing processes and process chains, Schönemann et al. proposed a simulation-based assessment of innovative composite manufacturing process chains., that has been extended in further works (Schönemann et al., 2016). Figure 13.8 presents the approach that is based on the modeling of manufacturing processes and their individual behavior with respect to cycle times, operation states, and associated electricity loads. The process models can be combined with process chains and evaluated in specific production scenarios. As a result, energy intensities for specific production runs can be determined, while simultaneously analyzing productivity and quality considerations. Primary inventory data can be fed to the models based on automated identification of state-based energy demands as well as quality information on manufactured parts, as for example presented by Gellrich et al. (Gellrich et al., 2019).

Another strategy to lower environmental impacts related to manufacturing of composites is the change of electricity sources used for manufacturing. Switching from an industrial average to green electricity from renewable resources could lower greenhouse gas (GHG) emissions in manufacturing by over 40%—a number being

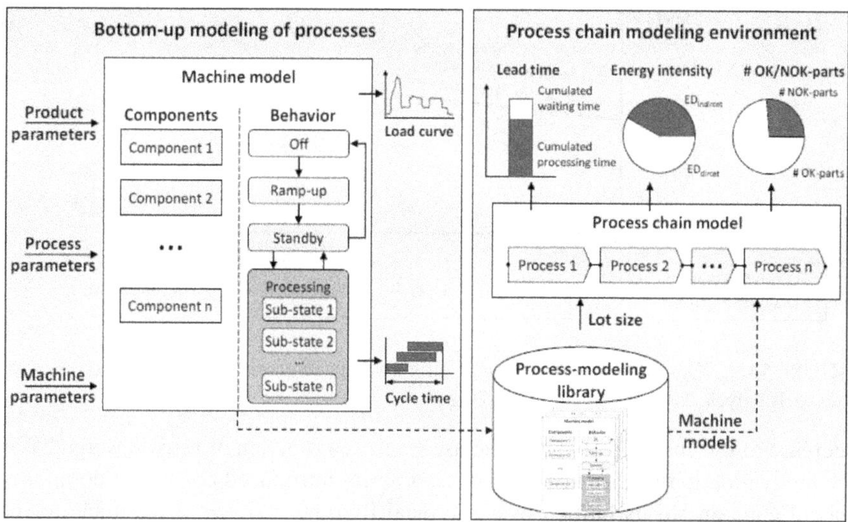

FIGURE 13.8 Simulation-based approach toward determining energy intensity of composite manufacturing chains. (Further developed based on Schönemann et al., 2016.)

equivalent to potential reductions due to technological improvements (Hohmann, Albrecht, Lindner, Wehner, et al., 2018).

13.5 USE

The following section discusses the evaluation of use stage environmental impacts of composites. As shown in Section 13.1, the use stage of thermoplastics and thermosetting composites can be manifold. Important applications include aviation industry, sports industry, automotive industry, machine tools, and buildings. Thereby, a composite part or structure typically is part of a larger product system and fulfills a specific function within that system (Herrmann et al., 2018). Complying with the LCA methodology, the influence of a composite part on the overall function of the larger product system needs to be determined for assessing its use stage. One example is the use of a composite part in a vehicle while saving weight compared to a reference design, as illustrated in Figure 13.9. For example, carbon composite lightweight structures tend to cause additional burdens compared to steel structures when used in automobiles (Herrmann et al., 2018; Egede, 2017; Delogu et al., 2018). During the use stage, a vehicle requires energy or fuel to realize the vehicle movement. This demand depends on various parameters such as vehicle weight, drivetrain characteristics, driving behavior, or traffic volume.

The introduction of a composite component as part of a lightweight strategy can reduce the energy demand in the use stage, as a lower weight decreases driving forces required for acceleration. Savings can be expressed by technical constants. A common denomination is the "fuel reduction value (FRV)" or "energy reduction value (ERV)," as for example applied by Koffler and Rohde-Brandenburger or Hofer (Koffler and Rohde-Brandenburger, 2010; Hofer, 2014). Those constants express the

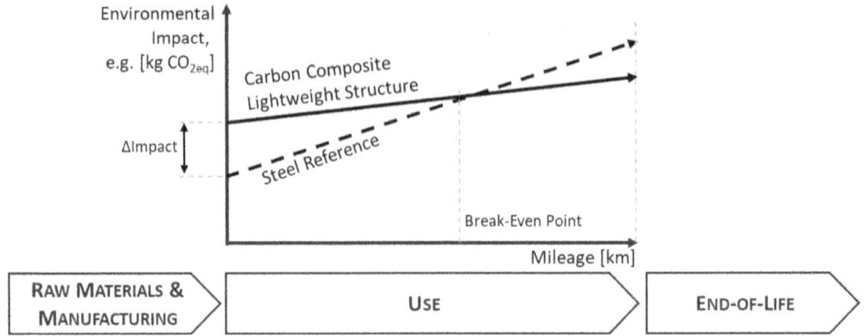

FIGURE 13.9 Qualitative comparison of a lightweight structure with a reference structure over its life cycle. (Based on Dér et al., 2018.)

decrease of the energy demand in the use stage as a function of vehicle weight. With the motivation to increase eco-efficiency, a newly introduced composite component should compensate its burden over a typical lifetime mileage of a vehicle toward achieving a "break-even point" (see Figure 13.9). In the case of electric vehicles, the break-even point regarding GHG emissions and other environmental impacts strongly depends on electricity mixes applied for vehicle charging. While a component could reach a break-even point in a market with large shares of fossil electricity carriers, it may fail in markets with a high share of renewables (Egede, 2017). Further variables need to be considered, e.g., in the course of vehicle automation.

If the motivation of introducing composites into technical products is not or only partly motivated by weight reduction, the use stage assessment needs to be adapted accordingly. For example, aircrafts show much longer life cycles in comparison to road vehicles. While the assessment of energy savings in comparison to outdated designs is not purposeful, composite use can be assessed regarding potential service efforts. This includes the non-destructive testing of composite components as well as repair or replacement scenarios.

13.6 RECYCLING AND END OF LIFE

Treatment of production and end-of-life wastes—incorporating reuse, remanufacturing, and recycling—is an important mitigation strategy for environmental impacts of industrial goods. Products or materials can be recirculated into the production stream and consequently to a secondary use stage. As illustrated in Figure 13.10, primary material production decreases by successful recycling, and landfill is avoided. However, efforts for collecting and reprocessing secondary products or materials need to be taken into account when assessing the overall environmental impacts of end-of-life processes (Geyer et al., 2016). However, in order to generate secondary material flows in required quantities, virgin material needs to be processed first (Herrmann et al., 2018).

Recycling efforts might also help to achieve compliance with legislative targets. For example, the European directive on end-of-life vehicles (ELV) states that at least 95% of the mass of end-of life vehicles should be reused and recovered, and at least 85% should be recycled (the rest may be incinerated with energy recovery, for

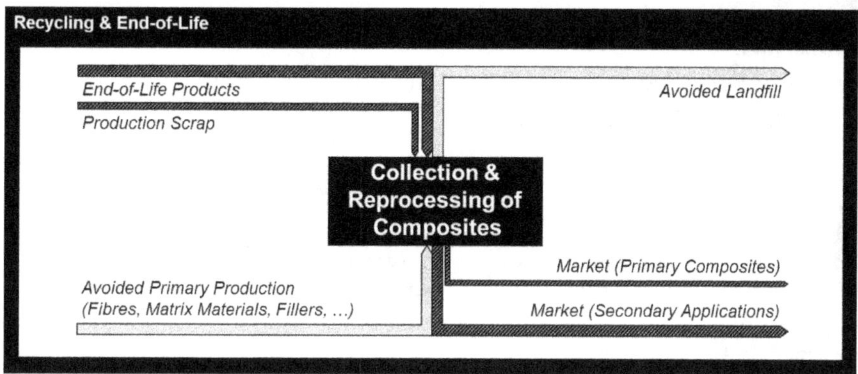

FIGURE 13.10 Environmentally successful recycling of composites avoids primary production and landfilling of material. (Adapted from Geyer et al., 2016.)

example) (European Parliament and Council, 2000). Coincidently, the relevance of recycling carbon fiber wastes has raised the attention of large players in the aviation industry. Airbus, for example, set a target of recycling 95% of its post-production carbon fiber waste by 2020–2025 (Airbus, 2014).

Composite waste streams can result from different stages of the composite life cycle, i.e., yield losses in manufacturing as well as end-of-life wastes. Rybicka et al. estimate that 30 to 50% of materials in aircraft production are scrapped during manufacturing (Rybicka et al., 2015). Nilakantan and Nutt performed an analysis on autoclave production scraps of thermoset materials and estimated that overall manufacturing waste lies between 6 and 19%, consisting of uncured prepregs, ply cutter scrap, and post-processing wastes (Nilakantan and Nutt, 2018). Different sources of waste streams are listed in the following.

- **Cut-offs from manufacturing processes** (dry and wet): Waste streams result from yield losses of manufacturing processes (the ratio of input material weight to part weight), which is subject to technical boundaries (Wong et al., 2017). Another source could be end-of-bobbins materials from manufacturing processes.
- **Prepregs that were not cured within their shelf life**: Materials cannot be used for structural applications without recertification (Wong et al., 2017). Mitigation could be achieved by testing methods to extend shelf life as well as treatment.
- **End-of-Life wastes**: Composite waste streams occur at the end-of-life of industrial goods. Major contributors are aircrafts and wind turbine blades but also automotive wastes (Shuaib et al., 2015).

Treating production waste and end-of-life composites is widely addressed by research and also by industrial implementation. Thereby, composite recycling technologies aim at separating fiber and filler reinforcements from matrix materials (Fleischer et al., 2018). This could be performed by different technologies, as listed in Table 13.4 based on works by Fleischer et al. and Mativenga et al. In general, these

TABLE 13.4
Review of Recycling Technologies for GFRP and CFRP Composites (Fleischer et al., 2018; Mativenga et al., 2016)

	Description	Implemented scale (TRL)	Energy demand	% Tensile strength (CFRP)
Biotechnological	Microorganisms used to degrade matrix	Limited availability even at laboratory scale	N/A	N/A
Chemical	Solvents used to break matrix at high temperature and pressure	Most research conducted at laboratory scale	61–93 MJ/kg	87–99
Electrochemical	An electrical current is applied through an electrolyte solution to degrade the matrix	Limited availability even at laboratory scale	N/A	80
Fluidized bed	Matrix decomposition through fluidized air at a given temperature	Process investigated at a pilot scale	N/A	74
High voltage fragmentation	High voltage creates pressure waves along plasma channels to disintegrate materials in water	Laboratory and pilot scale machine are available	N/A	N/A
Mechanical	Size reduction and separation into powder and fibrous fractions	Commercial scale with wide range of potential applications	0.17–0.27 MJ/kg; 0.35 MJ/kg	N/A
Microwave pyrolysis	Decomposition of matrix using microwave heating in an inert atmosphere	Limited availability even at laboratory scale	5–10 MJ/kg	79
Pyrolysis	Decomposition of matrix using conventional heating (oven) in an inert atmosphere	Industrial scale process available at a commercial level	23–30 MJ/kg	80–96

technologies can be distinguished as biotechnological, chemical, electrochemical, and fluidized bed processes, high voltage fragmentation, mechanical processes, microwave pyrolysis, and pyrolysis (Fleischer et al., 2018; Mativenga et al., 2016).

A major challenge is avoiding downcycling of the recyclates, meaning that secondary materials show an inferior quality compared to primary materials (Haas et al., 2015). This could be the result in both the fiber and other types of reinforcement, and from matrix or binders, e.g., the thermoset type (Yang et al., 2012). Another major challenge lies in the mixing of materials in waste streams, as well as in including designs where composites are combined with metal structures (Herrmann et al., 2018; Fleischer et al., 2018).

Mechanical recycling processes result in shortened fiber lengths and reduced mechanical properties, e.g., strength and strain properties, so that, for example, recycled carbon fibers are only available for lower value applications (Pickering, 2006; Shuaib and Mativenga, 2015; Li, Bai, and Mckechnie, 2015). Pyrolysis, fluidized bed methods as well as chemical or electrochemical routes show promising results in carbon fiber recycling by achieving tensile strengths of 74% and above when compared to virgin carbon fibers. Despite fiber length, length distribution, surface quality (impact on the adhesion to a new matrix) or the mixing of different fiber grades in a batch of recycled composites influence the quality and application potential of composite recyclates (Oliveux, Dandy, and Leeke, 2015).

Observed energy demands vary widely for recycling processes. Values listed in Table 13.4 are based on empirical studies and cover the full range between laboratory-scale processes and industrial implementations. When analyzing and comparing recycling process chains, LCA studies would need to investigate the specific processes of interest thoroughly. For example, efficiency gains could significantly influence the environmental impacts of different routes.

Overall, different sub-challenges with a major influence on environmental impacts of production waste and end-of-life treatment occur for composites and need to be taken into account when performing an LCA of end-of-life processes. These are listed below.

- Production waste streams of composite materials need to be directed based on the properties and qualities of the composite wastes. Mono-material streams could enable high-quality recycling. However, separation and logistics efforts need to be assessed to identify a potential trade-off to the added value due to higher-quality recyclates.
- End-of-life composites are more difficult to separate with existing disassembly and recycling technologies, compared to conventional engineering materials. Also, composites might be applied in multi-material designs side by side with steel and aluminum, as e.g., observed within the automotive industry. The state of the art of vehicle recycling incorporates a process chain consisting of shredding, sorting, and separation (Herrmann et al., 2018; Dalmijn and De Jong, 2007). For composite materials, the initial shredding process would lead to a shortening of fiber lengths and thus downcycling, as secondary applications of recyclates are inferior. Separation processes would need to be adapted for composites, e.g., prior to mechanical recycling, if feasible.
- Viable applications of secondary materials need to be identified and matched with waste streams from recycling processes. For example, carbon fiber recycling by pyrolysis can save up to 90% of the energy required to produce virgin material (Witik et al., 2013). However, this only applies to a high-quality secondary material that allows the replacement of virgin

material of the same grade. While being theoretically feasible, both quality and quantity requirements for the secondary application need to be fulfilled. Matrix material is lost during the process.

- Even if all lightweight structures could be produced from recycled material, the availability of recycled material might lag behind production needs. Lefeuvre et al. anticipate the use of stocks of CFRP for recycling. Only 34 tons of CFRP are estimated to be available for recycling from a total amount of 1.2 million tons in 2050. For the year 2015, the numbers are 13 tons of CFRP from a total amount of 125 tons (Lefeuvre et al., 2017).
- Most research efforts to date exclusively tackle the material recycling of composite production and end-of-life wastes, while reuse and remanufacturing are not in the focus (Pimenta and Pinho, 2011; Herrmann et al., 2018). However, direct use of production and end-of-life wastes in secondary applications could minimize efforts for collection and reprocessing.

13.7 MAIN CONCLUSIONS

The assessment of environmental impacts of composite structures relies on a systematic assessment of energy and material flows as well as the associated depletion of resources and emissions into water, soil, and air occurring along the life cycle of composite structures. A system model description complying with the life cycle assessment methodology is presented and single life cycle stages within the composite life cycle are discussed. Thereby, general and specific challenges for performing LCA of composites were observed.

General challenges for performing LCA comprise:

- The LCA methodology requires the definition of goal and scope of a study and the performance of inventory building, life cycle impact assessment and interpretation based on that definition. An LCA can be performed with different motivations in mind, that influence the definitions of system boundaries of a respective study.
- Depending on the goal and scope of a study, certain life cycle activities are in the foreground system, meaning the scope of influence of the target audience, e.g., to support a certain decision, whereas others are in the background system.
- In line with the scope of influence of a target audience and data availability, inventory building needs to be performed based on primary and secondary data sources. As shown by the example of manufacturing processes, simulation-based approaches could complement inventory data building from primary data acquisition, whereas secondary data is available through commercial and non-commercial databases.

Specific challenges for performing composite LCA include:

- The assessment of manufacturing and end-of-life processes should not be limited to value-adding process steps, but also link equipment of technical building services. This has been shown with the example of carbon fiber production.

- The introduction of matrix materials and fiber reinforcements from renewable resources might lead to a shift in environmental impacts. Whereas energy and resource demands from industrial processes can be mitigated, composites from renewable materials create burdens in other industrial sectors, that require a thorough analysis for any specific case.
- Sensitivities due to variable technical parameters need to be analyzed for single process steps as well as the entire life cycle. A large number of manufacturing and end-of-life processes for composites are still under development, meaning that efficiency improvements can be expected in scale-up scenarios. Energy provision can significantly influence environmental impacts, as well.
- The impact of composites in the use stage depends on the specific application. In the case of vehicles, a common approach is the comparison of a reference design to an innovative composite design. Whereas carbon fiber-reinforced composites are likely to carry a burden from materials manufacturing and manufacturing processes, weight reductions enable the saving of energy. Eco-efficient composites should fully compensate burdens and reach a break-even point within vehicle use.
- End-of-life assessment needs to take the technological maturity of technical processes into account. At the same time, it is highly relevant to analyze the quality and market availability of secondary materials over time. Only if supply of secondary materials is ensured, can avoided burdens be accounted for in environmental assessment of processes. Repair and remanufacturing efforts could improve the environmental impact of composites, but were not in the focus of past research activities.

REFERENCES

Airbus. 2014. "An Airbus Working Group Sets Out a Composites Recycling Roadmap."

Allwood, Julian M., Michael F. Ashby, Timothy G. Gutowski, and Ernst Worrell. 2011. "Material Efficiency: A White Paper." *Resources, Conservation and Recycling* 55(3): 362–81. doi:10.1016/j.resconrec.2010.11.002.

Arnold, Uwe, Andreas De Palmenaer, Thomas Brück, and Kolja Kuse. 2018. "Energy-Efficient Carbon Fiber Production with Concentrated Solar Power: Process Design and Techno-Economic Analysis." *Industrial and Engineering Chemistry Research* 57(23): 7934–45.

Boland, Claire S., Robb De Kleine, Gregory A. Keoleian, Ellen C. Lee, Hyung Chul Kim, and Timothy J. Wallington. 2015. "Life Cycle Impacts of Natural Fiber Composites for Automotive Applications: Effects of Renewable Energy Content and Lightweighting." *Journal of Industrial Ecology* 20(1): 179–89. doi:10.1111/jiec.12286.

Cerdas, Felipe, Alexander Kaluza, Selin Erkisi-Arici, Stefan Böhme, and Christoph Herrmann. 2017. "Improved Visualization in LCA through the Application of Cluster Heat Maps." *Procedia CIRP*: 732–37. doi:10.1016/j.procir.2016.11.160.

Dalmijn, W. L., and T. P. R. De Jong. 2007. "The Development of Vehicle Recycling in Europe: Sorting, Shredding, and Separation." *Jom* 59(11): 52–6. doi:10.1007/s11837-007-0141-1.

Das, Sujit. 2011. "Life Cycle Assessment of Carbon Fiber-Reinforced Polymer Composites." *The International Journal of Life Cycle Assessment* 16(3): 268–82. doi:10.1007/s11367-011-0264-z.

Das, Sujit, Josh Warren, Devin West, and Susan M. Schexnayder. 2016. "Global Carbon Fiber Composites Supply Chain Competitiveness Analysis. Report No. ORNL/SR-2016/100 - NREL/TP-6A50-66071." May 2016. https://www.nrel.gov/docs/fy16osti/66071.pdf.

Delogu, Massimo, Laura Zanchi, Caterina Antonia Dattilo, Silvia Maltese, Rubina Riccomagno, and Marco Pierini. 2018. "Take-Home Messages from the Applications of Life Cycle Assessment on Lightweight Automotive Components." doi:10.4271/2018-37-0029.

Deng, Yelin. 2014. *Life Cycle Assessment of Biobased Fiber-Reinforced Polymer Composites.* KU Leuven: Leuven, Belgium.

Dér, Antal, Alexander Kaluza, Denis Kurle, Christoph Herrmann, Sami Kara, and Russell Varley. 2018. "Life Cycle Engineering of Carbon Fibres for Lightweight Structures." *Procedia CIRP* 69(May): 43–8. doi:10.1016/j.procir.2017.11.007.

Dicker, Michael P. M., Peter F. Duckworth, Anna B. Baker, Guillaume Francois, Mark K. Hazzard, and Paul M. Weaver. 2014. "Green Composites: A Review of Material Attributes and Complementary Applications." *Composites—Part A: Applied Science and Manufacturing* 56: 280–89. doi:10.1016/j.compositesa.2013.10.014.

Duflou, Joost R., Yelin Deng, Karel Van Acker, and Wim Dewulf. 2012. "Do Fiber-Reinforced Polymer Composites Provide Environmentally Benign Alternatives? A Life-Cycle-Assessment-Based Study." *MRS Bulletin* 37(04): 374–82. doi:10.1557/mrs.2012.33.

Duflou, Joost R., Deng Yelin, Karel Van Acker, and Wim Dewulf. 2014. "Comparative Impact Assessment for Flax Fibre versus Conventional Glass Fibre Reinforced Composites: Are Bio-Based Reinforcement Materials the Way to Go?." *CIRP Annals—Manufacturing Technology* 63(1): 45–8. doi:10.1016/j.cirp.2014.03.061.

Egede, Patricia. 2017. *Environmental Assessment of Lightweight Electric Vehicles.* Sustainable Production, Life Cycle Engineering and Management. Cham: Springer International Publishing. doi:10.1007/978-3-319-40277-2.

European Parliament and Council. 2000. "ELV Directive."

Fleischer, Jürgen, Roberto Teti, Gisela Lanza, Paul Mativenga, Hans-Christian Möhring, and Alessandra Caggiano. 2018. "Composite Materials Parts Manufacturing." *CIRP Annals* 67(2): 603–26. doi:10.1016/j.cirp.2018.05.005.

Gellrich, Sebastian, Marc-André Filz, Johannes Wölper, Christoph Herrmann, and Sebastian Thiede. 2019. "Data Mining Applications in Manufacturing of Lightweight Structures." 15–27. doi:10.1007/978-3-662-58206-0_2.

Geyer, Roland, Brandon Kuczenski, Trevor Zink, and Ashley Henderson. 2016. "Common Misconceptions about Recycling." *Journal of Industrial Ecology* 20(5): 1010–17. doi:10.1111/jiec.12355.

Griffing, E., and M. Overcash. 2009. "Carbon Fiber HS from PAN [UIDCarbFibHS]. Contents of Factory Gate to Factory Gate Life Cycle Inventory Summary." *Chemical Life Cycle Database* 2009.

Haas, Willi, Fridolin Krausmann, Dominik Wiedenhofer, and Markus Heinz. 2015. "How Circular Is the Global Economy?: An Assessment of Material Flows, Waste Production, and Recycling in the European Union and the World in 2005." *Journal of Industrial Ecology* 19(5): 765–77. doi:10.1111/jiec.12244.

Hauschild, Michael Zwicky, Christoph Herrmann, and Sami. Kara. 2017. "An Integrated Framework for Life Cycle Engineering." *Procedia CIRP* 61: 2–9. doi:10.1016/j.procir.2016.11.257.

Hauschild, Michael Z., Ralph K. Rosenbaum, and Stig Irving Olsen. 2017. *Life Cycle Assessment: Theory and Practice. Life Cycle Assessment: Theory into Practice.* doi:10.1007/978-3-319-56475-3.

Herrmann, Christoph, Wim Dewulf, Michael Hauschild, Alexander Kaluza, Sami Kara, Steve Skerlos, and Life Cycle Engineering. 2018. "Life Cycle Engineering of Lightweight Structures." *CIRP Annals* 67(2): 651–72. doi:10.1016/j.cirp.2018.05.008.

Hofer, Johannes. 2014. "Sustainability Assessment of Passenger Vehicles: Analysis of Past Trends and Future Impacts of Electric Powertrains." *ETH Zurich* (22027). doi:10.3929/ethz-a-010252775.

Hohmann, Andrea, Stefan Albrecht, Jan Paul Lindner, Bernhard Voringer, Daniel Wehner, Klaus Drechsler, and Philip Leistner. 2018. "Resource Efficiency and Environmental Impact of Fiber Reinforced Plastic Processing Technologies." *Production Engineering* 12(3–4): 405–17. doi:10.1007/s11740-018-0802-7.

Hohmann, Andrea, Stefan Albrecht, Jan Paul Lindner, Daniel Wehner, M. Kugler, T. Prenzel, T. Pitschke, et al. 2018. "Recommendations for Resource Efficient and Environmentally Responsible Manufacturing of CFRP Products Results of the Research Study MAI Enviro 2.0."

Huang, Xiaosong. 2009. "Fabrication and Properties of Carbon Fibers." *Materials* 2(4): 2369–403. doi:10.3390/ma2042369.

IPCC. 2018. "Global Warming of 1.5°C. An IPCC Special Report on the Impacts of Global Warming of 1.5°C above Pre-Industrial Levels and Related Global Greenhouse Gas Emission Pathways, in the Context of Strengthening the Global Response to the Threat of Climate Change."

Kara, S., and S. Manmek. 2009. "Composites: Calculating Their Embodied Energy." http://www.wagnerscft.com.au/files/6813/4730/1312/sustainability.pdf.

Kelly, Jarod C., John L. Sullivan, Andrew Burnham, and Amgad Elgowainy. 2015. "Impacts of Vehicle Weight Reduction via Material Substitution on Life-Cycle Greenhouse Gas Emissions." *Environmental Science and Technology* 49(20): 12535–42. doi:10.1021/acs.est.5b03192.

Koffler, Christoph, and Klaus Rohde-Brandenburger. 2010. "On the Calculation of Fuel Savings through Lightweight Design in Automotive Life Cycle Assessments." *The International Journal of Life Cycle Assessment* 15(1): 128–35. doi:10.1007/s11367-009-0127-z.

Koronis, Georgios, Arlindo Silva, and Mihail Fontul. 2013. "Green Composites: A Review of Adequate Materials for Automotive Applications." *Composites Part B Engineering* 44(1): 120–27. doi:10.1016/j.compositesb.2012.07.004.

Lefeuvre, Anaële, Sébastien Garnier, Leslie Jacquemin, Baptiste Pillain, and Guido Sonnemann. 2017. "Anticipating in-Use Stocks of Carbon Fiber Reinforced Polymers and Related Waste Flows Generated by the Commercial Aeronautical Sector until 2050." *Resources, Conservation and Recycling* 125(April): 264–72. doi:10.1016/j.resconrec.2017.06.023.

Li, Xiang, Ruibin Bai, and Jon Mckechnie. 2015. "Environmental and Financial Performance of Mechanical Recycling of Carbon Fibre Reinforced Polymers and Comparison with Conventional Disposal Routes." *Journal of Cleaner Production* 127: 451–60. doi:10.1016/j.jclepro.2016.03.139.

Liddell, Heather, S. B. Brueske, A. C. Carpenter, and J. W. Cresko. 2016. "Manufacturing Energy Intensity and Opportunity Analysis for Fiber-Reinforced Polymer Composites and Other Lightweight Materials." *Proceedings of the American Society for Composites—31st Technical Conference, ASC 2016*, October 2017.

Liddell, Heather, Caroline Dollinger, Aaron Fisher, Sabine Brueske, Alberta Carpenter, and Joseph Cresko. 2017. "Bandwidth Study on Energy Use and Potential Energy Saving Opportunities in U.S. Glass Fiber Reinforced Polymer Manufacturing."

Mainka, Hendrik, Olaf Täger, Enrico Körner, Liane Hilfert, Sabine Busse, Frank T. Edelmann, and Axel S. Herrmann. 2015. "Lignin—An Alternative Precursor for Sustainable and Cost-Effective Automotive Carbon Fiber." *Journal of Materials Research and Technology* 4(3): 283–96. doi:10.1016/j.jmrt.2015.03.004.

Mativenga, Paul T., Norshah A. Shuaib, Jack Howarth, Fadri Pestalozzi, and Jörg Woidasky. 2016. "High Voltage Fragmentation and Mechanical Recycling of Glass Fibre Thermoset Composite." *CIRP Annals* 65(1): 45–8. doi:10.1016/j.cirp.2016.04.107.

Mohanty, A., M. Misra, and G. Hinrichsen. 2000. "Biofibers, Biodegradable Polymers and Biocomposites: An Overview." *Macromolecular Materials and Engineering* 276(1): 1–24.

Möhring, H.-C. 2017. "Composites in Production Machines." *Procedia CIRP* 66: 2–9. doi:10.1016/j.procir.2017.04.013.

Neitzel, Manfred, Peter Mitschang, and Ulf Breuer. 2014. "Handbuch Verbundwerkstoffe.". In: *Handbuch Verbundwerkstoffe*, Edited by Manfred Neitzel, Peter Mitschang, and Ulf Breuer, I–XXI. July. doi:10.3139/9783446436978.

Newcomb, Bradley A. 2016. "Processing, Structure, and Properties of Carbon Fibers." *Composites—Part A: Applied Science and Manufacturing* 91(December): 262–82. doi:10.1016/j.compositesa.2016.10.018.

Nilakantan, Gaurav, and Steven Nutt. 2018. "Reuse and Upcycling of Thermoset Prepreg Scrap: Case Study with Out-Of-Autoclave Carbon Fiber/Epoxy Prepreg." *Journal of Composite Materials* 52(3): 341–60. doi:10.1177/0021998317707253.

Oliveux, Géraldine, Luke O. Dandy, and Gary A. Leeke. 2015. "Current Status of Recycling of Fibre Reinforced Polymers: Review of Technologies, Reuse and Resulting Properties." *Progress in Materials Science* 72: 61–99. doi:10.1016/j.pmatsci.2015.01.004.

Pickering, S. J. 2006. "Recycling Technologies for Thermoset Composite Materials-Current Status." *Composites—Part A: Applied Science and Manufacturing* 37(8): 1206–15. doi:10.1016/j.compositesa.2005.05.030.

Pimenta, Soraia, and Silvestre T. Pinho. 2011. "Recycling Carbon Fibre Reinforced Polymers for Structural Applications: Technology Review and Market Outlook." *Waste Management* 31(2): 378–92. doi:10.1016/j.wasman.2010.09.019.

Plastics Europe. 2019. "Eco-Profiles." https://www.plasticseurope.org/en/resources/eco-profiles.

Prabhakaran, S., V. Krishnaraj, M. Senthil Kumar, and R. Zitoune. 2014. "Sound and Vibration Damping Properties of Flax Fiber Reinforced Composites." *Procedia Engineering* 97: 573–81. doi:10.1016/j.proeng.2014.12.285.

Rybicka, Justyna, Ashutosh Tiwari, Pedro Alvarez Del Campo, and Jack Howarth. 2015. "Capturing Composites Manufacturing Waste Flows through Process Mapping." *Journal of Cleaner Production* 91: 251–61. doi:10.1016/j.jclepro.2014.12.033.

Schönemann, Malte, Christopher Schmidt, Christoph Herrmann, and Sebastian Thiede. 2016. "Multi-Level Modeling and Simulation of Manufacturing Systems for Lightweight Automotive Components." *Procedia CIRP* 41: 1049–54. doi:10.1016/j.procir.2015.12.063.

Shuaib, Norshah Aizat, and Paul Tarisai Mativenga. 2015. "Energy Demand in Mechanical Recycling of Glass Fibre Reinforced Thermoset Plastic Composites." *Journal of Cleaner Production* 120: 198–206. doi:10.1016/j.jclepro.2016.01.070.

Shuaib, Norshah Aizat, Paul Tarisai Mativenga, James Kazie, and Stella Job. 2015. "Resource Efficiency and Composite Waste in UK Supply Chain." *Procedia CIRP* 29: 662–67. doi:10.1016/j.procir.2015.02.042.

Spierling, Sebastian, Eva Knupffer, Hannah Behnsen, Marina Mudersbach, Hannes Krieg, Sally Springer, Stefan Albrecht, Christoph Herrmann, and Hans-Josef Endres. 2018. "Bio-Based Plastics—A Review of Environmental, Social and Economic Impact Assessments." *Journal of Cleaner Production* 185: 476–91. doi:10.1016/j.jclepro.2018.03.014.

Stamboulis, A., C. A. Baillie, S. K. Garkhail, H. G. H. Van Melick, and T. Peijs. 2000. "Environmental Durability of Flax Fibres and Their Composites Based on Polypropylene Matrix." *Applied Composite Materials* 7(5–6): 273–94. doi:10.1023/A:1026581922221.

Suzuki, Tetsuya, and Jun Takahashi. 2005. "Prediction of Energy Intensity of Carbon Fiber Reinforced Plastics for Mass-Produced Passenger Cars." *Ninth Japan International SAMPE Symposium JISSE-9*, 14–19. http://j-t.o.oo7.jp/publications/051129/S1-02.pdf.

Swolfs, Yentl. 2015. *Hybridisation of Self- Reinforced Composites: Modelling and Verifying a Novel Hybrid Concept.*

Timmis, Andrew J., Alma Hodzic, Lenny Koh, Michael Bonner, Constantinos Soutis, Andreas W. Schäfer, Lynnette Dray, Andreas W. Schäfer, and Lynnette Dray. 2015. "Environmental Impact Assessment of Aviation Emission Reduction through the Implementation of Composite Materials." *International Journal of Life Cycle Assessment* 20(2): 233–43. doi:10.1007/s11367-014-0824-0.

UN Environent Life Cycle Iniative. 2019. "Life Cycle Terminology." https://www.lifecycleinitiative.org/resources/life-cycle-terminology-2/#f.

Vegt, O. M. De, and W. G. Haije. 1997. *Comparative Environmental Life Cycle Assessment of Composite Materials.*

Vo Hong, Nhan, Grzegorz Pyka, Martine Wevers, Bart Goderis, Peter Van Puyvelde, Ignaas Verpoest, and Aart Willem Van Vuure. 2015. "Processing Rigid Wheat Gluten Biocomposites for High Mechanical Performance." *Composites Part A: Applied Science and Manufacturing* 79: 74–81. doi:10.1016/j.compositesa.2015.09.006.

Witik, Robert A., Jérôme Payet, Véronique Michaud, Christian Ludwig, and Jan Anders E. Månson. 2011. "Assessing the Life Cycle Costs and Environmental Performance of Lightweight Materials in Automobile Applications." *Composites—Part A: Applied Science and Manufacturing* 42(11): 1694–709. doi: 10.1016/j.compositesa.2011.07.024.

Witik, Robert A., Remy Teuscher, Véronique Michaud, Christian Ludwig, and Jan Anders E. Månson. 2013. "Carbon Fibre Reinforced Composite Waste: An Environmental Assessment of Recycling, Energy Recovery and Landfilling." *Composites—Part A: Applied Science and Manufacturing* 49: 89–99. doi:10.1016/j.compositesa.2013.02.009.

Witten, Elmar, Volker Mathes, Michael Sauer, and Michael Kühnel. 2018. "Composites Market Report 2018: Market Developments, Trends, Outlooks and Challenges." 1–44. September.

Wong, Kok, Chris Rudd, Steve Pickering, and Xiao Ling Liu. 2017. "Composites Recycling Solutions for the Aviation Industry." *Science in China (Technological Sciences)*. doi:10.1007/s11431-016-9028-7.

Yang, Yongxiang, Rob Boom, Brijan Irion, Derk-Jan Jan van Heerden, Pieter Kuiper, and Hans de Wit. 2012. "Recycling of Composite Materials." *Chemical Engineering and Processing: Process Intensification* 51(January): 53–68. doi:10.1016/j.cep.2011.09.007.

Index

D

E

I

K

L

M

N

O

P

Q

R

S

T

U

V

W

X

Y

Z

For Product Safety Concerns and Information please contact our EU
representative GPSR@taylorandfrancis.com
Taylor & Francis Verlag GmbH, Kaufingerstraße 24, 80331 München, Germany

www.ingramcontent.com/pod-product-compliance
Lightning Source LLC
Chambersburg PA
CBHW060750220326
41598CB00022B/2392

* 9 7 8 0 3 6 7 5 4 1 6 2 0 *